D0706463

ETHOLOGY

JAMES L. GOULD
Department of Biology
Princeton University

ETHOLOGY

The Mechanisms and Evolution of Behavior

W · W · NORTON & COMPANY · NEW YORK · LONDON

Published simultaneously in Canada by George J. McLeod Limited, Toronto.
Printed in the United States of America

First Edition

Book design by Antonina Krass

Library of Congress Cataloging in Publication Data
Gould, James L.
Ethology: the mechanisms and evolution of behavior.

Bibliography: p.
Includes index.
1. Animals, Habits and behavior of. 2. Behavior evolution. I. Title.
QL751.G66 1982 591.5 81-16973
ISBN 0-393-01488-6 AACR2

W. W. Norton & Company, Inc. 500 Fifth Avenue, New York, N.Y. 10110
W. W. Norton & Company Ltd., 37 Great Russell Street, London WC1B 3NU

1 2 3 4 5 6 7 8 9 0

Dedicated to

Donald R. Griffin

whose high standards of research and
unflinching insistence on a comprehensive,
commonsense approach to the study of
behavior have given direction and discipline to
modern ethology.

Contents in Brief

Contents

Preface

Ethology is the study of the mechanisms and evolution of behavior. E. O. Wilson, in his influential book *Sociobiology: The New Synthesis,* predicted that ethology's older synthesis of neurobiology and behavioral ecology would split in time into two mutually exclusive camps, one mechanistic, the other evolutionary. As Wilson sees it, this division will be the unavoidable result of increasing narrowness of training as biology, which used to represent the unity of living things, becomes more and more specialized, and of the difficulty of applying both mechanistic and evolutionary points of view to the same animal. In short, ethology is in danger of an imminent demise. The irony, though, is that even now we are just beginning to appreciate the full power of ethology's classic ideas, and to understand the crucial link between neural mechanisms and the evolution of behavior. Now more than ever before ethology poses some of the most exciting intellectual challenges in modern biology while it offers the most powerful conceptual tools for answering them. The goal of this book is to illustrate the power of ethology's broad and integrative approach in unraveling how behavior, both simple and complex, is organized and orchestrated.

The book develops this theme by looking first at traditional ethology to establish familiarity with the models which will then be used to examine neural mechanisms, complex individual behavior, ecological mechanisms, social behavior and species interactions, and then finally our own species. In stressing the integrative nature and principles of ethology, the book is necessarily more selective than most. Suggested readings at the end of each chapter serve to broaden the book's base with examples of first-rate research treated in far more detail than would be possible in the text. Where such concise, semitechnical supplements are not available, short essays devoted to particular experiments or approaches accompany the text. Study questions at the end of each chapter are designed to stimulate creative thinking about the chapter's subject, and rarely have an obvious or even uniquely correct answer. Students wishing to delve deeper into the literature surrounding a particular subject may make use of the chapter-by-chapter selected bibliography at the end. Throughout, the common literary device of treating evolution as an actor is used, though in reality, as we all know, evolution is a passive process wholly lacking in purpose or goals.

This book owes a great deal to those advocates of the mechanistic approach to behavior under whom I have been fortunate enough to study, Seymour Benzer, Donald Griffin, Peter Marler, and Fernando Nottebohm in particular. Since one of my goals was readability, it has been impossible to mention by name most of the researchers whose work has contributed to the organizing principles of the field. Four reviewers, Peter Marler, William Quinn, James Simmons, and Charles Walcott, as well as Joseph B. Janson and James Jordan of W. W. Norton & Company, made especially helpful comments and suggestions. The publisher wishes to thank Christopher T. DeRosa, George B. Hooper, Philip Lehner, Richard E. Phillips, Michael Salmon, and Barry Schwartz for their comments and suggestions during the various stages of manuscript preparation. Kate M. Schenck made possible the impossible task of locating and organizing the illustrations, and Barbara Delanoy bore patiently with typing draft after draft. The most indispensable contributor of all, however, was Carol Grant Gould, who wrote about a quarter of the text, criticized and edited the entire manuscript, and gave it whatever grace it may possess.

ETHOLOGY

Part I

DEVELOPMENT OF ETHOLOGY

CHAPTER 1

The Biology of Behavior

Every spring pairs of pigeons bow and strut, bob and coo across campuses and in cities and parks the world over. Each sex has its own characteristic role to play in this elaborate courtship ballet, which begins in the spring regardless of temperature or of the availability of food and mates. From laboratory experiments we know that neither member of a courting pair need ever have observed courtship before, or indeed even another pigeon (Fig. 1–1). Nevertheless, the pair mates, selects a suitable nest site without needing to have encountered one previously, and formally agrees on the site through a distinctive nest call. Over a period of a week the two birds will construct a nest (again, perhaps the first either has ever built or seen), and late some afternoon a few days later the female will lay two eggs directly in the nest. Although neither bird may ever have come across an egg before or have shown the least inclination to sit, both will now begin regular, dutiful shifts of incubation. In about two weeks the developing chick will begin a precise pecking behavior until it hatches, and will then, upon seeing another bird for the first time, begin directing specialized begging calls and pecks at its parents. For their part the parents will begin feeding a liquid diet to perhaps the first newborn chicks

Fig. 1-1 A pair of laboratory-reared doves begin courtship. They will go on to build their nest in the glass dish and rear offspring in this artificial environment. Their remarkable willingness to continue this behavior outside of their natural environment has marked them out as one of the standard laboratory animals. The details of the underlying programming which permit parenting behavior to continue normally under such unusual circumstances will be examined in Chapter 4.

they have encountered, followed several days later by solid food.

The behavior of each of these performers is highly adaptive—they are all doing the right thing at the right time to survive and reproduce. How do they come by this knowledge? Who or what has choreographed the calls, the posturings, the building, the special recognition at just the right moment of objects and individuals never before encountered? In fact, all this knowledge and much more has been tucked away inside each bird prior to its birth, encoded in some mysterious way in its genes. When the information crucial to its well-being is too complex to be stored word for word in the genes or cannot be predicted in advance, the animal is born instead with detailed instructions about how to find out what it needs to know. This reservoir of information and instructions is, of course, *instinct,* and the scientists who seek to understand it fully in all its subtleties are *ethologists.*

TWO VIEWS OF BEHAVIOR

Ethology (from the Greek *ethos,* manner or behavior), is the study of what animals do, and how and why they do it. Ethologists, then, are interested in accurate observation and description of an animal's behavior, in the mechanisms and programming which underlie it, and most important, in the reasons an animal must behave as it does. Ethology is often defined as the study of an animal's behavior in its natural environment. Biology has shown in the last hundred years how the necessities of the environment have given rise through evolution to each animal's peculiar behavior. This explains why behavior, under natural conditions, is so well suited to the exigencies of each species' particular environment. For ethologists, then, the environment is of enormous importance, and includes not only the ecology of the physical surroundings, but the animal's social interactions with its own and other species, the sensory world in which the animal lives, and the many series of environments though which and to which the animal has evolved. To an ethologist, knowledge of an animal's evolution, ecology, social organization, and sensory abilities is crucial to understanding its behavior. As a result, ethologists are strongly motivated to keep these aspects of the animal's environment constant during the process of studying its behavior. Because of this concern, ethology has relied first on careful and inconspicuous observation, and only secondarily (and warily) on experimentation (Fig. 1–2).

Ethologists believe that the secrets of behavior lie largely in an animal's genes, in the ways in which evolution has prepared the animal to deal with its peculiar environment. This concern

Fig. 1–2 In their attempt to leave the environment undisturbed, ethologists try to make themselves as unobtrusive as possible. Here a blind, present since before these birds began to arrive on the nesting grounds, provides an opportunity to observe the social behavior of herring and lesser black-backed gulls.

with instinct as a driving force is a result of the realization that the unit of selection in the living world is the gene, and that all the rich variety of life we see about us is a reflection of the billion-year struggle of genes both with the environment and against other combative sets of genes, to ensure their own propagation into the next generation. Richard Dawkins puts it in its most extreme form when, in his book *The Selfish Gene,* he states that organisms are "survival machines—robot vehicles blindly programmed to preserve the selfish molecules known as genes." These genes ensure their survival by directing the construction of specifically tuned sense organs and neural circuits which control and generate adaptive behavior in that equally well-run corporation, the body. The neural mechanisms, programming, and evolutionary strategies which combine to produce a well-adapted animal provide ethologists with what we see as the most intriguing and important questions in science today.

Animal behavior is also the concern of a branch of psychology. Psychology is usually defined as the study of human and animal behavior, but animals have traditionally been studied not for their own sakes, but, as a current introductory text puts it, as human stand-ins whose "rudimentary forms" of behavior nevertheless permit researchers to do "many important experiments . . . that [they] cannot do with people because people cannot be treated as guinea pigs."

Although many schools of psychology exist, and the work of physiological, comparative, and cognitive psychologists will figure prominently in this text, the dominant school for at least the last fifty years—behaviorism—has maintained a point of view which makes behaviorists' work difficult and often impossible for ethologists to apply to their own concerns. Behaviorism, as developed by J. B. Watson and elaborated and popularized by B. F. Skinner, is based on three basic assumptions: (1) that the elements of behavior are not the largely prewired neural circuits imagined by ethology, but simple reflexes linked together by "conditioning" (a simple form of learning) to generate behavior; (2) that the ultimate basis of behavior therefore must be learning, and that instinct, even as the director of learning, need not exist; and (3) that worrying about what may be going on *inside* animals is irrelevant or even dangerous, since such a concern for the inner workings of the animals inevitably attributes to animals untestable concepts such as feelings, expectations, concepts, and other such "mental states." Hence, at least in its purist form, behaviorism could and did largely ignore evolution, genetics, instinct, and an animal's natural environment—the cornerstones of what we now call classical ethology. The most extreme form of behaviorism is illustrated by Watson's memorable quote: "Give me a dozen healthy infants, well-formed, and my own specified world to bring them up in and I'll guarantee to take any one at random and train him to become any type of specialist I might select—doctor, lawyer, artist, merchant-chief and, yes, even beggarman and thief. . . . " For Watson the genetic makeup of the individual was immaterial, the crucial factor in character, and even career, formation being the "laboratory" environment of life.

This purist view of men and animals as *tabulae rasae*, formless balls of putty to be shaped and conditioned by chance and experience, combined with an apparent disregard for what might be going on inside the subjects of their experiments, led behaviorists to take an approach to experimentation very different from that of ethologists.

Behaviorism presupposes that the study of animals is relevant to the study of humans because there are "laws" of behavior which apply to higher animals in general. As a result behaviorists could study, say, feeding in rats—perhaps something as simple as how making food better affects how quickly it is eaten—with the expectation that the laws of feeding would prove to be the same for humans. Behaviorists formulated these laws to relate changes in an animal's environmental "input" to the animal's behavioral "output," without offering any speculation about what might be happening in between. In the case of feeding, for example, the quantity and quality of the food offered might be the controlled input, and the rate and frequency of feeding might be the measured output. It seems now as though behaviorism treated its subjects as black boxes whose inner workings were unknowable (Fig. 1–3). The obsession of behaviorism with controlling the only variable admitted in behavior—the animal's physical environment—forced researchers to work in the laboratory rather than in what must have seemed the uncontrollable chaos of nature. Care was also taken to minimize any genetic differences among their animals to control experiments yet more thoroughly. While nature often goes to great lengths to prevent creatures from mating with their im-

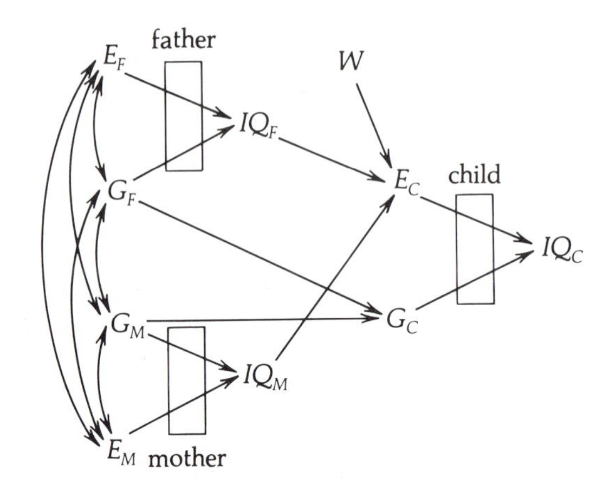

Fig. 1-3 A simplified black-box model for intelligence (IQ). IQ is treated as the measurable output, while the parents' genetics (G_M and G_F), parental environment (E_M and E_F), and the world (W) serve as the input to be varied. What is going on inside the boxes (mother, father, and child)—the workings of the genes, the nature and timing of the relevant environmental information—is not of primary concern in the behaviorists' model.

mediate relatives, experimental animals have been domesticated and inbred for dozens of generations. With genes and environment under careful control, behaviorism set the stage for precise and reproducible experiments. Since researchers avoided studying natural behavior, and concentrated instead on painstakingly training an animal through "operant conditioning" (Fig. 1-4) to perform a specific, arbitrary task for a reward, these experiments of early behaviorism seemed unbiological to ethologists.

CRITICISMS OF ETHOLOGY

Animal behavior, as we have seen, may be approached and explained from either of two distinct points of view, or indeed from anywhere along the classic continuum from nature to nurture. Although there has certainly been a generous element of misunderstanding, the basic disagreement between the two schools which developed in the first half of this century over the existence of instinct and the importance of genes and evolution in behavior made the two approaches essentially irrelevant to each other and generated a heated and often bitter debate between the two diciplines in the 1950s and 1960s.

It is now widely accepted that behaviorism is on the decline, its loss of vigor the result of its inability to come to grips even with the existence of innate behaviors, much less with their mechanisms and evolutionary origins. Emerging in its place is comparative psychology, a way of looking at phenomena that extends well beyond the behavior of the domesticated white rat, though it still retains the traditional focus on learning and controlled laboratory experiments. But in the behaviorists' criticisms of early ethology there was much truth: in some sense behaviorists unwittingly helped shake ethology loose from some fond but fatal misconceptions, hastening both the emergence of modern ethology and the ultimate demise of behaviorism itself.

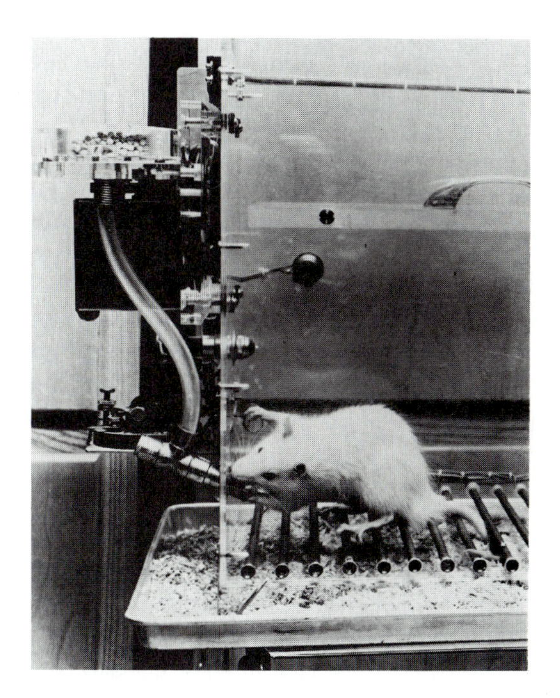

Fig. 1-4 A rat in a Skinner Box. When the rat presses the bar in response to the correct stimulus, a pellet of food is dispensed from the apparatus. This behavior is created through "operant conditioning": the rat's behavior is first shaped by rewarding ever-closer approximations of the desired performance. For example, it may first be fed for being in the correct end of the box, and then only for accidentally touching the bar. Then the reward threshold can be raised to require actually pressing, and finally the learning task can be added. At that point, the experiment can be turned over wholly to automated equipment.

CLEVER HANS

One of behaviorism's chief tenets rests on a stolid anti-Romantic foundation: attribute no internal process to your experimental animals which is inaccessible to experimentation, lest you risk being "tricked" into fallacies and disabling beliefs. The recurrent spectre of "mental processes," so alluring to those who get to know their animal subjects perhaps too well, has been a particular bugbear to behaviorists. That there is a great deal of validity in their caveat is illustrated by the infamous case of Clever Hans. The painstaking analysis of this remarkable creature's career showed definitively not only how easily our perceptions may be fooled by phenomena in the animal world, but how our very expectations may themselves *create* the phenomena.

Clever Hans was a horse owned and adored by Mr. Von Osten, a retired German mathematics professor who lived around the turn of the century. Von Osten had, he felt, been able to teach his friend Hans many extraordinary things, among them the abstract concepts of counting, color, spelling, musical theory, and arithmetical calculation. Hans was able to display his abilities largely by tapping correct answers with his hoof. Completely uninterested in monetary gain, Von Osten proudly displayed Hans to select crowds of viewers, openly inviting criticism and scientific evaluation. At one point a distinguished interdisciplinary group of scientists was assembled, really at the request of a public eager to be convinced either that the animal was conscious, or that Von Osten was a consummate liar and trickster. The group's unanimous verdict absolved Mr. Von Osten of any sort of knavery, intentional or, they thought, otherwise, and thus opened the way for the enthusiastic public to endow the horse with rationality.

Their findings, however, based largely on observation and minimal manipulations of Von Osten and his pupil, spurred Oskar Pfungst to delve more deeply into the matter. Pfungst, an early experimental psychologist, employed careful observation and meticulous screening methods to discover just what it might be that Hans was doing: Was it indeed burgeoning rationality, or was it something more, or less? If not real abstract conceptual thinking, what cues could Hans be using to arrive at his admittedly brilliant answers to abstruse problems?

After many months Pfungst arrived at an astounding conclusion: the clever animal, alert for the possibility of reward, was attending not to the question, but to the questioner, taking his hint from unconscious, minute, nearly inperceptible movements of his interrogators. He established that expectation on the part of the interrogator was the crucial factor by making use of what we now call "double-blind" experimentation. In this method of attacking a problem neither subject nor experimentor has access to the correct response. When Hans was asked a question that his interrogator did not know the answer to, he was almost infallibly wrong—even when the question was one which he had often answered correctly under different circumstances. Conversely, if the questioner knew the answer, Hans could handle extremely complex concepts, some almost metaphysical in their nature, so long as the answer could be signified by tapping, pointing, or fetching objects from an array.

Pfungst's next battlefront was that of modality: How might the questioner's expectation be expressed? The horse was known to have an extraordinary sense of hearing, so Pfungst tried in various ways to establish the role of audition in the process. He discovered that questions could indeed be whispered, ventriloquized, or even merely *thought,* and Hans still be miraculously able to produce correct responses.

When visual blocks such as curtains or tents were interposed between Hans and an "informed" questioner, however—even Mr. Von Osten himself—Hans

Fig. 1-5 Clever Hans.

became unable to perform. As the light failed in the evening or as his questioner moved away, error crept into Hans's responses.

Only Pfungst's combination of a critical mind and careful and painstaking observation allowed him to uncover Hans's true cleverness. The horse was cued by small, unconscious head movements or body alignments (almost imperceptible among the nervous fidgetings of his vivacious master, and only slightly more pronounced with calmer questioners) that almost everyone exhibited when Hans reached the correct number of taps, or the correct object in an array. That perfectly reliable, sincere investigators could overlook such an important phenomenon in themselves—even to the point of denying that they had in any way indicated outwardly their inward expectations—was a lesson of the greatest importance to behaviorists. Pfungst himself admits that he was completely unaware of his own movements until he had come to recognize those of Von Osten.

What if Pfungst had ceased his inquiry when his tests assured him that the phenomenon he was studying disappeared under double-blind conditions? Had he not gone on to the next step in the process, that of patient and deliberate observation, he would never have discovered the underlying explanation of the animal's very real behavior. It took weeks of tedious watching to discern—first in Von Osten, then in others, then in himself—the amazingly minute movements that the sensitive animal was detecting. Without this observation our whole understanding of the horse's sensory world and of our own unwitting participation in a psychological conspiracy would have been lost.

Beyond the question of instinct, behaviorists rejected the classical ethological approach because of its questionable methodology. Moving out of the laboratory to study unrestrained animals in a world with countless uncontrolled variables was bad enough, but once there ethologists simply observed and took notes. Behaviorism was born with a healthy distrust of just such subjective methods, and could point to a long list of cases in which the fond and overly romantic imaginings and expectations of observers had strongly biased the results of their observations. While behaviorists designed their own experiments to measure easily quantifiable behavior, employing either an unbiased machine or a "blind" (uninformed) observer to take the data, ethologists depended of necessity on more subtle and subjective measures which are not easily quantified (Fig. 1-6). And even when the experiments of early ethologists did yield quantifiable data, almost no effort was made to evaluate the results with that awesome battery of statistical procedures developed by behaviorists to remove as much of the human factor as possible from the process of data evaluation itself.

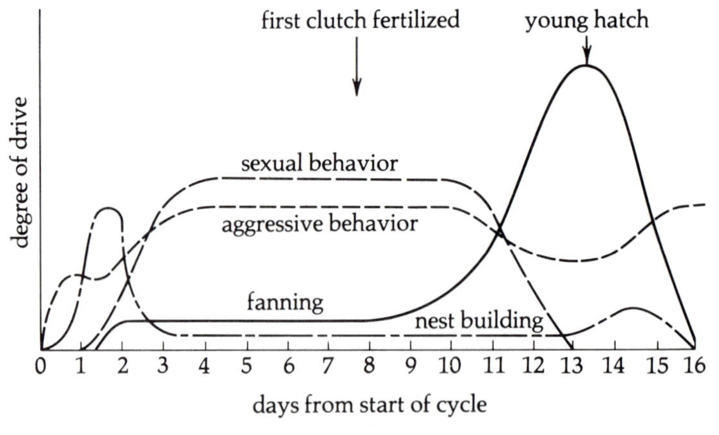

Fig. 1-6 Data from ethological experiments tend to be fairly subjective. This represents the reproductive behavior of a male stickleback (which is discussed more fully in Chapter 3). Notice the absence of numbers on the vertical axis and any discrete data points.

Moreover, although the early ethologists insisted that understanding the inner workings—the neural and physiological mechanisms—of behavior was crucial, they did precious little to further that understanding. As a result, behaviorists could with considerable justice assert that the standard ethological technique was to observe something, point out that it was adaptive, label it as innate, and leave it at that. Although we know now that the early ethologists were often right about which behaviors have instinctual bases and which do not, their glib applica-

tion of the term as though the very name were an explanation roused the just ire of behaviorists and probably postponed that analysis of the marvelous workings of instinct which is the foundation of modern ethology.

But in the end the issue goes beyond questions of method and scientific rigor. In a very real sense the traditional dichotomy between ethology and behaviorism reflects an underlying difference in world views, one based on the assumption that evolution is as strong a force in shaping behavior as it has been in morphology, the other (still a potent force in the social sciences) essentially pre-Darwinian. Is man, as this latter view contends, a special creation, so much more complex as to have become different in kind from the animals researchers must lamentably use to study him, freed in some magical way from "animal" instinct and genetic constraints? Or could it be that man occupies a point on the evolutionary continuum, differing only in degree from the rest of the animal kingdom in the particular course evolution took as our species began to diverge several million years ago, and so is best understood in the context of the mechanisms and evolutionary forces which shape the behavior of the rest of the animal world? In the next three chapters, this book will trace the evolution of the modern ethological viewpoint, which integrates neural mechanisms, behavior, and evolution. It will look in detail at the sensory and neural bases of behavior, and then at the sorts of complex individual behavior which these mechanisms generate. The focus will then shift to the genetic and evolutionary mechanisms which shape behavior, and then at their most impressive product: social behavior. This book will show not only that animals are themselves inherently interesting, but will argue in the last chapters that our place on the behavioral *scala naturae* can only be appreciated and understood by the rigorous (although admittedly ego-threatening) application of the ethological concepts of instinct, genetics, and evolution. Only then can we hope to understand our behavior and the many unrecognized forces at work within us.

SUMMARY

Ethology is the study of the mechanisms and evolution of behavior. The innate orchestration of behavior which causes animals to respond adaptively to situations important to their survival is known popularly as instinct, and is the distinctive focus of ethology. Since this programming represents a species-specific strategy optimized by natural selection for a particular environment and lifestyle, ethologists are vitally interested in leaving the environment as undisturbed as possible until they discover the features that are salient for the animal in question.

In their study of the mechanisms and evolution of behavior, ethologists hold to the standard biological assumption that the building blocks of behavior are best studied in simpler animals where they are probably fewer in number and less elaborately connected, and where the evolutionary and social constraints of the animal are more obvious. This biological bias leads to the further expectation that evolution will pragmatically have conserved successful behavioral strategies and building blocks just as it has conserved biochemical successes. Hence, just as the metabolism of a protozoan is very similar to, if less elaborate than, that of man, so too should the behavioral workings of simpler animals be similar to ours. As a result ethologists can study, say, learning in insects, with the hope that the underlying processes will turn out to be similar in humans. As we shall see, this viewpoint, which is opposed to the anthropocentric assumptions that permeate traditional American behaviorism and the social sciences, has enormous power when focused on the fascinating lives of the animals which share our world.

STUDY QUESTION

In a behavioristic study of stress (see Brady's "Ulcers in 'Executive' Monkeys," listed under "Further Reading"), monkeys were arranged in pairs to receive electrical shocks. One member of each pair, however, received a warning of the impending jolt, and inevitably developed serious or even fatal stomach ulcers. The unwarned members of the pairs, on the other hand, remained generally healthy. When the same experiment was run with rats, neither the warned nor the unwarned rats showed any obvious physical effects. What do you imagine might be an ethologist's objections to this study and to the conclusions it drew for ameliorating stress in human environments?

FURTHER READING

Beach, Frank A. "The Snark Was a Boojum." *American Psychologist* 5 (1950): 115–24. Reprinted in *Readings in Animal Behavior,* 3rd ed., edited by T. E. McGill (New York: Holt, Rinehart & Winston, 1977).

A prominent psychologist begins to worry about his field's narrow focus and perspective.

Brady, J. V. "Ulcers in 'Executive' Monkeys." *Scientific American* 199, no. 4 (1958): 95–100.

Details the experiment described in the study question.

Breland, Keller, and Breland, Marian. "A Field of Applied Animal Psychology." *American Psychologist* 6 (1951): 202–4.

Behaviorist theory suggests that any "higher" animal can be conditioned to perform any behavior of which it is physically capable. The Brelands put this expectation to a unique commercial test.

———. "The Misbehavior of Organisms." *American Psychologist* 16 (1961): 681–84.

Ten years later, the Brelands report in this witty article that their animals have instincts after all (reprinted here in Appendix B, page A15).

Lorenz, Konrad. Introduction to vol. 2 of his *Studies in Animal and Human Behavior* (Cambridge, Mass.: Harvard University Press, 1971).

Lorenz uses this opportunity to attack behaviorism, expound on the "black box" / "dissect into parts" dichotomy, and defend his dislike of experiments.

Skinner, B. F. "How to Teach Animals." *Scientific American* 185, no. 6 (1951): 26–29.

CHAPTER 2

Early Ethology

DARWIN

Darwin's revolutionary insights into evolution also demonstrated for the first time the inextricable link between an animal's world and its behavior. His theory of natural selection made it possible to understand why animals are so well endowed with mysterious instincts—why a wasp, for example, gathers food she has never eaten to feed larvae she will never see. Natural selection, Darwin hypothesized, favors those animals which leave the most offspring. Through countless generations the survivors of the unceasing struggle for a limited amount of food have had to be ever more perfectly adapted to their worlds, both morphologically and behaviorally.

Darwin's historic discovery arose from two clear sources. First, immersed in the Victorian preoccupation with utilitarian plant culture and selective breeding, he became aware that plants which were multiplied by cuttings produced offspring identical to the parent, while seeds, the products of sexual combination, could produce offspring radically different both from each other and from either parent. From this observation he predicted that there must exist in living things a genetic material which directs the development of a plant or animal, and that from the union of two such sets of material, variation emerges. Selective breeding capitalizes on the chance variation such a system makes possible, and has led to the establishment of lines so distinct from their ancestral stocks as to be hardly recognizable as members of the same species.

Second, Darwin had read Thomas Malthus's gloomy *Essay on Population* in which the pessimistic philosopher points out that our species invariably produces so many offspring that some must starve. Any increase in the food supply, Malthus observed, is quickly accommodated by an increase in population and a rapid return to a grisly balance between birth rate and death rate.

Darwin saw the same forces operating on the animal world: he could see individual animals competing with each other in the natural world for the necessities of life—food, mates, places to live—just as surely as people did in the London slums which had inspired Malthus's somber predictions. He saw that the inevitable variation among individuals in the natural world reflected in part differing ways of responding to environmental pressures, which meant that some individuals would naturally be better able to live and reproduce than others. Since most variation is heritable, their offspring could well perpetuate the competitive edge their parents had bequeathed to them. Thus nature selects the winners in each round of this omnipresent contest—those best adapted to survive. But with each generation the competition must grow keener, and surviving members of the species must be ever more perfectly adapted to the rigors of their environment. Adaptation, as Darwin realized from the first, was as much behavioral as it was morphological or physiological. Carefully programmed behavior like that of the wasp must provide an enormous competitive advantage for animals, eliminating wherever possible the need for the time-consuming and hazardous process of trial-and-error learning.

Darwin's ideas about instinct were rejected by many Victorian scientists. Then as now, many scientists went so far as to deny that instinct could exist, maintaining instead that everything must be learned, albeit very quickly in some cases. When a chick is born, for instance, it is some few hours before it begins to peck. Since the first pecks are amazingly accurate, skeptical scientists in Darwin's day maintained that the chick must learn to judge distances while standing around waiting for its down to dry. Douglas Spalding, a tutor of the family of

Bertrand Russell, appears to have been the first to put the question to a critical test. He reasoned that if visual learning were in fact involved, then blinding chicks until the time when they would usually begin to peck should delay the onset of pecking behavior until the chicks were able to "make up" the learning. Spalding put tiny hoods on the chicks before they opened their eyes and removed them several hours later. In case after case the chicks began pecking immediately with normal accuracy—one even captured a passing fly within seconds of being allowed to see. Spalding deduced from this that the ability to peck accurately is inborn in chicks, and that they need only wait until they have "matured" sufficiently before beginning to peck (Fig. 2-1).

Separating learning from simple maturation has been a persistent problem for ethologists. Although Spalding had convincingly demonstrated the existence of instinct and many other now-classic principles of ethology by similarly ingenious methods (see Apendix A), modern critics of the field have been equally ingenious in suggesting often untestable explanations of how learning might be responsible for seemingly innate behavior. Despite the pioneering work of men like Darwin and Spalding, and a series of famous naturalists like J. H. Fabre who accumulated a massive list of clearly innate behaviors in animals, the field did not catch on. What we now call ethology had to wait another sixty years until a small group of German naturalists began to attack the question of how and why animals behave as they do.

CLASSICAL CONDITIONING

An intellectual climate in which the concepts essential to a Darwinian approach to behavior could flourish was being prepared on the continent by scientists working in other fields, notably the great Russian physiologist Ivan Pavlov, and Karl von Frisch, an Austrian zoologist. While engaged in his Nobel Prize-winning research on the physiology of digestion, Pavlov discovered a phenomenon now called "classical conditioning." To his complete surprise, Pavlov found that salivation would begin in hungry animals at the mere sight of food (the cue); this he termed an "unconditioned response." The sight of food, the "unconditioned stimulus," released salivation. Pavlov found

Fig. 2-1 Eckhard Hess repeated Spalding's original experiment with an illuminating elaboration. Newborn chicks (A) peck at a target with fair accuracy (B), but their aim improves with age until at four days the pecks are tightly clustered (C). This improvement could be the result of some sort of maturation—better vision, perhaps, or strengthened neck muscles—or of learning, by which the chick recognizes and corrects its errors. Hess pitted these alternatives against one another by raising chicks with hoods which deflected their vision 7° to the right. As newborns, such birds produce the usual set of scattered pecks, but the pecking is well to the right of the target (D). By the fourth day, the pecks are tightly clustered but still misdirected (E), indicating that chicks are unable to learn to adjust their aim. The coordination of beak and eye involved in pecking must then be a wholly innate behavior which matures without benefit of learning.

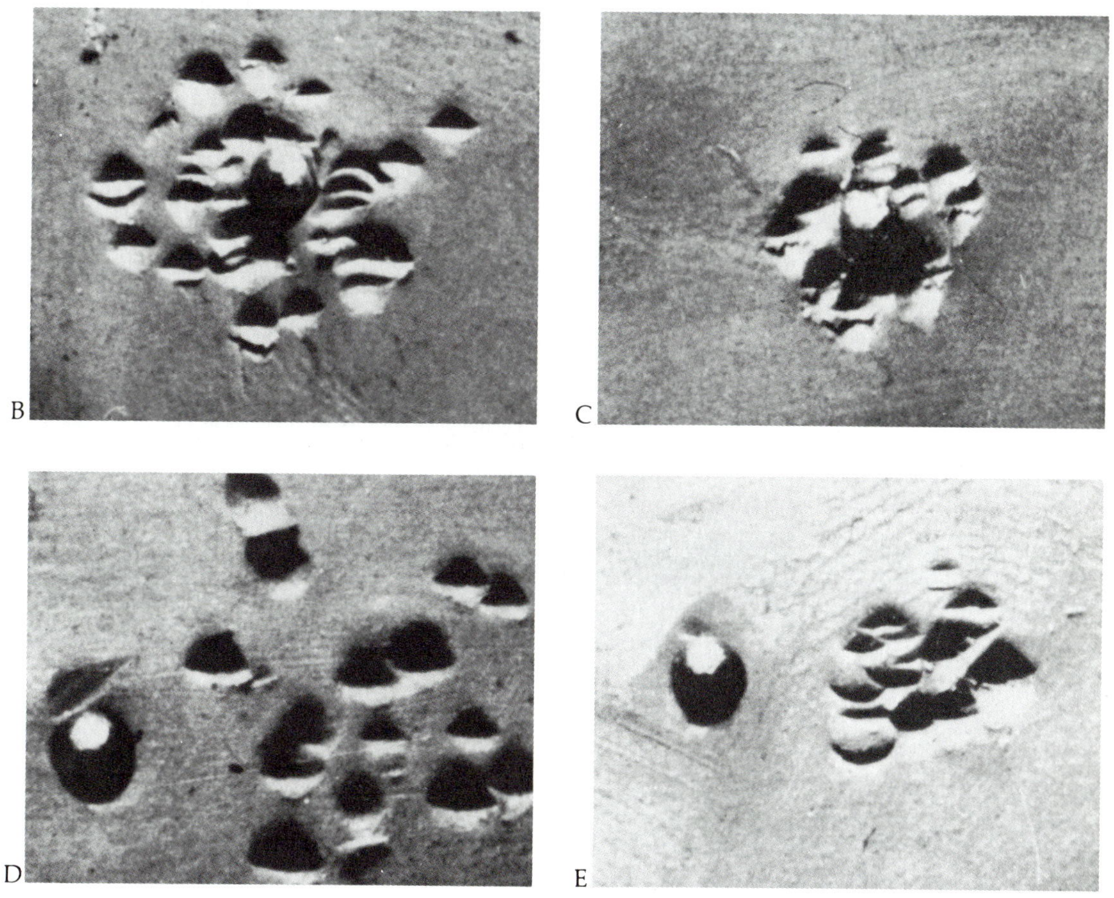

that if he presented an irrelevant stimulus—a light, say, or the famous bell—along with the food, the animal would in time come to associate the sound or light so strongly with food that this "conditioned stimulus" alone would release salivation (Fig. 2–2).

Classical conditioning (or "associative learning" as it is more appropriately known) is, in theory at least, quite different from the operant conditioning of behaviorism discussed in Chapter 1. Both cue recognition and the subsequent behavioral response are assumed to be innate, and for many years a substantial group of behaviorists denied that classical conditioning could involve "true" learning. (Whether there is any essential difference between the two forms of behavioral conditioning, and just where the phenomena fit into the ethological world view, will be taken up in Chapter 18.) Associative learning is important here because Pavlov had discovered what was to become one of the key experimental techniques in ethology: the natural stimulus in an innate stimulus-response sequence is replaced through learning with an otherwise unrelated ("conditioned") stimulus.

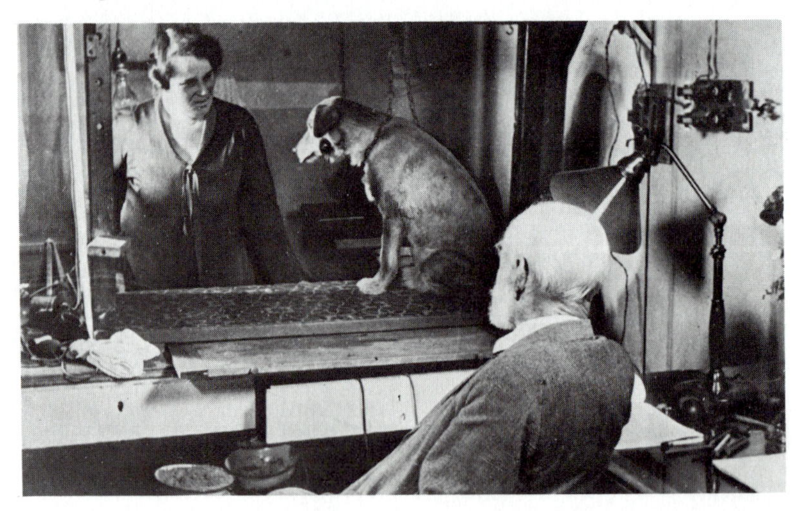

Fig. 2–2 The Russian physiologist Pavlov *(right)*. A dog is being conditioned to associate a light *(upper left)* with food. This procedure, known as classical conditioning, takes advantage of an innate or "unconditioned response" (UR) which is triggered by a releaser or "unconditioned stimulus" (US). By pairing a novel or "conditioned stimulus" (CS) with the US, Pavlov found that an animal would often come to associate the two so strongly that the CS would release the UR. The device on the dog's cheek is measuring salivation, the UR, while the dish at the left contains meat powder, the US. The CS is the light.

As it happens, Pavlov's discovery also provided ethology's early critics with what became their standard way of explaining behavior: by a sort of self-conditioning, animals must quickly link simple behavioral "reflexes" into complex chains which naïve ethologists and romantic naturalists interpret as innate.

SENSORY WORLDS

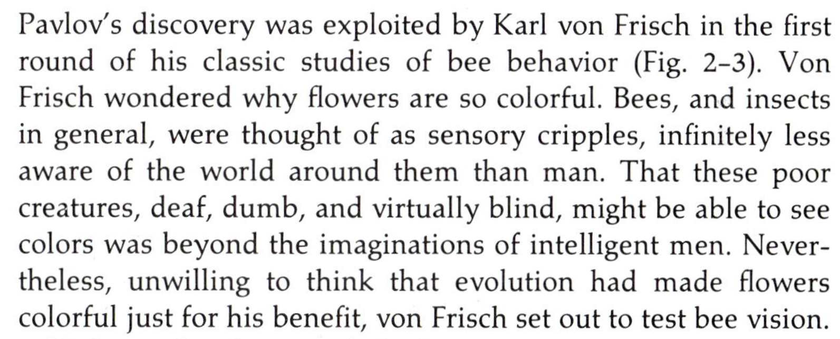

Fig. 2-3 Karl von Frisch with one of his observation hives. He is standing in the lower half of a prefabricated blind. The bees fly in and out of the "flight funnel" (which extends from the lower right of the hive) to the artificial flowers von Frisch has trained them to frequent.

Pavlov's discovery was exploited by Karl von Frisch in the first round of his classic studies of bee behavior (Fig. 2-3). Von Frisch wondered why flowers are so colorful. Bees, and insects in general, were thought of as sensory cripples, infinitely less aware of the world around them than man. That these poor creatures, deaf, dumb, and virtually blind, might be able to see colors was beyond the imaginations of intelligent men. Nevertheless, unwilling to think that evolution had made flowers colorful just for his benefit, von Frisch set out to test bee vision.

He began by placing a dish of sugar water at the entrance to a bee hive. As the bees began collecting the food, taking it inside, and returning for more, von Frisch began moving the dish by degrees to a nearby table. His bees quickly learned to fly to the table for the sugar solution. Von Frisch placed a piece of colored paper under the feeding dish and left it there until his bees had each made several trips. Then he removed the food and the paper, and set out new pieces of paper of the same color and of various shades of gray, each with an empty feeding dish. Von Frisch reasoned that having learned to associate the food and the color, if bees had color vision they would go to the colored sheet, but if they saw the world in black and white they would confuse at least one of the shades of gray with the colored sheet. The bees searched for the missing food exclusively on the colored paper (Fig. 2-4).

In the course of these experiments, von Frisch discovered two anomalies. First, bees confused red and gray, indicating that they are blind to red as a color. More interesting, he found that his bees would distinguish sheets with the same shade of gray made by different manufacturers. After having ruled out odor, von Frisch discovered that one of the sheets reflected more ultraviolet (UV) light than the other. He quickly confirmed that bees see UV light as a separate color, one to which we are blind. Applying his evolutionary logic in reverse, he wondered if

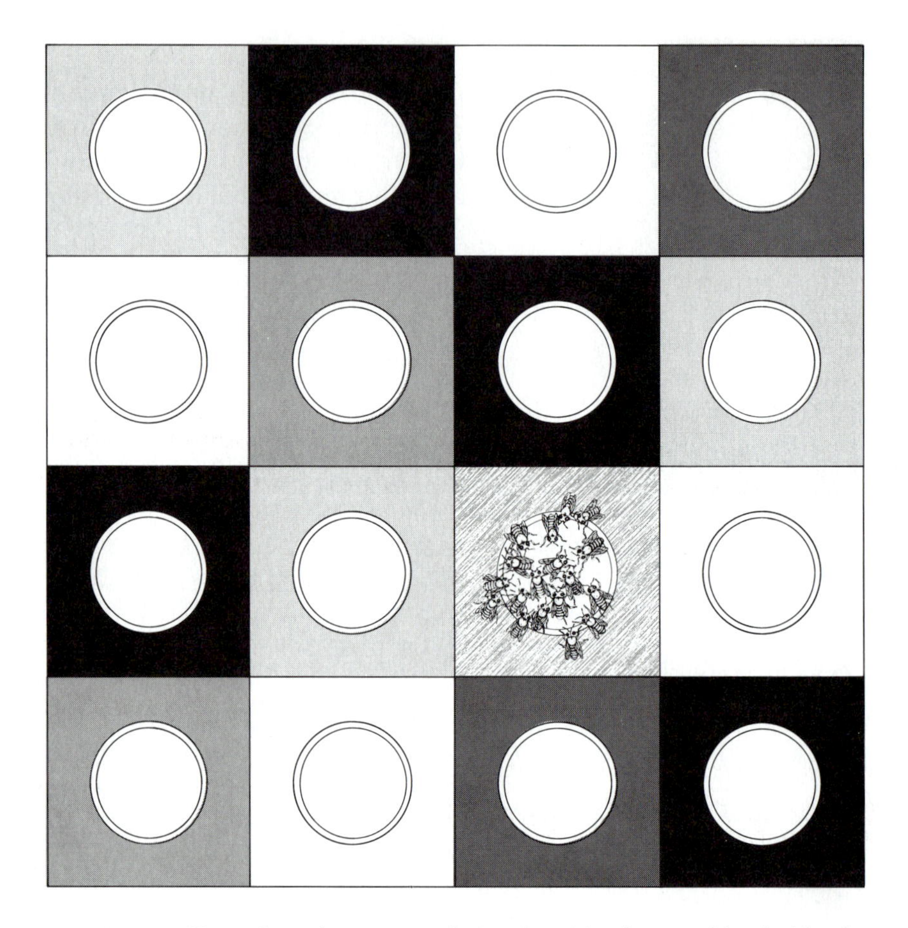

Fig. 2-4 Honey bees demonstrate their color vision by searching fruitlessly for the missing food on the one colored card in an array of cards of various shades of gray. [From "Learning and Memory in Bees," by Randolf Menzel and Jochen Erber. Copyright © 1978. Scientific American, Inc. All rights reserved.]

flowers, viewed in UV light, might reveal some secret that would justify that part of the bees' color vision. Amazingly, in ultraviolet light the flowers which rely on bees for pollination are transformed into striking bulls-eye patterns, dark in the center and light at the edges (Fig. 2-5). Von Frisch came to understand that bees innately recognize this general flower pattern, using the bright edge as a landing pad and walking to the dark center where the nectar is stored.

Von Frisch's discoveries revealed that not only are animals ideally adapted to their sensory worlds, but that far from being

Fig. 2-5 To honey bees, even the plainest flowers in visible light (A) are bull's eyes in the ultraviolet (B).

sensory cripples, they may sense features of the world to which *we* are blind. The problem of studying the behavior of animals sensitive to stimuli of which we are unaware is a serious one for ethology, for we are often blind to our own blindnesses. This melancholy truth is tellingly illustrated by John Dalton, the English physician and chemist who worked on the colors produced by the combustion of various elements and compounds for years before discovering that he was blind to red. Dalton became, in fact, the discoverer of the phenomenon of color blindness, a perceptual deficiency which must have afflicted hundreds of millions of people before him who never noticed their blindness.

Von Frisch and his students went on to show that bees inhabit a sensory world vastly different from ours. As we shall see in subsequent chapters, they are able to see patterns of polarized light in the sky, to sense CO_2 and relative humidity; they are enormously sensitive to a few special odors called "pheromones," can detect and orient to the earth's magnetic field, and even have an abstract and symbolic language to describe to their cohorts where food may be found. Indeed, much of this book will be devoted to exploring with the aid of technology the sensory worlds to which our evolution, in the absence of need, has denied us admission. Von Frisch's landmark work should forever alter our view of "lower" animals.

LORENZ AND TINBERGEN

The generally acknowledged founder (and namer) of ethology, the Bavarian naturalist Konrad Lorenz (Fig. 2–6), discovered most of the classic phenomena of ethology, and publicized both his results and his own controversial interpretations of them far more effectively than did Spalding. Unlike Spalding, whose work was unknown to him, he made his discoveries almost without ever doing an experiment. Lorenz's obsession with leaving the animal and its environment undisturbed and his keen powers of observation and induction combined to make experimentation in any modern sense both unaesthetic to him and often unnecessary. Such experiments as did take place were mostly unplanned. He discovered, for example, what we now call "releasers"—simple features in the natural world which trigger certain innate behaviors in animals—when, swim trunks in hand, he was attacked by his pet jackdaws (a sort of crow). A second, unrelated incident, which occurred while he was helping a jackdaw free itself from a screen, convinced him that anything black in the grip of a larger animal would release these otherwise uncharacteristic attacks. Lorenz discovered another releaser when, like Spalding before him, he saw even his youngest birds flee at the sight of a hawk overhead. Although he

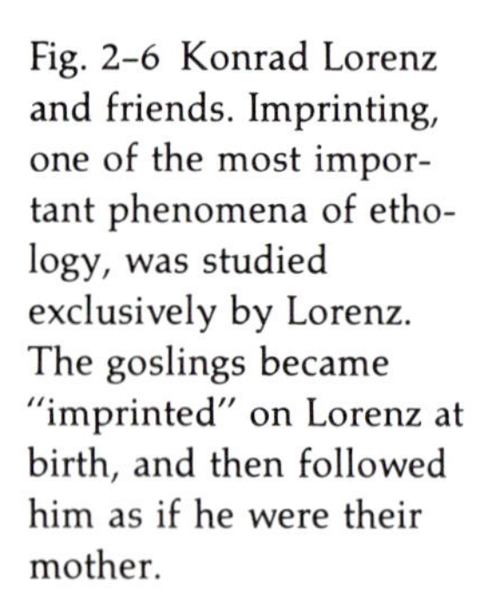

Fig. 2–6 Konrad Lorenz and friends. Imprinting, one of the most important phenomena of ethology, was studied exclusively by Lorenz. The goslings became "imprinted" on Lorenz at birth, and then followed him as if he were their mother.

Fig. 2-7 Niko Tinbergen, armed with the prerequisites for field ethology: binoculars, camera, note pad, and lunch.

realized immediately that this must be an innate response to the shape of hawks, it remained for others to make and test a variety of cardboard silhouettes in order to define the essential characteristics of "hawkness" which released this eminently adaptive fright behavior (see Chapter 3).

The founder of experimental ethology, the Dutch zoologist Niko Tinbergen (Fig. 2-7), blends Lorenz's concern for observation within the natural environment with the power of von Frisch's purely experimental approach. In this way Tinbergen has been able to refine Lorenz's initial observations into such general principles as early ethology may be said to have possessed. His experiments are so charmingly straightforward that virtually each one has become a classic. His very early work on digger wasps is a case in point.

EXPERIMENTAL ETHOLOGY

One species of digger wasp, the "bee wolf," makes its living by capturing and paralyzing honey bees. When the female wasp has dug an elaborate, chambered tunnel, she provisions each

VON UEXKÜLL'S TICK

Von Frisch's discoveries blazed a trail for serious exploration of the various and distinct sensory worlds of animals. The notion of private worlds was the inspiration of a little-known German biologist, Jakob von Uexküll, who coined the term "Umwelt" around 1905 to describe the unique world that each species, even each individual, inhabits. Not only are the worlds of sight and smell vastly different from creature to creature, but as von Uexküll realized, our senses of time and space differ as well.

As an illustration of his concept von Uexküll chose the lowly tick, and invited us through our imaginations, coupled with what we know of the life of the tick, to enter its "mind" and view the world through its sensory windows. Like the familiar mosquito, the female tick must have a blood meal before she can lay her eggs. Her entire behavioral repertoire consists of maturing to the proper form, singlemindedly acquiring the blood she needs to procreate, then laying her eggs and dying.

During this inexorable cycle, the tick's sensory organs are presenting her with a select range of stimuli which comprise her whole world. Her skin is receptive not to touch, as is ours, but to light; it leads her up from the darkness of the ground to a branch or grass blade. There she lurks, attentive now to nothing but the odor of butyric acid, a chemical common to all mammals. When that substance reaches her receptors, she drops. With her fine sense of temperature she searches for something warm, blind to all the other cues around her which might betray the presence or absence of her prey. Once she finds a source of warmth in her now dark and odorless world, a warmth which she takes as a sign of blood, she searches for a hairless spot, burrows in, and begins to suck. Lacking all sense of taste, she will drink any liquid of the correct temperature once she has perforated the membrane.

This is her first and last meal; she swells with blood, drops, lays her eggs in the earth, and dies. Thus her entire world is composed of only those few sensory inputs her system requires. Insensible to all stimuli but light, first she climbs skyward; then, impassive, she waits for the one odor in the world which tells her that prey is within her grasp. Even time, as we perceive it, stands still, perhaps for years, as she waits for butyric acid to start her clock again. The buzz and motion of the animate world is lost on these living receptor cells. Ticks have been known to wait for as long as eighteen years for this one magic odor to release them from the spell that holds them captive. But once freed from her state the tick loses all interest in the chemical which has held her in thrall and responds now only to heat. Her world becomes one not of smell but of temperature, and her whole body becomes tuned to that sense.

Fig. 2–8 Von Uexküll's tick.

These few stimuli, classic examples of the key ethological concept of sign stimuli (see Chapter 3) make up the sensory and aesthetic world of the tick. In von Uexküll's words, "the whole rich world around the tick shrinks and changes into a scanty framework consisting, in essence, of three receptor cues and three effector cues—her *Umwelt.* But the very poverty of this world guarantees the unfailing certainty of her actions, and security is more important than wealth." By abridging her indulgence in sensual luxuries, nature has ensured the tick as failsafe an existence as the natural world provides. For animals like us, bathed continually in our own set of sensory inputs, only our imaginations allow us entry into the world of such a creature, an entry essential to the ethological perspective.

chamber with a few paralyzed bees, lays an egg, and seals the chamber. The larvae which hatch devour the bees and emerge some time later as adult wasps.

Tinbergen's curiosity was aroused by two facts: the wasps never caught anything but honey bees, and each female always found her particular tunnel despite having to fly hundreds of meters to catch her prey. From observing wasps as they hunted, Tinbergen noted that they would fly around randomly, then move quickly upwind to within 15 cm of the bee, pause for a moment, and then strike (Fig. 2-9).

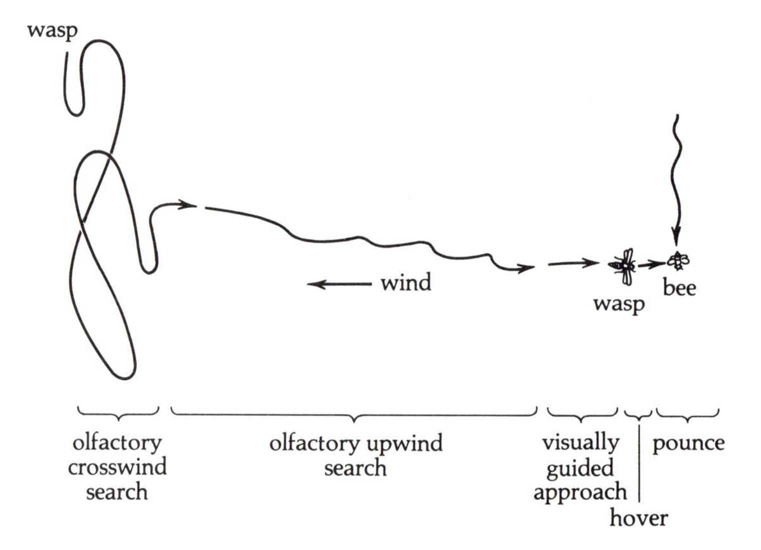

Fig. 2-9 Bee-hunting by the digger wasp. After a period of crosswind searching, the wasp flies slowly upwind, suddenly darting to within 15 cm of the bee. There the wasp pauses, then pounces on and stings the bee.

Tinbergen supposed that either vision or odor (or both) must be involved in the hunt. He tested this idea in a preliminary way by putting a paralyzed bee in the bottom of a clear test tube and then placing the tube across the wind so that an approaching wasp could both see and smell the bee. Tinbergen reasoned that the wasp, approaching always from downwind, would go to the closed end if vision were more important, and to the open end if odor were the cue. In fact, the wasp flew consistently to the open end of the tube.

Since his observations indicated that wasps accomplished the final "pounce" visually, Tinbergen performed another experiment in which separate odor and visual cues were provided. Wasps approached upwind toward the source of bee odor until they came near a tethered, odorless, bee-like object, then hovered, and finally pounced on the visual dummy (Fig. 2-10).

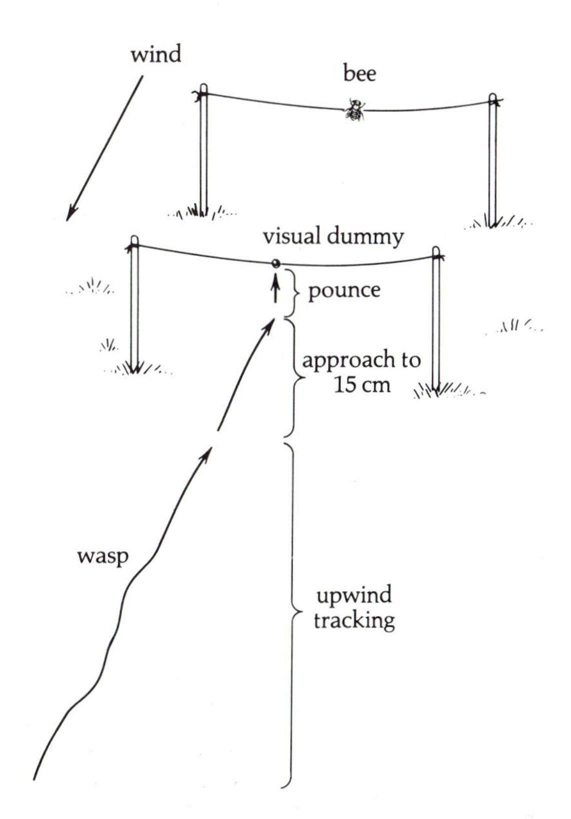

Fig. 2-10 Separation of the olfactory and visual phases of digger wasp hunting. A dead bee is suspended just upwind of an odorless dummy. The wasp flies upwind in response to the bee odor until it sees the dummy, whereupon it darts, hovers, pounces, but fails to sting.

Oddly enough, though, hunting wasps never stung the dummies. Tinbergen wondered if this meant that the dummies lacked the feel, the odor, or the appearance of real bees. He tested these possibilities in turn. Wasps did not sting dead bees made odorless by alcohol extraction, but the same bees, after being shaken with freshly killed, unextracted bees to renew any odors, *were* stung. Similarly, the wasps attacked and attempted to sting dead flies shaken with unextracted bees, but the wasps' stereotyped stinging behavior by which they stab the bee through a tiny, fleshy patch in the neck, seemed unequal to the task. On the other hand, wooden beads shaken with bees were not stung. Apparently foraging wasps require odor and some attribute of actual shape or texture of their prey.

These few, elegant experiments demonstrate how a wasp hunts bees. It searches for their odor, flies upwind until it can see one, then flies to a fixed distance from the target, pauses while taking aim, and then strikes. Finally, the odor and texture of the bee release stinging. As Spalding had realized half a century earlier, all of this is innate. The wasp is born "knowing" how to find and capture bees, and what to do with them. The superb organization of its mindless behavioral programs makes the outcome of the wasp's very demanding, unpredictable, and apparently intelligent search for prey a foregone conclusion.

Tinbergen's analysis of how a digger wasp finds its tunnel home is equally straightforward, though how they manage to return at all from hundreds of meters away after a long, irregular hunting flight over featureless terrain was a mystery left for von Frisch and others to tackle, as will be shown in Chapter 13. He could estimate a wasp's accuracy at locating its tunnel, however, to be on the order of 2 or 3 m. Tinbergen guessed that once back in the vicinity the wasps used local landmarks to guide them to their nests. To test this idea, he placed rings of pine cones around the nests, waited until each wasp flew out on a hunting trip, and then moved the circles a meter or so away. Invariably, the wasps would return to the center of the pine cone ring and search frantically for their nests (Fig. 2–11).

In the course of these experiments, though, Tinbergen discovered that not all of his artificial landmarks worked equally well, so he decided to investigate just what visual cues the wasps seemed best able to learn. His general technique was to encircle the nest with alternating and contrasting landmarks for the wasp to learn (Fig. 2–12), and then to test the wasp on its return by offering it a circle of each type of component. In this way Tinbergen was able to show that color and odor are irrelevant, but that three-dimensionality is critical. He showed, for

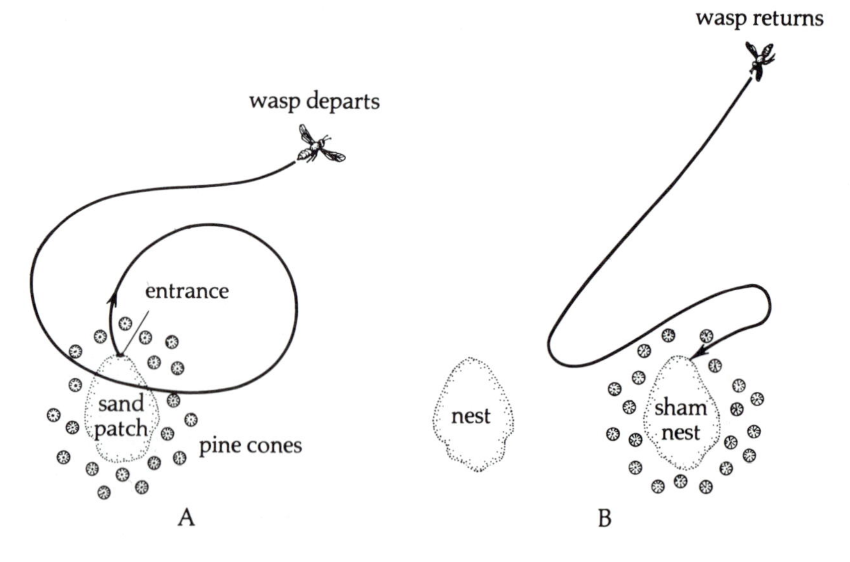

Fig. 2–11 Landmark learning by digger wasps. While the nest owner was underground, a ring of pine cones was placed around the nest (A). After the wasp flew out in search of a bee, the ring was moved to a nearby location (B). When the wasp returned, she always chose the pine cone ring in preference to her real nest.

Fig. 2–12 An alternating-landmark test. The wasp sees two sorts of landmarks in the ring around her nest as she flies out—discs and inverted cones in this case. When she returns, she finds two rings, each consisting of only one of the two classes of stimuli. Her choice reveals which class she learns most readily.

example, that cones worked far better than discs of the same size. Tinbergen began to worry, however, that the wasps might be reponding to the shadows on the three-dimensional objects rather than perceiving depth directly. To test this, he painted the appropriate "shadows" on his flat discs, but to no avail: apparently wasps prefer landmarks with real depth.

ETHOLOGY AND EVOLUTION

Tinbergen shared Lorenz's concern for the "why" questions of behavior, but unlike Lorenz he sought to put evolutionary guesses to an experimental test. As mentioned above, the two key elements of evolution are variation and selection. No two animals, cloning species and identical twins excepted, have the same set of genes, so limitless heritable differences in morphology, physiology, and behavior exist, even among members of the same population. Since, as Malthus noticed, animals inevitably produce more offspring than their environment can support, any genetic variation that gives an animal an edge in the contest that is survival in our crowded world is one that will slowly increase in frequency from generation to generation as it continues to confer an increased "fitness" on the animals carrying it, and on their offspring. The implication of this logic is that everything we see around us is "fit" and continually becoming more so, although just how this occurs is sometimes hard to discern. A favorite pastime of some ethologists is forming theories to justify the ways of evolution to man, theories that explain why species are what they are and do what they do. To be more than just an intellectual parlor game, however, Tinbergen realized that some sort of empirical constraints must be imposed on the hypotheses which result. In case after case Tinbergen did just this. Consider the stickleback's spines. Two species of these minute fish have evolved from a common ancestor: one, the ten-spined stickleback *(Pygosteus),* has ten small, stubby spines forming a comblike ridge along its back, while the three-spined variety *(Gasterosteus)* sports three long, sharp spikes on its spine and two ventral ones which, when erect, form a rigid, lethal crossbar across its body (Fig. 2–13). Behaviorally, the two species differ as radically as they do in their appearance. *Pygosteus,* with its ten small spines, is a shy, retiring fish, seldom straying far from the undergrowth it prefers, fleeing to cover at the hint of a predator. *Gasterosteus,* on the other hand, is bold and gregarious. It has no fear of open habitats, and will nest in the merest suggestion of vegetation. During courtship, male three-spined sticklebacks discard their protective coloration, assuming blatant red underbellies and bluish-red backs, while females school wantonly, inviting attention from males and predators alike.

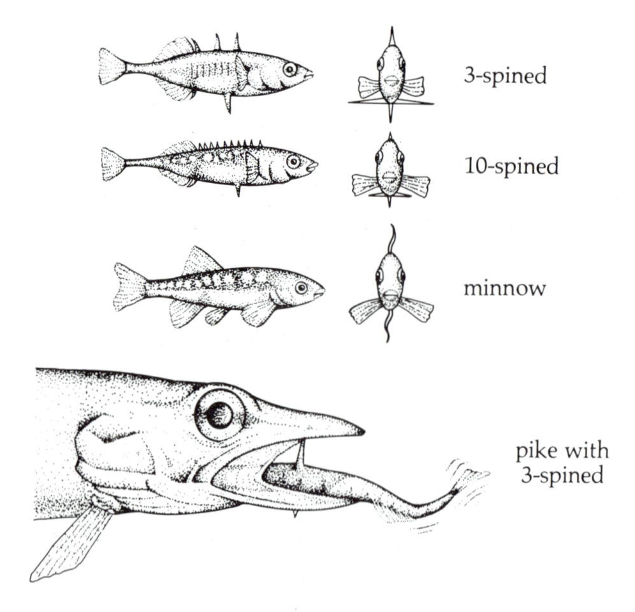

Fig. 2–13 A. In a test of the evolutionary "purpose" of spines, Tinbergen compared the ability of minnows, ten-spined sticklebacks, and three-spined sticklebacks to survive predation. B. The predators (in this case a pike) had particular difficulty with the three-spined fish.

The evolutionary implications of this morphological and physiological disparity between the species are clear: the awesome spines of *Gasterosteus* must somehow compensate for their reckless behavior. A reasonable explanation seems to be that fish with genes for slightly longer spines were less often eaten, bequeathing to the breeding population—fish which survived long enough to mate—an ever-higher fraction of long-spine genes.

The genes of the bolder fish in the parental population, with their enhanced sexual and nutritional fitness, were able to take advantage of longer, deadlier spines to repel the predators which their showy behavior would inevitably attract.

The parsimony and aesthetics of this surmise would in themselves have been enough to content Lorenz, but Tinbergen required a test. With two of his students, Desmond Morris and R. Hoogland, he designed a series of experiments, classic in their simplicity, which demonstrated conclusively the function of the spines. They offered a Chinese menu of prey, minnows and three- and ten-spined sticklebacks, to the fishes' main predators (pike and perch) in laboratory tanks, and "scored" how many of each group remained at set intervals. The minnows disappeared rapidly, demonstrating definitively that they were the predators' first choice. *Pygosteus,* with its bumpy ridge, came in a close but clear second. Not until there was nothing left for it to eat but the three-spined sticklebacks did the pike and perch reluctantly—and with a great deal of difficulty—down the prickly mouthfuls. Indeed, even if they seized the three-spined fish the predators would gape, strain, and "choke" in an effort to reject or at least reorient their prey, which remained determinedly motionless, spines erect, until either swallowed or set free. Some of these encounters lasted as long as an hour, but except with a starving predator seldom resulted in the sticklebacks' being killed, eaten, or even harmed; and on subsequent rencontres the pike and perch were unlikely to show more than a passing interest in *Gasterosteus,* preferring to save their energies for less well-armed prey.

These results are suggestive, certainly, but not quite conclusive to an inquiring mind. It might be that sticklebacks come equipped with some failsafe defense mechanism, possibly chemical—a bad taste, perhaps—that is their redemption. To explore this possibility, Morris and Hoogland carefully removed the spines from the fish and replaced them in the tank, where they were summarily eaten. In this way Tinbergen and his colleagues demonstrated that evolutionary guesses may often be put to stringent tests, and something more solid and intellectually satisfying distilled from airy speculation.

Tinbergen's "observe, then experiment" approach has become the most popular and valuable strategy in ethology. His experiments have again and again unraveled a string of simple components which together comprise a seemingly complex behavior, and have set a rarely equalled standard for care, simplicity, and clarity. His insistence on putting speculations about behavior's adaptive value to a direct experimental test has also been profoundly influential. Not the least of Tinbergen's achievements has been to bring ethology to the English-speaking world, where it has been enthusiastically received.

SUMMARY

Once Darwin, Pavlov, and von Frisch had prepared for its reception, ethology—the study of the mechanisms and evolution of behavior—was born. Darwin showed how natural selection would favor specialized behavioral programming for survival, and so focused attention on the adaptive quality of behavior. Pavlov's discovery of associative learning opened the way to controlled intervention in natural, innate behavior, and proved especially valuable in elucidating the sensory capacities of species. Von Frisch's discovery of the rich, highly specialized sensory world of a mere insect opened the eyes of scientists to the unique perceptual worlds of animals.

The uniqueness of the studies of Lorenz, Tinbergen, and the later work of von Frisch was their focus on the mechanisms and evolution of behavior. Each sought to dissect an animal's behavioral repertoire into discrete units or programs, to determine what cues were being used and in what context, how the information was being processed and behavioral "decisions" being made, and how the response was organized. Finally, each sought to interpret the behavior in the context of natural selection and the contingencies of the animal's world. From these analyses emerged the first organizing principles of ethology, which are the subject of the next chapter.

STUDY QUESTIONS

1. Von Frisch noticed, as had Aristotle before him, that a bee on a particular foraging flight tends to remain faithful to one kind of flower,

bypassing other species during her nectar- or pollen-gathering expedition. Presumably the bee is remembering one or more cues by which the correct species can be identified. Using Tinbergenian techniques, how could you go about discovering the salient cues and placing them in the appropriate hierarchy?

2. Pavlov's original discovery of the unconditioned stimulus focused on the sight of food, most often meat powder. Is it likely that this is a true unconditioned (i.e., innate) stimulus? If not, how would you reinterpret Pavlov's work and its implications for ethology? How would you test your ideas?

3. Why should Tinbergen's wasps—clearly able to see colors, smell odors, and respond to two-dimensional objects—be programmed to ignore all these cues and focus on three-dimensionality when learning landmarks?

4. How do you suppose that Tinbergen came by his estimate of 2–3 m for the accuracy of the homing navigation system of wasps?

FURTHER READING

Blough, D. S. "Experiments in Animal Psychophysics." *Scientific American* 205, no. 1 (1961): 113–22.

Examples of the use of conditioning to define the sensory limits of animals.

Lorenz, Konrad. "Contributions to the Study of the Ethology of Social Corvidae." In vol. 1 of his *Studies in Animal and Human Behavior* (Cambridge, Mass.: Harvard University Press, 1970), pp. 1–9.

In the first pages of this classic study, Lorenz's techniques become clear as he discovers "releasers."

Spalding, Douglas A. "Instinct, with Original Observations on Young Animals." *MacMillans Magazine* 27 (1873): 282–93 (reprinted here as Appendix A).

Spalding discovered many of the phenomena of classic ethology, and performed the first experiments to separate learning from maturation of instinct. His work, however, was forgotten.

Tinbergen, Niko. "On the Orientation of Digger Wasps." In vol. 1 of his *The Animal in Its World* (Cambridge, Mass.: Harvard University Press, 1972), pp. 103–27.

von Frisch, Karl. *Bees: Their Vision, Chemical Senses, and Language.* Ithaca, N. Y.: Cornell University Press, 1971, pp. 1–34.

Von Frisch describes his classic experiments on color vision.

CHAPTER 3

Principles of Early Ethology

Perhaps the single most famous paper in the history of ethology is that of Lorenz and Tinbergen on the egg-rolling response of the greylag goose. It combines for the first time the concepts of releasers, "motor programs," and drives, which together allow us to make sense of so much of animal behavior. Egg-rolling behavior is striking: when an incubating goose notices an egg near the nest, its attention is suddenly riveted. It fixates on the egg, slowly rises, extends its neck over the egg, and with the bottom of its bill painstakingly rolls the egg back up into the nest (Fig. 3-1). With the egg safely back in the nest, the goose nestles down to incubate.

At first sight this looks like a thoughtful and intelligent piece of behavior on the part of the bird: the goose has perceived the problem and solved it. Lorenz and Tinbergen, however, were struck by the stereotyped nature of the whole sequence—the goose performed as if it were a machine. They wondered if the rolling might be some sort of program which, once triggered, would inevitably run to completion. To test this notion, they tried removing the egg once the goose had begun its neck extension. Perversely, the animal went through the rest of the behavior, gingerly rolling in and settling down on the nonexistent egg.

Fig. 3–1 The egg-rolling response of the greylag goose. The behavior begins when the brooding goose notices an egg outside the nest (A) and fixates on it. The goose rises, extending its neck to touch the egg (B). The goose places its bill carefully over the egg (C) and gently rolls it back into the nest (D).

RELEASERS AND MOTOR PROGRAMS

Lorenz and Tinbergen termed the goose's remarkable behavior a "fixed-action pattern" (FAP). The distinguishing characteristics of the behavior are the innate and stereotyped coordination and patterning of several muscle movements which, when released, proceed to completion without requiring further sensory input. In terms of its almost total independence of feedback, the fixed-action pattern represents an extreme class of prewired behavioral performances which have come to be known as "motor programs."

Tinbergen went on to discover a phenomenon more remarkable still: an incubating goose can be stimulated to perform egg rolling by a wide variety of only marginally egg-like objects—beer cans and baseballs, for example. From further experiments it became clear that the egg-rolling behavior could be triggered by virtually any large, nearby convex object with smooth rounded edges—objects some of which, once in the nest, clearly felt wrong to the goose, and which upon investigation were disdainfully discarded as obviously not goose eggs. Lorenz and Tinbergen realized that geese must possess an innate and highly schematic filter which, when stimulated by anything satisfying its crude criteria for "eggness," releases the fixed-action pattern. They called the filter-trigger complex an "innate releasing mechanism" (IRM) while the features of the stimulus essential to triggering it were termed "releasers" or, because the geese responded only to one aspect of the stimulus object, "sign stimuli." In nature, of course, the simple but diagnostic criteria of the IRM are sufficient to exclude almost everything the goose is likely to encounter which is not an egg.

The egg rolling then is a behavioral unit. It has a specific trigger which is satisfied by an egg (among other things), and a specific response, curiously independent of feedback, which results in the recovery of the egg. As we shall see in Chapter 14, this entire unit is turned on and off by another class of circuitry known popularly as "drive" or "motivation," which ensures that eggs are rolled only from the onset of incubation until hatching is due to begin. The recognition of foreign objects in the nest which results in the discarding of a beer can which the goose has just gone to great pains to acquire is a separate program or behavioral unit with its own special cues and motor responses.

In fact, there is yet a third program for dealing with eggs, this one aimed specifically at ridding the nest of broken eggs and getting the shells several meters away. After showing that broken-egg removal depends on sign stimuli for brokenness (sharply defined or jagged edges and concavity; see Fig. 3–2), Tinbergen began to wonder why the black-headed gulls he was studying would *want* to remove broken eggs in the first place. Observation indicated to him that the cause was unlikely to be danger of injury or disease, since the cliff-dwelling kittiwake gull should face the same problems and yet is indifferent to its own empty eggshells. Tinbergen then guessed that predation

Fig. 3–2 A. Objects shown by Tinbergen to trigger the eggshell-removal program generally have sharp edges and flat or concave contours. B. Objects which are convex and lack sharp edges (as long as they are small enough to be comfortable) are tolerated in the nest and dutifully incubated. The dropper and skull are intriguing anomalies.

might represent a strong selective pressure for ground-nesting birds, and that the bright-white interiors of the eggs, when exposed, destroyed the camouflaging effect of the speckled, earth-toned exteriors. Hence, a broken egg would be likely to attract the attention of sharp-eyed predators, and lead them to a nest of eggs and chicks.

Tinbergen tested this hypothesis by setting out an array of both normal gull eggs and gull eggs painted white. Crows and even other species of gulls began to prey on the eggs almost immediately, even though they were set out well away from the nesting area. The predators took more than 60 percent of the white eggs, but found fewer than 20 percent of the normal ones. That evolution does a superb job of camouflaging eggs and wiring predators to spot them was indicated by the ability of predators to find the hand-colored mimics of gull eggs that Tinbergen set out in another experiment almost as well as the white ones. Trying the same experiment with only real, unpainted eggs, with and without broken eggshells 5 cm away, confirmed Tinbergen's guess (Table 3–1). Two-thirds of the intact eggs which were near broken ones were taken, while predators found only one-fifth of the other eggs. Of course in a real colony where there are adults to defend against predation, losses would be lower, but a three-to-one improvement in the rate of loss to predation is far more than is needed to drive evolution and lead to eggshell removal. Even the distance to which broken eggs are removed is important: the farther away, up to 2 m, the better. Hence, the removal behavior is almost certainly a product of predation pressures.

TABLE 3–1 *Survival Value of Eggshell Removal*

Distance from egg to eggshell (cm)	*% eggs taken by predators*
5	65
15	42
100	32
200	21
no eggshell	22

Source: N. Tinbergen et al., "Egg Shell Removal by the Black-headed Gull," *Behaviour* 19 (1963): 74–117.

SUPERNORMAL STIMULI

Tinbergen's insistence that evolutionary guesses must be put to the test brought to ethology that essential rigor which separates modern science from the armchair speculations of its Aristotelian and medieval precursors. Moreover, his informed and restrained experimentation uncovered phenomena of sweeping significance which Lorenz's purely observational approach, though powerful in its own way, could never hope to have touched. For example, in the process of asking just how unegg-like an object of human manufacture could be and still trigger the egg-rolling releasing mechanism, Tinbergen found that incubating geese would even attempt to roll volleyballs into their nests. This observation led immediately to one of Tinbergen's classic choice experiments. Allowed to decide between a goose egg and a volleyball, geese inveterately chose to recover the volleyball. In some way volleyballs seemed to be better stimuli for the goose's IRM than real eggs. In other words, there are "supernormal stimuli" (Fig. 3–3).

Fig. 3–3 A supernormal stimulus. Given a choice between its own egg *(left)* and a giant egg, a brooding oyster catcher chooses the larger of the two.

Supernormal stimuli, although inherently unnatural, provide a crucial clue to the process of natural selection. It is clear that the natural releaser, the goose's egg in this case, is only one of many stimuli which could satisfy the animal's simple criterion and release egg rolling. Since variation exists in nature, and since changes in some stimulus features like the egg's size can result in an increased ability to trigger the response, selection should favor the genes leading to more potent releasers. In other words, goose eggs should eventually become the size of volleyballs.

Of course, evolution must balance conflicting demands. The goose's egg, for example, must represent a balance between the goose's physiology—how large an egg it *could* lay and incubate—the pressure of sharp-eyed predators, the optimum number of offspring, and so on. The most dramatic examples of this balance are found in sexual selection. Although there are rarely two eggs outside the nest for an incubating goose to choose between, animals quite often have several members of the opposite sex to compare. Species recognition usually involves releasers, and the role of the releaser can be quite complex. Animals frequently must compete for the opportunity to pass on their genes by mating. It is a fact of life that an animal which does poorly in sexual competition leaves no offspring, and so its genes are not perpetuated. Animals whose genes code for features which most stimulate the opposite sex, on the other hand, will leave disproportionately more offspring. As a result, sexual sign stimuli should become increasingly supernormal, though again under the restrictions imposed by the natural world.

Because mating is so crucial, sexual selection can have bizarre results. Darwin, in his *Descent of Man,* gathered together a long list of examples of species in which the males possess some dramatic sexual dimorphism, a structure whose existence seems exclusively the product of sexual competition among males, and whose metabolic cost, attractiveness to predators, and general unwieldiness must be a substantial burden (Fig. 3–4). Fiddler crabs are a compelling example of this system. Male fiddler crabs spend most of their time "beckoning" to females with their one oversized claw (Fig. 3–5). The claw-waving display is a simple, species-specific sexual releaser. The pattern of waving is a code which is unique to, and therefore an indicator of, the species; and as such is used by female fiddler crabs to sort out which males are which. Once male fiddler crabs had the temporal pattern of the display pretty well perfected, the conspicuousness of the waving must have become very important in attracting the attention of females in the presence of other waving males. Natural selection—which is to say, the female fiddler crab—has favored ever larger and more conspicuous claws. The enormous claws of modern fiddler crab males are useless for feeding or for digging burrows, and are known by predators to contain a very tasty meal; furthermore, they pose a great impediment to movement and escape. Clearly the morphological arms race has gotten out of hand, but the IRMs of the females

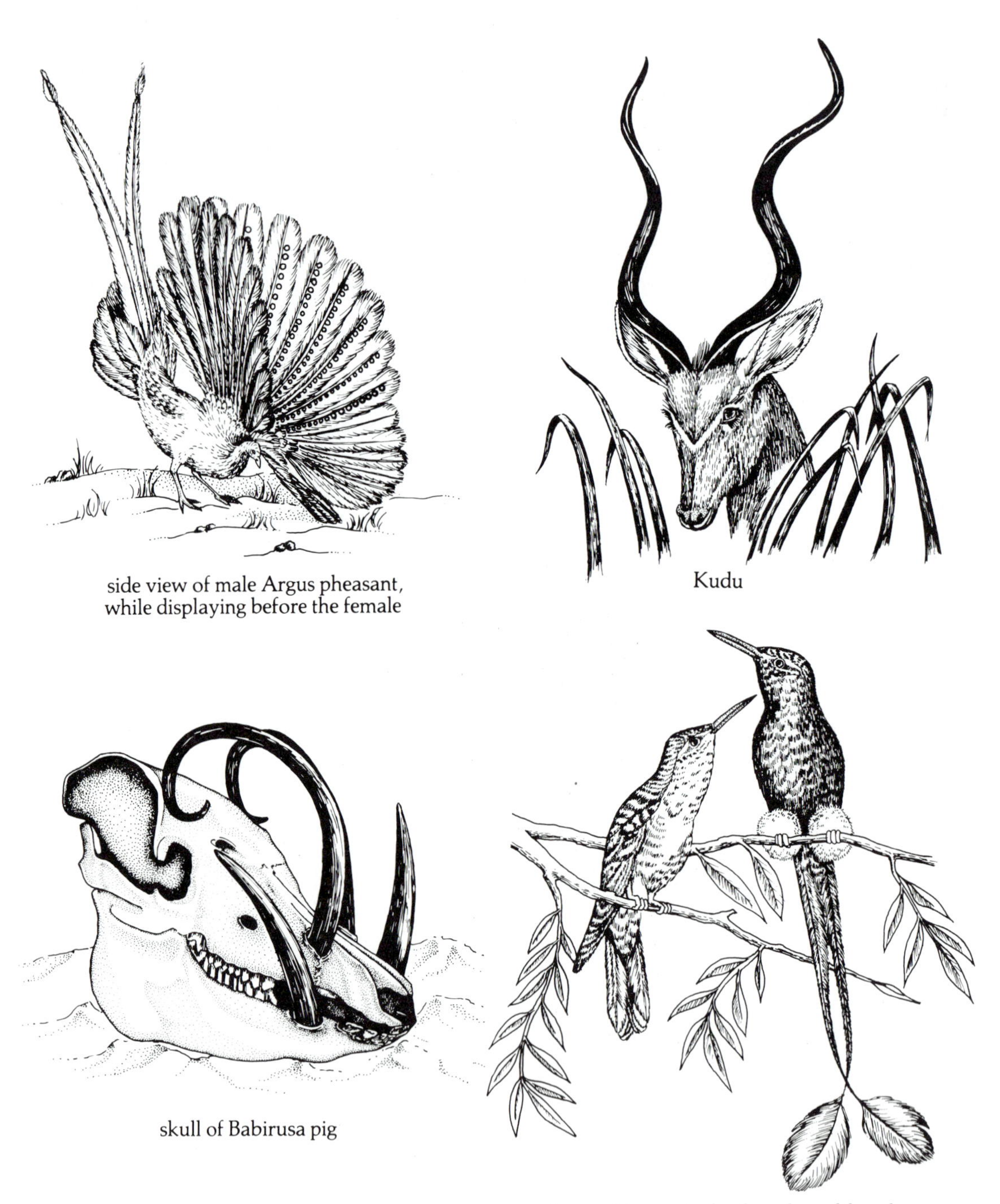

Fig. 3-4 Examples of sexual dimorphisms used by Darwin to illustrate sexual selection.

make disarmament out of the question. Evolution seems here to have operated like a ratchet with its one-way movement, and the sexual advantage of even larger claws is almost certainly still there, awaiting perhaps only a reduction in predator pressure to remove the present constraints. Only extinction or the establishment of some selective advantage for females without this releaser circuitry could bring an end to the supernormal claw of male fiddler crabs.

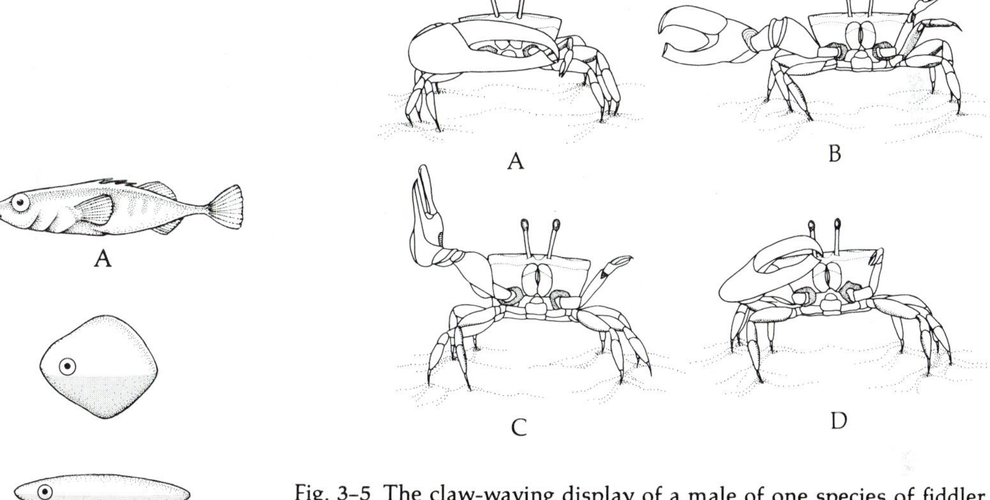

Fig. 3-5 The claw-waving display of a male of one species of fiddler crab.

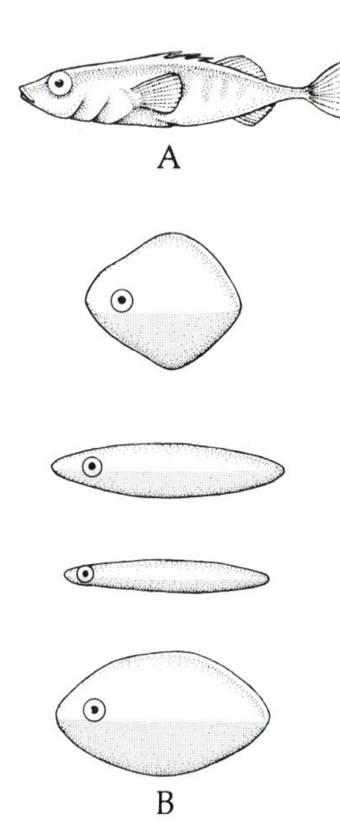

Fig. 3-6 Stickleback models. An accurate model of a stickleback without any real coloration (A) fails to elicit attacks from territorial males, while any of a variety of unfishlike models with red undersides (B) effectively release attack behavior.

Perhaps the most famous example of an IRM other than the egg rolling of geese is the territorial and mating behavior of the three-spined stickleback mentioned in Chapter 2. The males of this versatile species of fish divide their world into territories which they guard jealously. During this territorial phase the underside of each male becomes bright red, and the approach of a neighboring male in territorial garb releases an aggressive display or even an attack at the invisible boundary between the two males' territories. From the first Tinbergen suspected that the red belly was the releaser for this behavior. When a passing red postal truck elicited attacks from the males, there was little doubt that the sign stimulus was the color, and not the body shape or odor of other males. In typically Tinbergenian fashion, a series of increasingly unfishlike models were presented to territorial males (Fig. 3-6), and the red stripe emerged as the one necessary criterion. In fact, the red stripe conveys a double message: when the male is swimming in its normal, horizontal orientation, it indicates a fish in full aggressive mood; but when the loser in an encounter adopts a vertical posture, it signals submission and an intention to retreat. That evolution can get away with such a simple sign stimulus as a red bar can be taken as pretty good evidence that there is a dearth of red objects in the natural world of the stickleback.

Almost the same story may be told of European robins: the red feathers on the breasts of territorial males are the sign stimulus for aggressive interactions. Just as with the sticklebacks, a red object of almost any shape is sufficient to release aggressive behavior—a tuft of red feathers on a wire, for example. And again, as with sticklebacks, we must suppose that the trees of spring contain little else that is red.

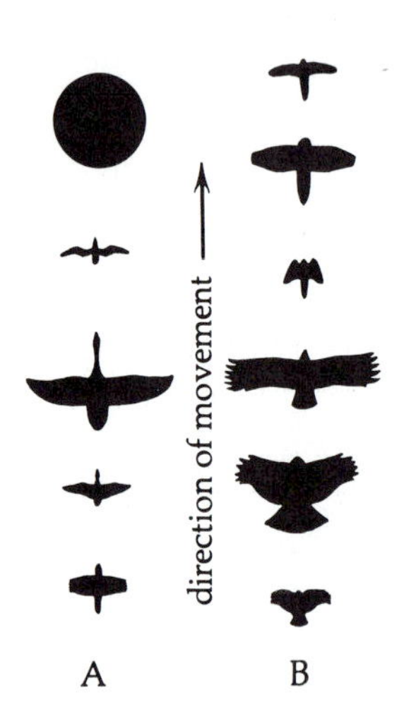

Fig. 3-7 Some bird silhouettes evoke escape responses (B) in naïve chicks, while others do not (A). "Hawkness" to birds seems to be mostly a matter of relative neck and tail length.

A famous releaser reported by Spalding, rediscovered by Lorenz, and systematically investigated by Tinbergen, involves the escape response of naïve chicks when they are shown the silhouette of a flying hawk. After numerous model experiments, it now seems clear that a short head and a long or broad tail are the features necessary to trigger the escape behavior (Fig. 3-7). The most elegant test utilized the silhouette of a flying goose. Presented normally the model had no effect, but when flown in reverse so that the broad, short tail became the head, and the long neck and head became the tail, the escape response was triggered (Fig. 3-8). The adaptive value of this behavioral unit is clear: it is crucial that young birds "know" to hide from hawks from the first, and the simple but diagnostic requirements of the IRM should be satisfactory for even the most myopic chick.

The same adaptive value for an IRM holds true for snake-eating birds: in many species, the sight of the deadly coral snake releases alarm behavior. Like predatory birds, the various species of coral snakes share only a few features, and it is from these that natural selection has fashioned the IRM. Birds which had never before seen a snake were presented rods painted with various patterns. Stripes running lengthwise, regardless of

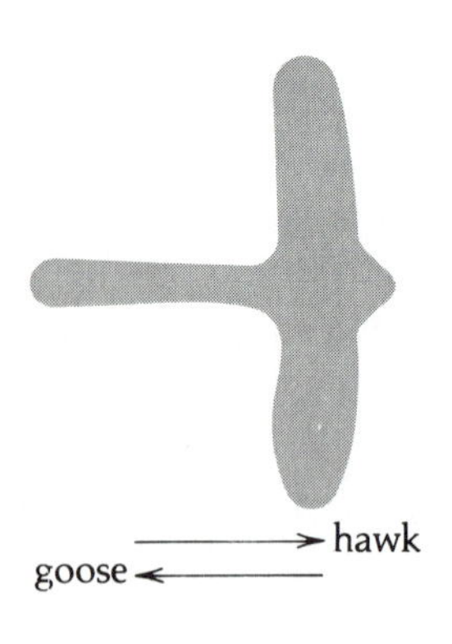

Fig. 3-8 The same bird silhouette has very different meanings to naïve chicks depending on its direction of travel. When "flown" to the right with the short neck leading it releases escape behavior, but when flown to the left with the long neck leading it is ignored.

color, never upset the birds. Rods with alternating blue and green rings were likewise ignored. But a clumsy, hand-held wooden rod with just a pair of red and blue rings at one end—quite obviously a dowel and not a snake—inevitably elicited alarm from the hatchlings. Clearly a bird which does not recognize and avoid coral snakes from the outset may not survive the experience of learning.

INTERLOCKING RELEASERS

Very often, particularly in courtship, IRMs and FAPs are arranged in serial fashion. For example, in butterfly courtship a behavior on the part of the female releases a behavior in the male, which then elicits the next behavior in the female, and so on until mating is achieved. The first releaser is the flight pattern of the female. Tinbergen showed through the usual model experiments that three factors trigger pursuit by males: the contrast between the dark female and her lighter background, the "bobbling" flight pattern, and the rapid alteration of her apparent size which is a consequence of her wing flapping. Tinbergen demonstrated that neither the details of the elaborate wing patterns nor even the butterfly shape are important. On the other hand, higher rates of wing flapping—presumably impossible for physiological or aerodynamic reasons—are more attractive to males than the natural rate.

Males of the queen butterfly respond to these sign stimuli through pursuit, followed by an elaborate behavior known as "hair penciling." The males possess fine brush-like structures at the tips of their abdomens which they extrude and wave in front of the flying females. These hair pencils release a special odor which the females detect with their antennae.

Females respond to the male odor by alighting with wings spread. The male responds by hovering in front of the female, sweeping his hair pencils across her antennae until she responds by closing her wings. This is the signal which causes the male to alight next to her, where he begins drumming her antennae with his own until copulation begins.

The elaborate series of releasers in butterfly courtship serves to ensure that mating occurs only between members of the opposite sex of the same species, both of whom are physiologically ready to mate (Fig. 3–9). The flight-pattern releaser informs the male of neither the sex, the species, nor the physiological readiness to mate of his intended partner. The hair penciling of the male, on the other hand, informs the female of all three. If the courted butterfly lands and folds its wings, the male can afford to proceed in the certainty that the species, sex, and readiness are all appropriate.

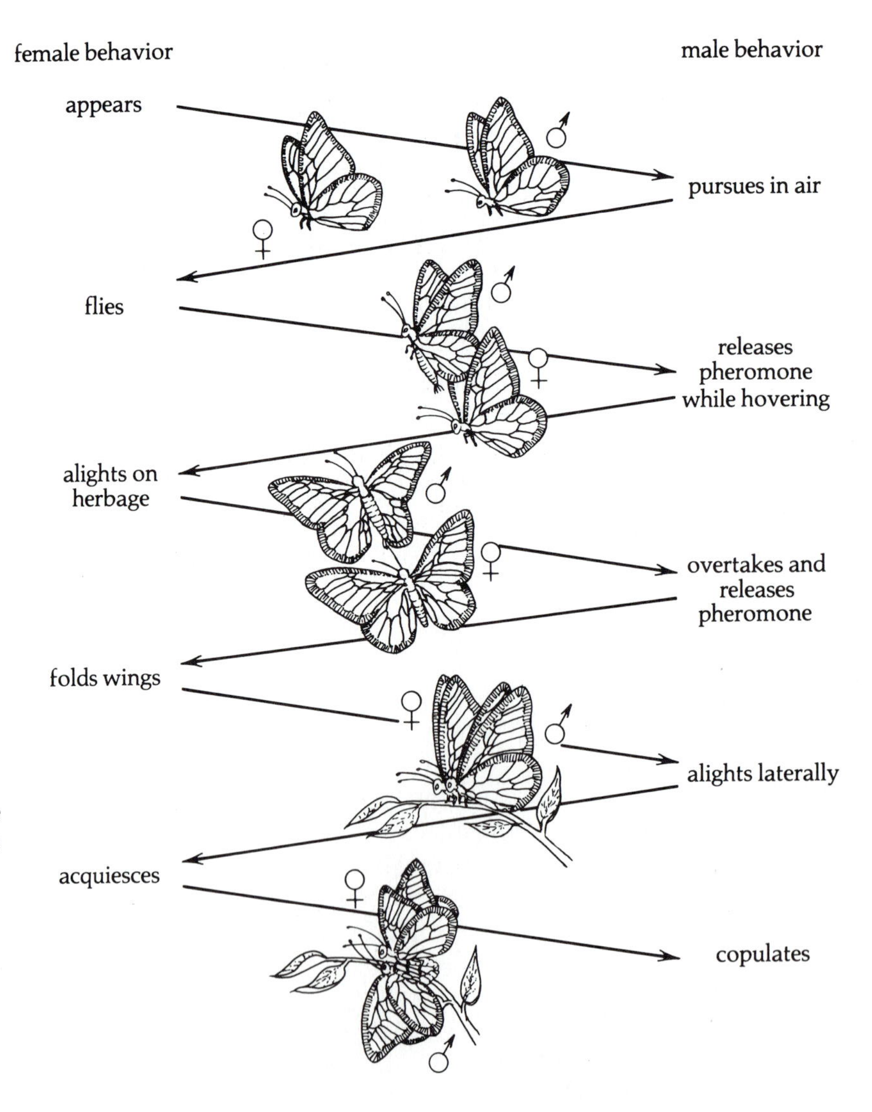

Fig. 3–9 Courtship sequence of queen butterflies. The female's flight pattern releases pursuit and hair penciling by the male. The hair-pencil odor releases settling by the female. This causes the male to hover and sweep his hair pencils across her antennae. The female responds by folding her wings, the signal which causes the male to alight and begin mating. This interlocking sequence of releasers assures that only queen butterflies in reproductive condition and of opposite sex actually mate.

THE "DO-LOOP"

One of the most frustrating things about working with computers is their consummate stupidity. The slightest error on our part, even so much as a misplaced comma or extraneous space which leaves the meaning of the instruction clear to us, confounds these simple-minded computational giants completely. This inflexibility is one crucial feature which distinguishes man from even the most complex of the machines he creates—at least from the ones that work. The computer lacks even any rudimentary insight or imagination, and so must be told not only what the problem is that it is going to be called upon to solve, but exactly how to go about solving it—and even that in agonizing mechanical detail.

Suppose, for example, we ask our computer to calculate something so simple as the number of doublings necessary to generate a million offspring. That problem requires fewer than a dozen English words to state, but for the computer the problem must be put more explicitly:

Step #	*Instruction*
1	Let $A = 1$
2	Let $N = 0$
3	Do steps 4 through 6 until A is greater than or equal to 1,000,000; then go to step 7
4	Let $A = 2 \times A$
5	Let $N = N + 1$
6	Return to step 3
7	Print N, the number of doublings
8	Stop

Obviously, this program requires that the grammar, the functions (addition and multiplication), and the complex conditional cycling instruction of step 3 be "hardwired" or already stored in the machine by some previous programmer. Step 3 is known to users of FORTRAN (one of the most widespread of the computer "languages") as a "do-loop": the machine cycles mindlessly through the steps until the contingency ($A \geq 10^6$) is met. Typical programming errors might include requiring A to *equal* 10^6, which, as the calculation leaps from 524,288 to 1,048,576 in one step, never happens; or to mistype "A" as "S" which, as far as the machine knows, is nonexistent, and so cannot reach 10^6. The result in either case is that the computer, slavishly following its instructions, gets hung up in an endless loop of perfectly exact but pointless calculations. Such errors of programming, when they occur, must be painstakingly culled out by the wary programmer through trial and error in a process known as "debugging."

We might say that evolution, the wiliest of programmers, has been elaborating and debugging its programs for millions of years. One of its most intriguing debugging solutions for keeping its creatures from getting hung up in do-loops is the phenomenon of habituation or, more familiarly, boredom (see Chapter 5). Nevertheless, some behavioral programs, never having gone awry under the orderly auspices of nature, have never needed to have this safeguard built in. These cases, more plainly than any others, illustrate the machine-like nature of animal behavior.

Take, for example, the species of wasp which builds those familiar tunnel-like mud chambers on the sides of houses. The wasp carefully seals the bottom of each chamber and drops prey she has paralyzed in from the top. Then she lays an egg, closes the top, and begins another nest. This seems reasonably clever at first sight. However, if we cut off the bottom of the nest so that it is open at both ends, the wasp becomes caught in a do-loop. She captures prey and drops them in from the top, but they fall out the bottom. The wasp may even explore the nest and emerge from the open bottom without repairing it. Instead, she continues to try to stock the chamber. It seems that the only contingency which nature has programmed in to instruct her to "exit the loop" is a full chamber.

Apparently wasps in general are remarkably resistant to boredom. One species of digger wasp which, as we know, builds its nests in the ground, preys on crickets. This wasp typically returns from its hunt to the burrow, sets the cricket down about an inch from the entrance, goes briefly into the nest (apparently to check on things), and then returns to the surface and takes the paralyzed cricket back down. Again this seems sensible enough. If, however, we move the cricket away from its original spot even a bit, the wasp must search a moment for it. Then she moves the cricket back to the exact spot on which she had placed it before, and reexamines the nest. As you can guess by now, this sets up a loop. As long as we keep moving the cricket, the wasp cannot go on to the next step, and will replace the cricket and reinspect the nest at least as many times as human patience can endure. J. H. Fabre, the eminent French naturalist, tried forty times before throwing up his hands in exasperation.

These examples illustrate not only the machine-like nature of some behavior, but also how we can go about understanding the programming instructions the genes have generated. The essential element in these cases was the lack either of some contingency plan—boredom, for example—to terminate the loop, or of one of the "backup routines" so obvious in animal navigation (see Chapters 13 and 14) to which the whole problem may be referred when the first-order routine fails to generate an answer. In examining other cases we should be looking for simple manipulations for which evolution could not have prepared the animal, and which ought to affect the presumptive program in revealing ways. In Chapter 13, for example, we will see how merely closing a beehive for two hours at midday causes the bees' superficially complex sun-movement compensation behavior to fall apart or, in Chapter 14, how attaching a magnet to a pigeon on a cloudy day destroys its mysterious ability to get home. Nature never performs such irrational experiments, so the animals' programming is confounded in revealing ways.

WHY RELEASERS?

Releasers, then, and the accompanying physiological arrangements which produce the IRM and the associated motor program, represent the single most general strategy in animal behavior from prokaryotes to primates. The concept of releasers dramatically illustrates the innate, programmed nature of much of behavior. At the same time, a releasing mechanism is not an explanation but rather another ultimately divisible component of behavior, and one which has proved especially troublesome. It was not easy for early ethologists to account for the selective advantage of releasers. They proposed that IRMs might have evolved before animals developed the intellectual capacity to "reason" for themselves, and have been preserved by that most pragmatic and conservative force, evolution. It is undoubtedly the case that IRMs enable animals to react more quickly in a variety of situations and moods, without interference from the time-consuming and error-prone process of thinking. Perhaps they serve to focus an animal's attention, allowing it to ignore irrelevant and confusing stimuli in situations that are particularly important to its genes. They may enable animals to perform perfectly some crucial piece of behavior when there is no room for learning because even a small mistake may be fatal. Or just as often IRMs might show an otherwise distracted animal exactly what information it needs to acquire. It might be, too, that IRMs simply began as a way to compensate for relatively crude sense organs—eyes and the like—and have proved useful despite the enormous technical refinements evolution has accomplished along the way.

Against all these alleged advantages stand two very obvious objections: IRMs occasionally trigger responses to cues which are obviously inappropriate, responses which are maladaptive and which, with only a slight increase in IRM specificity, might have been avoided; and IRMs in sexual selection can force species such as fiddler crabs (and perhaps the sabre-toothed tiger and the giant elk before it) down evolutionary paths which are potentially debilitating or even fatal. As we shall see in Chapter 4, one of the most impressive consequences of the discoveries which led to modern ethology is a convincing, mechanistic explanation of the evolution of IRMs.

IMPRINTING AND PROGRAMMED LEARNING

In addition to releasers, motor programs, and drives, the early ethologists discovered one more dramatic and controversial example of behavioral programming: imprinting. Originally thought of as a remarkable curiosity, imprinting has become the classic example of programmed learning, a widespread phenomenon whose sweeping importance makes it one of the cornerstones of modern ethology. Lorenz, following up an earlier observation by Oskar Heinroth, found that orphaned baby geese and ducks would begin to follow him as they would a parent of their own species as long as he "adopted" them before they were two days old. Later, they would ignore members of their own species, evidently having accepted Lorenz as their model of the perfect parent (see Fig. 2–6). Still later, the geese would court humans instead of other geese. Young geese isolated until two days of age, however, failed to imprint on anything. Lorenz referred to this age of susceptibility as a "critical period" (known more commonly now as the "sensitive period"). He concluded that the birds would imprint on anything, that imprinting etched indelibly a general species configuration in the birds as well as enabling them to identify their particular parents, and that the process was irreversible.

But as with releasers, motor programs, and drives, the early ethologists had two sorts of problems with imprinting. The first, far more severe with imprinting than with the other three, was that it was hard to repeat Lorenz's observations under controlled conditions. Chicks simply did not imprint well for other people, partly because these other scientists used the wrong techniques (Lorenz had failed to report many of the essential details, the importance of which even he was mostly unaware) and partly because certain of Lorenz's untested interpretations were wrong. The other problem is more familiar: imprinting, like releasing mechanisms, seems curiously unselective and maladaptive. Mistakes seem inevitable, mistakes fatal for the chick which must identify its parents correctly to receive their protection and care, and which must later select an individual of the proper species with whom to mate, on the basis of a "follow-whatever-moves" program. Where is the decisive selective advantage of this remarkable behavior?

As with IRMs, one of the triumphs of modern ethology has been the unraveling of this conundrum. As we shall see, the mechanistic and evolutionary insights which have resulted help explain nearly the whole realm of adaptively flexible behavior in animals as simple as bees or as complex as birds, primates, and perhaps even ourselves.

SUMMARY

Animals come neurally wired to recognize important stimuli in their environments on the basis of one or two simple but diagnostic features. These "sign" stimuli are then used to trigger appropriate behavioral responses. The most obvious of such behavioral reactions are motor programs—self-contained neural circuits which in the most dramatic cases produce a coordinated muscle performance (a fixed-action pattern) wholly independent of feedback. Two other much more subtle but equally critical reactions to releasers are learning and changes in drive. Drive or motivation controls an animal's responsiveness to releasers and other stimuli, while programmed learning as exemplified by imprinting directs an animal to acquire particular information at an appropriate time and from a correct source. Together, these four classic phenomena form the basis of modern ethology.

STUDY QUESTION

Look back at the hunting sequence of Tinbergen's digger wasps (Chapter 2) and interpret it in terms of releasers and motor programs. How does it compare with butterfly mating? Why should the wasp, having caught an insect to sting, care whether it is actually a honey bee or not?

FURTHER READING

Lorenz, Konrad, and Tinbergen, Niko. "Taxis and Instinct in the Egg-Rolling Response of the Greylag Goose." In vol. 1 Lorenz's *Studies in Animal and Human Behavior* (Cambridge, Mass.; Harvard University Press, 1970), pp. 328–42.

Tinbergen, Niko. "The Curious Behavior of Sticklebacks." *Scientific American* 187, no. 6 (1952): 22–26.

CHAPTER 4

The Emergence of Modern Ethology

RELEASERS AND PATTERN RECOGNITION

If the discoveries of early ethologists had so much truth in them, and the ethological view of animals as machines brought so many intriguing insights, how is it that behaviorism thrived while ethology all but withered away in the 1950s and 1960s? In fact, early ethology suffered from a conceptual and physiological roadblock: What were those all-important IRMs in neural terms; how could they be wired to accomplish all that they were supposed to be responsible for; and how could they, imperfect as they are, have been favored by natural selection? Neurophysiologists could offer no help—none of the nerve cells they listened in on could be coding for "eggness" or the other pattern-recognition functions attributed to IRMs. As we shall see in Chapters 7 and 9, the specialized neurons they were finding in the visual and auditory systems coded for very simple and discrete features; and so, in the absence of that mechanistic link so basic to ethological thinking, the whole notion of IRMs began to fall into disfavor, and ethology with it.

Ethologists, meanwhile, could only suppose that neurobiologists were somehow missing the crucial cells; but in fact, the problem lay with ethology's conception of the IRM. Let's consider, for example, the releasers involved in the feeding of nestlings. There must be one set to tell chicks, which for their own safety ought to be quiet and inconspicuous, when to begin their noisy begging and at what to direct their pleas. These auditory supplications, along with the brightly colored throat patches of the young, release and direct feeding by the parents. In most species of birds the nestlings are born with their eyes closed, and so depend initially on nonvisual cues to tell them when to begin their vocal, highly conspicuous and competitive begging.

In an elegant series of experiments, Tinbergen showed that the releaser for gaping in thrushes (Fig. 4–1A, B) was the impact

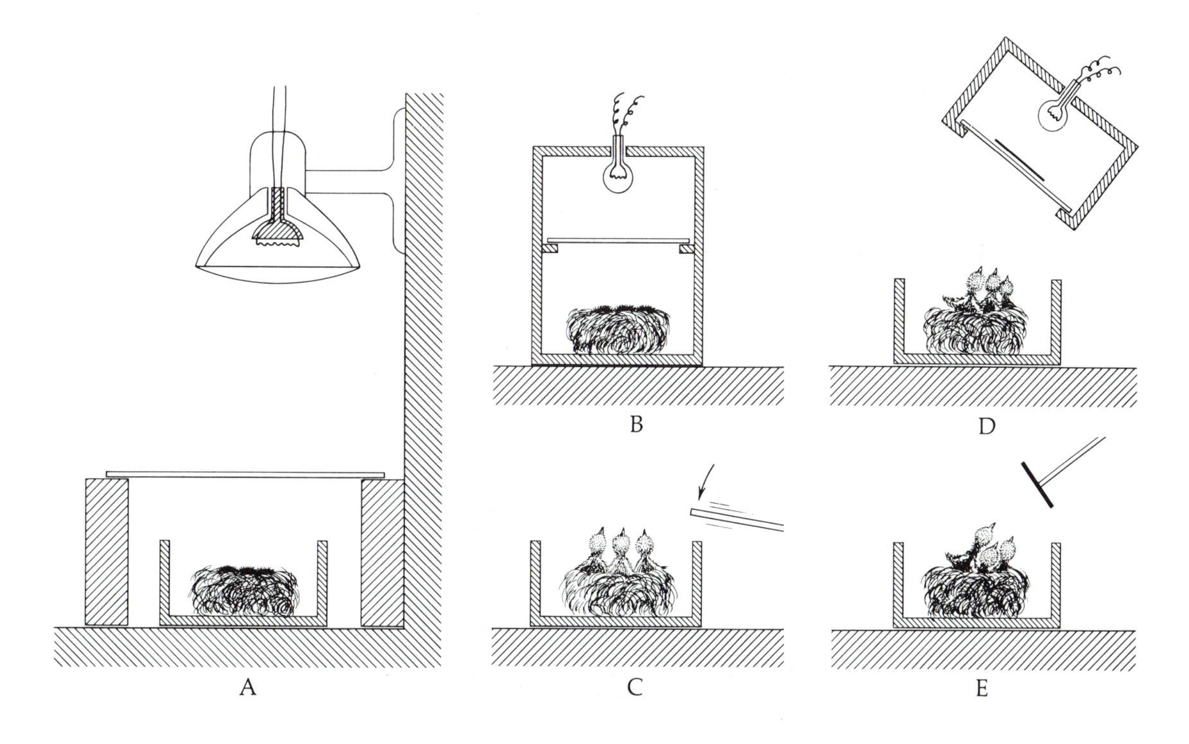

Fig. 4–1 Tinbergen tested the various stimuli which could conceivably release begging in baby thrushes. With a variety of experimental arrangements, he showed that the change in ambient light level (A) and temperature (B) occasioned by a parent's arrival were ignored, as was the sound; whereas the impact alone (C) proved sufficient to rouse the huddled chicks. Once the nestlings opened their eyes, Tinbergen offered various visual stimuli (D, E) to see which elicited gaping, and where the nestlings' efforts were directed.

of the parent on the nest edge and not the air turbulence it stirs up, nor the changes in light intensity or temperature created by a parent's arrival, nor the sound of the arrival itself. Despite all the cues available to the sightless chicks, the one which releases gaping, which in turn begins the feeding sequence, is a classic example of a sign stimulus which depends on a single stimulus characteristic: the impact (Fig. 4–1C).

Once hatchlings open their eyes, visual cues transform the blind, upward gaping into a highly directed behavior. Tinbergen set out to discover what innate knowledge allowed the chicks to identify their parents. His basic technique was to offer cardboard models and observe which elicited gaping or pecking, and just where on the model these behavioral requests were directed (Fig. 4–1D, E). Tinbergen's most complete study was on herring gulls, a species whose chicks are born able to see from the outset. He concluded from this study and others that baby birds have some sort of vague, shadowy picture of their parents to guide their efforts. For example, each herring gull parent has a red spot on the lower tip of its bill (Fig. 4–2A), and when it has food for the chicks it waves its downward-pointing bill slowly

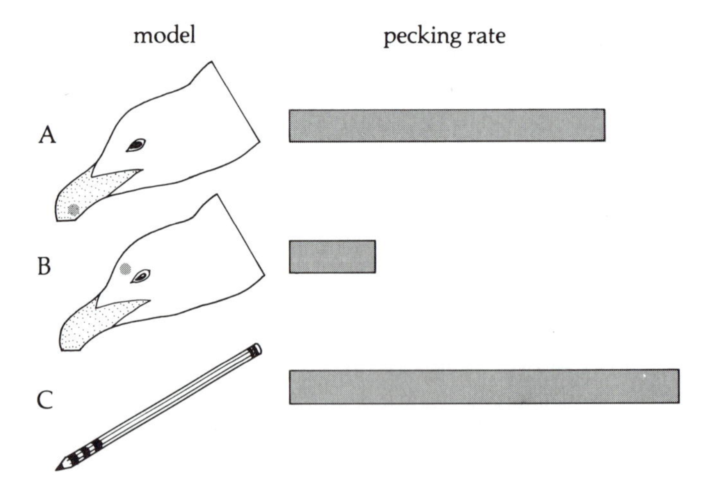

Fig. 4–2 Three models used by Tinbergen to elicit pecking by chicks. Model A corresponds to a normal herring gull with its red spot near the tip of the bill. Model B had the red dot moved to the forehead, where it was only one-fourth as effective. From this, Tinbergen concluded that the chicks "know" that the spot belongs on the tip. Despite this apparent selectivity, chicks prefer a pencil with three rings to anything else (C).

Fig. 4–3 A parent herring gull waves a bit of food in front of its chick. The food itself is irrelevant since the chick will peck at the red spot on the bill whether there is food there or not.

Fig. 4–4 Herring gull chicks are more stimulated by a stick waved in front of them than by a parent's beak.

back and forth in front of them (Fig. 4–3).

If the chicks' pecking were released by a single stimulus element (a classical sign stimulus) we would guess that the moving red spot would be the key. According to Tinbergen, though, the *relative configuration* of head, bill, and spot are important. For instance, if he placed an artificial spot on the forehead of the model of the parent bird rather than on the bill (Fig. 4–2B), the chicks' pecking rate was only one-fourth as high. Clearly, Tinbergen might argue, if the spot itself were the releaser, either position ought to work equally well. The chicks, he felt, must be using an innate picture, an inborn idea of what its parents ought to look like.

This elusive notion of releasers as being organized patterns such as "eggness," "hawkness," or what have you, innate mental pictures to be compared against the objects around an animal, pervades early ethology. And yet Tinbergen also found that a simple stick with a red ring was an even more potent releaser to herring gull chicks than an accurate model of the parent's head, spot and all, despite the stick's obvious configurational shortcomings (Fig. 4–2C and Fig. 4–4). A careful comparison of the various models (Fig. 4–5) reinforces our doubts about innate pictures.

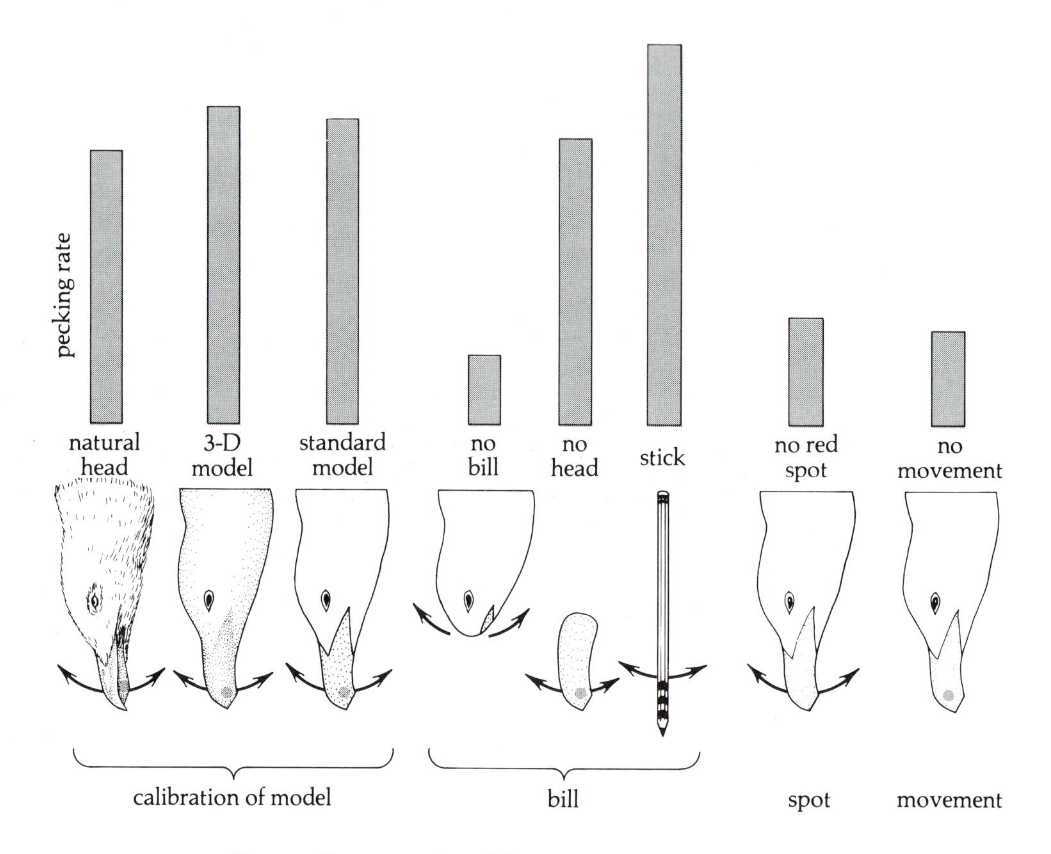

Fig. 4–5 Summary of model experiments with herring gull chicks. The set of three models on the left serves to calibrate the standard model—a flat, white head and yellow bill with a red spot on the tip held beak down and moved from side to side. Most of the remaining models are independent variations of one potentially important feature. The model without a bill was not used by Tinbergen, the data coming instead from Hailman's later work. The crucial features seem to be a vertical horizontally moving bill with a spot. Only the model on the far right (the "Configuration" model) favors the innate-picture hypothesis.

RELEASERS AND SIGN STIMULI

Clearly something is wrong here, and in this contradiction is that stumbling block which separated early ethology from its modern counterpart. Although the emergence of modern ethol-

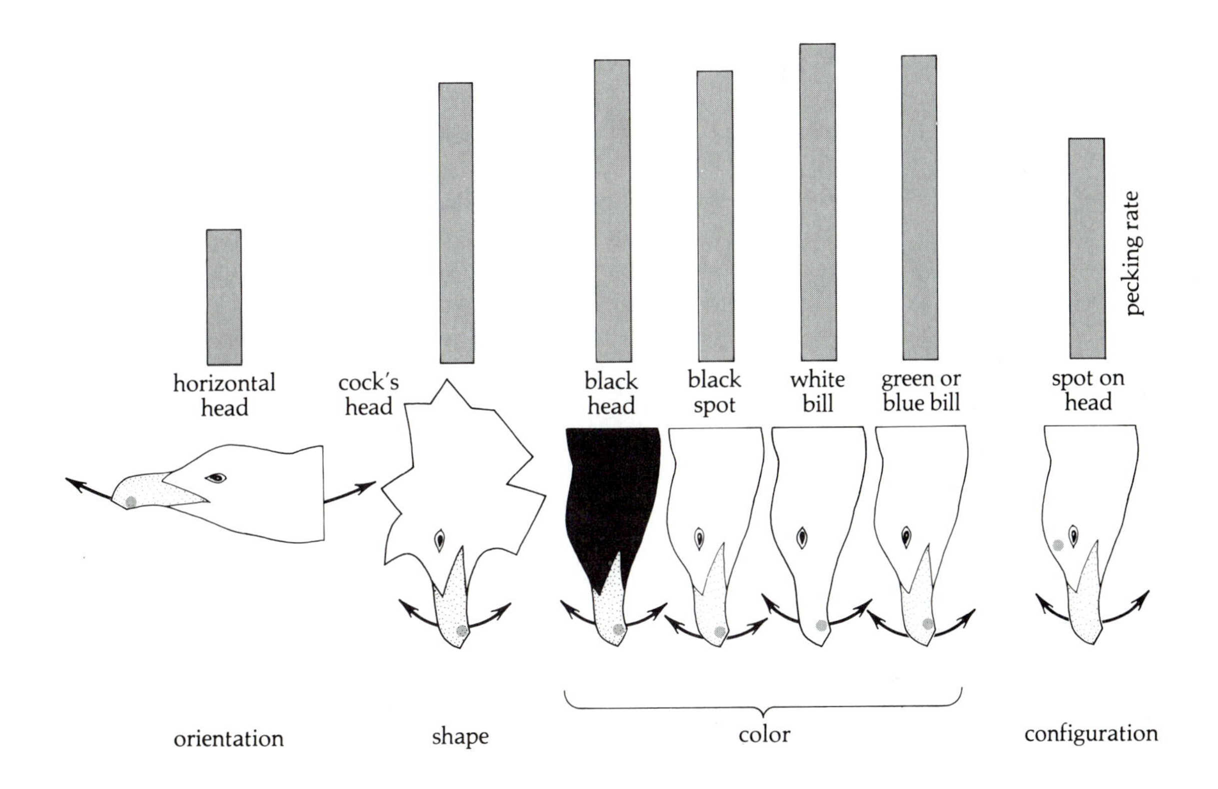

ogy has been painfully gradual—indeed, is still going on—the credit for isolating the insidious misconception which blocked the way for so long goes to the American ethologist Jack Hailman. Hailman, like so many of ethology's greatest contributors, began as a skeptic. He saw nothing in Tinbergen's results which could not be explained by conditioning or trial-and-error learning. For example, suppose that chicks have a pecking "reflex" and simply learn to associate a food reward with pecking at the parent. In this case no IRM in the chicks may be necessary and the reactions of chicks toward models is simply a measure of how good a picture the chicks have formed through experience. Hailman found, though, that the parents do indeed have releasers which trigger and direct the pecking of their chicks; but at the same time, he demonstrated that learning is a crucial element in this process.

Hailman began by showing that the gulls' pecking accuracy improved with age, presumably just as Spalding's chickens' pecking matured (see Fig. 2–1). Then, using laughing gulls (red bill, black head), he asked what *their* releaser was. By comparing a series of increasingly schematic models, Hailman established unambiguously that for newly hatched chicks of this species, only the parent's bill was imporant—indeed, a disembodied bill worked just as well as the whole head, bill, and neck; whereas a bill-less head was virtually ignored. Moreover, color was irrelevant: a white head and bill was almost as effective as the naturally colored one. Hailman also showed that the orientation and specific movement of the bill was crucial: a stationary vertical bill is twice as effective as a horizontal one, and horizontal movement is still more stimulating while vertical motion is ineffectual (Fig. 4–6). Of course, these criteria accurately describe the position and movement of the parent's bill during feeding.

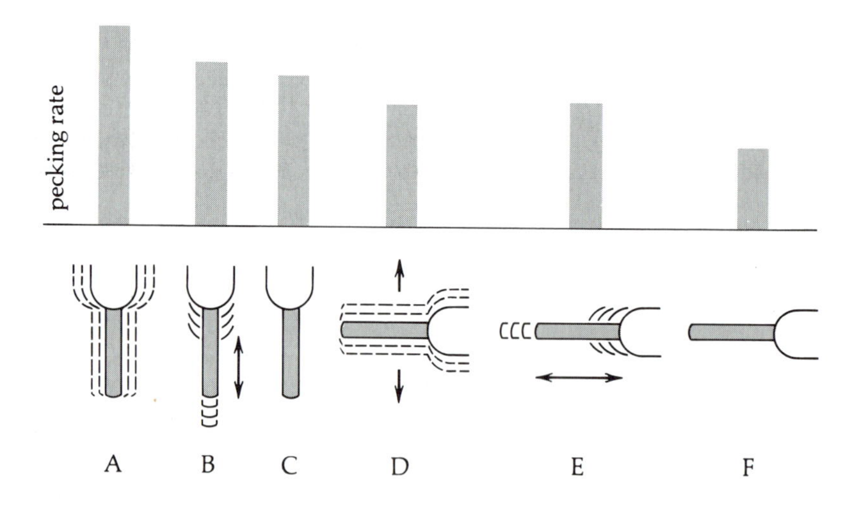

Fig. 4–6 When Hailman tested the preferences of naïve gull chicks with wooden dowels, he found that the normal vertically oriented "bill" moving horizontally (A) was more effective than a vertical bar moving vertically (B) or not at all (C). Horizontally oriented "bills" were even less effective (D–F).

When Hailman turned to herring gulls he found virtually the same story: both the bill *and* the red spot were important, though not their relative configuration. Using naïve chicks, Hailman demonstrated that it is the *movement of the red spot per se*, and not its location, which is the releaser (Fig. 4–7). Tinbergen's misinterpretation arose from two sources. First, the dummy heads were mounted on sticks which were moved pendulum-

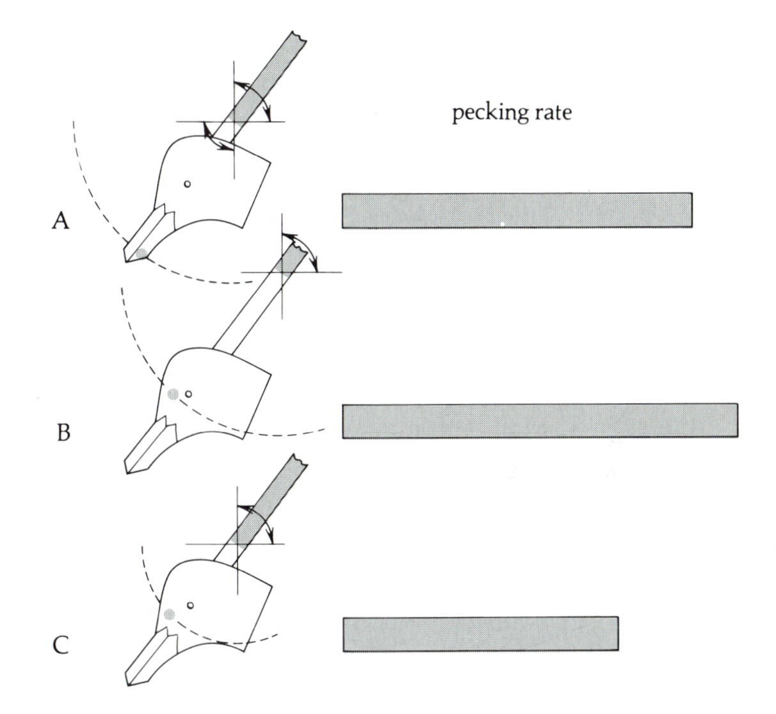

Fig. 4-7 Lack of preference for a normal model over a model with a forehead spot moved at the same apparent rates. Lengths of the bars indicate mean pecking rates. Chicks prefer models whose spots move the most, regardless of the location of the spot.

like so that the ostensible bill-tip spot swept out a larger arc and moved faster than the forehead spot of the other model. When Hailman arranged his models so that the spots, regardless of their location, moved at the same rate, newborn chicks actually preferred the unnatural forehead configuration. This perverse result is a consequence of increased movement of the bill in the forehead model: the bill is a secondary releaser (an isolated bill with no spot elicits some pecking) which in this arrangement moved more, and so added its exaggerated releasing value to that of the spot. In short, newborn chicks are attending to two independent, perfectly discrete features rather than to any patterns of head configuration and spot placement.

Tinbergen's second oversight involved the precise age of the chicks. Hailman found that after two days baby gulls began showing a clear preference for just the sort of correct configuration that Tinbergen had concluded was innate. Since both Tinbergen and Hailman had left the chicks with their parents except for testing, this implied to Hailman that learning had taken place, and that the chicks had replaced the innately recognized releasers with a picture. Indeed, since chicks and parents learn to recognize each other as individuals well before the end of a week, it would seem foolish if either were programmed to continue to depend on releasers.

Hailman's classic experiments, the full significance of which is still being appreciated, point the way to two of the cornerstones of modern ethology. The first is that releasers are discrete features rather than "schemata" or patterns, features which are often the optimum stimuli for well-known classes of feature-detector neurons in the visual and auditory systems of higher animals. As Chapter 7 will illustrate, the moving red spot of the herring gull, for example, is the ideal stimulus for one class of cells in the visual system while recognition of horizontally moving vertical bars (i.e., swinging beaks) is the special task of another group. The correspondence is even stronger since moving-bar detectors inevitably respond best to some particular rate of movement, and Hailman found that gull chicks are similarly tuned (12 cm/sec being the optimum). Examples of the tuning of particular neural cells to stimuli are endless. The bright-center/dark-surround pattern of flowers, for example, is a perfect stimulus for a class of cells in honey bees, while the vertical or horizontal red stripes resulting from a stickleback's behavioral posturing each stimulate very different sets of cells and corresponding sets of moods and behaviors.

The discovery that IRMs depend on simple features leads us to suspect that releasers are used by animals to recognize the salient features of their world because the genes, master instructors though they are, simply cannot encode innate neural photographs. Obviously, an ability to conjure up elaborate pictures of its prey, its predators, or its own species from birth would be of great service to any animal. That the recognition of each of these potentially crucial entities must depend at least initially on relatively crude, disembodied releasers—bars and spots, for example—argues that the genes in most animals, try as they will, simply cannot encode anything so complex as pictures, and so must gear the machines they control to make use of such sketchy cues as they *can* specify.

The realization that releasers are discrete features finally accounts for the evolution of IRMs which posed such a problem for early ethologists (Chapter 3). The oddly indiscriminate nature of IRMs when confronted with unnatural and obviously inappropriate stimuli offers no advantage over more demanding and discriminatory criteria based on a detailed innate pattern that might be imagined. The innate releasing mechanism and the behavior that ensues reflects the genes' inability to do any better: an IRM is the best system possible given the available hardware.

RELEASERS AND LEARNING

The second insight we derive from Hailman's work is the realization that although genes may not be able to encode pictures, they *can* instruct an animal to memorize the appearance of important individuals and objects as soon as possible. The gull chicks substitute a rapidly acquired mental picture of their parents for the innate releasers they must depend on at first, and this image takes over the IRM's function in releasing and directing begging behavior. In some sense, then, gull chicks imprint on their parents. Indeed, these two principles—that discrete features act as releasers and that they are replaced by pictures through internally directed learning—make classic imprinting seem less curious and more of a piece with behavior in general.

As we mentioned in Chapter 3, Lorenz saw imprinting as a program by which precocial young—newborn animals that must be mobile almost from birth to keep up with their ever-moving parents—memorize the appearance of the first moving objects they see. They then direct their stereotyped requests for food and comfort to this individual, and later in life use this image as a "schemata" for identifying potential mates. Lorenz's unequaled powers of observation and intuition were insufficient in this instance to unravel the full story, and later experimenters, who came from different intellectual backgrounds and had different expectations, obtained predictably conflicting results when they attempted to reproduce the phenomenon under their more rigorous conditions. Ultimately, all aspects of Lorenz's original models were rejected.

Even now the actual picture of what goes on during imprinting is only beginning to come into focus. It is now clear that more than one imprinting program is involved: not only do offspring imprint on their parents and their species *separately*, but parents imprint on their offspring as well. As fate would

PARENTAL INSTINCTS IN PIGEONS

Much of the conflict between classical ethology and behavioristic psychology grew out of the combatants' failure (or inability) to imagine or take seriously the alternative world view. Although the most common scientific technique is to test the predictions a specific model makes, the result is all too often a piling up of data consistent with what is later found to be a totally fallacious hypothesis. A better approach is often to imagine an opposing hypothesis, and then to work out an experiment which distinguishes between the two. Another all-too-frequent problem is our blindness to the logical extensions of our own theories. These scientific problems and the lessons they teach are illustrated by the debate over parental behavior in birds.

The ethological view of "parenting" is clear enough: chicks are programmed to get food from their parents, and parents are programmed to provide it. All that remains to be discovered are the relevant releasers. The behavioristic view, enunciated by Lorenz's most literate critic, Daniel Lehrman, sees parental behavior as the result of ordinary physiological constraints and trial-and-error conditioning of reflexes. Lehrman attacked the question experimentally with ring doves, a species of pigeon. His hypothesis was that prolactin, a hormone produced by incubating birds which stimulates the production of the "crop milk" pigeons feed their newly hatched young, leads also to an increased irritability of the crop and a painful distension as it fills with the "milk." The chicks, provided with a random pecking reflex, would by chance strike their parents' open throats and induce a reflexive regurgitation of the milk. The chicks would come to associate pecking at their parents' throats with food, while the parents, finding that the decreased distension of the crop after disgorging felt good, would associate relief with their chicks and thus present open mouths to them. In support of this hypothesis Lehrman reported that hormone-injected birds which had raised brood before, having supposedly already learned the appropriate reflex association, will feed seven-day-old chicks presented to them by the experimenter. Uninjected birds and injected birds which have not previously raised brood will not. Moreover, some birds with anaesthetized crops (and therefore no discomfort, no matter how full of milk they might be) failed to feed their adopted chicks normally.

Although the results in this experiment are consistent with the initial hypothesis, the hypothesis was wrong. The correct mechanism—a classic mélange of releasers and imprinting—was discovered quite incidentally a decade later by ethologists Erich Klinghammer and Eckhard Hess. The normal pattern of pigeon parenting consists of courtship and pairing, followed a week or so later by the appearance of two eggs in the nest. These two eggs hatch after sixteen days of incubation when, of course, the parents feed the young. What the ethologists discovered, first, was that crop milk was produced sixteen days after eggs appeared in the nest, even if the eggs were placed there by the experimentors days before the nesting pair produced their own. Clearly the eggs themselves serve as a releaser which must start an internal clock. The scheduled release of prolactin causes the crop milk to be produced. Second, by putting partially incubated eggs in the nest so that hatching occurred before the sixteen-day clock could have run its course, they were able to show that parents begin attempting to feed chicks *whether or not* their crop milk is ready—that is, regardless of any "painful" distension of the crop—independent of any previous experience with chicks. Similarly, parents begin providing solid food on schedule a few days after their eggs hatch, regardless of whether or not the eggs hatched on time or whether the crop milk ever came in. Another systemic clock must control this process, this time set by the sight of the chicks.

The realization missed by Lehrman was that seven-day-old chicks lack the necessary feeding releasers. What Lehrman's critics failed to grasp was that parents must imprint on their young. This difficulty is, in retrospect, an obvious extension of imprinting: how are parents, if they lack innate gestalt pictures of their young, to distinguish them in the nest from rocks or mice or other intruders, or outside the nest from the chicks of other doves? The chicks' releasers must direct both feeding and learning, just as the sounds and movements of parents direct both following and learning in the classical case of imprinting in water fowl. Within a week the parents' imprinted picture must take over, just as by then imprinted chicks must be able to recognize their parents without the need of any releaser. Hence Lehrman's unimprinted first-time parents could not recognize the seven-day-old chicks as anything special, whereas experienced birds could call upon carefully orchestrated "memories" from their previous broods. In the end, the basic assumption which allows ethologists to construct realistic explanations of behavior is that although animals sometimes seem painfully stupid, their genes, for the most part, are excellent teachers.

have it, only geese, the species Lorenz observed in detail, run their parental- and species-imprinting programs simultaneously. In all species the parental program always comes in the first day or two and engenders the "following response," while for most species the drive to learn the general features of the species comes days or weeks later and is evidenced by choice of companions and future mate selection. Most students of imprinting saw the chicks' following response as evidence that imprinting was underway, but viewed mate and companion selection as a measure of when the learning took place, its selectivity, and its irreversibility. Needless to say, confusion reigned.

In addition, baby animals turn out not to be quite so naïve about what their parents ought to look, sound, and smell like as Lorenz had thought. There seems to be a hierarchy of sign stimuli by which parents and offspring alike identify each other and which initiate imprinting; in the case of young birds, these appear to be both auditory and visual. In general, the first stimulus the newborns attend to is the species-specific following call produced by their parents. One species of duck, for example, attends solely to the repetition rate of a certain syllable, while another responds only to the frequency sweep peculiar to its species' call. As Chapter 9 will indicate, both of these features correspond to the optimum stimuli picked up by well-known classes of acoustic feature detectors lodged in the brains of higher animals. Only after a chick has located the source of the acoustic sign stimulus (or in the case of experimentally deprived animals, given up on ever hearing them) does it begin to learn and imprint on an object it sees moving away from it. Lorenz's uncanny talent for mimicking the calls of birds, which he invariably exercised as he worked to imprint his chicks, may in large part account for his success in eliciting the following response where others regularly failed.

As with the herring gull chicks, this use of two discrete releasers vastly enhances the specificity and increases the chances that the bewildered chick will correctly identify its parent under natural conditions. In short, the genes in the newborn, unable to encode a picture of its species or to predict identifying characteristics in the parents, program the chick to memorize the appearance of individuals which display appropriate releasers during a specific "sensitive period," and to substitute these pictures for the sketchy releasers in directing their subsequent behavior. Nor need the pictures thus formed be visual: when the contingencies of a species' lifestyle make it appropriate, imprinting on the individual idiosyncracies of calls or odors is just as effective in forging the crucial link between parent and offspring.

As we shall see, the discovery of a programmed sort of learning based on releasers and modulated by drives, and the realization that releasers are recognized by specific feature-detector cells in the nervous system, allow ethologists to understand much of the adaptive but mysterious flexibility of behavior and its evolution. With these evolutionary and mechanistic insights, the behavior of animals is just beginning to fall into sensible, predictable patterns. The emergence of modern ethology has brought with it a growing appreciation of the marvelous, rococo complexity of behavioral programming that genes can concoct from such primitive components as IRMs, motor programs, drives, and directed learning.

SUMMARY

The major obstacle to ethology's development was the assumption that releasers must depend on innate pictures. Ethologists had been greatly influenced by experiments which showed that recognition of familiar objects or individuals could result from seeing or hearing a very limited number of their features, provided that the features were in the correct configurational relationship to each other. A few notes in the proper order call up in our minds a whole melody, whereas the very same notes in another order are meaningless. In fact, as Jack Hailman's experiments with gulls proved, the configuration of stimulus elements seems *not* to be recognized innately by most animals. Instead, they attend only to the elements themselves, which constitute neural sign stimuli. This discovery has led to the realization that nothing more esoteric than the simple feature-detector cells in the nervous system is required to explain most IRMs. This insight in turn has led to the gradual appreciation that the documented cases of recognition based on stimulus configuration are made possible only by releaser-directed learning. Releasers based on feature detectors, motor programs, drives, and programmed learning, then, have become the conceptual tools which modern ethology uses in its endeavor to understand the mechanisms and evolution of behavior.

STUDY QUESTIONS

1. In Hailman's experiment as summarized in Figure 4-5, the pecking preferences of young birds (raised by their parents but removed briefly and tested with models) changed as they grew older. The baby birds came to prefer a realistic model of their own species over one of another species or a model with a misplaced bill spot. Three interpretations come to mind: (1) Hailman suggests that the birds, although innately programmed to peck at a releaser, become classically conditioned to the object that reinforces them with food, namely the parent, and so begin to distinguish the models; (2) Tinbergen might suggest that no learning is involved, but rather that the birds have a gestalt picture of the parent all along, and that their crude vision at birth improves as their eyes or visual processing mature; hence, as the chicks grow older and better able to see, they are able to make more exact distinctions between models; and (3) Lorenz might suggest that the young birds are learning, but that the process is imprinting rather than conditioning; as the imprinting period progresses, the baby birds form an ever more exact picture of their parents. How could these three interpretations be distinguished and tested?

2. In the Klinghammer and Hess experiment on parental instincts in pigeons, how could you demonstrate conclusively (and as dramatically as possible) that parents really do imprint on their offspring?

3. Many birds are able to recognize and discard the eggs of cuckoos added after their own clutch is laid. The ability to recognize and discard the cuckoo egg before the baby cuckoo hatches and proceeds to discard the eggs of its host could be a case either of imprinting or innate recognition. How could the two be distinguished?

Fig. 4-8 A newly hatched cuckoo, its eyes still tightly shut, prepares to roll the eggs of its host out of the nest.

FURTHER READING

Hailman, J. P. "How an Instinct Is Learned." *Scientific American* 221 no. 6 (1969): 98–108.

Lehrman, D. H. "A Critique of Konrad Lorenz's Theory of Instinctive Behavior." *Quarterly Review of Biology* 28 (1953): 337–63.

Lorenz, K. Z. "Analogy as a Source of Knowledge." *Science* 185 (1974): 229–34.

Tinbergen, N., and Perdeck, A. C. "On the Stimulus Situation Releasing the Begging Response in Newly-Hatched Herring Gull Chicks." *Behaviour* (1950): 1–38.

Part II

NEURAL MECHANISMS

CHAPTER 5

Nerves and Circuits

Innate releasing mechanisms and motor programs, like all behavior, must result from underlying physiological events. The nervous system not only generates information but receives it, filters it, processes it, and activates the appropriate responses. The strategy of modern ethology demands that IRMs and motor programs be treated not as black boxes, but as new systems made up of smaller parts which in turn must be dissected and analyzed. The parts, in this case, are composed of receptor, nerve, and muscle cells. They can be studied either as groups—circuits whose specialized wiring processes information and generates behavior—or as individuals, each with its own peculiar properties.

REFLEXES VS. MOTOR PROGRAMS

Different people study the nervous system at different levels, and there was a time when these levels reflected two very different sets of assumptions about behavior. Ethologists, convinced as they are that behavior is organized into subsystems such as releasing mechanisms and learning and motor programs, naturally set about looking for the wiring of those key circuits at the group level. However, the alternate and dominant view until the 1960s assumed that behavior grew out of chains of reflexes rather than prewired central programs processing complex sensory information and orchestrating coordinated movements of several muscles. A reflex is a simple stimulus-response circuit by which a particular input is inevitably followed, with little or no intervening processing, by a unitary output. A typical example is the familiar knee-jerk reflex of humans (Fig. 5–1). Behaviorists hypothesized that complex behavior could be constructed out of reflexes combined in particular sequences, and could be fine-tuned or modified by alterations in the "threshold" or "excitability" of single nerve

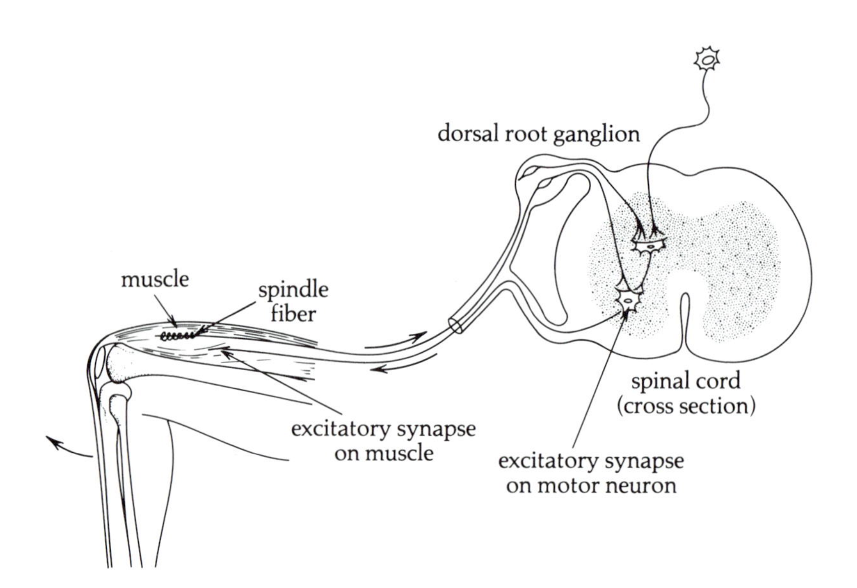

Fig. 5–1 Classic monosynaptic stretch reflex. When the muscle spindle is stretched (due normally to increased load on the limb which the muscle serves) it sends a signal into the spinal cord to, among other things, a motor neuron going to the very same muscle. The interaction at this single synapse causes the nerve which serves the muscle to signal the muscle to contract—a message which results not only in contraction of the muscle, but also in reduction of the stretch on the spindle and, finally, a cessation of the spindle's signal to the spinal cord. In its simplest form, this circuit is a servo mechanism, a system which is self-adjusting to maintain a constant setting regardless of load. In this case the servo system holds a limb in a constant orientation in response to various pressures.

cells along these chains. Supporters of the reflex-chain theory felt that alterations in the behavior of single cells could lead to changes in the "branching" of these hypothetical chains and thus could explain the mechanics of almost any behavior. This latter model, an obvious but mistaken outgrowth of Pavlov's work on the conditioning of reflexes, was consistent with the behavioristic view of behavior, and would eventually lead its students to examine single cells.

The turning point in the study of nerves and behavior came in the early 1960s when Donald Wilson published his studies on locust flight. Here was an archetypal reflex behavior: when the upstroke muscles get the wings to the top of their arc, a downstroke reflex must be triggered which brings the wings to the bottom, thereby setting off the upstroke reflex, and so on ad infinitum. All that the system requires are trigger elements to provide some way of knowing when the wings are all the way up and all the way down. Wilson found the sensory organs, the stretch receptors which must be responsible for triggering the response, and cut them. To his surprise, the locust kept right on flying, although a little slower than before. As he discovered, the flight pattern is generated by an innate, prewired ("endogenous") motor program circuit located in the locust's thorax which is switched on and off by a separate decision-making circuit in the head, and merely fine-tuned by the feedback from the stretch receptors. With this unshakably ethological discovery at the neural level, the supposed reflex nature of other motor behaviors began to dissolve under scrutiny and a new generation of "neuroethologists" started actively searching for and discovering motor program circuits in a wide range of behaviors.

And yet reflexes like the familiar knee-jerk reflex in humans do exist, providing indispensable, simple, quick responses or automatic adjustments, though perhaps they may most usefully be thought of as the limiting case of a single nerve cell motor program. The years of work under the "reflex" hypothesis were not lost, however; the resulting single-cell level of analysis has proven important to our understanding of the mechanics of receptors and other nerves, and how changes in their properties affect behavior. Some of these details are crucial in understanding how animals perceive, analyze, and act in and on their environments.

PHYSIOLOGY OF NERVES

The underlying mechanism of neural activity is ionic. In the usual case, an ion pump in the membrane of nerve cells actively moves at least one class of ion (usually sodium) out of the cell and transports another type of ion (generally potassium) in. This creates both an osmotic pressure and an electrical potential (difference in charge) across the membrane: the interior of the cell holds an elevated concentration of both positively charged potassium ions and negatively charged organic ions, while the outside fluids are rich in positively charged potassium (Fig. 5-2A). Hence the inside of a neuron is simply more negative than the outside by about 70 mv. The membrane itself is relatively impermeable to organic ions but slightly permeable to potassium, so it requires the charge difference generated by the unequal distribution of sodium and organic ions to balance the osmotic pressure, and to keep the concentration of potassium higher within the nerve than without.

This delicate electrical and osmotic balance is regularly upset when the neuron carries information. Input from other cells (or in the case of receptors, energy from the environment) causes a change in the cell's membrane potential at isolated spots. When the membrane is depolarized to a cell-specific threshold (which usually happens because of signals being received from other cells) a self-regenerating electrical event—the "action potential" or "nerve spike"—is propagated down the membrane. This event, triggered by the original depolarization, is the result of a stereotyped and coordinated opening and closing of first one set of ion gates and then another. The membrane contains a vast array of poorly understood, ion-specific pathways. The sodium gates are voltage-dependent—that is, they are opened when the membrane potential falls below the cell's active threshold. The result, of course, is a rapid influx of sodium ions and a further lowering of the local potential (Fig. 5-2). Since this molecular fixed-action pattern results in complete depolarization of the affected piece of membrane, it triggers the same event in the adjacent patch of membrane, which then triggers the next, and so on, thus propagating the spike. All that remains now is for the cell to recover and clean things up so that further action potentials can be propagated (see Fig. 5-2).

The unitary, all-or-nothing nature of these nerve spikes

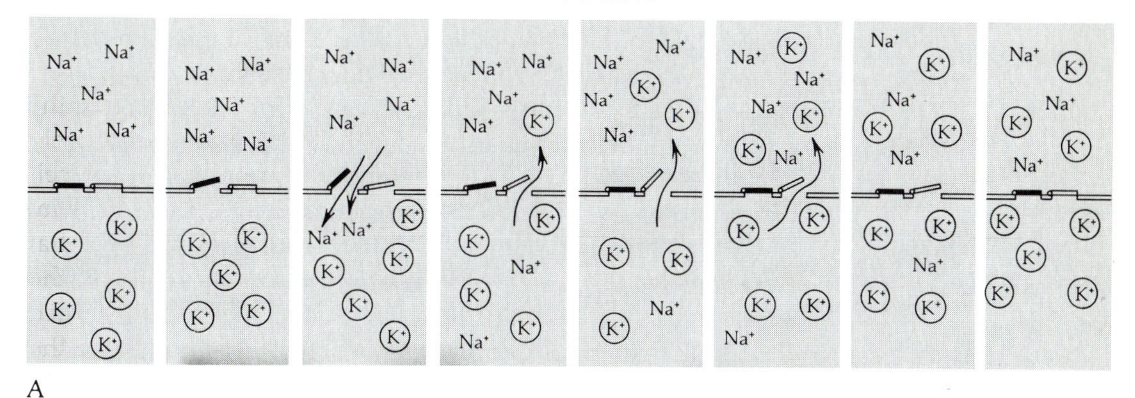

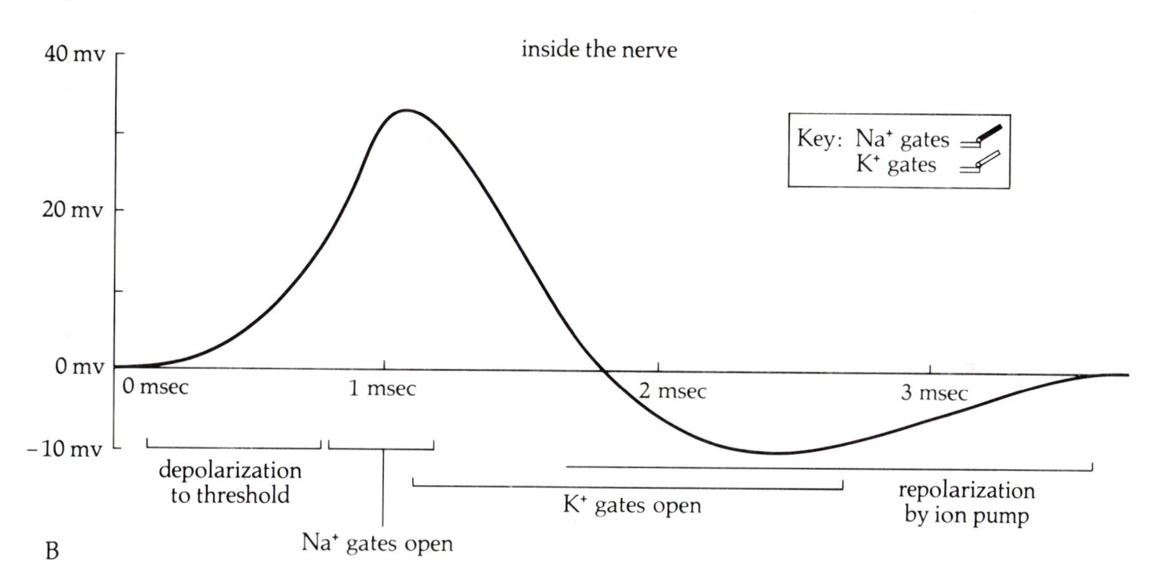

Fig. 5-2 Transmission of an action potential along a nerve. In A, we see that the nerve membrane consists of, among other things, gates or channels which specifically pass either sodium ions (Na^+) or potassium ions (K^+). Normally these gates are closed (as at 0 msec), but when an action potential approaches (0.5 msec) the voltage inside the cell begins to change. This change triggers the quick opening of the sodium gates and the slower opening of the potassium channels (1 msec and 1.5 msec, respectively). The concentration gradient causes the ions near the membrane to cross. The sodium channels close quickly (by 2 msec) while the potassium gates close much more slowly. The effect of the channel openings and closings is relatively brief since the sodium ions in the nerve—an enormous volume compared to the thin layer next to the membrane—quickly diffuse in as the potassium is diluted out, while just the opposite is happening outside. An active pump is running all the while, moving stray Na^+ inside and K^+ outside. The electrical consequences are shown in B.

means that information in the nervous system is carried in pulses, and only the frequency or pattern of those pulses is important to the cells which receive the messages.

The archetypal nerve possesses a central cell body which nurtures the cell, a set of short finger-like extensions called "dendrites" which collect the output of other cells; a long axon which carries the signal to some distant destination; and a set of synapses which pass the signal to the dendrites of other nerves or to muscles (Fig. 5-3). Communication between cells at synapses is usually accomplished by the release of transmitter chemicals from the incoming cell which affect the potential of the receiving cell. Specific receptors on the receiving cell alter the neuron's local membrane potential briefly when the appropriate transmitter is received. This information is therefore graded, depending as it does on the number and kind of synapses stimulated as well on temporal factors. Each active synapse creates a small membrane depolarization; a weak electrical stimulus will spread, but being below the threshold of an action potential it will die out slowly some little distance away.

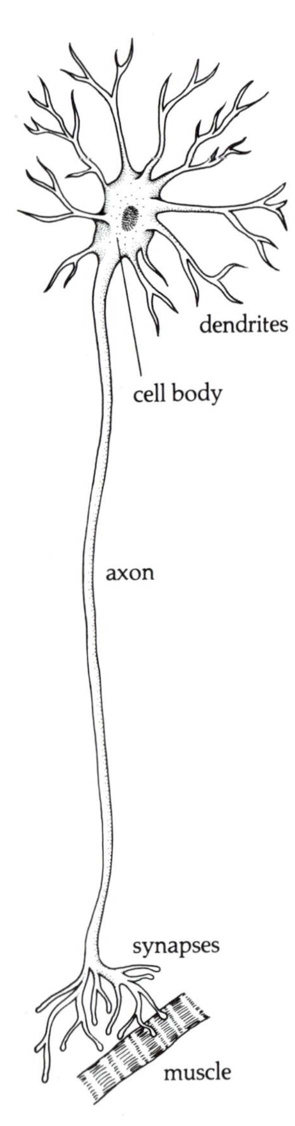

Fig. 5-3 A generalized nerve cell synapsing, in this case on a muscle.

The interaction of these weak spreading electrical stimuli causes the input to be processed and refined in a way characteristic to each cell. The input, usually coming in through the synapses of several hundred other nerves, is detected by the dendrites. No single synapse can stimulate the dendrites sufficiently to discharge the cell, but each incoming signal starts a brief, spreading electrical charge which may be joined by other inputs arriving nearby at the same time, or even by a rapid "buzz" of impulses from one cell with many synapses. If the overlapping electrical spreading in the dendrites is of sufficient strength, the cell will discharge and carry a spike down the length of the axon to its synapses. Some dendritic inputs are inhibitory, and thus block the electrical spreading; obviously, nerves are active electronic processing devices rather than passive wires. Many different kinds of nerves exist, and depending on their anatomy and physiology, they can perform addition, subtraction, differentiation, or integration.

ADAPTATION AND HABITUATION

Moreover, nerves have properties which depend on recent "experience." They may appear to "fatigue," responding only to the initiation of a stimulus, or they may respond continuously. Those at the end of the spectrum which respond briefly at the initiation of a stimulus are called *phasic* cells, and are ideally suited for detecting change, while those at the other end, known as *tonic* cells, report absolute values (Fig. 5-4). *Receptor* cells typically fall into the middle of this range, adapting only slowly to changes in sensory input.

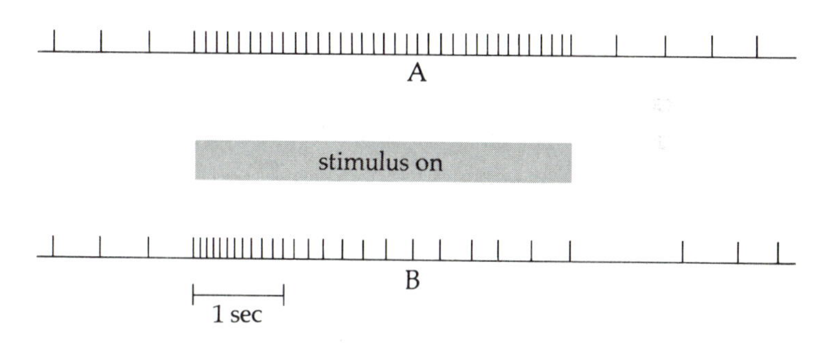

Fig. 5-4 Response of two different classes of sensory cells. A tonic cell (A) reports the exact stimulus strength—the degree of stretch in a joint, for example—while a phasic cell (B) emphasizes the degree of change, exaggerating the onset of a stimulus, slowly adapting to it, and then exaggerating the end of the stimulus.

The intermediate strategy of most receptor cells is a compromise which permits relatively accurate measurement of stimulus levels at any given time while allowing the midpoint of that sensory range to slide up or down the sensitivity scale as conditions change. Hence we are nearly blinded leaving a movie theater on a sunny day, but we quickly adjust. Bath water seems warm or even hot when we first get in, and swimming pools are notoriously chilly, yet in time we no longer sense that the water is unusually warm or cold. That the water itself has not changed is easily demonstrated by putting a dry part of our bodies into the water. This phenomenon is referred to as *adaptation.* The sensory system of one limb can become accustomed to a stimulus—in this case to hot or cold water—while the rest of the body remains sensitive.

This example also serves to illustrate another distinction. Some sensory systems adapt in this way, while others are said to *habituate.* In the latter case the sensory "numbing" takes place centrally so that if habituation were involved, exposing one arm to cold water should desensitize us to cold water applied to the other arm. Obviously we do not habituate to temperature, we adapt; but we do habituate to many other stimuli. For example, one loud noise, a shot perhaps, may startle, even "sensitize" us, making us more alert to other sensory modalities, but repeated shots usually cause the startle behavior to wane. Except in the most extreme cases our ability to hear is not affected, but the habituation of our startle response to shots heard by only one ear will continue even if the same sound is presented to the other. The receptors themselves are not involved.

The ability to become instantly more sensitive, to lose our apparent sensory numbness and become alert when another sensory modality is stimulated, is the single most diagnostic characteristic of habituation: such irrelevant inputs have no effect on the state of adaptation of receptor cells. Indeed, different parts of the same sensory modality may habituate separately. Responsiveness to one acoustic releaser can be virtually abolished without affecting others mediated by the auditory system.

We may think of habituation, then, as a central process of some sort, but independent of sensory fatigue. As a higher level of sensory "boredom," it is mediated by cells which must be averaging or integrating over all the sensory inputs affecting any one behavior—the proboscis-extension response of flies to the taste of sugar, for example—to control the general responsiveness of the animal toward that one special set of stimuli. In general, habituation controls the probability of specific behavioral responses, and thereby represents the basic level of an organism's behavioral plasticity or learning, as we shall consider calling it later. Sensitization is the converse, a generalized alerting reaction which serves to *turn off* habituation.

The cellular bases of adaptation, habituation, and sensitization are of great interest to neurobiologists (see especially the Kandel article in "Further Reading"), but considering our present state of knowledge, these mechanisms are often taken by ethologists for the time being to be irreducible components of behavior. As we try to understand how the circuitry works as a whole to generate behavior, the details of calcium conductances and cyclic-AMP metabolism at synapses seem elegant but extraneous luxuries. Indeed, as neuroethologists have discovered, dissecting the IRM–FAP chain even at the circuit level is very difficult. The difficulty arises both from the technical problems of working with such small neural elements, and from the nervous system's propensity to hook nearly every cell to a vast and formidable array of other cells.

NEURAL CIRCUITS AND BEHAVIOR

And yet when the modern ethological perspective has risen to the challenge of looking at the multicellular interactions which underlie behavior, not only has the expected circuitry been found, but our understanding of the physiological constraints and potentials has been greatly advanced. Consider, for example, how the intricately beautiful sea slug *Tritonia* makes its escape from its archenemy, the starfish. One touch from this predator and *Tritonia* instantly contracts into a cowering posture, from which it begins a rhythmic swimming movement that propels it up off the bottom and into the weak ocean currents that carry it out of harm's way (Fig. 5-5). In setting up a laboratory situation conducive to making detailed electrical recordings of the behavior, neurobiologists quickly discovered that *Tritonia* could be made to perform its escape routine when merely touched with a small bit of starfish, or with a dropper full of starfish extract. Apparently, *Tritonia* comes wired to recognize the "smell" of starfish, which it takes as a sign stimulus. Inside this saltwater slug is a now well-understood circuit that generates the rhythmic swimming motions which constitute the escape motor program.

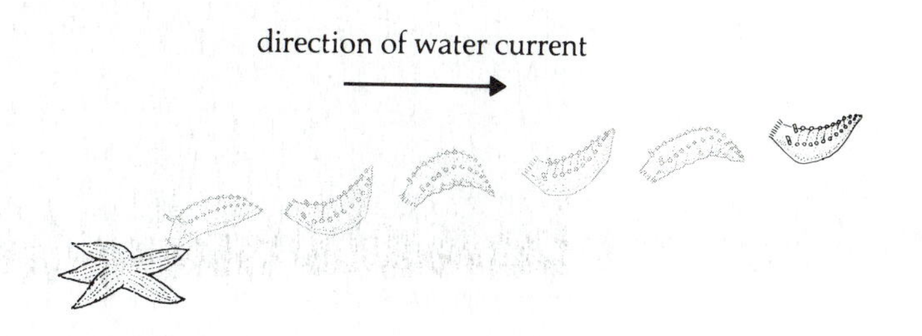

Fig. 5-5 Escape response of *Tritonia*. When a *Tritonia* is touched by a starfish, it responds first with a dorsal contraction which pulls its extremities back from danger, followed by a rhythmic swimming motion that carries it up off the bottom into the water currents and away from danger. [From "Giant Brain Cells in Mollusks," by A. O. D. Willows. Copyright © 1971. Scientific American, Inc. All rights reserved.]

Finally, systematic and painstaking recordings from inside single cells turned up an "executive neuron" which when artificially stimulated caused the motor program to begin, and which when artificially inhibited from firing prevented the animal from reacting to the ordinarily repulsive touch of a starfish. This cell looks very much like it might be the Prime Innate Releasing Mechanism postulated by early ethologists.

The first (and perhaps the best) example we have to date of a satisfactory explanation of a nonrhythmic behavior at the level of neural wiring is feeding in blowflies. Blowflies make their living by finding and eating decaying or fermenting food. They fly about essentially at random until they smell an appropriate odor, then fly upwind to the source in much the same way that digger wasps search for honey bees. Upon finding the source of the odor, the blowfly lands and walks around it until it literally steps into the food. Taste receptors on the front feet alert the brain, which turns the fly toward the leg that found the food, and then the fly extends and inflates its strawlike mouth or proboscis (Fig. 5–6). When the sensory hairs on the proboscis detect the substrate they cause the lips of the proboscis to open, exposing another set of taste receptors. The taste receptors of the proboscis signal the brain to turn on the pumping muscles. The blowfly pumps for a bit, stops, pumps some more, pauses, and so on until it deflates its proboscis and flies away.

Blowfly behavior is an obvious series of sign stimuli and motor programs: the foot receptors release turning and proboscis extension; the proboscis touch receptors release lip opening; the lip receptors release pumping, which is a protracted motor program. Clearly, there must be feedback to stop the pumping or the animal might overeat or even explode. Several

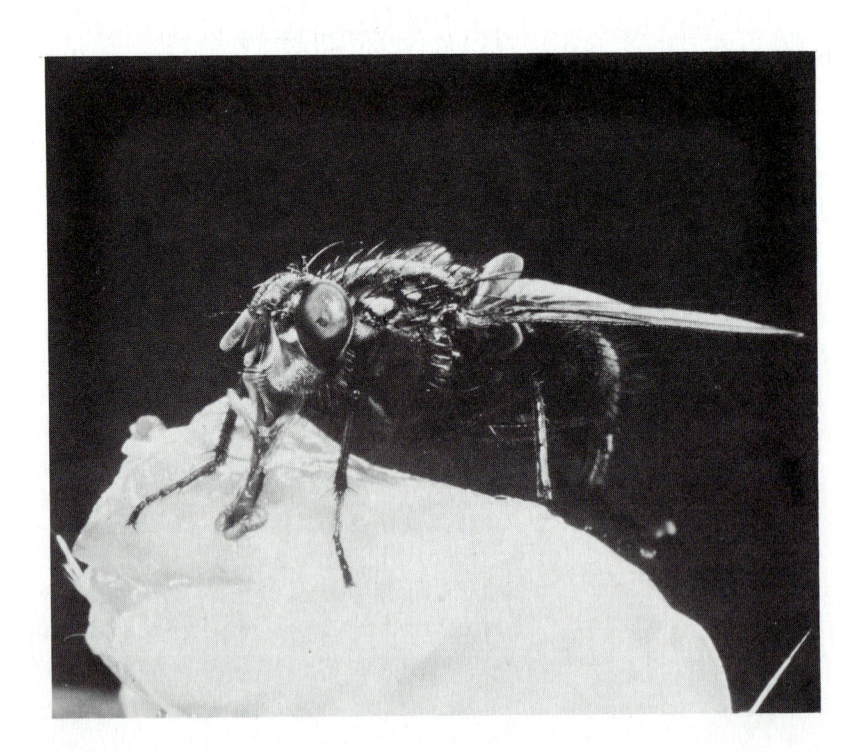

Fig. 5–6 A blowfly with its proboscis extended.

possibilities are open at this point: the taste sensors, for example, might adapt, or the fly habituate to the food. In fact, though, the blowfly judges when it has had enough to eat in quite another, much more exact way. It has a stretch receptor in its stomach (crop) which measures just how full and swollen that organ is. When the stomach gets too full, the stretch receptor turns off the pumping until the stomach can begin to empty. Obviously, the stretch receptor must be a tonic cell to work in this circuit since it measures the strength of the stimulus. Once the stomach has emptied into the gut, pumping can begin again. A second stretch receptor in the gut tells the fly when the second half of the digestive system is also full. This stretch receptor turns off the proboscis inflation, and thereby brings an end to the meal. Cutting the nerves to either receptor results in a fly which will eat indefinitely (Fig. 5-7).

In the case of the blowfly a knowledge of the wiring, the workings of stretch receptors, and the effects of cutting small pieces combine to explain much of the feeding behavior. But the hope that to know the anatomy of the wiring diagram is to understand behavior has, as yet, never been fully realized.

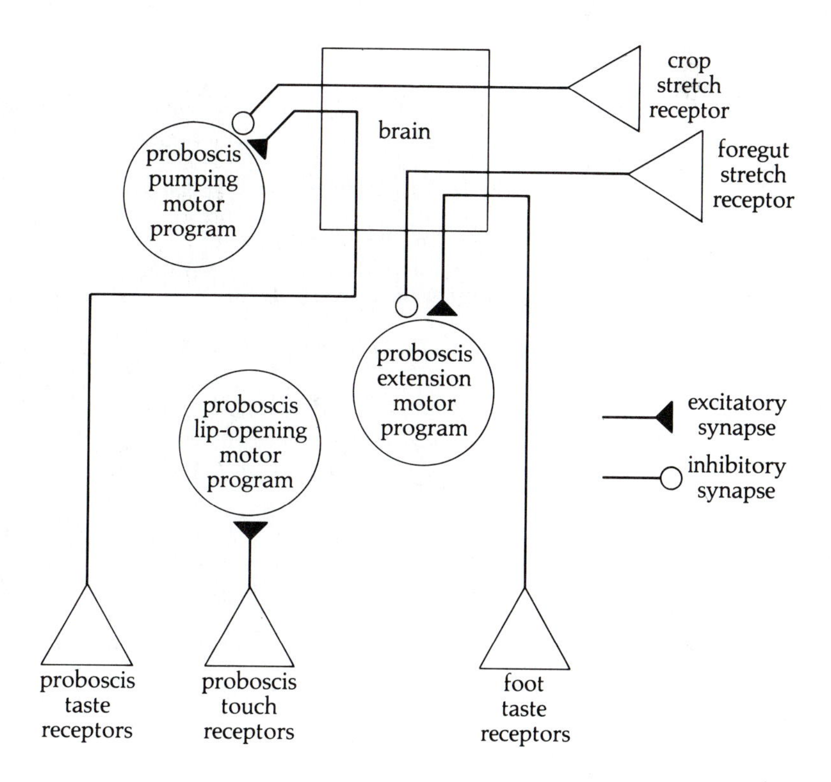

Fig. 5-7 Simplified wiring diagram for the post-search phase of blowfly feeding behavior.

Perhaps the most dramatic example of the anatomical dissection of behavior is the ongoing attempt to map the nervous system of nematodes. The human parasite *Ascaris,* for example, is known to have only 257 nerve cells—almost the smallest number possessed by any animal with discrete neurons. Nematodes have an adequate behavioral repertoire despite their diminutive supply of nerves: free-living nematodes track their food, mate, orient to heat and moisture, and of course swim. Early in the century a German biologist, Goldschmidt, painstakingly mapped the entire nervous system of *Ascaris.* Goldschmidt's studies revealed a daunting complexity even in this least complex of organized nervous systems (Fig. 5–8). More recently, Sidney Brenner has mapped another 257-celled nematode somewhat more quickly using an electron microscope and a computer.

Although much has been learned from attempts to map the nematode nervous system, one set of questions has been asked and quite another set answered. This heroic mapping effort was begun despite the discouraging knowledge that the animal is too small to allow neurobiologists to put electrodes into the cells

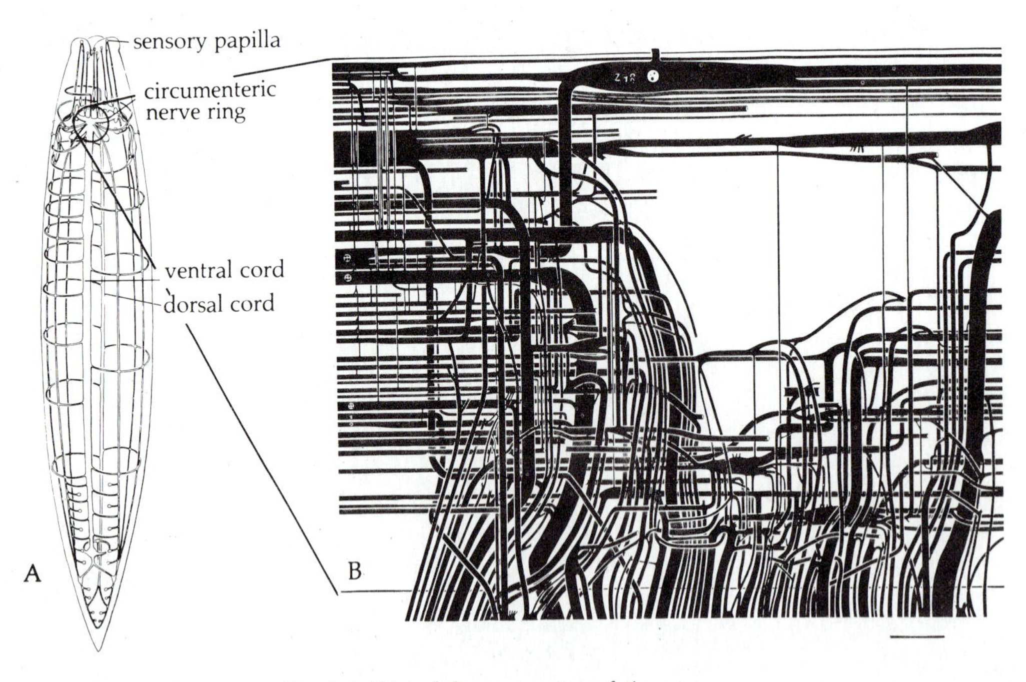

Fig. 5–8 Part of the nerve ring of *Ascaris.*

to listen to what they are doing. Hence, unlike the blowfly investigation, we must rely on anatomy *alone* as our source of information. Since processing occurs on the dendrites, and since the strength and speed of electrical spreading depends on the local dimensions of these irregularly shaped processes, anatomical studies can yield only limited information; and the interaction of the dendrites of a cell with hundreds of cells, both inhibitory and excitatory, adds to the confusion. There is as a result no way in practice to predict on the basis of anatomy alone just how a particular cell will process its output. The nervous system's mystery is all the more complete since the cells seem to have little difficulty in wiring themselves during development and keeping things straight.

With nematodes we can only guess what the nerves are doing, and so we must either adopt the Lorenzian strategy of trusting our guesses, or try something else.

The "something else" has proven to be an investigation of the small, relatively independent systems of nerve cells in larger animals which control well-defined and convenient behaviors—the circuits responsible for insect flight, cricket song, or crayfish swimming, for instance. Each of these is a classical motor program. Ideally the mutual interactions of the cells in the circuit would be analyzed, and rules to explain the neural behavior formulated. This approach has unfortunately turned out to be something like the three-body problem of physics: mastering the interactions of more than a very few cells seems almost beyond comprehension. Very recently, however, the oscillator circuits controlling rhythmic movement in these relatively simple animals have begun to yield up their secrets.

Pessimists maintain that the neurophysiological approach has no future either because of the enormous and incomprehensible complexity of nonrhythmic neural circuits, or because behavior is somehow an "emergent" process—the whole being in some mysterious way more than the sum of its parts. Optimists assert that progress, though slow, has nevertheless been steady, particularly when biologists have begun with the sensory receptors and tracked the information into the brain. The basic assumption of ethologists, as of all modern biologists, is that nature is conservative with things that work. Hence all nerves work in basically the same way, and an understanding of the sensory windows and motor programs of convenient animals might provide the clues needed to tackle more complicated systems.

In the case of vision, as we shall see in Chapters 6 and 7, the lowly, antediluvian horseshoe crab gives us the necessary models for understanding the visual system of frogs, which in turn provides the key to comprehending the more elaborate system of cats, and the cat work itself has been crucial to our ability to sort out what is going on in primate vision. The next few chapters will begin the search for the neural bases of behavior (releasers, motor programs, and learning) with an investigation of the mechanisms of sensory reception, and we will follow the paths of species-specific sensory processing as far into and back out of the central nervous system as has yet been possible. After that, we will turn to the behavioral systems which combine many separate "programs" to construct such (at least superficially) complex behaviors as navigation and social organization.

SUMMARY

Nerves underline virtually all that is behavior: the detection of sensory information, its integration, the processing which leads to behavioral decisions, the motor-program circuits which orchestrate movement, and since muscles are modified nerve cells, even the movements themselves. Although the once-dominant school of thought held that complex behavior developed out of simple reflexes through a process of trial-and-error conditioning combined with the intrinsic plasticity of neurons, we now know that prewired circuits exist which manage much of the integration of processing of inputs, as well as the patterned coordination necessary for adaptive behavioral responses.

Nevertheless, the plastic and dynamic characteristics of neurons are important to our understanding of how these building blocks of the nervous system actually combine and interact to create behavior. A complete understanding of the actual processing of information through the submicroscopic electrical interactions of the thousands of synaptic inputs on the dendrites of a cell at present seems beyond our grasp for technical reasons.

On the other hand, adaptation—that process by which a receptor cell adjusts its responsiveness to a change in stimulation, either very quickly as in phasic cells, or very slowly if at all as with tonic neurons—is relatively well understood. Other dynamic properties are typical of whole circuits. The two most common are habituation (the mechanism by which the responsiveness of each specific behavioral pathway to its particular integrated input is lowered) and sensitization (the process which alerts an animal by erasing an existing habituation).

Beyond these relatively general properties, our understanding of the workings of specific behavioral circuits will ultimately depend on coordinated, cell-by-cell anatomical and neurophysiological analyses.

STUDY QUESTIONS

1. In Kandel's article, he describes the phenomenon he is studying as habituation. How does he know it is not adaptation instead?
2. A blowfly extends its proboscis when a foot is touched with sugar. Eventually this response goes away. Is the fly adapting or habituating? How could you find out? How could you deal with the contention that the proboscis extensors are just fatiguing?
3. Many cells respond only to the *onset* of a continuous input. Their brief output, often important in circuits responsible for timing things, could be a consequence of a highly phasic construction. Recently, however, cells have been found which are basically tonic in physiology, but turn themselves off. Draw a circuit diagram to explain this behavior.
4. Most of what goes on behaviorally and biochemically in animals may be viewed in terms of the principle of negative feedback or servo control, in which the system attempts to cancel any changes which affect it. Take the temperature in a centrally heated house, for example. When the temperature falls in the winter, the change is sensed by a thermostat which activates the furnace. The heat from the furnace serves to cancel the original temperature change, returning things to their former state. When the set-point of the thermostat is reached, the furnace switches off. By contrast, positive feedback, the *amplification* of change, is relatively rare in organisms. How do the action potentials of nerves fit into this dichotomy? How about egg rolling in geese? Courtship in sticklebacks? The proboscis-extension reaction of blowflies?

FURTHER READING

Gelperin, Alan. "Regulation of Feeding." *Annual Review of Entomology* 16 (1971): 365–78.

The classic description of the neural control of blowfly feeding.

Kandel, Eric. "Small Systems of Neurons." *Scientific American* 241, no. 3 (1979): 66–76.

Sheperd, G. M. "Microcircuits in the Nervous System." *Scientific American* 238, no. 2 (1978): 92–103.

Stevens, Charles. "The Neuron." *Scientific American* 241, no. 3 (1979): 54–65.

Werblin, F. S. "Control of Sensitivity in the Retina." *Scientific American* 228, no. 1 (1973): 70–79.

Wilson, Donald M. "The Flight Control System of the Locust." *Scientific American* 218, no. 5 (1968): 83–90.

CHAPTER 6

Visual Design

The physical world is full of energy—light and sound, for example—which, if organisms could detect and analyze it, would provide them with important information about their world. Animals are faced with two problems: how best to collect the energy, given its physical properties and peculiarities; and then how best to extract information from what is received. In each independent solution of the first problem, the process of evolution has incidentally defined each animal's peculiar sensory world, and in deciding how to process this sensory information, evolution has also created the necessary neural circuits which are responsible for releasers and perception in general. This chapter deals with the problem of gathering visual information while the next describes how it is processed.

LIGHT

Sunlight is the basis of vision, but more than that it is the basis of life on earth. It warms the planet and provides the energy which plants turn into more plants. In the process sunlight also provides the energy to make and run higher animals—both those that eat plants and those that eat the animals that eat plants. Vision evolved from the necessity of light for life. Most plants, of course, live in a world dominated by sunlight. They enhance their ability to use the sun as an energy source by aiming their leaves to capture more light. Using pigments similar to (and doubtless derived from) plant photochemicals, most living things began to use light as their major source of information about the world. Simple animals regularly orient to (or away from) light, sometimes using it, as *Euglena* does, merely to find out which way is up. Both groups of organisms must be able not only to sense light but to determine the direction from which it is coming. As we shall see, the strategy for detecting light originally hit upon by animals—shading—led directly to the eyes most creatures now use to form actual pictures of their world.

Light has direction, intensity, color, and polarization. Perplexingly, it behaves both like a particle and like a wave. Particles of light, called "photons," give up energy when they collide with receptor pigments, and this energy initiates the biochemical reactions that ultimately result in the sensation of vision. Very few of the photons that reach our eyes are actually detected; in fact, nocturnal animals such as cats even have a reflective layer behind the receptor pigments which serves to bounce the uncaptured photons back through for another try. That cats' eyes shine so brightly at night testifies to the low capture rate of even two trips through the receptor layer. Receptors nevertheless make very efficient use of those photons they do capture, so that even a single photon out of the billions entering the eye each second under daytime illumination is sufficient to initiate the biochemical reaction which underlies reception.

Light also has a wave nature, the most familiar aspect of which is wavelength, or color. The visible portion of the spectrum for humans runs from red at about 750 nm (nm = 10^{-9} m) to violet at about 400 nm (Fig. 6–1). Since the energy of a photon is related to its wavelength, specialized receptor molecules that react to specific amounts of energy provide the basis for color vision. As a wave, light undergoes a variety of wave-specific interactions including diffraction and refraction. These two properties, respectively, have provided the major limitations and potential for animal vision. Diffraction is the tendency of waves to bend around objects. As a result of this property, light passing through a small hole will spread out rather than project

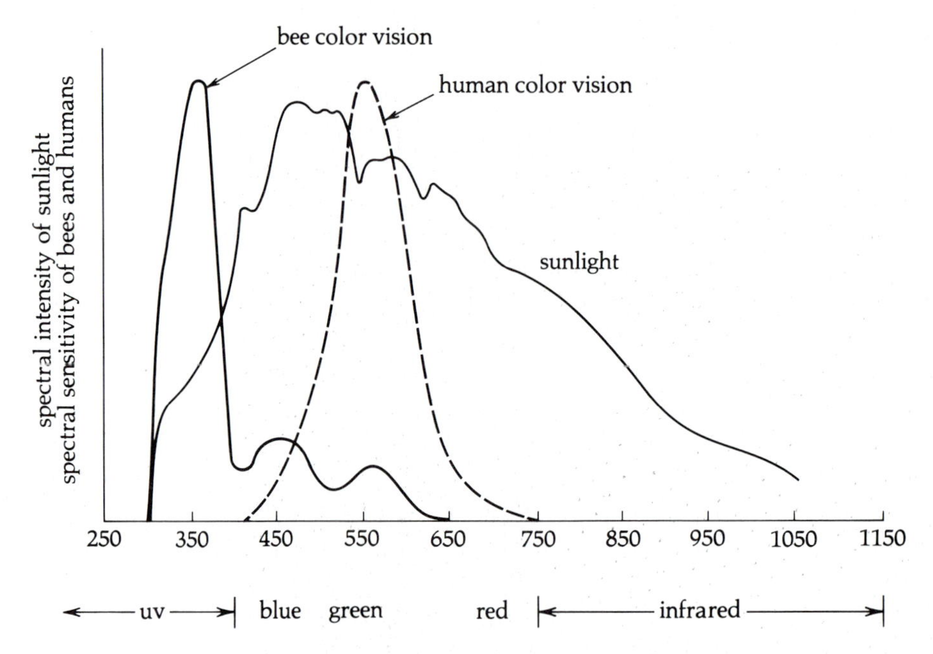

Fig. 6–1 The distribution of color and intensity of light coming from the sun, as well as the very different color sensitivities of bees and humans. Most vertebrates make a special screening pigment to filter out the UV.

as a simple spot of light (Fig. 6–2A). The failure of light to travel in straight lines near objects serves to blur images, and so diffraction is the major factor limiting visual resolution. Refraction is the process by which light is bent as it passes from one material or medium to another of different density—from air to water for instance, or from air into a glass telescope lens (Fig. 6–2B).

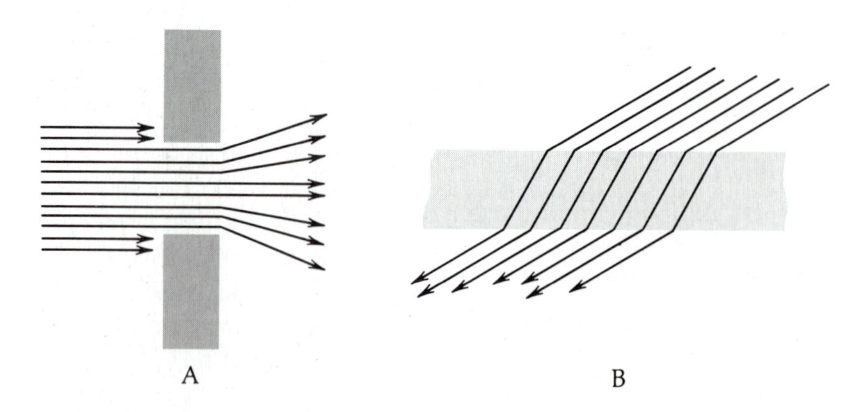

Fig. 6–2 A. Diffraction of light passing near an edge. B. Refraction of light passing into an optically dense material such as water or glass.

PHOTOTAXIS

Animals which detect light do so for a reason, and generally use light to control or orient some part of their behavior. Photosynthetic bacteria reverse the direction in which they are swimming, for instance, when they encounter darkness or even a slight decrease in illumination. This tactic results not only in the bacteria's remaining in patches of that all-important resource upon which they "graze," light, but in a general movement toward the light as an incidental consequence of simply never being allowed to swim away from it.

The movement of photosynthetic bacteria is one of a class of behaviors known as *kineses,* a set of strategies by which animals merely alter their rate or direction of movement in a random manner in response to external stimuli. Wood lice, for instance, move about only when their current environment proves too dry, and stop when they again find that welcome damp essential for their survival. As such, their behavior is not actually oriented.

By far the most common strategies for animal movement result in orienting animals with regard to a relevant stimulus, and so do not depend on chance to turn the organism in the right direction. Such oriented behaviors are called *taxes. Euglena,* for example, a simple photosynthetic protozoan which confounds any strict plant or animal dichotomy (Fig. 6–3), is able to move directly toward the light by means of a true "phototaxis." *Euglena* has a small, opaque "stigma" in front of its light-sensitive pigments. When a *Euglena* is pointed toward a light source,

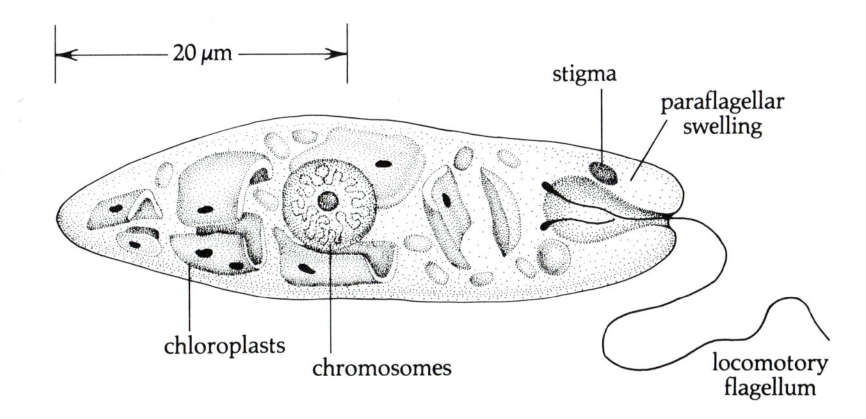

Fig. 6–3 *Euglena* pulls itself along with its flagellum. Just behind the flagellum is the opaque stigma whose shadow is the basis of *Euglena's* phototaxis.

the stigma shades the receptor, thereby providing the animal with directional information. In order to remain oriented, however, *Euglena* has to compare the light coming in from two or more directions. It accomplishes this by simply swimming in a helix so that its stigma-receptor axis points systematically in a circle about its direction of travel. If one part of that circle is brighter, the *Euglena* turns slightly in that direction. When the organism is perfectly oriented, all directions should be equally shaded and thus appear equally bright, and in this way it swims directly toward the light.

The strategy of shading light-sensitive pigments so as to provide a clue to the direction of a light is the key to the function of eyes in general. The simple eye cup of flatworms is perhaps the first step toward the elaborate camera eyes of vertebrates. The eye cup has receptors arranged on the interior surface, while the cup itself is opaque. Light falling on the cup casts a shadow from the lip into the receptors, and the shadow line reveals the direction of the light (Fig. 6–4).

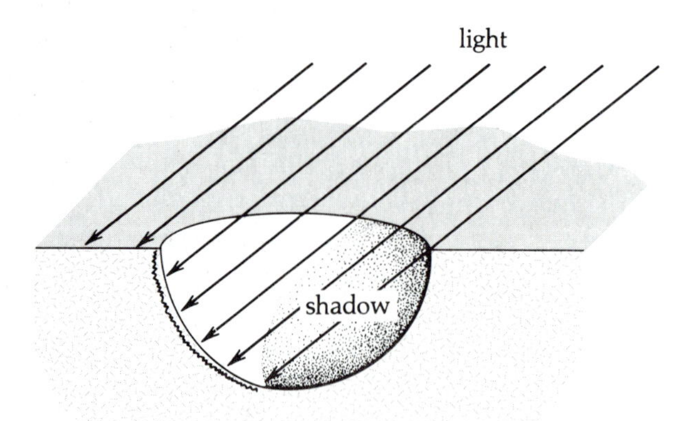

Fig. 6–4 Any eye cup in cross section. The lowest receptor receiving light indicates the direction and elevation of the sun.

EYES

In order actually to *see* their world, however, animals must somehow project an image onto an array of receptors. As early scientists discovered, there are only three ways to do this, and evolution has tried each one. The first is the pinhole camera—

essentially an eye cup with a very tiny opening. Light is reflected from objects in the environment in all directions. Only light scattered in precisely the right direction will pass through the tiny hole in an otherwise opaque barrier, but once having passed through, the light ought to project to a unique point on the retina behind the hole. As a result, an array of receptors will receive a precise (although inverted) image of the outside world (Fig. 6–5). This strategy (used by the chambered nautilus, for example) has two serious drawbacks: only a tiny amount of light comes through the hole and falls on the receptors; and diffraction or bending of the light at the edges of the hole reduces the quality of the already-dim image.

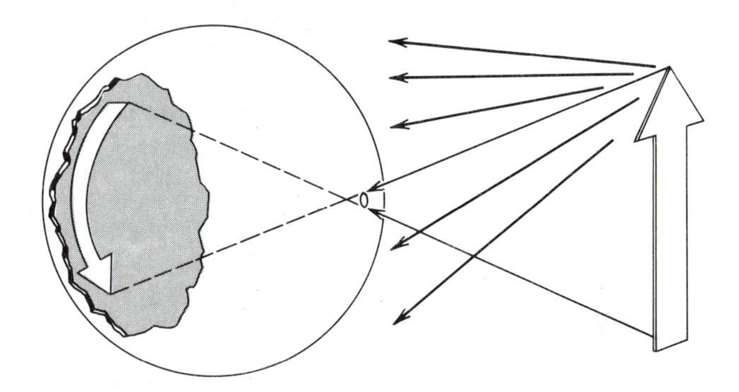

Fig. 6–5 A pinhole eye. Very little of the light scattered by the arrow passes through the hole.

A second strategy, elected by cephalopods and vertebrates such as man, is to use a lens to form the image. The lens, by being denser than air, bends and focuses the light arriving over a much larger area than the pinhole (Fig. 6–6). As a result, much more light falls on the receptors and diffraction becomes only a minor problem. Other difficulties, however, arise with lenses. For one thing, lenses have a unique focal point—that is, for any distance from the lens to an object in the environment there is only one distance behind the lens at which the image will be in focus. Given that the environment provides things to see at all possible distances, eyes with lenses ought, in the simplest case, to be able to change the distance from the lens to the retina in order to focus. In fact, lens eyes do not do this at all. Instead, muscles stretch the lens, altering its curved shape and thereby its focal distance (Fig. 6–6C). So crucial is this operation in many species (including ours) that the muscles which "aim" the eyes

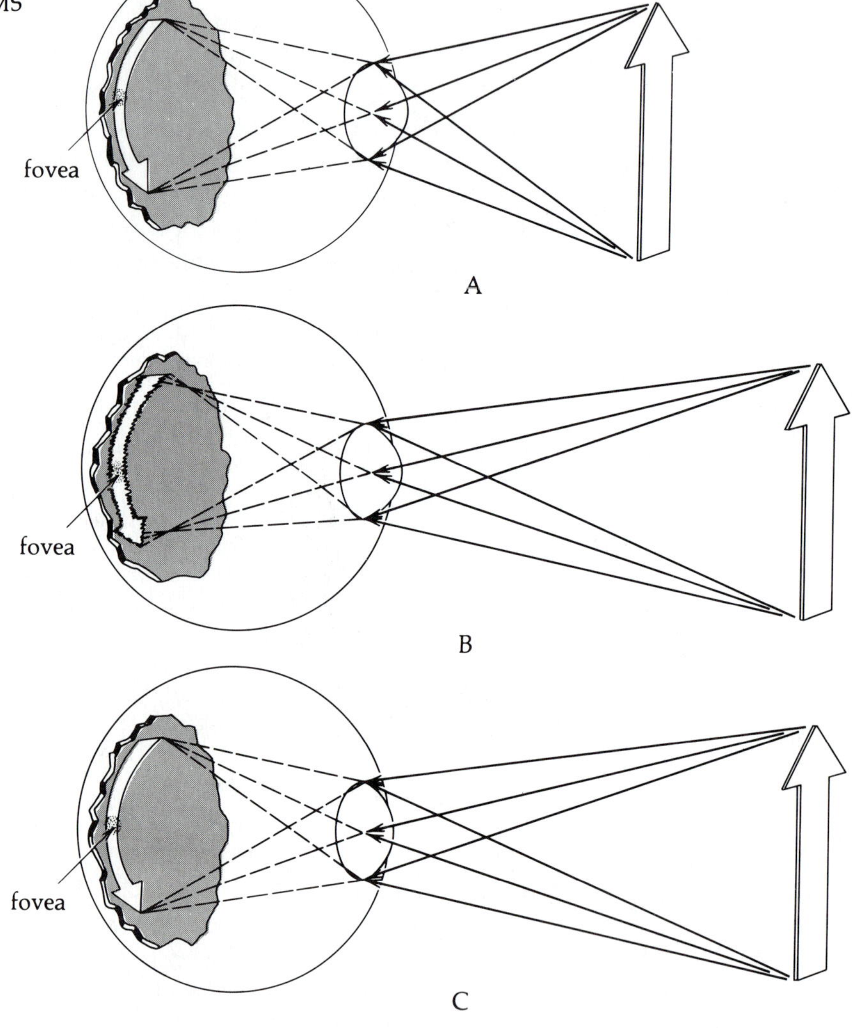

Fig. 6-6 A camera eye. A. The lens admits more light, but must bend it to form an image. B. When the distance to the object changes, the image is lost unless (C) the shape of the lens is changed to compensate.

and control the shape of the lens contain the highest density of stretch receptors in the whole body. This need to control the precise shape (and thus the focus) of a lens, however, requires complicated neural wiring not needed for the more economical pinhole eye. Moreover, the lens strategy is less forgiving of any anatomical defects in the shape of the eye or lens which may generate systematic nearsightedness, farsightedness, and astigmatisms.

The third strategy is the compound eye typical of insects. While the camera eye is basically an eye cup with a lens, compound eyes are vast arrays of eye cups, each pointing out in a unique direction (Fig. 6–7A). The result is a mosaic image of the world which must be something like a needlepoint picture (Fig. 6–8). The compound eye appears to be a marvelous compromise negotiated by evolution between good vision, small size, low weight, and Newtonian optics. Each element or ommatidium is actually a tube with a lens at one end and layer upon layer of receptors at the other (Fig. 6–7B). Each ommatidium looks out at a unique point in space. Of course, using more but smaller ommatidia to subdivide the visual world into more and

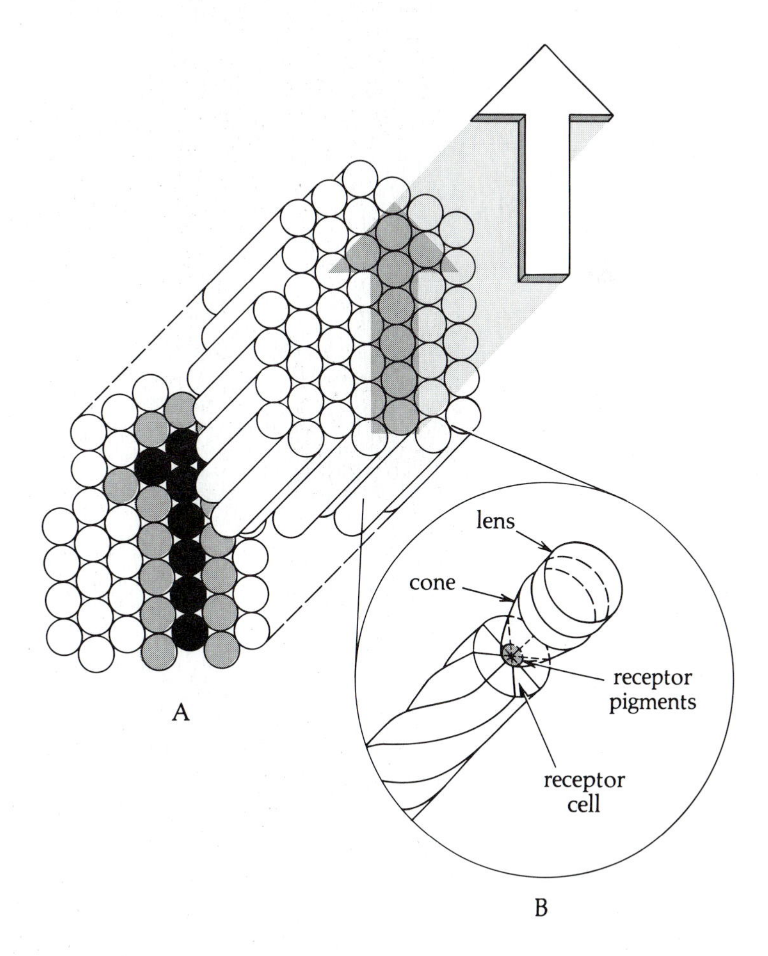

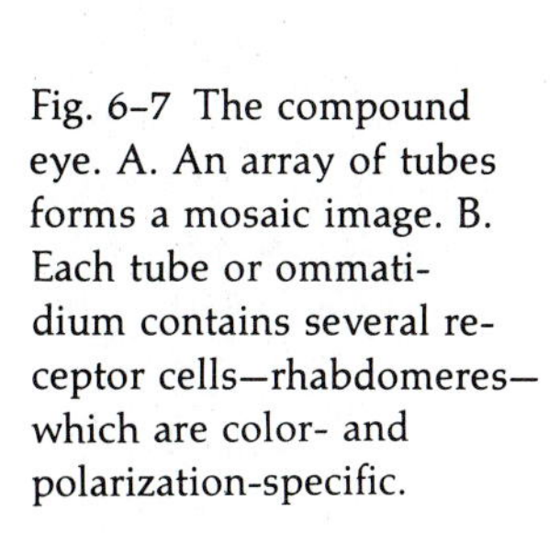

Fig. 6–7 The compound eye. A. An array of tubes forms a mosaic image. B. Each tube or ommatidium contains several receptor cells—rhabdomeres—which are color- and polarization-specific.

Fig. 6–8 A St. John's wort from 10 cm as seen by a lens eye (A) and as it might appear to a compound eye (B).

smaller parts should increase the resolution of the eye (Fig. 6–9).

Diffraction, however, puts a strict limit on the number of ommatidia. As the opening gets smaller diffraction increases, bending light from other directions into the ommatidial tube. The exact angle of bending is roughly proportional to λ/d, where λ is the wavelength of light and d is the diameter of the

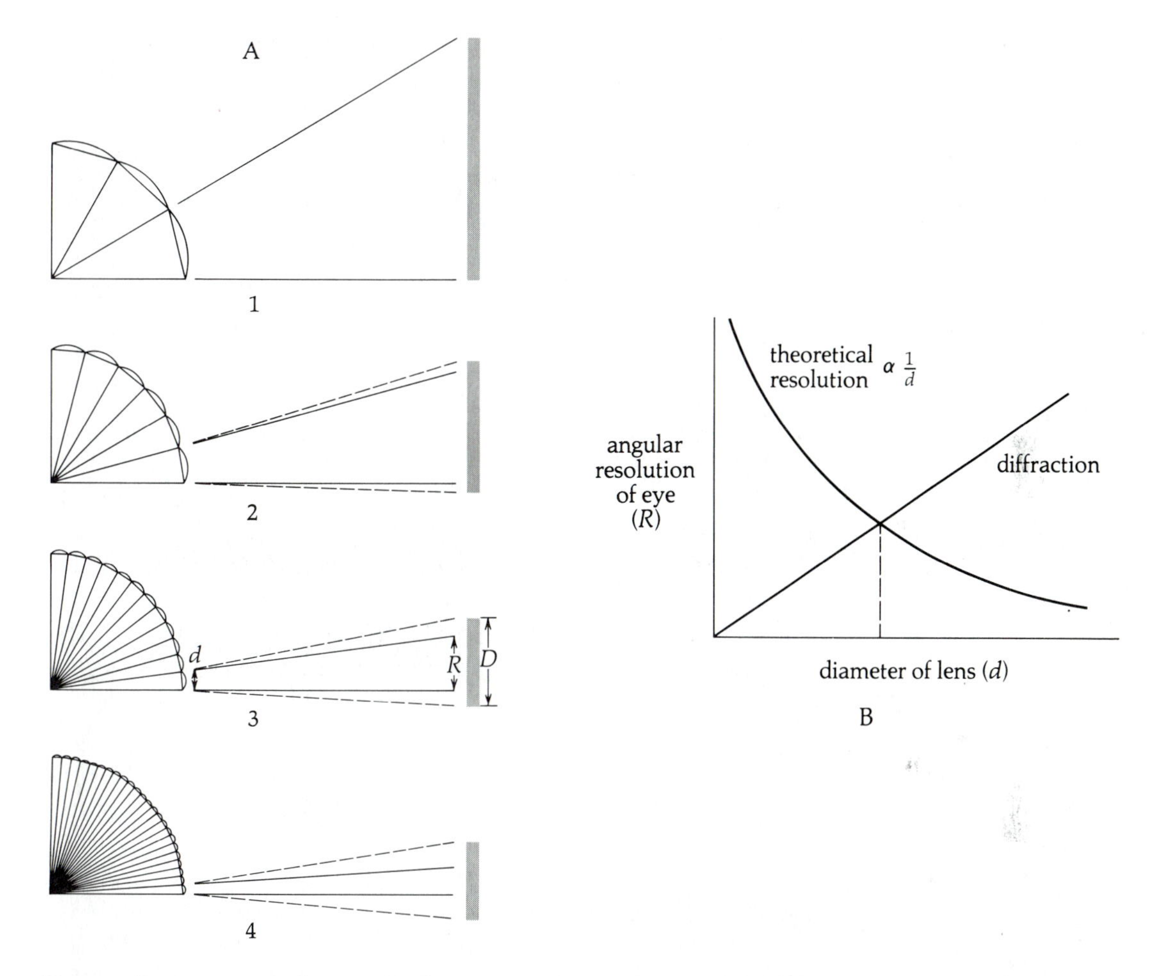

Fig. 6–9 Optimum size of an ommatidium. A. As the number of ommatidia is increased, each becomes narrower and looks out at a smaller section of the visual world. In theory, as the diameter *(d)* of each element gets smaller, the resolution *(R)* improves. However, diffraction *(D)* becomes important as light from other directions is bent into the increasingly narrow ommatidia. In this example, the actual resolution in the fourth case is considerably worse than the third despite the twofold increase in facets. B. As a result of the interaction of lens diameter and diffraction, an optimum size exists for ommatidia.

ommatidium. As d becomes smaller, the angular field of view of each ommatidium necessarily increases, and resolution decreases (Fig. 6–9).

With the field of view (Θ) increasing because of diffraction and decreasing by the addition of more elements, there must be some best compromise value for d so that Θ is as small as possible. Since a smaller λ allows a smaller d and thereby increased resolution, it is perhaps no surprise that insects in general have abandoned red light and pushed their visual range into the shorter wavelengths of the ultraviolet, right to the point at which atmospheric absorption removes the UV coming in from the sun (Fig. 6-1).

Evolution has favored the compound eye in insects because of its low weight and volume. The honey bee eye, for example, is said to have 1/50th the average resolution of a human eye while it weighs only 1/5000th as much.

Although vertebrates, because of their size, can afford to pay a high price to gain resolution by using the bulky camera eye, it happens that they cannot afford the brain space to process the highly detailed information which that eye can gather. As a result, vertebrates pack their receptors densely and only process the output of individual receptors in a tiny part of the retina known as the fovea (Fig. 6-6). The receptors are less densely placed over the rest of the eye, and the output of many receptors may be added together for economy. A cat eye, for example, contains 130,000,000 receptors, but sends only a few million fibers to the brain. Indeed much of the retina, although it is actively processing the image as we shall see, seems to exist primarily to tell the brain when there is something worth pointing the fovea at.

Vertebrate retinas usually consist of some mixture of two classes of receptors: rods and cones. Rods are high-efficiency photon counters, good in dim light, while cones are low-efficiency wavelength-specific detectors which make color vision possible while they sacrifice sensitivity to light. The relative proportions of these two classes of detectors usually reflect an animal's evolutionary needs. Nocturnal animals—cats, for example—have mostly rods. Nature is willing in this case to sacrifice wavelength information for high sensitivity in dim light. Owls are said to see as well by starlight as humans do at sunset. Diurnal animals, on the other hand, have mostly cones in the fovea. In the course of evolution our genes have been willing to sacrifice some sensitivity to light in order to make fine color distinctions. Since the human retina has rods as well, it represents something of a compromise.

Insects, too, have had to make the choice between color vision and sensitivity in dim light. Bees are a good example of the color-vision option. Their ommatidia are each divided into eight cells or rhabdomeres (Fig. 6–7B). Each ommatidium in the bee eye has two cells with green receptors, two with blue receptors, and two with UV receptors. The remaining two rhabdomeres are green receptors if they are situated in the bottom half of the eye, and UV or blue receptors if they are in the top half—a specialization appropriate for what a bee is likely to see in flight. Although insect eyes are smaller and so cannot take in as much light as a lens eye, they are very efficient at capturing the incoming photons. Each ommatidium has such a dense and deep layering of receptor pigments that it is said that nearly 90 percent of the incident light falling on them is absorbed—a far cry from the low capture-efficiency of vertebrate retinas.

The "primitive" eyes of invertebrates manage to extract a wealth of information denied to vertebrates. For one thing, their flicker-fusion rate—the ability to distinguish two separate images closely spaced in time—is about ten times higher than that of more complex eyes. Hence bees can see fluorescent lights flashing on and off 120 times a second, and live in such a world of slow motion that their own wings must appear not as a blur as they do to us, but rather as distinct flappings, much as the flight of a bird is seen by us. So fine is this temporal resolution that crabs can even see the sun move.

The morphology of arthropod eyes opens yet another world. The long rhodopsin pigment molecules are aligned in parallel on the ordered folds of the rhabdomeres. This arrangement gives rise to a sensitivity to the polarization of light. The ability to perceive polarized light is a great advantage to animals, for the sky, when analyzed for polarized light, becomes a giant map which points to the sun. For an animal which needs to navigate in order to find its way home, the sky becomes its compass (see Chapter 13). With some odd exceptions, vertebrates are regrettably blind to polarized light.

Evolution, then, has fashioned eyes which take advantage of all the known properties of light—direction, intensity, color, and polarization. Now that it has all of this information the nervous system must decide what to do with it. Regardless of the kind of eye, any image on the array of receptors is completely without meaning until the enormous amount of processing which follows sorts and interprets the visual world. The mechanisms by which animals manage to recognize and extract the information relevant to their particular species, and to discard the rest, are the subject of the next chapter.

SUMMARY

Sensory receptors transduce environmental energy into the electrochemical energy which is the common currency of the nervous system. Visual receptors use the energy of photons to begin a biochemical chain reaction which ultimately alters membrane potential. A variety of tricks permit animals to know more than the intensity of light. Simple shading reveals the direction of a light source, energy-specific receptor molecules can determine its color, and the parallel arrangement of the molecules yields a sensitivity to its polarization. The most conspicuous triumph of evolution, however, is the image-forming eye—the lightweight, mosaic eyes typical of invertebrates and the high-resolution camera eyes of vertebrates.

STUDY QUESTIONS

1. Insects, though they come equipped for the most part with a pair of image-forming compound eyes, cannot perform normal phototaxis with only one eye. Instead, they behave as if each eye were a separate single intensity detector and try to balance the two intensities. What path should such an animal with one eye blinded follow? What might be the advantage of such a dual comparison strategy in the natural world?
2. To what extent could insects make up for their poor spatial resolution with their superior temporal resolution? What behavior, if any, would you expect to see if they were attempting this trade-off?

FURTHER READING

Horridge, Adrian. "The Compound Eye of Insects." *Scientific American* 237, no. 1 (1977): 108–20.

Wald, George. "Eye and Camera." *Scientific American* 183, no. 2 (1950): 32–41.

CHAPTER 7

Visual Processing

The retina and the brain are often thought of as passive neural arrays which receive and project an accurate representation of the visual world. Nothing could be farther from the truth. The moment a photon is absorbed by a receptor, the eye and brain begin the systematic process of discarding one piece of information and exaggerating another. Just what gets thrown away or emphasized, however, depends on what you are: the visual world, far from being absolute, is species specific. To a frog the tiniest movement, the most delicate flickering of an insect's wing, is blown out of all proportion while brightly colored and conspicuous but stationary objects fade into muted, uninteresting shades of gray. Approaching objects stand out in bold relief while those moving away are virtually erased from the frog's picture of the world. On the other hand, the human visual world might seem to a frog to be dominated by colors, vertical and horizontal lines, exaggerated edges, detailed patterns in the center of the visual field, and overemphasized contrasts between light and dark. This chapter will focus on some of the better understood systems by which animals interpret their visual worlds.

COLOR

Color discrimination is the most familiar example of visual processing. The three classes of color-sensitive pigments in bees and humans, for example, are broadly tuned to UV (bees only), blue, green, and yellow (humans) (Fig. 7–1). We discriminate colors by comparing the outputs of the receptors. Because of the overlap in sensitivity, any particular wavelength elicits a response from three classes of receptors in a ratio unique to that

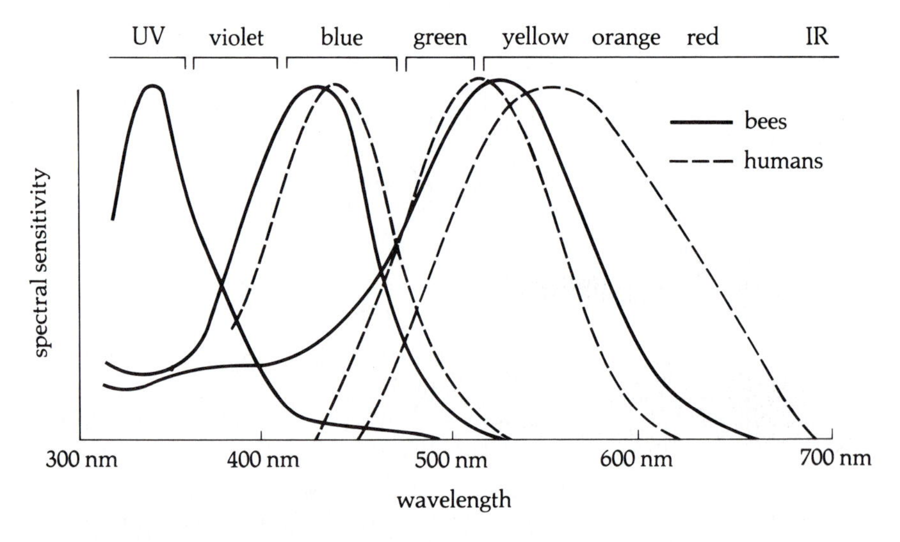

Fig. 7–1 Both honey bees and humans have trichromatic color vision. In the case of bees, the three color receptors are neatly tuned to UV, blue, and yellow-green. Color is then determined by the relative firing rates of the three types of receptor. Humans also have three types of cones, but their mutual tuning appears less inspired. One absorbs blue light best, another yellow-green, while the last is most efficiently stimulated by yellow light. Despite the "crowding" of these latter two pigment-response curves, we seem to see and appreciate red and orange perfectly well.

color and independent of intensity. This same processing strategy is used by the tonic stretch receptors in our joints which keep us informed of the positions of our limbs (Fig. 7–2). For color vision the CNS breaks the resulting visual continuum into arbitrary categories (see Chapter 30).

Fig. 7-2 The body uses a group of differently tuned tonic stretch receptors (proprioceptors) to measure the angle of a joint. Just as with vision, any single cell has only a limited range of sensitivity and inherent ambiguity—any output could represent two separate joint angles. The relative firing rate of any two active receptors, however, uniquely specifies the angle.

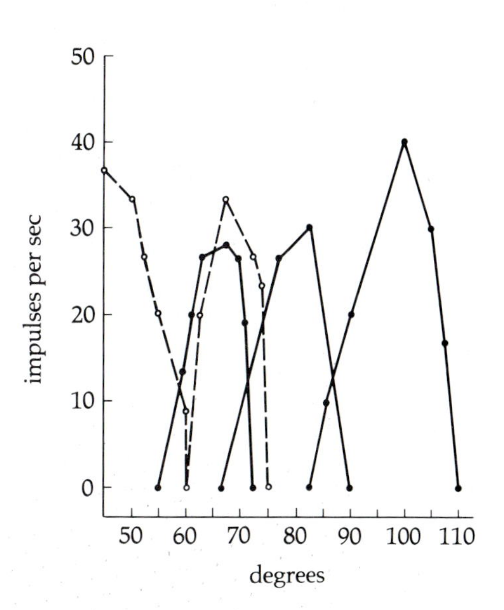

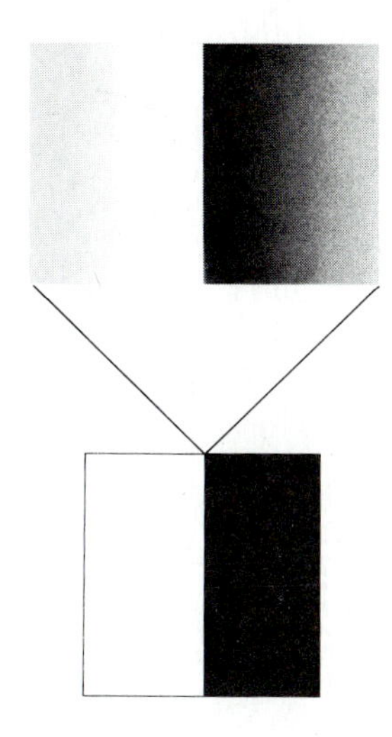

Fig. 7-3 Mach bands. On each side of the black/white boundary, a narrow stripe can be seen which is even blacker in the one case and whiter in the other. Lateral inhibition in the retina underlies this illusion.

LATERAL INHIBITION

Much of the deliberate distortion of what our visual receptors actually see is accomplished right in the retina. The simplest example, and perhaps a grand organizing principle of both sensory processing and innate releasing mechanisms, is lateral inhibition. When we look at a border between light and dark, for example, the light side becomes brighter and the dark side darker just where the two areas meet (Fig. 7-3). The narrow strips along the edge which seem brighter and darker than the larger areas to which they belong are called Mach bands. They are, of course, optical illusions, but like most such tricks which our eyes play on us they reveal something about the way we are wired.

Mach bands are the result of lateral inhibition—an interaction between receptors in the retina. When hit by light, the so-called horizontal cells (driven directly by the receptors and named for their morphology rather than any selection or sensitivity) simply inhibit or prevent their neighbors from firing quite as often as they would otherwise. We might expect this to result merely in a slight darkening of the visual image which is passed on to the brain. But along any edges between areas of different illuminations, a curious thing happens. The cells just on the bright side of the edge are less inhibited by their neighbors on the dark side than their next-door neighbors on the bright side are by them and by each other. As a result they fire more often. The cells on the dark edge, on the other hand, are very strongly inhibited by their neighbors on the bright side, and so fire infrequently, if at all. The consequence of this interaction is that cells on the edge between two areas of differing brightnesses exaggerate that difference to the brain (Fig. 7–4).

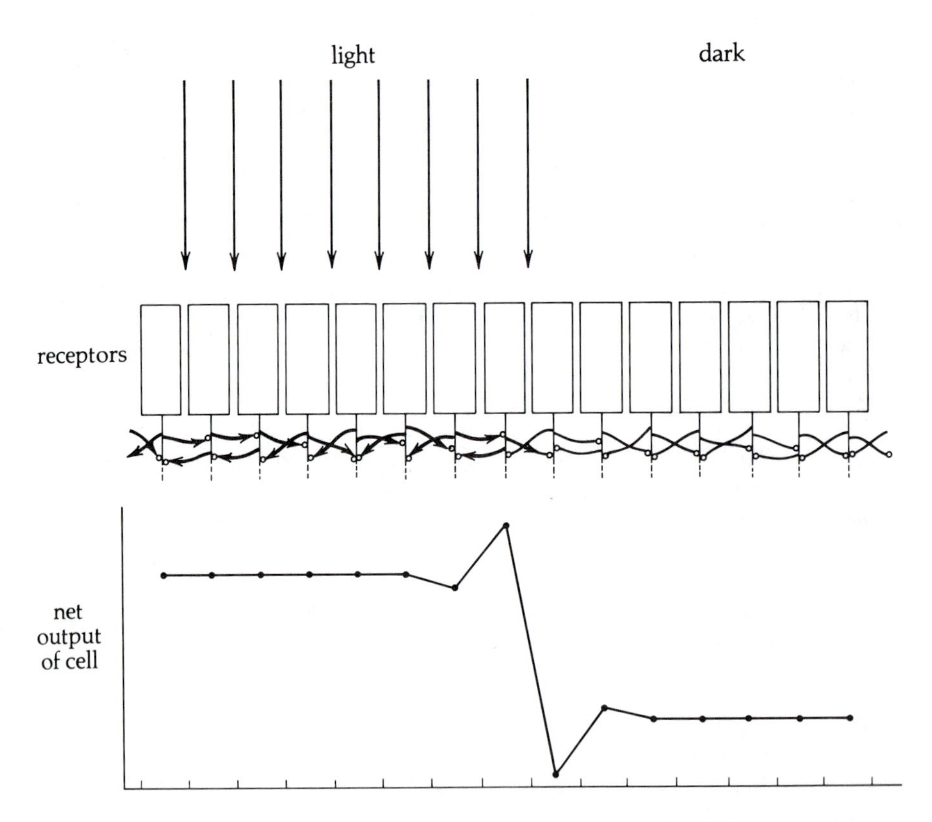

Fig. 7–4 Lateral inhibition. This hypothetical array of receptors illustrates the edge-enhancing effect of lateral inhibition. The horizontal cells are omitted for clarity. Each cell inhibits its neighbors to the extent it is active. This inhibition is indicated by the arrows. The result is an image which is artifically enhanced along the edge.

Lateral inhibition serves, then, to enhance our perception of edges. It is the most basic element in the repertoire of processes carried on by all visual systems. The retina and brain are wired to single out lines of specific orientations, simple shapes such as spots, or even movement in particular directions and at particular speeds from a vast world of visual stimuli; and they do this through minor alterations in the pattern of who inhibits whom (Fig. 7-5). The ultimate results of this trick of wiring are the so-called hypercomplex cells of the visual cortex—cells which are basically "feature detectors." Our visual world is a direct consequence of the sorts of these very specific detectors evolution has selected to best meet our species' needs.

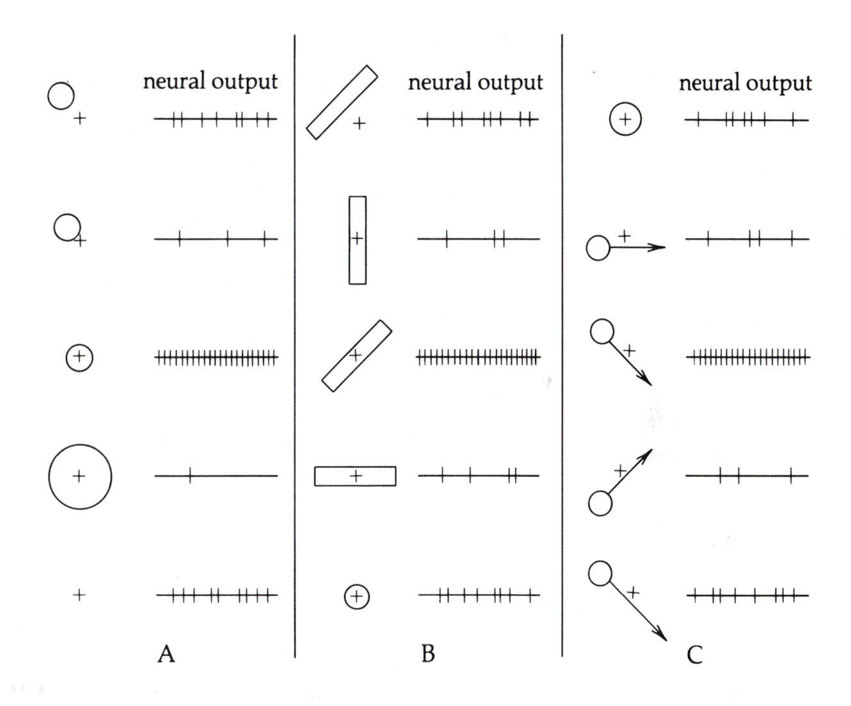

Fig. 7-5 Examples of feature detectors found in the visual system of higher vertebrates. The neural output is shown at the right in each case, and the third example shows the optimum stimulus. Spot detectors (A) are seen in the retina. Some respond best to a bright spot of a particular size centered on a particular point in the visual field, while others prefer a dark spot (the "t" shows the center of the visual field). Both are obvious adumbrations of lateral inhibition. Another familiar class (B) are found in layer 4c of the visual cortex of the brain's line-orientation detectors. These cells prefer a linear edge of a particular orientation. The third example (C) is a hypercomplex cell which demands an edge of a particular orientation moving in a particular direction at a particular speed. Vast numbers of far more complex cells are known to exist whose tastes in visual stimuli have yet to be defined.

FEATURE DETECTORS

Feature detectors are without a doubt the key elements behind sign stimuli, which are in turn the neural bases of classical innate releasing mechanisms. As we showed in Chapter 4, the behavior of gull chicks as they aim their begging at the red spots on their parents' bills is probably triggered by both a spot detector (perhaps wired together from red-sensitive receptors) and a moving-vertical-bar detector. Certainly, the basis of a frog's ability to track and capture flying insects is a dark-center/bright-surround spot detector wired into a movement-detector circuit. In both cases the animals are using feature detectors to make these simple but important discriminations quickly and reliably, without having to "think."

In fact, feature detectors operate at such an early stage in neural processing, so long before information actually reaches the higher CNS for further consideration, that they could account for that most striking feature of sign stimuli and innate releasing mechanisms: their heretofore inexplicable "irrationality." Under unnatural circumstances IRMs can cause an otherwise intelligent animal to act in ways which look entirely inappropriate and nonsensical. Feature detectors probably explain why European robins with perfectly good eyes attack shapeless tufts of red feathers, why a gull chick will peck wildly at a red spot on a stick, why a male stickleback will try frantically to attack a red mail truck passing in the distance, and why a goose will carefully roll beer cans into her nest and later do a "double-take" and kick them out again. It seems to observers that animals perform these and other similar actions by reflex, before they realize how silly they are being. As usual, though, evolution knew what it was doing: these intellectual "short circuits" control behaviors which are vital to the survival of the organisms or of their offspring. The almost one-to-one correspondence between the sometimes outrageously schematic stimuli that elicit appropriate responses in model experiments and familiar (or easily imagined) feature detectors strengthens the hope that we will soon be able to understand the physiological basis of this cornerstone of ethology.

The use of visual feature detectors to control behavior is especially obvious in invertebrates. Flying insects in particular require lightweight brains, and so seem to have elected wherever possible to use simple feature-detector circuits to control behavior. Bees, for example, face a severe test: they must fly to distant flowers and back home without getting lost along the way. They know just what to look for at each end of the journey, and the proper direction and distance to fly, but all this information is rendered useless by the slightest gust of wind. How can a bee judge distance in a headwind or a tailwind? How can a bee know how much to turn into a crosswind in order to move in the correct direction?

Honey bees come equipped to solve these problems not as

human pilots do, with charts and radio bearings, but with an impressive collection of feature detectors. The green receptors in the lower half of the bee's eye are wired as direction-specific and speed-specific motion detectors. As a result, a honey bee knows instantly which direction the ground is moving below it. The bee has only to turn until its direction-specific motion detectors tell it that the ground is moving by from the direction of the goal.

Other animals have feature detectors specially designed for a single particularly important distance. The digger wasp, for instance, seems to have chosen 15cm—the point from which it launches its attack on honey bees—as the most important distance to be measured exactly. Similarly, the distance detectors of the praying mantis are specialized to judge just one range with exquisite precision—the length of its arms, and therefore the precise distance it must be from its next meal before it strikes.

VISUAL PATHWAYS

As part of a strategy to reduce the volume of insect nervous systems, nature has reduced the size of each neural cell to a minimum. As a result it is exceedingly difficult to record neural activity in the insect brain. For detailed analyses of just where feature detectors are located and how they work we must depend on larger animals such as monkeys or the familiar house cat. In primitive vertebrates the retina simply sends its fibers directly to the thalamus. In mammals what was formerly the olfactory lobe has expanded to overlay the thalamus with large areas of involuted cortex. The thalamus, then, which is the seat of higher level processing in simple vertebrates, is thought to have become in mammals primarily a relay point on the way from sense organs to the cortex. Information from the retina, for example, reaches the visual cortex by way of the lateral geniculate nucleus (LGN) of the thalamus. In animals with binocular vision—usually predators as opposed to their ever-watchful prey—nerves from parts of the eye which look out at the same part of the world are brought into register in the LGN (Fig. 7-6). This process of aiming and focusing the two eyes on a single target, of course, informs the brain of the distance and actual size of the target—important data for a hunter.

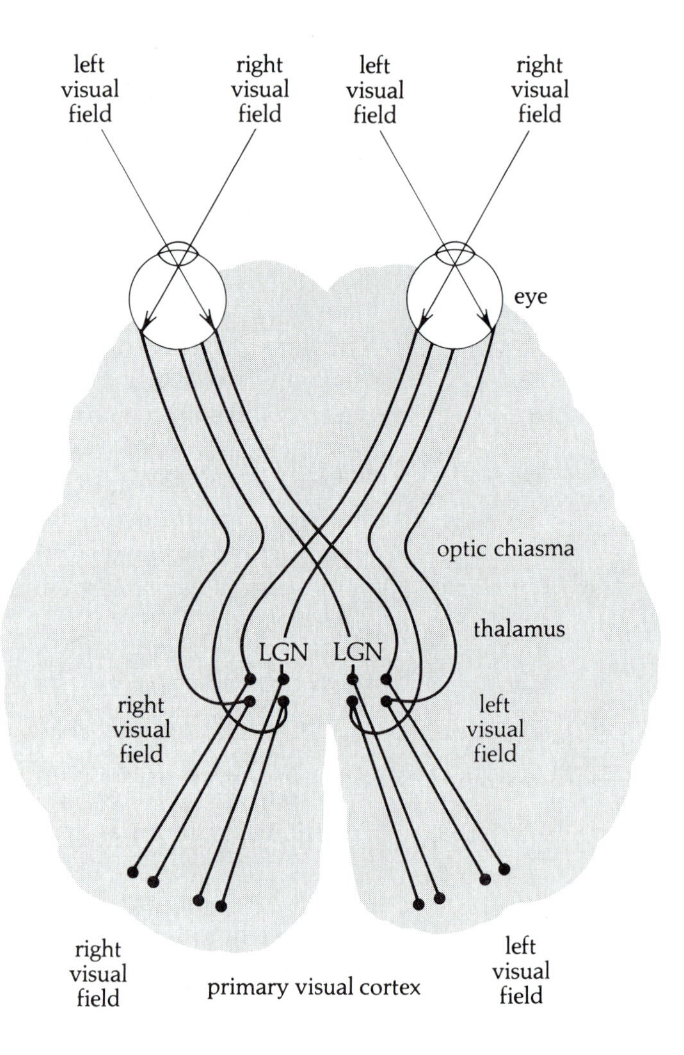

Fig. 7-6 Visual pathway in higher vertebrates (in lower vertebrates the visual information goes only as far as the tectum of the thalamus). The pathway shown is for a highly binocular animal. The fibers from the left side of each retina (which see the right half of the visual field) meet and segregate at the optic chiasma, from whch they proceed to the right LGN. The LGN serves to bring the receptors from the two eyes into register. From the LGN the visual information proceeds to the left primary visual cortex. The fibers terminate in layer 4c.

The visual fibers from the LGN make their way to the primary area of visual cortex (also known as the striate cortex), where the whole of what the retina sees is represented. Mapping these pictures in the cortex, however, has made it clear that the primate brain is something like a hall of mirrors where the visual world is projected in at least seven different places (Fig. 7-7). The most circuitous path shown here proceeds from the LGN to the striate cortex, on to the superior colliculus (a very old structure to be in such a new pathway), then to an area in the temporal lobe, and on to a doughnut-like area surrounding it. Neuroanatomists are currently busy mapping still more complicated pathways between various visual areas, and the

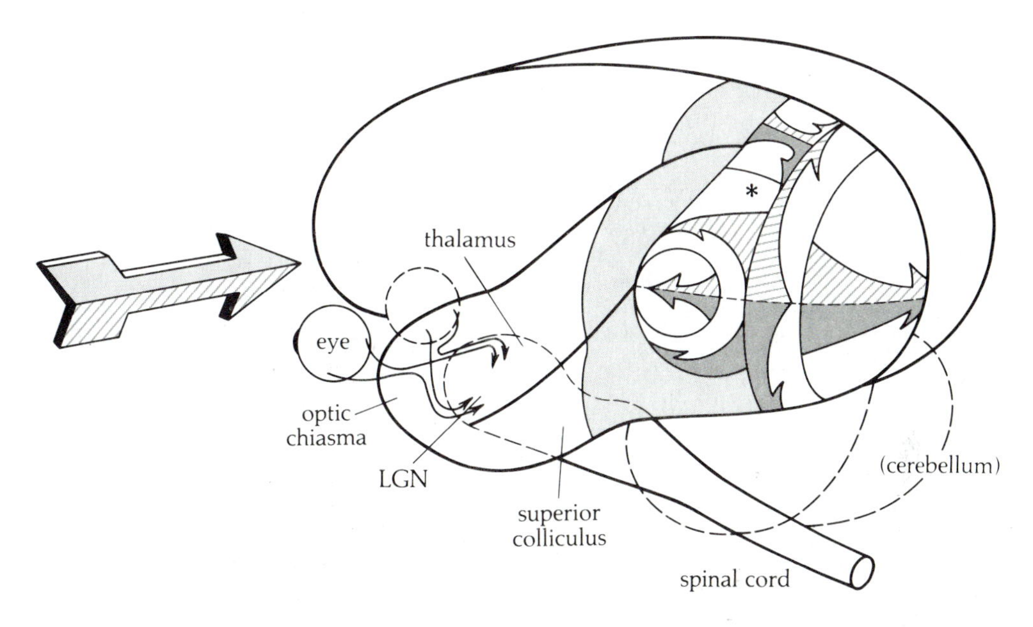

Fig. 7–7 The visual world of the owl monkey is mapped out in at least seven different places on each side of the brain. The primary target is the striate cortex at the very back. Note that in each case the image is inverted and the center of the field, the fovea, is highly magnified. Only six maps are shown since the geometry of the seventh area (indicated by "*") is not known. The shaded areas are also visual, but unmapped as yet.

end is not in sight. Just what all this shunting about is accomplishing is not yet known, although it is worthwhile to wonder whether the flow of information among the visual areas, often in loops, might have something to do with decision making or pattern recognition.

If we look at one cortical area in isolation, the striate cortex, intriguing and somehow satisfying patterns are becoming clear. The striate cortex gets its name from its layered appearance in cross-section. The projection of visual fibers from the LGN is to the fourth of these layers (Fig. 7–6). Careful recording reveals that layer 4 reproduces the general organization in the lateral geniculate; that is, fibers from the parts of the two retinas which look out in the same direction project side by side in layer four (Fig. 7–8 and Fig. 7–9). Fine-scale recording in this layer, however, reveals an unsuspectedly functional organization. An electrode moving across layer 4 encounters a thin stripe of cells

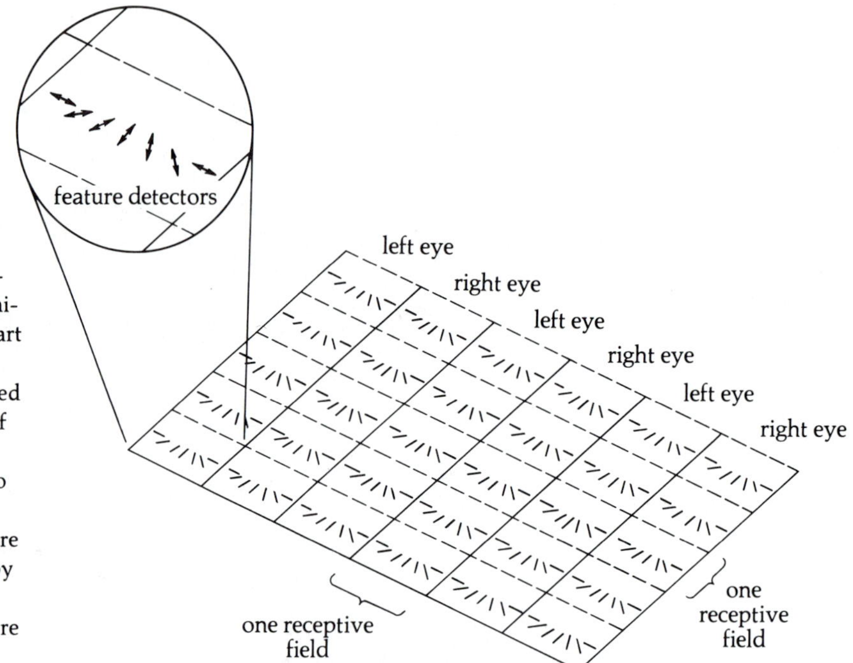

Fig. 7-8 Detail of idealized striate cortex organization. In layer 4 any part of the visual world of one eye will be presented next to the projection of the same part from the other eye. In addition to spot detectors, orientation-specific detectors are found in this and nearby layers, with orientation preferences varying more or less systematically across the cortex.

responding to the left eye, then a stripe responding to the right eye, then one to the left, and so on all the way across the visual field. This patterning of cells responsive to input from only one eye is known as "ocular dominance." But such recordings also reveal that something else is happening. Besides simple lateral inhibition, the only features emphasized in the retina are spots. The LGN of higher vertebrates adds no new emphases. Layer 4, on the other hand, has millions of orientation detectors built into it. These orientation-specific cells are also organized in a neat, functional pattern on the cortex. As an electrode moves across each tiny cortical stripe, the first brain cells respond preferentially to lines of one specific orientation, say horizontal, but then as the electrode moves the orientation preference shifts counterclockwise through vertical and back again to horizontal (Fig. 7-8).

Many details of the organization have yet to be worked out. It is not clear how the stripes of cells with the same orientation specificity are organized relative to the stripes of ocular dominance. Nor is it obvious just how the simple lateral inhibition spot detectors of the retina which arrive by way of the LGN are

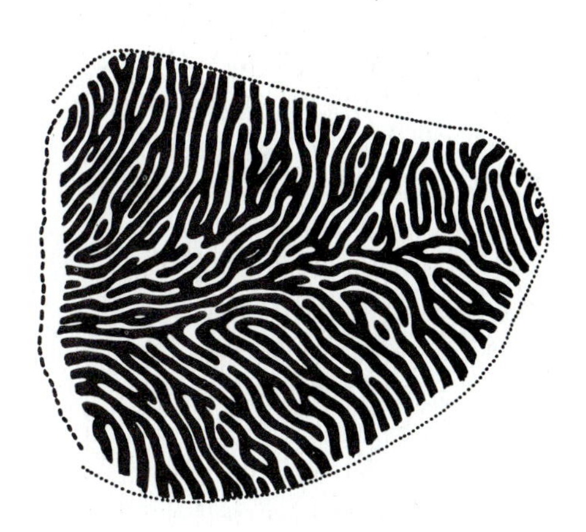

Fig. 7-9 An actual map of striate cortex from a primate shows ocular dominance columns (here indicated by the alternating bands of black for one eye and white for the other) which resemble a zebra hide more than a TV screen. This reconstruction is from layer 4c of the right side.

physically connected to give rise to the line-orientation detectors in the cortex, or how these cells then combine to produce the hypercomplex cells of adjacent layers which sort for angular size, direction and rate of movement, binocular disparity, and all the rest. Yet all these detectors and many more yet to be deciphered ones are organized there in a geometrically orderly array to make crucial sense out of the visual world.

ONTOGENY OF VISION

Some neurophysiologists have challenged the assumption that feature detectors are innate, prewired elements of the visual system. Instead, they hypothesize that in some way the brain "learns" to see. Early results seemed to support that view. In one classic experiment, kittens were raised in garbage cans whose interiors were painted with either vertical or horizontal stripes. Although normal cats have feature detectors for lines of all orientations, cats which as kittens had seen only vertical stripes had feature detectors only for vertical lines, while the horizontal-stripe cats had only horizontal detectors. This experiment was taken as evidence that cats *learn* to see—that feature detectors are not innate, but rather are the result of visual experience.

More careful repetitions of the garbage can experiment, however, were unsuccessful, and slowly led to the realization that the *age* of the kittens controlled the results. It was discovered that the orderly array of feature detectors is present at birth, but persists only if a cat is allowed to see between six and eighteen weeks of age. For example, if one eye is kept closed and the other eye open until six weeks, and then the situation is reversed until eighteen weeks, and then reversed again until testing some months later, the eye which saw from six to eighteen weeks will be normal, while the other will be blind. Precisely the same sort of result is obtained if a kitten is permitted to see lines of only one orientation during this period of time: feature detectors for other orientations will be rare. It is safe to say then that there is a *critical period* for vision in cats, and apparently the same thing is going on in humans. Very young children with cataracts which are not removed before one and a half years never see normally, while children and adults who develop cataracts after that age will have normal vision after their removal regardless of how many years they have been affected. Neurally, monocular deprivation—blindness in one eye during development—causes the alternate stripes in layer 4 assigned to the blind eye to become very narrow, perhaps even to disappear. Since the output of the retina is basically from spot detectors, it is probably not the absence of light per se in a blind eye which is responsible for the massive collapse of cortex organization, but the absence of effective stimuli to be encoded by spot detectors: spots, lines, indeed forms of any sort. In fact, if the eye of an experimental animal is covered with a translucent shield which permits light to pass through but destroys any image, the cortex becomes blind to input from that eye.

A similar pattern is observed in binocular cells, nerves which combine incoming visual information from both eyes. Most cortical cells are genetically programmed to be binocular, but only remain binocular when both eyes see the same thing at the same time during the critical period in early development. If both eyes are left open normal binocular detectors are found in the cortex, but when only one eye is left open throughout the critical period for development binocular detectors are absent. Manipulations during the critical period reveal further complications. For example, when both eyes are allowed to see during the critical period but not at the same time, no binocular cells develop. On the other hand, kittens which wear special goggles throughout their critical period generally fail to develop binocularly driven cells when the lenses present the animals with different images: horizontal versus vertical stripes, for example.

A perverse but potentially illuminating twist to this story is that if *neither* eye sees anything during the critical period, the cells remain binocular.

None of these experiments really resolves the nature versus nurture question for feature detectors. Either the brain "learns" to see by experience, or the detectors are there but are lost when they are not used. This question was answered conclusively when a baby monkey was delivered by Caesarian section and raised in the dark until it was old enough for experimentation. Its first visual experience occurred with electrodes in place in the striate cortex. Bright, moving visual stimuli were presented in a dark room, and all classes of feature detectors were found to be already there.

Apparently the brain comes knowing how to see at birth, but can "forget." The predominance of one eye or of one class of stimulus permanently destroys or attenuates the other(s) during the critical period for visual development. There is more and more evidence that this predominance effect is the result of competition among incoming cells for synaptic space on the target cells in the cortex. For example, if both eyes are kept closed so that neither has a competitive edge, the brain continues to listen to both eyes, and the cortex, although affected by this major disturbance, remains largely normal.

Analogous things seem to happen in other sensory areas, even in insects. In crickets, for example, nerves from the two cerci—touch and vibration receptors in the tail—converge in a particular part of the nervous system. If one cercus is kept from "hearing" during its critical period, nerves from the other side take over the cells in the abdomen which relay information to the brain. When the treated cercus is again allowed to function, however, it is too late. Because its nerves were unable to compete with those of other cerci for synaptic connections on the relay cell, its receptors fire normally but the brain remains deaf to their output.

Among many ethologists there is a growing sense that the general principles of lateral inhibition and neural competition may ultimately explain many aspects of sensory processing and behavioral control. Indeed, similar neural strategies pervade the auditory system as well. This emerging picture of neural organization has brought to those in the field a great sense of relief. The brain seems to have organizing principles and, what is more, we can hope to understand them.

In quite a different way, these same discoveries have removed some of the tension between advocates of nature and nurture by establishing definitively that most systems for neural processing examined so far are innate, but that proper (i.e., natural) sensory experience during ontogeny is required to make everything work, and may even serve to fine-tune some of the elements. Both sides of the conflict regard this interpretation as a clear victory for their traditional interpretations—a felicitous situation indeed.

SUMMARY

The sensory worlds of animals are species specific, finely tuned to the needs they are likely to encounter. The visual world in particular is a consequence of highly evolved and adaptive genetic design. Beyond the issues of whether to invest in color vision or high sensitivity, what wavelength range to concentrate on, and whether or not to be sensitive to the polarization of light, each species brings to the incoming sensory information a unique battery of processing devices. The basic processing trick—lateral inhibition—out of which the first-order repertoire of spot, motion, and line detectors are fashioned is the same from horseshoe crabs to primates. Species-specific elaborations and variations in these networks of feature detectors sort out and even enhance the crucial information for animals—releasers being the most specialized cases—and attenuate or discard the rest. The conscious experience of the visual world for other species rarely has much relation to ours, being at once rich where we may be blind, and dull where we are the most acute. Hence our manifest blindness, when added to our very incomplete understanding of any of the visual processing going on outside of the one layer of the one visual projection in the mammalian brain that we know anything about, once again require us to look to the overt behavior of animals and to our own limited imaginations for clues to the ultimate sensory worlds of other creatures.

STUDY QUESTIONS

1. How could the notion of feature detectors as the basic elements from which releaser circuits are created explain the existence of supernormal stimuli?
2. If the peak sensitivity of our so-called red cone is actually yellow-orange (Fig. 7–1), how do we come to sense and distinguish red?

3. About 8 percent of males are color-blind, lacking most often the "red" visual pigment. What is the world like for a red-blind person? For example, what color is the sky? "White" is our perception when all three pigments are equally stimulated. What is white for a red-blind male? For that matter, what color is the sky and what is white for bees, or for the rare blue-blind and green-blind people?

FURTHER READING

Ewert, Jörg-Peter. "The Neural Basis of Visually Guided Behavior." *Scientific American* 230, no. 3 (1974): 34–42.

Hubel, David H. "The Visual Cortex of the Brain." *Scientific American* 209, no. 5 (1963): 54–63.

———. "The Visual Cortex of Normal and Deprived Monkeys." *American Scientist* 67 (1979): 532–43.

———, and Weisel, Torsten. "Brain Mechanisms of Vision." *Scientific American* 241, no. 3 (1979): 150–62.

Michael, C. R. "Retinal Processing of Visual Images." *Scientific American* 220, no. 5 (1969): 104–14.

Pettigrew, Jack. "Binocular Vision." *Scientific American* 227, no. 2 (1972): 84–95.

Ratliff, Floyd. "Contour and Contrast." *Scientific American* 226, no. 6 (1972): 90–101.

CHAPTER 8

Auditory Design

SOUND

Sound is simply the vibration of molecules, whether in air, water, earth, or whatever. Hearing involves the filtering and decoding of the vibrations the various auditory organs are designed to detect.

In many ways hearing and seeing are analogous. Both systems seem as sensitive as possible: dark-adapted rods can respond to single photons, while quiet-adapted ears are at the threshold of hearing molecular vibrations and are able to detect a movement of the delicate basilar membrane in the ear which is hundreds of times less than the diameter of a hydrogen atom. Both light and sound are wave phenomena, and so have color or pitch. As a result, both are subject to diffraction, reflection, refraction, and wavelength-specific filtering. Like light, sound has a particle nature so that, as we shall see, two quite different strategies of hearing have evolved, one based on particles and the other on waves.

The sensory environment for hearing is even more species specific than the visual environment: men, bats, and bees each live in entirely separate worlds of sound. While animal vision runs from the infrared in snakes to the ultraviolet in bees—

barely a factor of two in wavelength—animal hearing runs from the 100,000 Hz of bats to the 0.1 Hz of pigeons—a factor of a million. Moreover, light passes through very few substances while sound can travel in almost everything. Nearly everywhere we look, whatever the wavelength, there is something listening—listening for prey, for predators, for the opposite sex, or for auditory landmarks.

Sound is created by molecules set in motion. In the case of a loudspeaker, for example, an AC signal put through a coil of wire generates an alternating magnetic field which in turn interacts with a nearby permanent magnet to attract and repel the coil. The coil is attached to a speaker cone which moves the adjacent air to create sound (Fig. 8-1). The vibrations of the cone serve to generate alternating bands of high and low pressure which appear to move out from the speaker at the speed of sound. On closer examination we find that the molecules whose movements underlie sound transmission merely move back and forth from the high-pressure bands into the low-pressure ones,

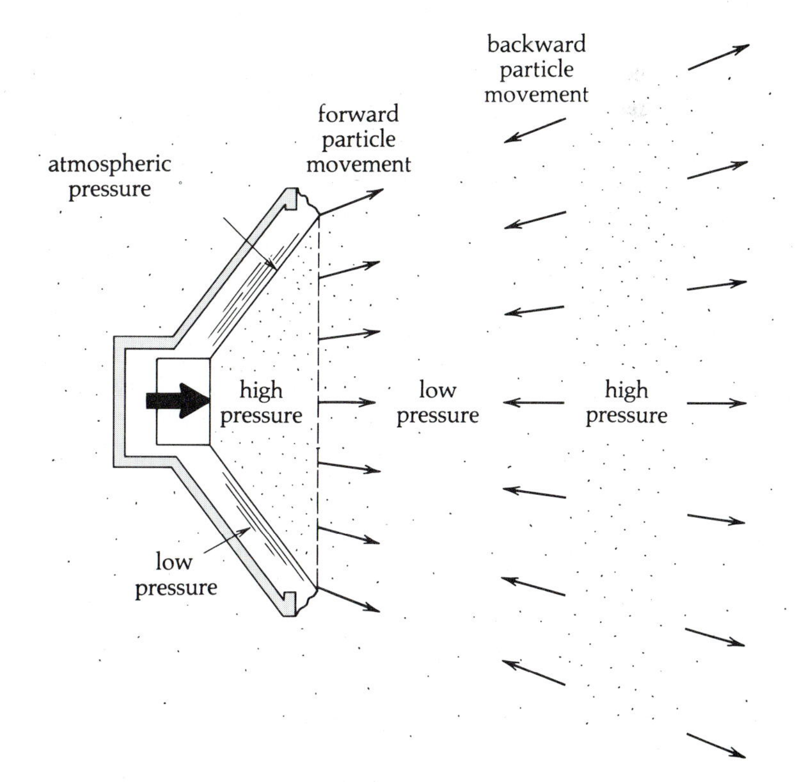

Fig. 8-1 Production of sound by a loudspeaker. The movement of the cone creates alternating bands of high and low pressure. The motion of air molecules in response to these pressure differences is indicated.

thereby creating new bands of high pressure out of which they subsequently rush (Fig. 8–1). As a result, sound may be thought of either as alternating bands of high pressure and relative vacuum, or as regions of alternating directions of particle movement. Evolution has designed ears for both perspectives.

Rabbits and many other animals transmit warning messages merely by stomping hard on the ground. Similarly, some species of temperate zone fiddler crabs communicate at night by rapping their claws on the ground. Other animals receive valuable information by means of sounds produced incidentally by movement. Mosquitoes, for example, find each other by listening for flight sounds of a particular wingbeat frequency. Other animals, like male *Drosophila*, use these same incidental sounds, but have arranged them into pulse-coded messages. *Drosophila* court with a "song" which consists entirely of one-wing bursts of flapping according to a species-specific code (see Chapter 15). Worker bees use the same strategy to communicate about food, while queens exchange quite different messages in the same way (see Chapter 24). As we shall see, in some animals these signals are the results of prewired motor programs, and carry acoustic sign stimuli which trigger the appropriate responses in their listeners.

All these examples involve the use of something already on hand, and are predictably inefficient. Despite their numbers, our world is notably free of the din of thumping rabbits and rapping fiddler crabs. The animal sounds which fill the world are the result of elaborate specialization. Cricket calls and bird songs, both of which are surprisingly loud considering the size of the animals generating them, are perhaps the best illustrations of this.

Crickets draw a file over a row of teeth (Fig. 8–2) in a species-specific pulse code. The teeth are designed to resonate like a row of cheap tuning forks to channel acoustic energy into a particular narrow frequency band. Crickets would be even louder if the wings on which the teeth are placed were closer in size to the wavelength of the frequencies that carry the sound. They would need wings about three to six inches long for the 2–4 Hz necessary to carry the sound most efficiently. The efficiency of what is called "loading" a sound into the air is very low when the source is much smaller than the wavelength, but at least two species of crickets have overcome this problem behaviorally. A bush cricket will carefully select a stiff leaf of

Fig. 8-2 Stridulation apparatus of crickets. The file (shown here) sets the teeth into motion. The teeth vibrate, causing the wing to vibrate as well. The movement of the wing sets the surrounding air into motion, thereby creating the sound we hear.

Fig. 8-3 A bush cricket and its leaf. Choosing a stiff leaf of the proper size and then cutting a hole to accommodate itself, this cricket sets the whole leaf into resonance, thereby producing a louder sound.

the proper size, cut a cricket-shaped hole in its center, and climb in. Keeping its wings firmly pressed against the leaf, the cricket constructs a resonant sounding board for its song (Fig. 8-3). Mole crickets, on the other hand, dig resonant cavities of superb acoustical design to amplify their signals. The result in either case is a very loud sound constrained to a single frequency band. Any information in the sound must be transmitted in the pattern of pulsing—an animal version of Morse code.

Birds have a very different apparatus for producing sound, and are consequently able to use a wider range of strategies in communication. Like most vertebrates, birds pass the air from their lungs over a thin tissue attached to muscles whose resonant frequency depends on muscle tension (Fig. 8-4). The passing air sets the muscle into motion much as the wind ripples a pond or a leaf. The movement of the vibrating muscle produces sound of the frequency chosen by the animal. Humans and other vertebrates have one of these devices (called a larynx) in their throats, while birds have two of more precise design—one

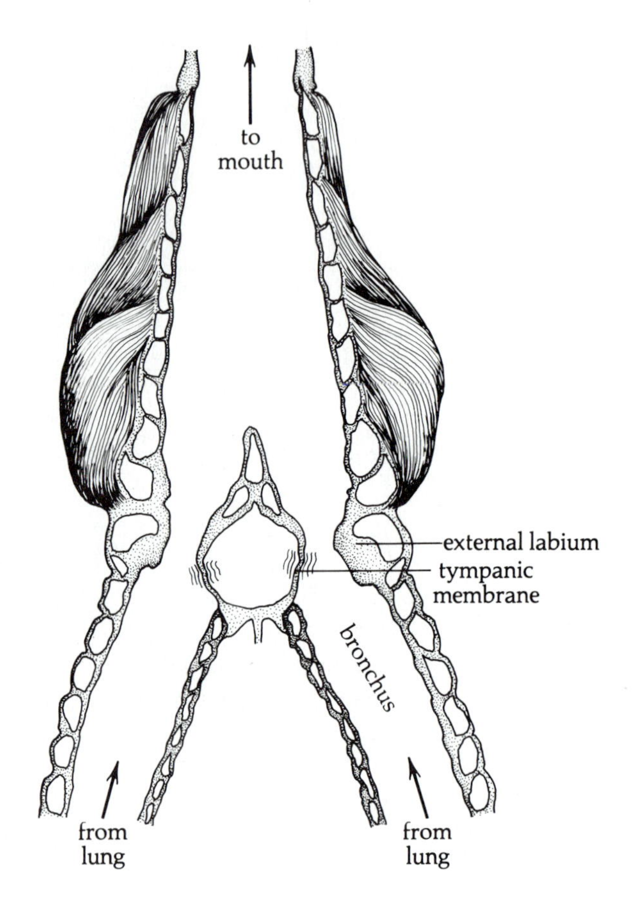

Fig. 8-4 The syrinx of a bird. The tensions on the two thin muscles attached to the tympanic membranes are controlled independently by the brain, allowing the bird to produce two different tones at once.

in each of the two branches of the trachea going to the lungs. As a result, birds can produce two independent pure tones at once, which is what makes it impossible for humans to produce realistic imitations. Singing birds add frequency variations to the amplitude modulations or pulse coding already available to create an enormous and complicated repertoire.

The vocal apparatus of humans is at once simpler and more complex. The sounds that emerge from the human larynx are not pure tones, although they are produced by a single vibrating element, the vocal folds, or cords. Unlike avian cords, the vocal cords of humans are sloppy oscillators, generating not only a fundamental frequency but a variety of harmonics or overtones. By changing the size of the throat and mouth and the relative accessibility of the nasal passage, we "shape" our sound output. Every time we speak we are selectively creating and modulating the physical resonances of these three chambers in a way which leads to the simultaneous production of several harmonically related tones (the "formants" of Chapter 30). It is more the manipulation of these harmonics than the production of the fundamental frequency by the vocal folds which controls the information in our vocalizations.

SOUND TRANSMISSION

Once sound has been produced, evolution has had to deal with a variety of problems in getting sound to its intended receiver. As with that of light, sound transmission is subject to two kinds of natural attenuation. The first is the simple loss resulting from the spreading out of the energy over distance: the intensity of sound falls off as the square of the distance. The second is the loss occasioned by the absorption of energy by a medium. With both light and sound, shorter wavelengths are absorbed more readily by almost everything. Thus when the sun is near the horizon and sunlight has more of the earth's atmosphere to travel through, the shorter wavelength blues are lost and the sun appears red. So, too, short-wavelength (high-frequency) sounds are absorbed more readily by air, water, and earth.

The constraints on sound transmission have helped shape the physiology of auditory communication. Small animals, for example, cannot produce low frequencies efficiently (i. e., loudly). For very short-range communication this is not a problem. Over long distances, however, the difficulties for small animals become enormous. For bats using 100,000 Hz, the air must seem like solid cotton batting: for them, hearing must be very much like trying to see in a thick fog, and only large or nearby targets are perceptible. This limitation, however, also adds information: the ratio of high to low frequencies in a familiar sound is used by many animals, humans included, to judge the distance over which it must have traveled. It is the loss of high frequencies which gives sound a faraway quality. When we hear a voice, for instance, our auditory systems automatically judge how distant the speaker is by measuring the proportion of low and high frequencies. Low frequencies, on the other hand, can travel tremendous distances. Were there not so much noise from ship engines in the sea, the 20-Hz sounds of some whales would be audible for the thousands of miles of water these solitary creatures need to call through in order to find each other and mate. As it is, evolution's failure to anticipate the deafening noise of heavy shipping may have doomed the legendary leviathan to extinction.

EARS

Evolution has developed three strategies for detecting sound, one based on the particle-movement nature of sound and two others on its pressure-wave nature. Each strategy has profound effects on an animal's auditory world.

Particle-movement detectors are usually thin, low-mass projections attached to solid, high-mass objects. Molecules rushing back and forth strike the detector and in turn push it back and forth. In male mosquitoes, for example, antennal hairs are wingbeat detectors whose resonant frequency is precisely tuned to the all-important flight sound of the female. The male mosquito is deaf to all other sounds in the world, but since his one purpose in life is to mate, this sort of crude detector system suits him well. Like von Uexküll's tick, security is more valuable than wealth or knowledge. Many other examples of particle detectors exist in nature. The lateral-line organs of many fish, for example, are small hairs designed to bend with moving water molecules, and some caterpillars have rows of short hairs which are tuned not to the sounds of other caterpillars but to the wingbeat frequencies of the species of wasps which hunt them.

An important variation on the particle-movement strategy is the *subgenual organ* of many invertebrates. It consists of a thin membrane stretched across the nearly hollow cylinders which insects and crustaceans, for example, use as legs. Although these organs are isolated from and therefore deaf to airborne vibrations, sounds carried through the ground and transmitted up the stiff legs set the low-mass subgenual organs into motion. These organs are used by fiddler crabs to detect coded rappings, and by queen bees when they send messages to each other through the honey comb. Simple as they are, subgenual organs can be enormously sensitive. Roaches, for instance, are some 100,000 times more sensitive to ground vibrations than humans—a sensitivity which, by alerting them to the approach of predators, has no doubt served to "preadapt" them to their present urban habitat.

Particle-movement detectors, however, have many limitations. Since they rely on the limited resonance of the detector they cannot hear most sounds, and cannot distinguish between a relatively quiet sound at their peak resonance and a loud sound close enough to the frequency of peak resonance to stimulate the detector, although less efficiently.

Of course animals with particle-movement detectors could circumvent this difficulty if they were to equip themselves with several detectors of diffent resonances. If the ranges of elements in this hypothetical array of tuning forks overlapped, absolute frequency determination would be possible in just the manner described in Chapter 7 for color vision (Fig. 7–1). Any sound, regardless of intensity, would produce a unique response ratio between two or three detectors which would specify the frequency. Although no clear case of overlapping ranges has yet been found, instances of animals provided with two or three separate detector resonances are beginning to be uncovered. Bees, for example, seem to use their whole antennae to detect sounds in the 0- to 50-Hz range, the *distal* ends of the antennae to cover the 250- to 350-Hz range, and their subgenual organs to listen to the auditory world between 1500 and 3000 Hz (Fig. 8–5). (Humans, by way of contrast, hear from 20 to 20,000 Hz, though the upper limit declines with age.) For bees, each of these ranges has its own set of messages which the bees distinguish by their characteristic pulse coding (see Chapter 24).

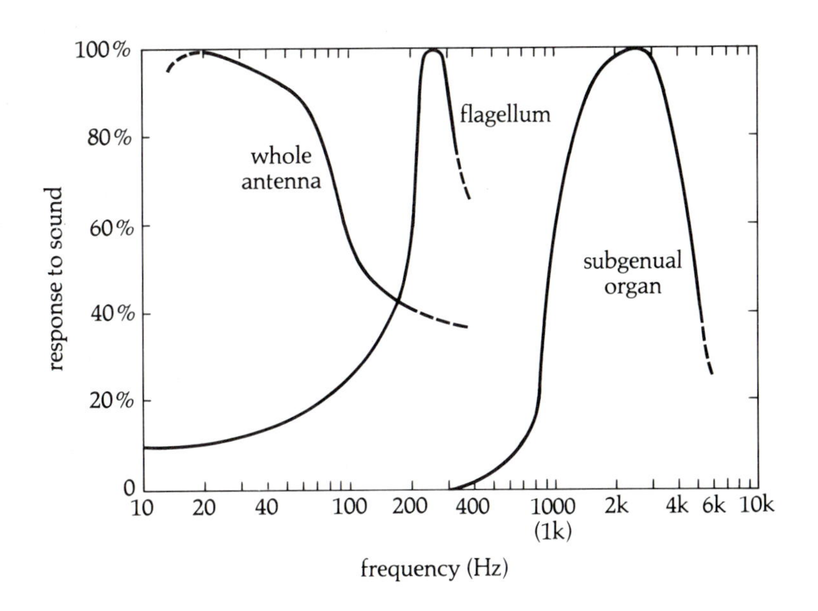

Fig. 8–5 Honey bees listen to the world in three frequency bands: 0–200 Hz, 200–500 Hz, and 900–5000 Hz. The function of the lowest band is unknown, though it may be involved in detecting the 13-Hz "waggles" of the bee dance or the various sorts of antennal drummings bees employ in many of their interactions. The intermediate frequency band is tuned to the wingbeat frequency, the pulse-coded distance message in the dance, and the much longer pulsed "stop" signal of dance attenders. The highest band is sensitive to the pulse-coded messages which pass between developing and adult queens. These various auditory behaviors are discussed in Chapter 24.

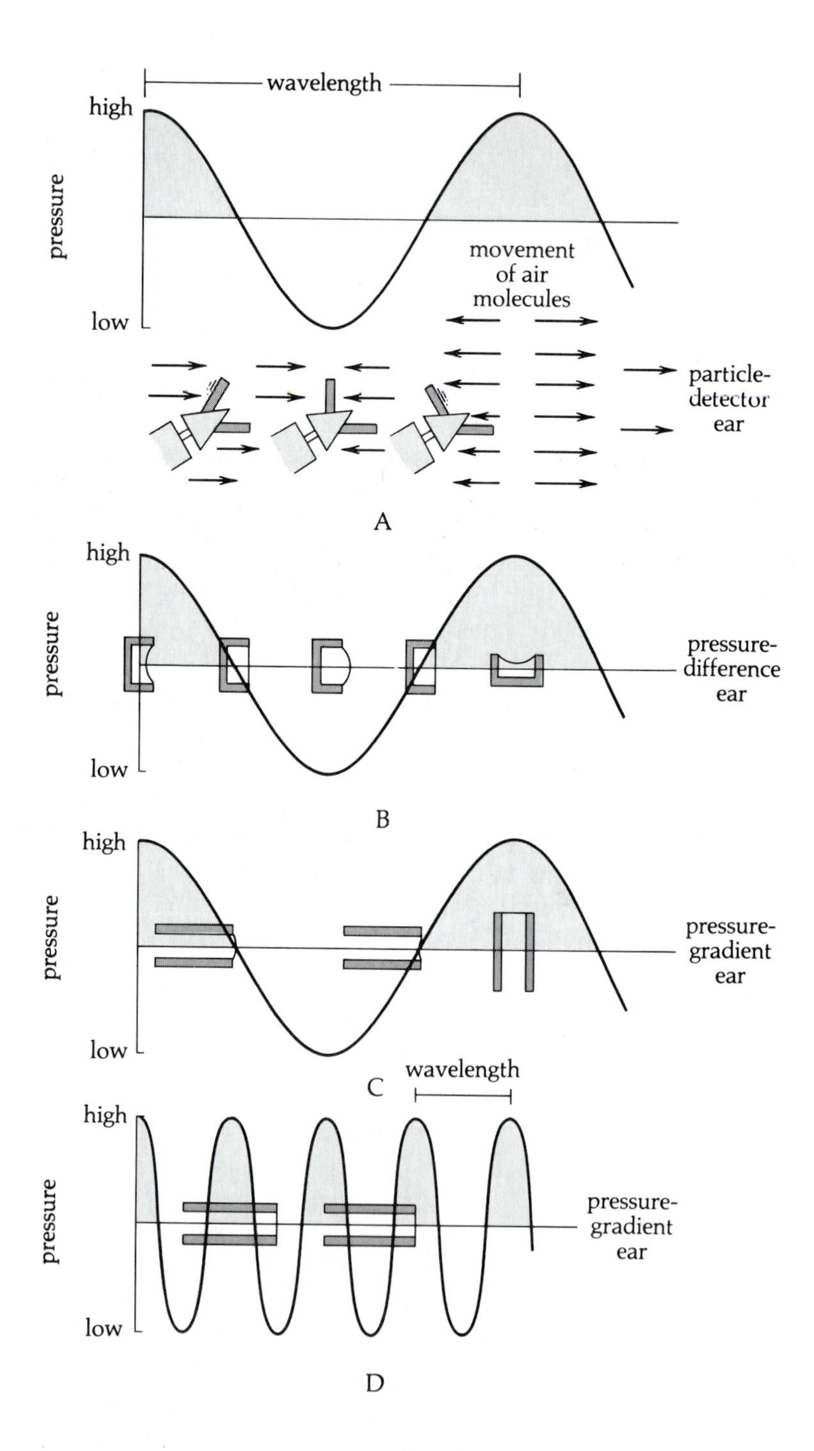
wavelength
high
pressure
low
movement
of air
molecules
particle-
detector
ear
A
high
pressure
low
pressure-
difference
ear
B
high
pressure
low
pressure-
gradient
ear
C
wavelength
high
pressure
low
pressure-
gradient
ear
D

Fig. 8–6 The three major types of ears respond to sounds in different ways. In these examples, sound is moving from left to right and the successive responses of the various detector types are illustrated. A. The particle-detector ear of bees, mosquitoes, fish with lateral-line organs, and so on, is insensitive to the waves of high and low pressure (indicated by the graph of pressure), but is instead physically deflected by air molecules rushing from the high-pressure zones to low-pressure areas. A particle detector aimed into the direction of sound propagation (the right antenna in this example) is not moved by the air motion, and so is deaf to the sound. B. The pressure-difference ear of mammals, some birds and fish, and even many insects, depends on a sealed chamber to supply a reference pressure. (As altitude or weather changes alter the normal air pressure, these ears can usually equalize the reference pressure by opening a tiny Eustachian tube or its equivalent.) The pressure fluctuations produced by sound waves passing by displace the membrane in and out, faithfully reproducing the sound. The orientation of the chamber is irrelevant, as the right-hand illustration indicates. C. The pressure-gradient ear of many insects, birds, fish, amphibians, and reptiles (and perhaps even some mammals) measures the difference in pressure between the two ends of a tube. This difference is registered by the displacement of a membrane placed anywhere in the tube. Unlike pressure-difference ears, this arrangement is very much affected by the direction from which the sound is coming: as illustrated, the ear is deaf to sounds arriving from the side of the tube, but maximally responsive when aimed along the axis of sound transmission. D. Another characteristic of pressure-gradient ears is their differing sensitivity to frequency. When the wavelength matches the effective tube length, the ear is deaf. A simple case in which the tube is along the sound axis and exactly one wavelength long is illustrated here. These ears, then, essentially have a negative resonance—a tone-deafness whose exact frequency depends on the direction of the sound.

Particle detectors have still other potential limitations. For example, they are unresponsive to sounds coming from the direction in which the detectors happen to be pointing (particles striking end-on do not move the detector) while they are most sensitive to sounds striking them from the side (Fig. 8–6A). As a result not only do they confuse frequency with intensity, they further confound intensity and direction. In order to sort out how loud a sound is, and therefore how far away the source must be, particle detectors must turn to try other orientations for comparison, or use a second detector with a different orientation and then perform some trigonometry on the two inputs.

An advantage of this calculation is that it also determines the *direction* of the source, an often crucial piece of information for which those of us with classical pressure ears must work a good deal harder.

Pressure detectors such as the human ear are moved by alternating bands of high- and low-pressure sound. A thin, low-mass (and hence easily moved) membrane is stretched across a high-mass cavity (Fig. 8–6B). Two superficially similar kinds of pressure ears give their owners dramatically different representations of the auditory world. When we speak of pressure detectors, we usually refer to the pressure-difference ear, in which the high-mass cavity is sealed, thereby trapping inside it a volume of air which serves as a reference pressure. When the high-pressure portion of a wave passes the membrane, the higher air pressure on the outside of the membrane causes it to bulge in, while when the low-pressure peak arrives the pressure difference causes the membrane to bulge out. The movement of the membrane, then, accurately reflects the characteristics—frequency and intensity—of the sound passing by (Fig. 8–6B). Like particle detectors, membranes have resonances, but these can be made relatively small. As a result the frequency sensitivity of a pressure ear can be broad. This is an advantage for moths, for example, which are vitally interested in hearing and avoiding predatory bats which use ultrasonic sound pulses of 20–100 kHz (depending on the species) as sonar signals to detect flying insects. Since moths do not greatly care just which species of bat is after them, they use a very broadly tuned membrane. Unlike particle detectors, the direction from which a sound is coming is irrelevant to this kind of pressure-difference ear: high pressure outside the ear pushes the membrane in regardless of direction from which the sound arrived.

In recent years it has become obvious that some ears with membranes, however, like those of locusts and crickets, do *not* work like classical pressure detectors. For one thing, this second kind of pressure-detecting ear is highly directional. The crucial difference in these curious ears is that the cavity behind the membrane is simply not closed. The membrane responds not to the difference in pressure between the outside and some sealed reference chamber, but to the difference in pressure between the two openings of the cavity (Fig. 8–6C). Hence, taking the simplest case of the ear as a tube with a membrane over one end, if the tube lies *across* the direction of sound propagation each end is exposed to the same part of the same wave: no pressure difference results and the ear is deaf. This is analogous to a particle-movement detector pointing *along* the axis of sound propagation. When the tube is along the sound axis, however, it responds to the pressure difference between the two ends. The magnitude of that difference depends both on the intensity and the wavelength of the sound. If the wavelength is long compared to the length of the tube the difference will be slight, while if it is exactly twice the length of the tube it will be very great. On the other hand, if the tube is precisely one wavelength long, it will hear nothing. Hence, just like particle detectors, the frequency and intensity information from a single ear are thoroughly muddled (Fig. 8–5D).

The advantage of this "gradient" ear is substantial. It can be far more sensitive and have a broader frequency range than particle detectors, while taking advantage of the same very simple direction processing used by particle ears. This makes it ideal for small animals with minimal brains—animals such as locusts and crickets, many fish, and probably all frogs. In the case of frogs, the wider resonance of the membrane allows them to distinguish the two frequencies present in their calls and deal with them independently. The mechanism for this separation of pitches is wonderfully simple: inside the fluid-filled inner ear to which the pressure-gradient membrane has faithfully transmitted the sounds of the external world are two more membranes, the amphibian and basilar papillae, each with its own specific resonance. One is tuned to the high-frequency component of the call while the other responds only to the low note. As a result, frogs not only can hear the species-specific pulse-coded signals of their species, but can use the ratio of high frequencies to low in the pulses to judge the age and distance of the sender. (Robert Capranica, a neurophysiologist at Cornell, has discovered that juvenile frog sounds lack the low-frequency portion of the signal, while according to University of Missouri ethologist Carl Gerhardt, distant frog sounds have had much of their high-frequency component absorbed by intervening air and vegetation.)

Locusts and crickets too use a dual-resonance version of the pressure-gradient ear to hear in two frequency bands at once. The dual resonance arises from a two-section membrane, one part of which is four times thicker than the others. In crickets, for example, one set of cells responds to sounds in the 2- to

4-kHz range—the frequency of their stridulations—while the other set covers the band 20–40 kHz. When female crickets hear these frequencies in flight they will turn toward stridulations but away from the higher frequencies. Crickets probably take any stimulation of the high-frequency part of their membrane as a sign stimulus indicative of bats.

The type of ear an animal possesses, then, profoundly affects the nature of the auditory world to which a species is privy. The means by which the ear turns incoming sound into nerve impulses, however, is just the first step in the fascinating process by which the nervous system extracts the species-specific messages that lie hidden in the cacophony of the animal's natural world.

THE DISCOVERY OF ECHOLOCATION

The inexorable reality that ordains that we shall be blind to our own blindnesses presents to modern science a challenge to be openminded and imaginative in the face of apparent mystery. But often imagination is not enough if we lack the instrumentation necessary to let us see vicariously beyond our senses. The discovery of bat echolocation provides a telling illustration of this caveat.

As early as 1793 Lazaro Spallanzini discovered that blinded bats flew perfectly well. This discovery, contradicting as it did all commonsense views of how animals navigate, spurred him to write letters to scientific colleagues and learned societies all over Europe, to help verify his experiments and to help him wrestle with the intellectually unmanageable results. Try as he might—and some of his experiments were amazingly ingenious—he could not escape from the surprisingly modern concept that "in the absence of sight there is substituted some new organ or sense that we do not have and of which consequently, we can never have any idea."

The following year Charles Jurine in Switzerland proved unambiguously not only that blinded bats could navigate, but that deafened ones could not. Two factors, however, one mechanical, one merely an idea, kept the two experimentors from uncovering the bats' secret. The idea they were denied was a notion of *Ümwelt,* a world view that could allow that animals might be not the sensory cripples our egocentric species deemed them but rather creatures uniquely adapted to life in their own worlds. Mechanically they lacked the technology and instrumentation to supply the senses denied us by evolution.

Both the thinking and the instrumentation, however, were lacking; and the scientific establishment of the time saw to it that Spallanzini's and Jurine's work was buried in obscurity, with the skeptical epitaph that their theories required "more faith and less philosophic reasoning than can be expected of the zoological philosopher." As Donald Griffin points out in his elegant updating of the controversy *Listening in the Dark,* "the experimental evidence contradicted common sense and reasonableness based on normal experience, and so the experimental evidence was laid aside." The prevailing theory remained that a highly sophisticated tactile sense such as is often exhibited in blind persons could account for the bats' uncanny abilities.

Indeed, in the early 1900s scientists again turned to experimentation, ignorant of the groundbreaking work of their eighteenth-century forebears. In 1908 an American, W. L. Hahn, with a leap of imagination tried some ear-plugging experiments, and came quite close to the real solution with the idea that bats have a "sixth sense," that of "direction," located somewhere in the inner ear. What was missing here, however, as Griffin points out, was "any clearly formulated realization that there could exist sounds inaudible to human ears but yet useful to a bat for obstacle avoidance."

Acoustical research was then a young but growing field, and scientists and engineers were experimenting with the nature of sound. One English physiologist, Hartridge, combined a knowledge of physics and of the nervous system to suggest, after watching bats maneuver in the dark, that they might use high-frequency sound waves, perhaps generated by the flapping of their wings, to locate obstacles.

Hartridge was not an experimentalist, however, and it was not until 1938 that Griffin, then an undergraduate at Harvard, became intrigued by the problem while studying bird migration. A Harvard physics professor, G. W. Pierce, had developed an apparatus that could both generate and detect a wide range of sounds above the audible range. With Pierce's ultrasonic detector they could "eavesdrop" on the bats to see if they were in fact emitting inaudible sounds.

The results were gratifying. In Griffin's own words, "the pen [on the chart recorder] responded, or the loudspeaker popped, almost regardless of the setting of the tuning dial; evidently these inaudible sounds of bats contained considerable energy throughout the frequency range to which the Pierce instrument was sensitive, although the maximum response was obtained at about 50 kc—more than double the highest frequency audible to human ears."

Griffin and Pierce's joy at their discovery, however, was tempered by disappointment that they were unable to detect any particular patterns of emission from free-flying bats, and when their short note was published it suggested merely that bats might use their high-pitched sounds as calls. Again in Griffin's words, though, "excessive caution can sometimes lead one as far astray as rash enthusiasm." Griffin and a fellow student, Robert Galambos, repeated Spallanzini's, Jurine's, and Hahn's experiments with the additional finesse of the Pierce "horn" to eavesdrop, and discovered that its highly directional sensitivity, combined with the sharply focused nature of the bats' cries, had masked the unremitting pulses of the flying animals. Test after test served to corroborate unequivocally what Spallanzini had surmised before 1800, that bats "see with their ears." Griffin's discovery of what Spallanzini had not imagined, that the bats *emit* vocally the sounds which they hear as reflections from surfaces of obstacles, was the crucial piece of the puzzle. Years later he also confirmed what no one, himself included, had previously wanted to believe could be true: that the animals could also "echolocate" their tiny insect prey. We now know that the bats' preternatural achievements do not stop even here. They are able to pick insects off vegetation as they fly, and to return to a particular "toehold" among hundreds of others in the darkness of their cave homes.

Thus bit by bit the pieces came together, but until the possibility that animals might have senses that we lack became part of researchers' view of their world, the most clear-cut results could find no reception; and until the equipment had been developed that could detect the hypothetical medium, no amount of theory could be convincing to a critical mind. In this slow and painstaking way, then, the myriad unsung abilities of animals are being discovered, from the infrared "vision" of snakes to the electrolocation that some fish use to detect prey hidden beneath the sand; from the ultrasonic communication between mother rats and their young, to the magnetic sensitivity that is even now becoming recognized as a force in the lives of many animals. Ironically, even with all our modern technology, people may, as Dr. Griffin writes, "look back on the present period as a time when fertile scientific fields lay incredibly untended. The only question is whether these lookers back will be ourselves a few years hence or our remote descendants."

SUMMARY

Ears transduce the rhythmic motion of molecules which we call sound into nerve impulses. All ears do this by detecting the displacement of long, thin projections either directly, as in subgenual organs, antennae, or lateral line systems; or indirectly, as the motion of a membrane which has been set into motion by rhythmic pressure waves. Two classes of ears—the pressure-difference ears of higher vertebrates with their vast arrays of hair cells, and the pressure-gradient ears of many vertebrates and invertebrates—use the pressure strategy. The pressure-difference ear, as we shall see, has the potential advantage of perfect frequency reproduction and analysis as well as directional localization, but at enormous neural cost. The other two types of ear, on the other hand, though limited and quirky in their responsiveness to both frequency and sound, yield directional and a certain amount of frequency information with a minimum of processing. Like the compound-eye/camera-eye dichotomy, the choice of strategy correlates well with size. The potential behavioral complexity which evolution has exploited in the auditory world is a secondary consequence.

STUDY QUESTION

At what angle to each other should two particle detectors be placed in order to hear sounds with no ambiguity about direction? Try 180°, 90°, 45°, and 30°. What processing tricks might resolve any remaining ambiguity? If ambiguity is inevitable, what angle yields the least ambiguity and maximum sensitivity? If practical, see how the antennae of bees, mosquitoes, and fruit flies compare with your calculation.

FURTHER READING

von Bekesy, Georg. "The Ear." *Scientific American* 197, no. 2 (1957): 66–78.

Sundberg, Johan. "The Acoustics of the Singing Voice." *Scientific American* 236, no. 3 (1977): 82–91.

Michelson, Axel. "Insect Ears as Mechanical Systems." *American Scientist* 67 (1979): 696–706.

CHAPTER 9

Auditory Processing

When a sound is detected by an ear, most of what we call "hearing" has yet to occur. At every point from an animal's ear to its brain the nervous system processes auditory data to extract the information that is important for its species. This intricate network is usually designed to identify incoming frequencies, localize the direction of the source (and often to calculate the source's approximate distance and rate and direction of movement), sort for sign stimuli, or otherwise to identify the source of the emission. The mechanisms by which these analyses are accomplished are just beginning to be understood.

TONE DISCRIMINATION

The pressure ears of higher vertebrates transfer incoming sound to a second (basilar) membrane in a fluid-filled sac. Just as the eye has rods for low-light sensitivity and cones for wavelength (color) discrimination in bright light, the ear has two classes of receptors: numerous outer hair cells which seem to pick up low-level sounds, and inner hair cells which are probably responsible for determining wavelength (pitch) accurately.

The basilar membrane in mammals is some 20–30 mm long,

and is so constructed that each tone causes a particular part of it to be vibrated more than any other (Fig. 9–1). Tuning for sound is quite broad (as it is for visual pigments), so that fine tone discrimination requires special processing. This could be accomplished in either of the two ways used so successfully by the visual system. The inner hair cells could be wired to perform lateral inhibition to exaggerate the peak frequency of a tone. Alternatively, the ratio of response of neighboring whole groups of cells with different broadly tuned peaks could be compared. This second strategy is exactly what the visual system does with three classes of broadly tuned pigments to span the "visual octave." (An octave is a factor of two in wavelength—from the 380 nm of violet to the 750 nm of red, for example, or from the

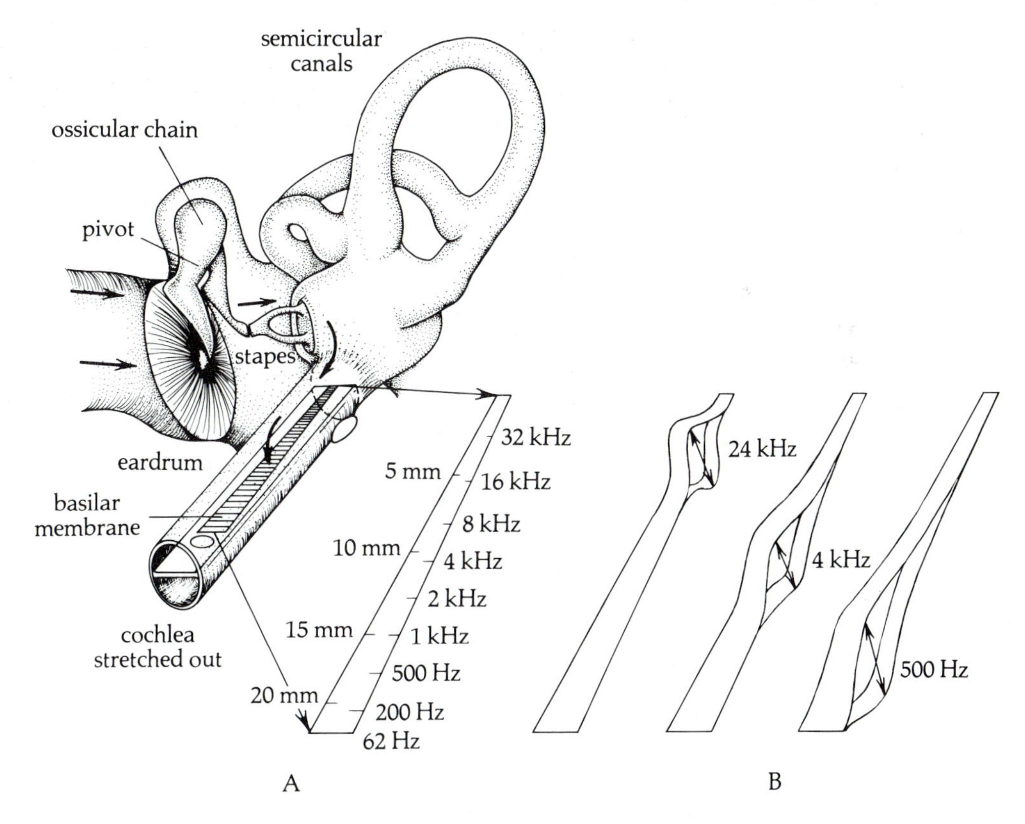

Fig. 9–1 A. The basilar membrane of mammals, normally coiled in the cochlea, is shown here stretched out. B. Its mechanical properties cause it to resonate at different places for different frequencies. [From *Experiments in Hearing,* by G. von Békésy, translated and edited by E. G. Wever. Copyright © 1960 by McGraw-Hill. Used with the permission of McGraw-Hill Book Company.]

540 Hz of middle C to the 1080 Hz of high C.) Indirect evidence suggests that the auditory systems of higher animals might be organized on the same three-filters-per-octave strategy over much of their range—ten octaves in humans for example. However, matters are not quite so alluringly simple. The inner and outer hair cells interact with each other to sharpen pitch discrimination, and at frequencies below 100 Hz the system seems designed actually to count the sound waves. In fact, counting waves is probably the sole mechanism of fine-frequency discrimination outside of birds and mammals.

Exact frequency analysis obviously requires a lot of hardware and processing. The human basilar membrane alone is larger than the bodies of most insects, and of course the auditory areas of our brains are larger still. The pressing need for physical and neural economy explains why smaller animals must use only one, two, or at most three elements—membranes, hairs, antennae, or subgenual organs—whose physical resonances serve to accomplish relatively crude frequency discrimination. Frequency considerations are also important in determining which kind of element to employ. The subgenual organs of the last chapter, although quite sensitive, are generally restricted to frequencies around 2000 Hz, and are naturally deaf to airborne sounds. Particle-movement detectors—hairs and antennae—are limited to rather low frequencies (below 500 Hz) and are relatively insensitive. Pressure-gradient ears, those tubes with membranes, on the other hand, can detect sounds that are quite faint, and depending on the membrane can be tuned to anywhere from the 100 Hz of bullfrogs to the 10,000 Hz of the Orthoptera. Beyond these ranges, however, severe problems arise. In both cases the difficulty is in the distance between the outside of the membrane and the rear opening of the cavity. Recall that at low frequencies any attempt to measure pressure differences across this distance becomes practically hopeless. At 100 Hz, for example, the wavelength is about 3 m, so that a separation of openings of, say, 1 cm, would be exposed to a maximum intensity difference of only 2 percent. By contrast, the just-noticeable difference for humans is about 10 percent. As a result of these limitations, insect pressure-gradient ears rarely produce any behavioral response below 2000 Hz, and bullfrogs manage to detect 100 Hz only by virtue of their long Eustachian tubes.

At high frequencies the opposite problem arises. When the

two openings to the membrane cavity are only one wavelength apart, the pressure at each end is the same and the ear becomes deaf. Hence, the larger separations in the bullfrog ear lead to high frequency failure above 1 or 2 kHz, while the shorter separations of Orthopteran ears are good to perhaps 10–20 kHz.

SOUND LOCALIZATION

Sound localization, an ability which yields information crucial to many animals, is relatively simple for particle ears. Particle detectors are, as we have seen, deaf to sounds coming straight down the axis of the detector (Fig. 8–5A), and maximally sensitive when oriented across the direction from which the sound is coming. The method the nervous system employs for analysis of this information is a familiar one: overlapping response ranges. Two detectors at different orientations generate two sets of responses whose ratio, independent of intensity and frequency, specifies the direction of the sound source.

Determining the direction of a sound with a pressure-gradient ear is done in much the same way as with a particle detector: two tubes with different orientations must be used. The threefold interaction between intensity, frequency, and direction means that the accuracy of localization will decline severely with frequencies near and higher than the "resonance" of the tubes. Both of these simple ears, then, localize sound by means of very simple processing within a constrained frequency range. Localization with pressure-difference ears, however, is another matter.

The basic difficulty with isolated pressure-difference ears lies in their fundamental insensitivity to direction (Fig. 8–5B). Hence, to a first approximation any sound affects this sort of ear in the same way no matter how the pressure wave got to it. Happily, our pressure-difference ears are not isolated, but rather are mounted on two sides of an acoustically opaque head. For short wavelengths—small relative to the size of an animal's skull—the head casts an acoustic shadow, creating a difference in sound intensity between the two ears. Diffraction causes the sound waves to bend around the head, so that the exact intensity difference between the ears depends on the angle from which the sound is coming *and* its wavelength (Fig. 9–2). The processing required to sort all this out must be formidable.

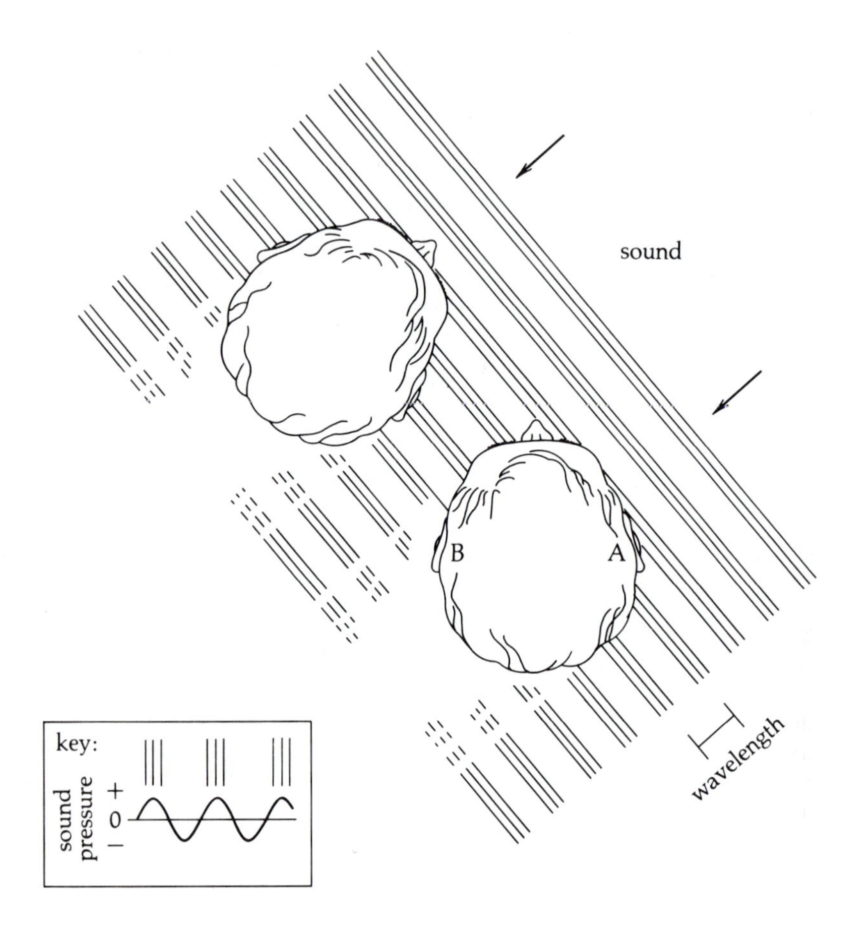

Fig. 9-2 Localization of sounds whose wavelengths are small compared to the diameter of the head is accomplished by attending to differences in intensity between the two ears. The upper head faces into the sound so that both ears hear the same sound intensity, while the lower head faces to the side, placing the ear away from the sound (B) in an acoustic shadow. Diffraction allows some of the sound to bend around the head to reach the shadowed ear. The exact amount, and hence the intensity difference between the two ears, depends on the angle of the head with respect to the sound source and the wavelength of the sound.

Many (indeed, probably all) newborn animals are able to localize sounds at least crudely, though. How have the genes encoded the frequency/intensity function which permits naïve animals to turn toward and look for hidden sound sources? Since vertebrate heads grow, it is not clear that evolution could completely prewire the high-frequency localization system without some provision for self-calibration from time to time.

Low-frequency localization is tougher still. Because of diffraction, sounds bend around the head at wavelengths larger than one head diameter. As a result, the two ears hear nearly the same intensity of sound. The only difference in perception between the ears is the moment of the sound's arrival. For any source angle there is a unique time difference between the ears which runs from a maximum of 0.5 msec when the sound comes directly from one side to a maximum distinguishable delay of about 0.02 msec (Fig. 9-3). To make this distinction, the two ears must each judge when the sound has reached them according to some equally precise criterion. For example, if the cue were the peak of maximum pressure in the sound wave, that peak would have to be judged to within ±0.01 msec. For very long wavelengths pressure changes are gradual, and therefore the peaks would be broad and difficult to compare. This is essentially the same difficulty pressure-gradient ears have at low frequencies in trying to judge the pressure difference between two points when the wavelength is very long. As a result, localization at very low frequencies should be less accurate, but this has never been measured.

Neither acoustic shadows nor delay can tell an animal anything about the elevation of a sound source, yet we are all sensible of whether a sound is coming from near our feet or out

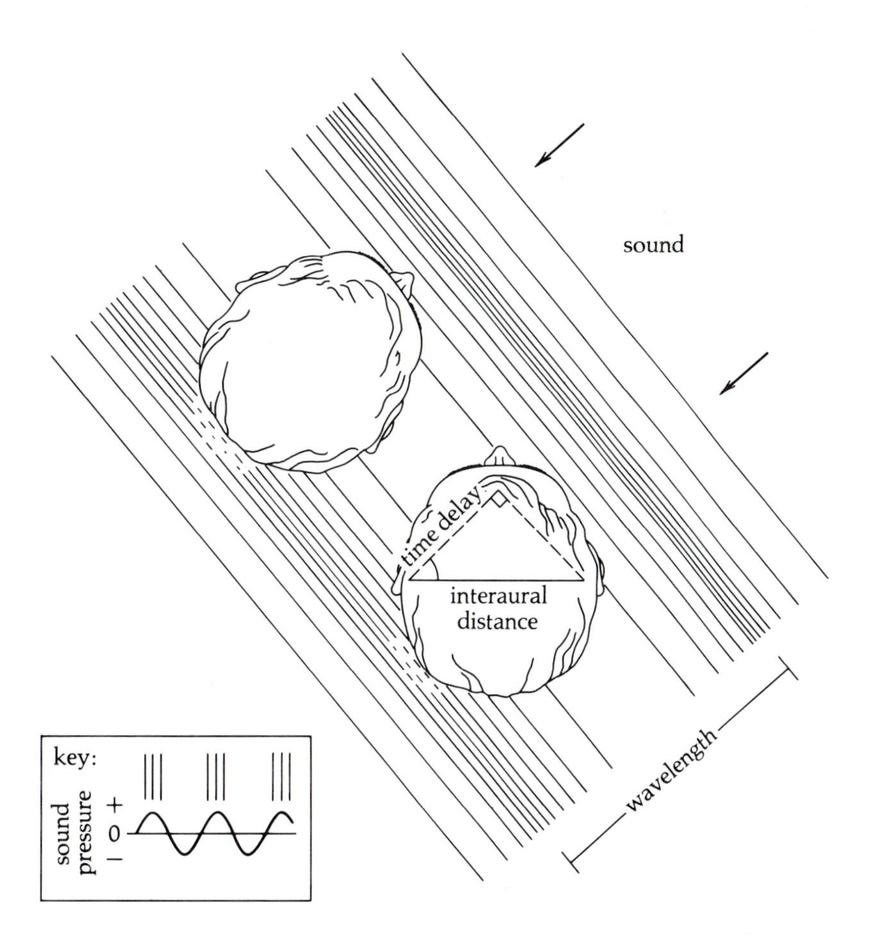

Fig. 9-3 Localization of sounds whose wavelengths are large compared to the diameter of the head cannot depend on intensity differences since low wavelengths are so strongly diffracted that the head is essentially transparent. The direction of the sound in these cases is calculated from the delay between the arrival of a wave at the nearer ear and its arrival at the farther ear. This means that the nervous system must "know" the distance between the ears—a value which changes with growth—and be able to count time. This system would not work for high frequencies (Fig. 9-2) because of the ambiguity in determining which wave is which: after a sound wave hits the nearer ear, it may be the first, second, third, or indeed thirtieth wave to pass the farther ear, depending on the wavelength and the angle of the head.

of the sky. The consensus of scientific opinion now is that the external ear may provide the final cues necessary for auditory localization. Perhaps the most powerful argument in favor of the external-ear hypothesis is that no one can think of a more plausible one. In humans, according to this theory, incoming sound would reflect off the various lumps and ridges of our ears, causing each reflected bit to arrive at the eardrum at a slightly different time. These delays would vary in a complicated way with the direction of the source, and must involve time differences of 0.005 msec or less. The delays ought, according to playback experiments, to depend on ultrasonic sounds not consciously perceived. If we can actually manage that sort of discrimination, it is surprising that our ability to localize sounds is no better than it is. A further complication is that since ear shapes differ, each individual would have to calibrate himself. At present, we have no idea how any of this intricate sorting and processing could be accomplished by the nervous system.

It has been known for years that there exists in the brain an auditory area in which sensitivity to frequency is mapped out sequentially from low to high. Recent work has demonstrated that the owl brain has an area which represents the auditory world spatially very much like the striate cortex represents the visual world. This map even includes a central area which is overrepresented: an acoustic fovea (Fig. 9–4). Whether this map exists in the auditory systems of other animals is not yet known, but it would be satisfying if the brain employed such general strategies.

There are also reports of cells on one of the half-dozen visual maps in the brain which respond to both light and sound from the same direction. Could this sort of convergence from two sensory systems be innate, or could it be that the calibration of acoustic space might depend on vision? Questions of this sort are not unprecedented. For instance, as mentioned in Chapter 8, the nervous system judges the distance of a sound source on the basis of how much of its high-frequency component has been lost to atmospheric absorption. This means that to be judged accurately, the sound must be familiar so that the brain knows how much high frequency it starts off with. Whether animals come completely programmed to calculate distance or must calibrate themselves is not known. If they calibrate themselves the role of vision might be important. Experiments to resolve this question are waiting to be performed.

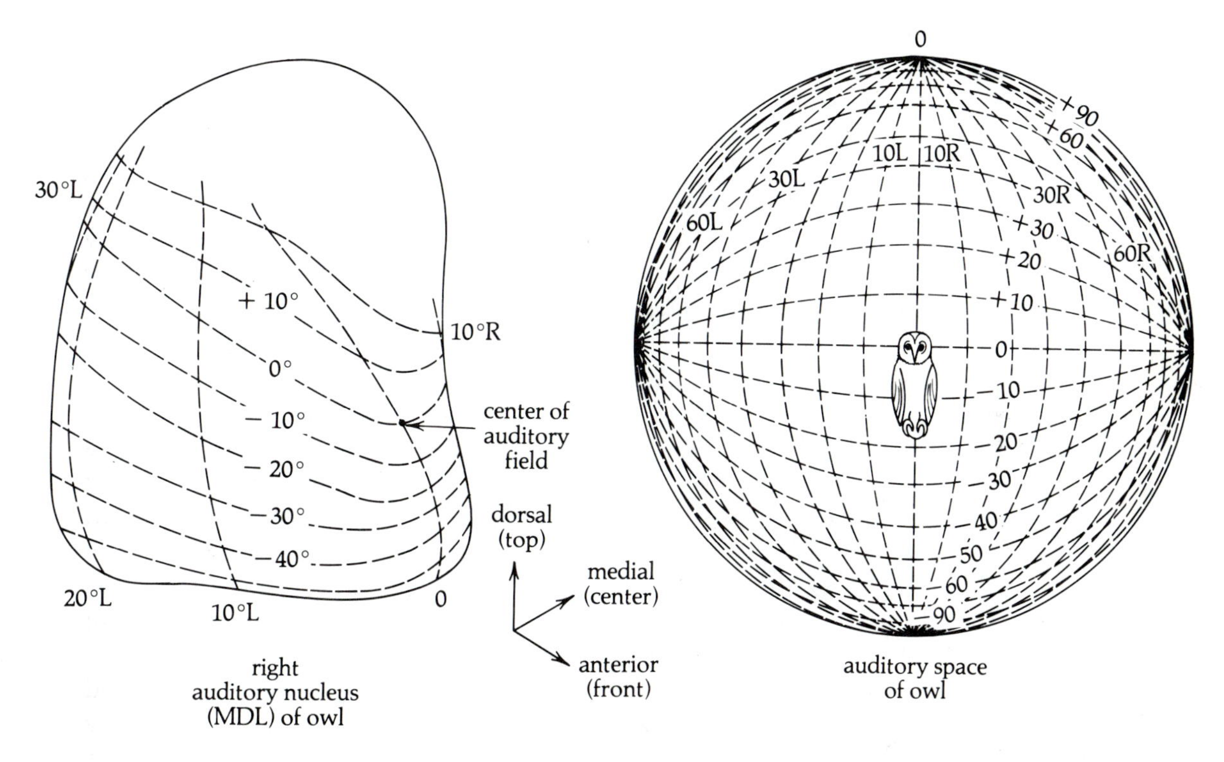

Fig. 9-4 The auditory world of the owl is displayed on the brain as a bicoordinate acoustic map. Appropriately enough for an animal which typically listens for terrestrial prey from a perch, the map of auditory elevation concentrates predominantly on the world below eye level. Note also the very large amount of brain space devoted to the first 20° of azimuth—the area directly in front of the owl—versus the vanishingly small bands for azimuths beyond 40°. This acoustic fovea for azimuth must serve to increase the owl's accuracy in auditory localization.

Many animals are able to judge the relative motion of sound sources by Doppler shift. The motion of a sound source toward or away from a listener artificially shortens or lengthens the pressure waves—that is, raises or lowers the pitch of the sound. This principle is used by bats and police to measure the speed of potential victims. Of the various kinds of acoustic analysis, this is perhaps the easiest. The auditory system compares the true frequency of the sound being emitted with the altered frequency of the sound as it arrives at the ear, and determines the difference. This is possible if you know something about the original sound, and is especially simple if you happen to be making it yourself and then analyzing the echoes. An upward shift of 1 percent in pitch, for example, means that the target is moving toward you (or you toward the target) at 1 percent of the speed of sound—about 10 ft/sec in this case. For bats pursuing insect prey this is crucial information, since it allows them to distinguish between the animate and the inanimate in their world of darkness.

ECHOLOCATION

Auditory processing has traditionally been thought of as the stepchild of vision research. Given our species' overwhelming visual bias, this sort of hierarchy is understandable. We speak of bats as able to "see" in the dark, and we find the discovery of the spatial auditory map in the owl brain to be somehow comforting. And yet the perceptual world of bats must be concerned with identifying auditory features—precise distance, size, rate and direction of movement, and "texture" of targets—that are almost wholly outside our range of experience.

How are such acoustic features recognized and mapped by the nervous system? As auditory specialists, bats devote an enormous portion of their brains to these tasks. Evolution has been forced to use every possible processing trick to enable bats to eat, as they must, their own weight in insects each day. And yet so obvious and specific is the bat's task, so carefully wired are all of the neural elements, that this highly complex acoustic animal is one of the simplest to analyze neurophysiologically.

The best known of these auditory machines is the mustache bat. Its sonar signal consists of a long constant-frequency (CF) component followed by an essential downward, frequency-modulated (FM) sweep (Fig. 9-5C). The functions of the CF component are clear: by listening for one single frequency for a relatively long time, the bats can detect smaller prey further away; and by measuring the Doppler shift of the echo of its cry the bat can determine the relative motion of that target. It makes this latter determination by adjusting the exact pitch of its own cry to produce an echo of one special frequency: 61.5 kHz. The frequency map on the bat's brain reflects this behavioral quirk: although the range of frequencies displayed in the brain runs from 20 to 120 kHz, fully 50 percent of the receptors are devoted to the narrow range of calls from 61 to 63 kHz (Fig. 9-6C). This acoustic fovea permits not only a precise measure of target

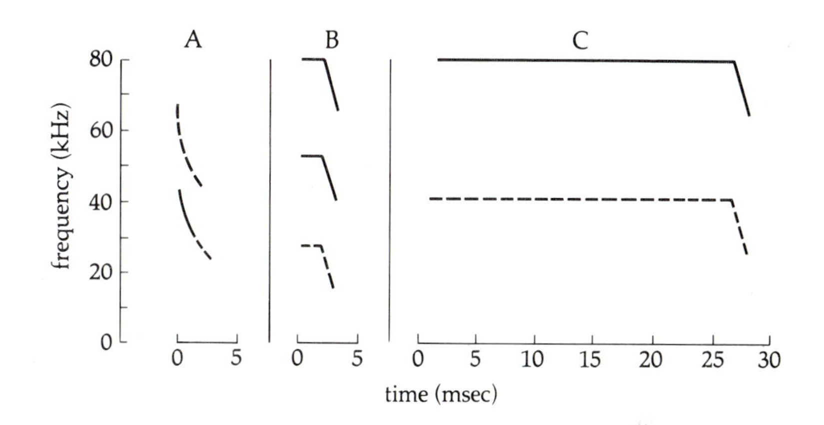

Fig. 9–5 At least three general classes of bat cries exist: (A) pure frequency-modulated (FM) sweeps; (B) short constant-frequency (CF) bursts followed by an FM sweep; and (C) long CF cries with an FM sweep at the end. Each requires its own specialized processing strategy. Best understood at the neural level is the long-CF/FM signal (C). The CF echo is Doppler-shifted in frequency as a result of the relative motion of the bat and its target. This not only tells the bat the speed of the object, but calibrates its subsequent analysis of the FM echo. The FM sweep provides precise target distance information, and since the wavelength at which objects become acoustically transparent depends on size, the target's diameter. Behavioral experiments indicate that bats extract more information about potential prey, but the neural bases of these discriminations are yet to be understood.

speed, but provides the basis for an equally precise analysis of the FM sweep.

Versatile as it is, the CF echo is nearly useless for distance analysis since it is difficult to judge just when a CF echo returns. For one thing the cry begins softly, only gradually reaching its full intensity. And then, the returned echo is a much-distorted version of the bat's cry—attenuated by atmospheric absorption and often modulated by the changing cross section (acoustic size) of the target as it turns with respect to the bat, or beats its wings. The result is that matching a return echo with the output wave is fraught with temporal ambiguities. Since all the individual CF waves look alike, and the chance of hearing the return of the first is nil, there is no accurate way of measuring echo delay. This problem does not plague the FM echo. Any particular frequency is swept through once and once only, and so the times of output and return of a particular frequency are unique—providing, of course, the bat allows for the Doppler shift of the FM sweep. Indeed, mustache bats use the shift of the CF echo to adjust the starting frequency of the FM sweep so that the echo will fall precisely on the array of FM feature detectors waiting to analyze the target.

The neural mechanism for analyzing distance has become clear from both behavioral experiments and neurophysiological recordings. The bat is comparing the fundamental of the FM sweep (a frequency not actually broadcast by mustache bats but perceived by their ears through bone conduction) with each of the three harmonics in the return echo, then determining the delay and mapping the result on a grid (Fig. 9–6A). (Incidentally, the point at which the target becomes acoustically invisible to the lowest harmonic gives the bat a clue about the target's size.) James Simmons has conducted elegant behavioral experi-

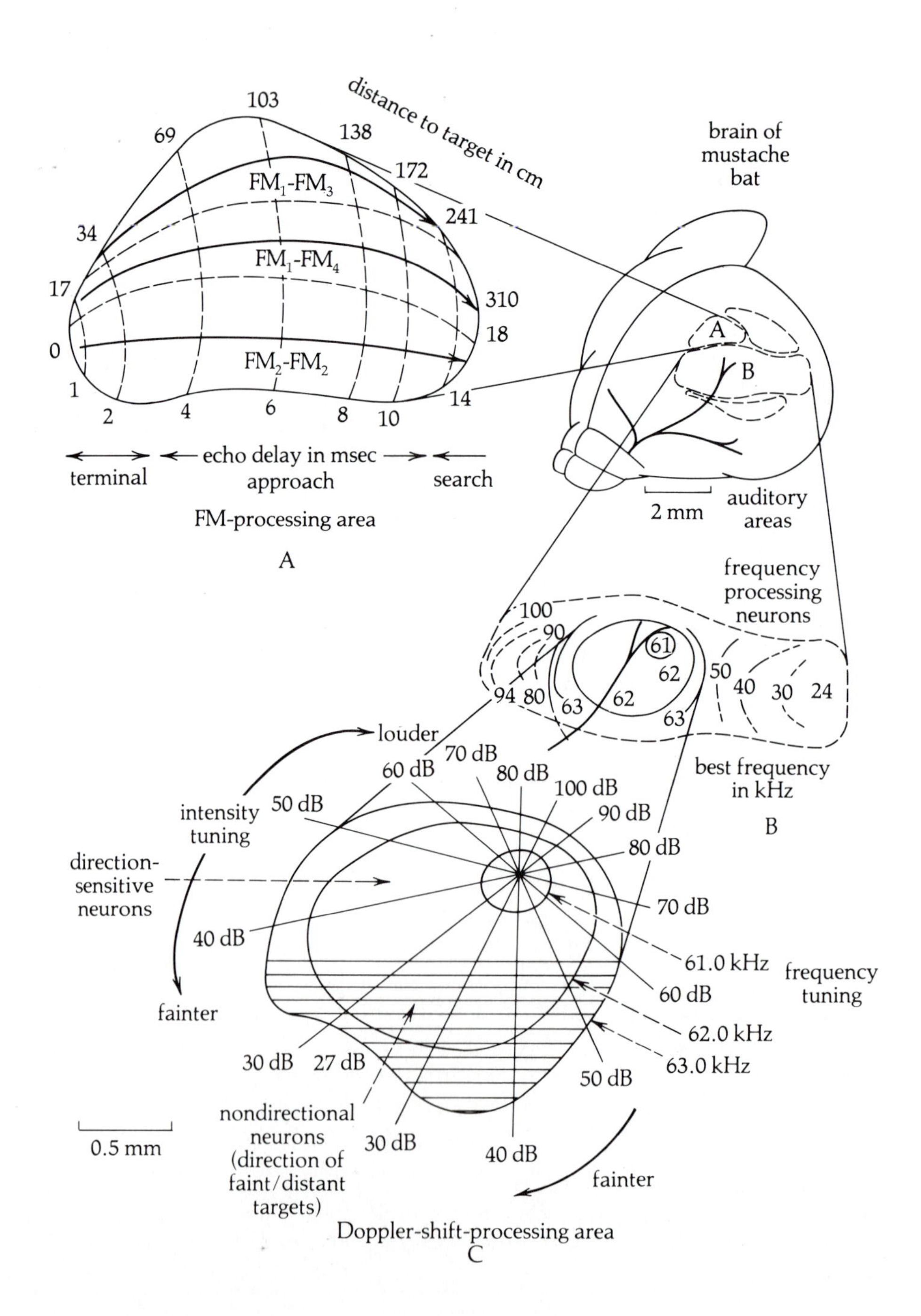

ments demonstrating that the mapping strategy is carried on by a process known as "cross correlation," which achieves a temporal resolution on the order of .0005 msec which translates into a distance of about 0.2 mm. What the bat does is store a copy of the outgoing signal, and continuously test the match between it and the returning echo. As the echo returns there is an instant at which the match, although never perfect, is best, and this defines the travel time of the sound and hence the distance of the target. Apparently the cross correlation is performed separately for each harmonic of the echo, thereby providing three simultaneous but independent estimates.

The auditory cortex of the mustache bat turns out to be a machine very much like the visual cortex of cats, full of highly specific feature detectors organized into logical arrays on the surface of the brain. It has the equivalent of the visual system's "hypercomplex" cells, tuned to several parameters at once—to some unique combination of frequency, rate and direction of frequency change, and delay.

The success of efforts to understand the auditory processing of bats has come from a brilliant battery of behavioral and neurophysiological techniques precisely designed to answer the ethological "how" questions. Simmons's behavioral experiments have determined what information is being gathered by bats, how the processing is being accomplished, and the overall accuracy of the system. Nubuo Suga's exacting neurophysiological studies have added clarity by locating the cells responsible and suggesting how the harmonics might be manipulated and compared, and the final output "displayed." The resulting picture is a clear answer to Donald Griffin's question of two

←

Fig. 9-6 The auditory areas of the mustache bat brain. Two specialized regions are shown in detail. The FM-processing area (A) analyzes, one by one, three harmonics of the echo against the fundamental frequency and maps out target range sequentially. Note that the resolution is highest at close range and worst at long range, and that the maximum range for the lower harmonics is about 3 m. The highest harmonic is most strongly absorbed by the air, and so is mapped out only to about 2.5 m. The frequency-processing area (B) maps out the bat's range of hearing from 20 to 120 kHz. Half of this area is devoted to the band 61–63 kHz (C). This is the frequency fovea into which the bat attempts to have the echo fall. Here the bat is able to analyze the precise Doppler shift (and hence the speed of the target) and adjust the starting frequency of the FM sweep accordingly. The area also maps out intensity, and thus the relative target size. In conjunction with the FM area, an absolute target size can be determined. The Doppler-shift area also is sensitive to the direction of the echo provided it is loud enough.

decades ago: How is it that with our 2 kg of brain, humans cannot match the remarkable sensory abilities locked inside the 1½-g brains of bats? The answer is now clear: Our genes have no need for—and so do not construct—the exquisite processing array that provides the bat with its unique *Ümwelt.*

Although the auditory system has proven to be more difficult to analyze than the visual system, it may ultimately be more important. Examples of acoustic releasers in imprinting and song recognition are accumulating, and impressive evidence is beginning to suggest that much of human language recognition may be based on innate acoustic releasers (see Chapter 30). This, and the many exciting unanswered questions about how evolution has designed animals to process and extract information from what they hear, make the auditory system the most exciting of the sensory pathways for behavioral and neural analysis.

CROSS CORRELATION IN BATS

Bats confront an extraordinary problem in attempting to discern the faint echoes of their sonar cries in the noisy world around them. What they must hear is an auditory clutter: the ultrasonic signals of other bats, louder sounds produced by the rustling of leaves, and other useless auditory information. To make matters worse the echo's intensity, already greatly attenuated by both spreading and atmospheric absorption, varies wildly as its prospective target flaps its wings or turns. Indeed, one part of the echo off the same object may be ten times louder than another. The result is a crazily distorted, almost inaudible echo embedded in noise perhaps several times as loud. How does the bat manage to extract from this acoustic jumble the information about distance, size, direction of movement, texture, and so on which is crucial to its survival?

The answer to this question has come out of a remarkable series of behavioral experiments by James Simmons. Simmons simply asks his hungry bats to stand on a platform, "look" at two targets, and crawl to a second platform in the direction of the correct choice. Their reward for distinguishing, say, the nearer of two targets, is a mealworm. Simmons can even remove the targets altogether, supplying the bats with computer-processed echoes to probe the system in greater detail. As usual, the solution worked out by evolution turns out to be, as Simmons discovered, at once both clever and charmingly simple: bats

store a copy of their original signal and attempt to correlate it continuously with what they subsequently hear.

Conceptually, the correlation process is a way of measuring the degree of similarity between two numbers or patterns of numbers. In this case the pattern is the temporal alternation of high- and low-pressure waves. Imagine two copies of a signal such as the one in Figure 9-7A, one on a transparent sheet which may be shifted left and right over the other. When the two signals are exactly in register, they correlate perfectly. The correlation with a completely different sound (or, indeed, no sound at all) is zero (Fig. 9-7B), while the correlation with the same signal one-half wavelength out of phase is negative (Fig. 9-7C). This latter curiosity is a consequence of how correlations are calculated mathematically. The value of every point on the master copy is multiplied by the corresponding point in the sound pattern to be tested. Then all the products are added together. If the two signals are in register and the echo is a perfect, undistorted copy, the product—the "cross correlation"—is essentially the

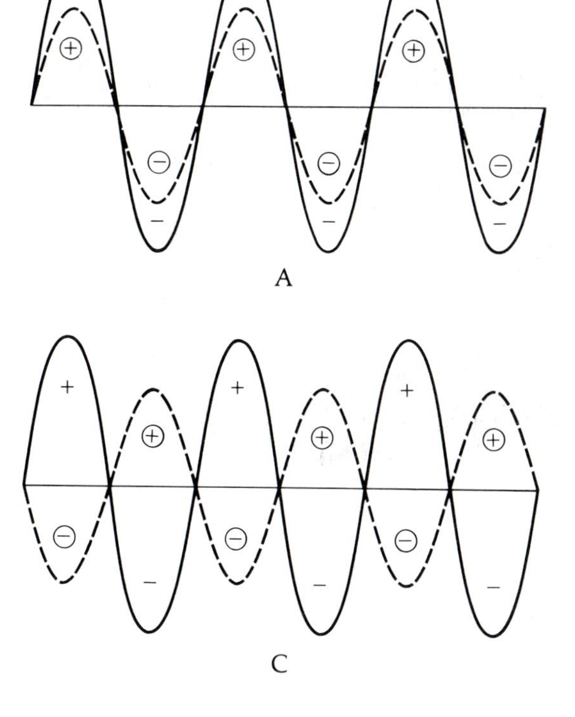

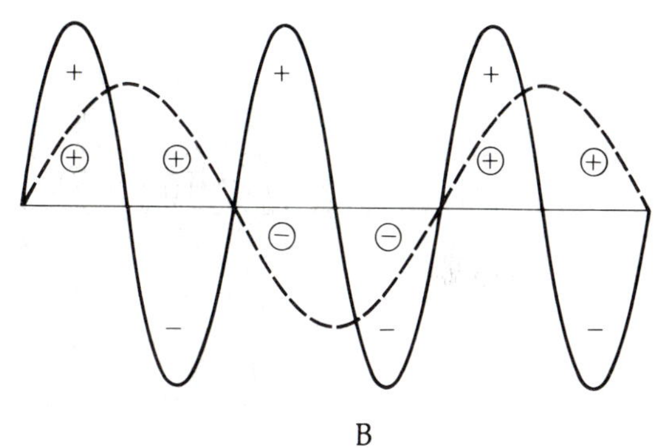

Fig. 9-7 Cross correlation is a process by which two patterns, signals, or sets of numbers are compared and their degree of similarity determined. Mathematically, each pair of values, one from each signal, is multiplied together one by one, and the resulting products summed. In (A), the summed product of these two identical signals in phase with each other will be positive and very large, indicating nearly perfect correlation. The same pair out of phase (C), however, always produces negative products, and adds to a large, negative number, signifying a negative correlation. When two unrelated signals are compared (B), the resulting mixture of positive and negative products will add to approximately zero.

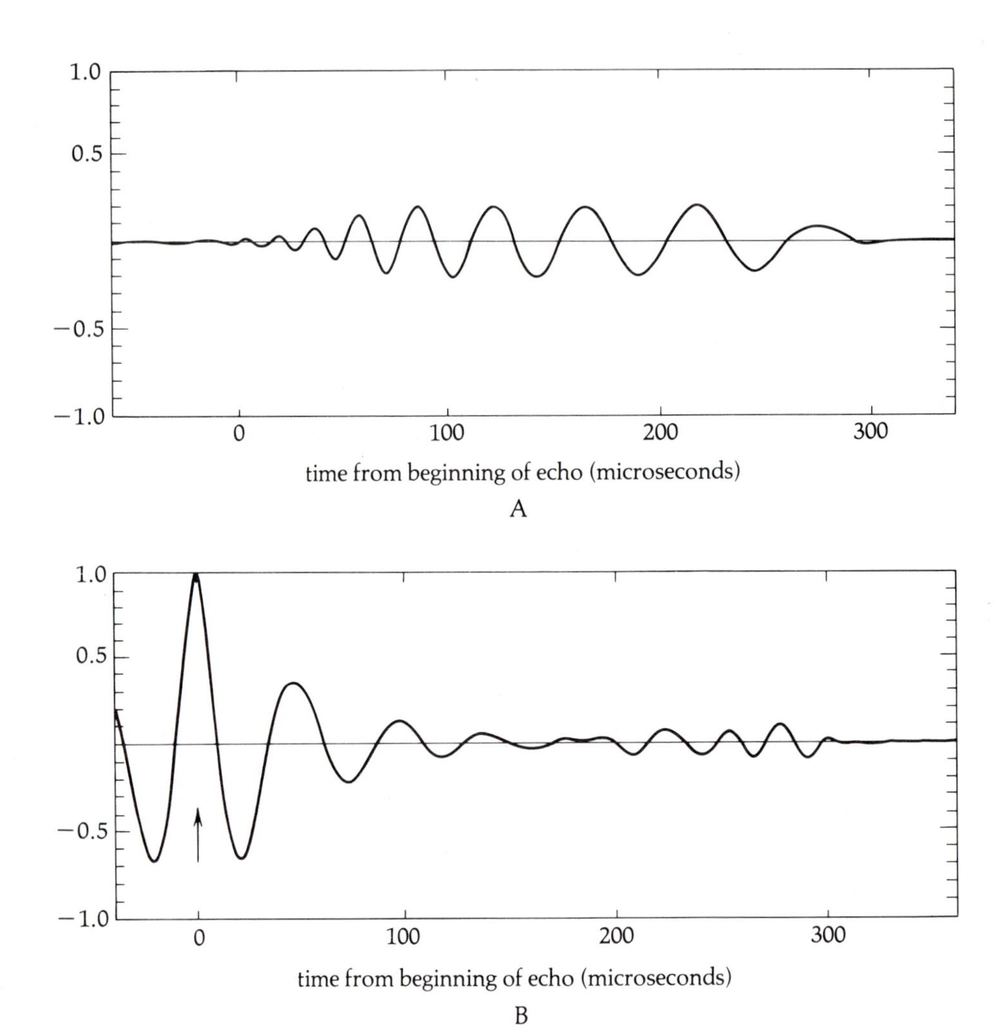

Fig. 9–8 When two signals are compared repeatedly over time, their phase relationship varies. The cross correlation will go positive when they are in phase, negative when out of phase; and these excursions away from the no-correlation value of zero will be the greatest when the two signals match the best. Bats produce two types of signals: FM sweeps (A) and CF tones (described in Fig. 9–9). When an FM sweep and its idealized echo are cross correlated (B) there is an obvious time of best fit (a zero, shown by an arrow, on this scale) as well as a pair of fair matches, one just before and one just after the true match. The time of best fit specifies the target distance. But echos are rarely perfect copies of the cry; often they are wildly distorted, and the signal is invisible to our eyes (C). Even so, the cross-correlation process for an FM sweep (D) ignores the noise, and is therefore able to pick out the correct echo travel time.

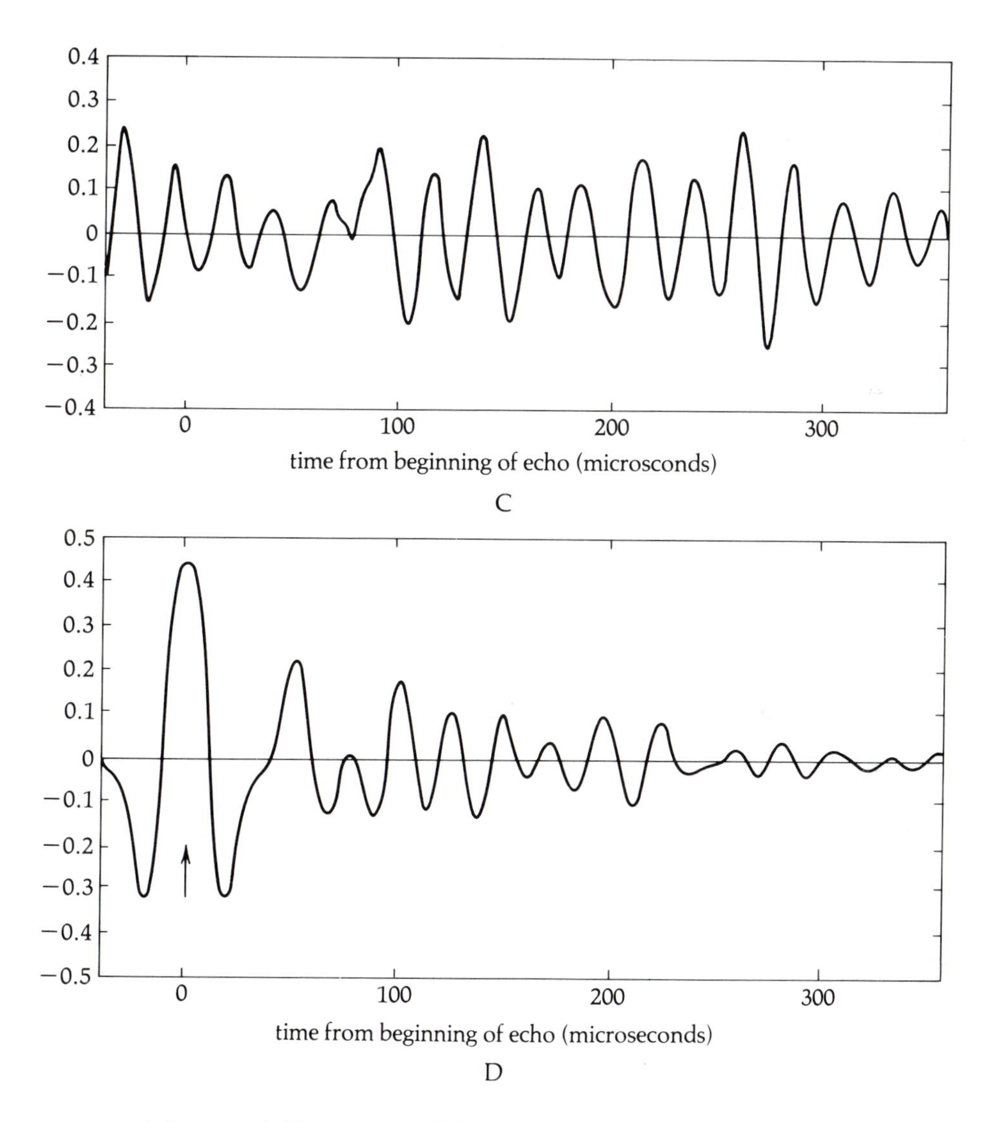

square of the signal (Fig. 9–7A); if they are one-half wavelength out of phase, the product is the negative square (Fig. 9–7C). The power of this analysis in defining the moment of best match is remarkable. Figure 9–8A is a simplified 0.03-msec (30-microsec) FM signal, unlike those of bats in that it is shorter, sweeps through more than an octave, and has no harmonics. Figure 9–8B is the cross correlation for an undistorted echo free of noise. The correlation assigns the travel time (equivalent to the target distance) within 1 microsec (± 3 mm) with a probability of about 50 percent. (That is, the bat sees three targets—one distinct and two fairly dim—but the prey is twice as likely to be in the center peak as either of the two side peaks.) Figure 9–8C is the same echo distorted by changes in target size over a range of 8:1 and embedded in noise averaging twice the intensity of the echo. Against this challenge the cross correlation still locates the target within ±17 mm with a probability of about 50 percent (Fig. 9–8D). And, of course, as the bat closes in and gets an ever-stronger echo, the precision of the correlation rises accordingly. By testing the bats systematically, Simmons has shown that the bat's errors exactly match the predictions of the cross correlation, even to the point of faithfully reproducing glitches in the theoretical curves.

Cross-correlation analysis also demonstrates the problems in temporal resolution inherent in constant frequency signals. Figure 9–9A shows an imaginary CF signal, and Figure 9–9B shows the cross correlation with a pure, noiseless

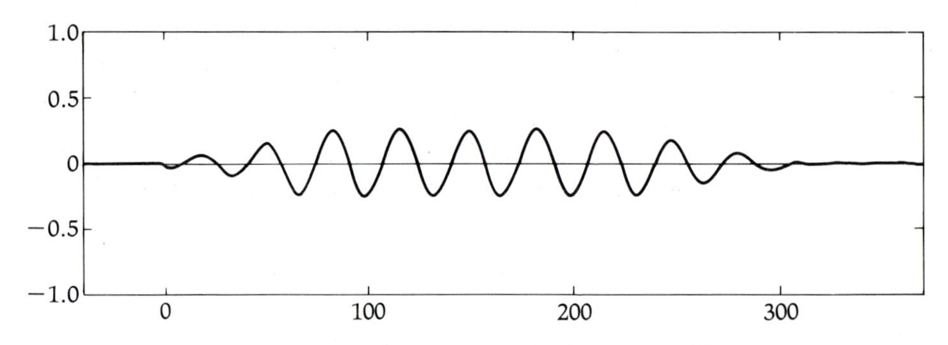

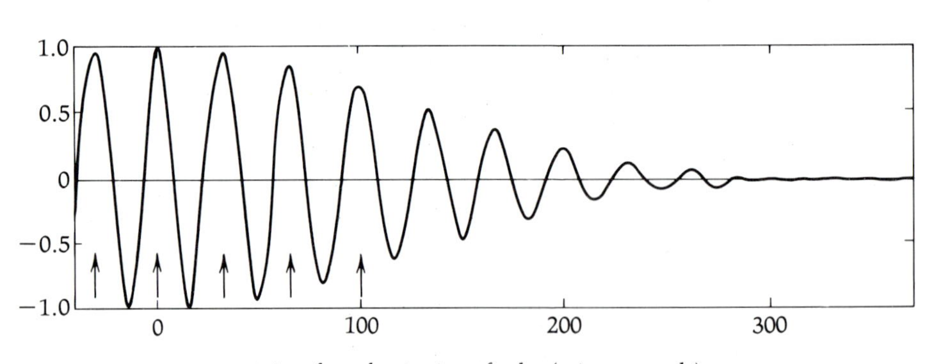

echo. The target could be any number of places and now the 50 percent probability of target location under the best of circumstances is smeared out over 22 mm—roughly three times the diameter of the bat's mouth. When the echo is subjected to the same shaping and factor-of-two noise used for the FM echo (Fig. 9-9C) this auditory hall of mirrors gets worse, and the cross correlation is

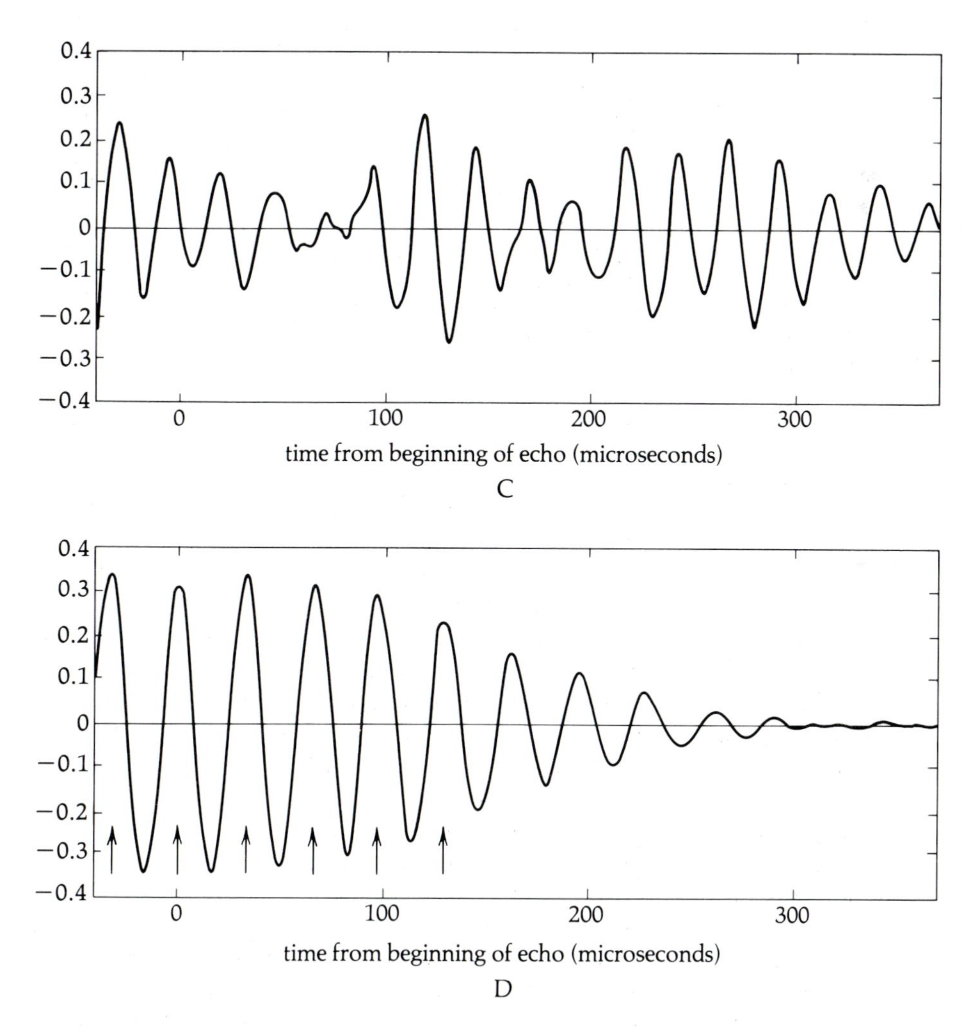

Fig. 9-9 The CF cry of bats (A) produces an echo which matches the outgoing signal nearly perfectly once per cycle (B). As a result, the travel time of the signal is impossible to determine precisely. (Times of best match are shown by arrows.) A more realistic, garbled echo (C) yields a cross correlation (D) which is even worse. Clearly bats cannot use CF cries for accurate target-range judgments.

almost useless (Fig. 9-9D). The FM signal does significantly better than this embedded twenty times its own depth in noise.

Perhaps the most significant thing about Simmons's discovery is that it provides a possible (and *testable*) mechanism by which a poor-quality or partial image of something can be compared with some array of master copies to determine the best match. A snatch of a melody, a piece of a puzzle, a partially hidden object, a distorted voice on the telephone, a crude sketch of an object, a blurred photograph are all quickly and accurately identified by our nervous systems if the original is familiar. Indeed, this is just the distinguishing characteristic of the gestalt process so dear to Lorenz—the ability to call to mind the whole on the basis of some part. It would be ironic, indeed, if those "silent" creatures of the night with their curious auditory double vision turned out to hold the key to the second part of that mystery of mysteries, learning and memory retrieval in animals.

SUMMARY

The three most fundamental tasks for the auditory system are frequency, direction, and pattern analysis. Frequency can be determined by resonances, either of the tuned projections or membranes of simpler ears, or of the basilar membrane of birds and mammals with its broad, location-specific tuning. When the frequency of a sound is low enough, it may be detected and analyzed by a system which actually counts its waves.

The sort of complex direction processing which higher animals employ depends on the type of ear, requiring temporal and intensity comparisons which must involve massive processing arrays with specialized acoustic feature detectors. Simple ears, though severely constrained in frequency and intensity ranges and accuracy, can manage the task with little more than simple trigonometry.

Owls, and perhaps most species, map the acoustic world on their brains according to Cartesian coordinates in very much the same way the visual system plots the visual world. Although simple auditory pattern recognition also depends on feature detectors, more complex signals require stored or memorized models as a standard against which incoming sounds are matched. This "cross-correlation" system, most obvious in the highly specialized echo analysis of bats, may turn out to be the basis of complex pattern recognition whether acoustic, visual, or even olfactory.

STUDY QUESTIONS

1. If high frequencies are so strongly absorbed by air, why don't bats use lower pitched cries?

2. The FM and CF areas in the mustache bat brain are mapped out in one dimension—target range and frequency, respectively. In most neural displays, a bicoordinate system is used. What might be the other axis in each case? How would you test the possibilities? Why might the Doppler-shift region have each of its bicoordinate points represented twice?

3. Bats also determine the wingbeat rate and surface texture of potential prey. What would flapping do to the pattern of excitation on the auditory map of the bat brain? How might texture be analyzed?

4. Some moths have an ultrasonic click which they produce upon hearing a bat. What do you suppose might be its function and the bat's response?

FURTHER READING

Griffin, Donald. "More About Bat 'Radar.'" *Scientific American* 199, no. 1 (1958): 40–44.

Konishi, Masakazu. "How the Owl Tracks Its Prey." *American Scientist* 61 (1973): 414–24.

Roeder, Kenneth. "Moths and Ultrasound." *Scientific American* 212, no. 4 (1965): 94–102.

Simmons, James, et al. "Information Content of Bat Sonar Echoes." *American Scientist* 63 (1975): 204–15.

CHAPTER 10

Other Senses

There are a variety of sources of information about the world around us from which our species has been largely or wholly excluded. Even our primary sensory modality, vision, which so manifestly shapes our perceptions, is paltry compared to the visual systems of some animals: birds of prey, it has been estimated, receive eight times the visual information about their world that we do. Hearing, the sense we rely on next to vision, is a distant second at best, when compared to the richness of the acoustic world for other animals. What, then, must be our poverty in regard to the other senses which enrich the lives of lower animals?

Taste and smell, for example, are two familiar senses which provide us with third-rate information at best. For most animals the all-important tasks of recognizing their own species or finding edible food depend on these systems. Both senses rely first on the interaction of molecules with receptors, and then on making neural judgments from that interaction. Making judgments about the sources of these molecules often involves a very real and powerful olfactory memory.

TASTE

The categories of human taste which have traditionally been denominated—salt, sweet, sour, and bitter—were determined from both introspection (a questionable form of verification) and psychophysical experiments. (The "purest" stimuli have been, respectively, NaCl, sucrose, H^+ acids, and quinine.) When neurophysiologists like Vincent Dethier finally succeeded in recording directly from insect taste organs, exactly these four classes of receptors were found. However, essentially nothing is known about how insects judge taste and identify food on the basis of the information from these receptors.

Anatomically, we know that insects have four receptors (one from each class) in each hair on the feet and on the opening of the proboscis. Like the hungry blowfly of Chapter 5, an insect typically finds food by stepping in it, whereupon it turns, sends its proboscis down, and pumps in the food. Vertebrates, on the other hand, have no such peripheral system, and so must make a special (and often awkward) effort to sample anything in their environment. Vertebrate receptors are located in "taste buds" on the tongue. Disappointingly, physiological recording from these receptors has failed to find the neat "one-taste, one-receptor" system previously found in insects. Instead, vertebrate taste receptors have wide responses and merely display "preferences." Essentially nothing is known about how "sweetness," for example, is extracted from broadly tuned receptors. Perhaps taste perception is analagous to that of color—to perceiving orange from the overlap of the red and green responses—so that sweet and sour, say, may represent the two ends of a continuum. Beyond the difficulty of understanding how the four tastes could be extracted from broadly tuned receptors—or why the cells shouldn't have been narrowly responsive in the first place—the sensation of taste results from more than just taste buds. In fact, it is mostly the *odor* of food which imparts the sensation of taste, a phenomenon which explains the otherwise puzzling but familiar observation that food is tasteless when our noses are plugged during a cold.

OLFACTION

But there is little comfort in lumping the question of taste processing together with that of olfactory analysis: although the biology of taste is poorly understood, that of olfaction is even worse. Attempts to break odors down into classes have met with little success. From studies of animal and human behavior and physiology, it is now clear that much of our inability to comprehend the evidently rich world of odor results from our species' relative insensitivity to smell, further compounded by the existence of literally dozens of receptor classes. Moreover, like Spallanzini trying to grapple with echolocation before ultrasonic sounds could be detected, we lack the machines, the technological crutches, to measure odor.

Human olfactory acuity is, by mammalian standards, poor. For example, we can detect the foul odor of butyl mercaptan—the substance to which we are, for reasons best known to evolution, most sensitive—only at concentrations of 10 million molecules per cc, while our pet dogs and cats regularly respond to the same odor at $\frac{1}{1000}$ of that level. Animals commonly detect odors that have special, species-specific significance at low concentrations. Moths, for instance, can detect 200 molecules per cc of their sexual scents. The world is a very smelly place, but we can smell it only about as well as most legally deaf people can hear.

Nevertheless, humans have five million olfactory receptors, and with talent, training, and suitably high concentrations of scent can recognize and name several thousand odors. On the other hand, equipped as they are with 100–300 million receptors, many dogs seem able to sniff out an unlimited number of humans, to find the stick we've thrown into a pile of similar sticks, or to follow a trail left by the odor of our feet which has passed through our shoes onto the ground. Who can imagine what the olfactory world is like for such an animal? Their sensitivity must be the product of natural selection, and therefore must play an important part in their lives. We know, for example, that animals regularly leave individually recognizable scent marks to advertise their territorial boundaries or social status. Other animals use odors to synchronize ovulation in females, and the discovery of substantial synchrony of menstrual cycles in college dormitories suggests that something similar may have evolved in humans. In insects, special odors ("pheromones") are used as sign stimuli to specify species and sex. The hair-pencil odor of the butterflies in Chapter 3 is a well-known and fascinating example of this phenomenon. Ants leave trails of specific odors to guide other ants, and have special odors for alarm and even for panic. In fact, some predators have broken these olfactory codes: assassin bugs lay false ant-odor trails and eat the hapless recruits. Slave-making ants spray a "panic" odor to clear out the guards at a nest they intend to rob, and bolas spiders fish for prey, their lines baited with sticky blobs redolent of female moth pheromone.

Very little can be said about how olfactory receptors work. Many animals possess receptors sensitive only to their own species-specific pheromones. These receptors probably derive from an ancient evolutionary strategy: very specific binding proteins. Similarly, certain animals have specific receptors for naturally occurring molecules, as bees have for CO_2 and H_2O (humidity). But aside from these specialized cells, any olfactory receptor shows some response to almost *any* odor—a complicated response (Fig. 10–1) with a unique time course for each

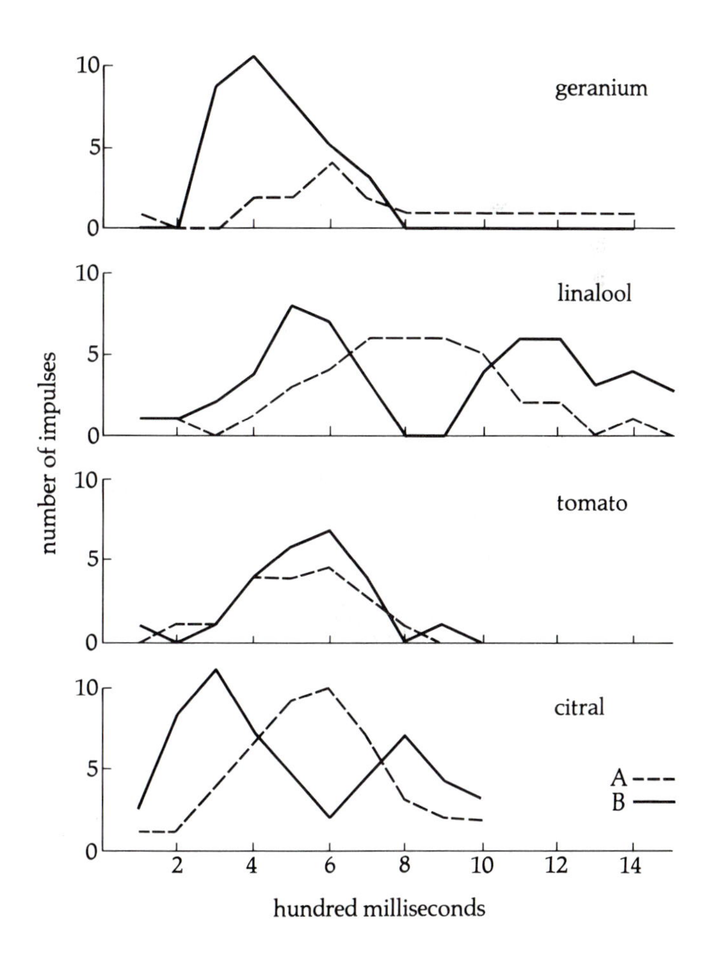

Fig. 10–1 The responses of olfactory receptors to odors are extraordinarily complicated. In this illustration of the output of two cells ("A" and "B") from a moth antenna, a unique temporal pattern of nerve impulses is produced by each odor, even though the concentration of the odor remains constant. Where in these patterns does the information used by the moth lie?

compound. In certain well-studied species, however, a slight hint of order has become apparent. In roaches, for example, olfactory receptors can be divided into a finite number (about thirty-five) of discrete classes—a far cry from their three of vision and four of taste.

Finally, nothing can be said about the processing of olfactory information—how, for example, we distinguish oranges from onions. Is there a "map" on our olfactory lobes, and if so, what are its coordinates? Does the distinction depend on two receptor classes or twenty? Does it depend on learning or "calibration"? How, for example, does evolution arrange for newborn garter snakes in one part of the U.S. to recognize and attack earthworms, while in another part the same species ignores worms and concentrates innately on fish? Are the receptors different, or is it a difference in processing? How does evolution wire female rodents to abort their pregnancies spontaneously when presented with the odor of a strange male, or make it possible for guard bees to recognize their own hive's foragers on the basis of their odor? And to what extent has our own poor olfactory acuity blinded us to what may be going on in the rest of the animal world? That shortcoming, combined with our inability

---

THE DISCOVERY OF ELECTROLOCATION

Sharks, rays, and catfish sense the electric fields which their prey generate by means of an array of specialized structures known as the ampullae of Lorenzini. These curious organs consist of small innervated bulbs, each of which is attached to a long, jelly-filled canal running to an opening in the skin. Although the anatomy of the bulbs offers no clue to their function, their arrangement lends itself to neurophysiological analysis. In fact the ampullae were first recorded from in 1895. Despite their accessibility it took seventy years to sort out even the primary functions of these tiny organs.

In 1909 Howard Parker found that a blunt glass rod pressed against the skin would fire the nerves. The ampullae of Lorenzini were therefore first identified as mechanoreceptors. By 1915 this response was interpreted not as a touch response but as the output of a receptor designed to read hydrostatic pressure (water depth). In 1938 the ampullae were identified as temperature receptors

---

with a demonstrated sensitivity of 0.05°C. In 1960 came the news that these organs were actually salinity detectors, measuring with great precision the concentration of sea water.

The first observation of what is now thought to be the true function of the ampullae actually came in 1917, when Parker discovered that blindfolded catfish could sense a metal rod at 7 cm, but were unable to detect glass rods until they bumped into them. Although Parker correctly guessed that an electrical sense must be involved, he saw no link between the behavior and any particular sensory system. The whole phenomenon was shelved until 1962, when Sven Dijkgraaf and Adrianus Kalmijn noticed that the sharks they were studying were sensitive to a rusty wire. Kalmijn went on to show that the electrical field produced by the wire was the stimulus, and that the ampullae were the detectors. He determined the behavioral threshold (the point at which sensitivity to the stimulus began to affect the behavior) of sharks and rays to be a voltage gradient of only 0.005 μv/cm. This is equivalent to the field produced by an otherwise camouflaged flounder (a favorite prey of sharks and rays) at 30 cm. Indeed, in a behavioral experiment on the ocean floor sharks consistently attacked buried electrodes rather than open tubes of ground-up flounder "extract."

The astounding sensitivity of the ampullae of Lorenzini to electrical fields has led Kalmijn to propose adding yet another page to the lengthy saga. When a shark swims in sea water its motion through the earth's magnetic field should induce a detectable electrical field. Could these organs serve as compasses as well?

It is difficult to imagine a practical and decisive test in the animals' natural environment, but sharks in a laboratory tank with an artificial magnetic field can be conditioned to seek food in a particular direction using only magnetic cues. The contingencies of the shark's natural habitat—the darkness of the ocean depths—might render such a compass sense quite useful. Distinguishing the steady induced field generated by the earth's magnetism from the alternating fields of prey should be a simple matter for these animals.

The long search for the function of the ampullae of Lorenzini illustrates the unyielding nature of physiology when it is pursued in isolation. Without any real sense of the natural world of sharks and the sensory systems that world dictates, we can come by few cogent answers from laboratory experiments. As long as those who worked on them had no knowledge of or interest in their respective sensory worlds, even pigeons and mice, those workhorses of neurophysiology, successfully kept to themselves their astounding "superhuman" sensitivities to ultrasonic sound, ultraviolet light, infrasonic sound, polarized light, air pressure, and magnetic fields. As Chapter 14 will illustrate, pigeons still gamely resist our attempts to uncover the basis of their celebrated and seemingly magical map sense.

to assign any meaningful parameters to odors (analogous, say, to frequency), makes olfaction the most difficult of our senses to analyze.

OTHER SENSORY WORLDS

Beyond the world of odors, there are other worlds in which we lack even the status of sensory cripples. Pit vipers such as rattlesnakes, for example, have separate infrared eyes whose 150,000 heat-sensitive receptors form an image which is mapped directly onto the regular visual display in the optic tectum (Fig. 10–2). As Chapters 13 and 14 will document, pigeons and honey bees sense the earth's magnetic field with

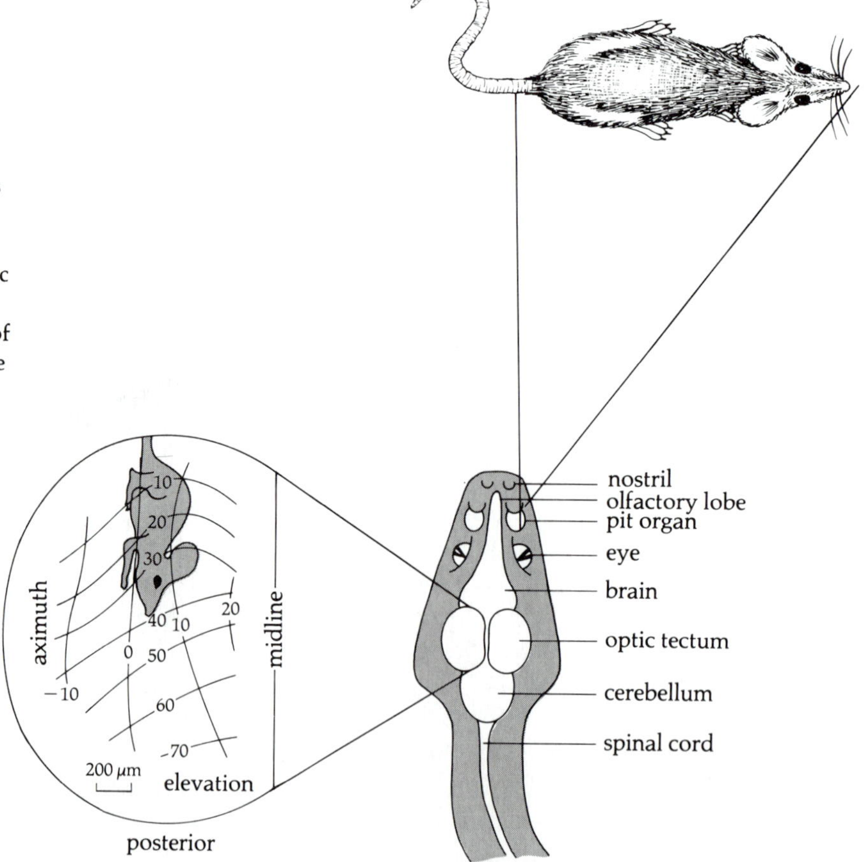

Fig. 10–2 A rattlesnake's view of the world in the infrared is mapped out on the surface of its optic tectum in a geometrical fashion. A similar map of visual space exists on the tectum as well.

considerable accuracy. But by far the best understood of these alien sensory windows is the world of electric fields. There are three classes of electrosensitive animals: strong, weak, and passive. The first group, exemplified by the electric eel and ray, use highly modified muscle cells like electronic capacitors to store the deadly electric charges they wield against predators and prey alike. The third class, in contrast, merely "listen" to the world. Sharks, for example, explore the dark and murky depths while listening for the telltale electric currents which leak through the skin of their otherwise camouflaged or well-hidden prey. The sensitivity of the electric organs of sharks is incredible: 10^{-8} V/m, the equivalent of one-fifth of a volt on the other side of the world. This sensitivity is so extreme, in fact, that sharks and (nonelectric) rays can be taught to distinguish the direction of the earth's magnetic field from the tiny electric currents the animal's own motion incidentally generates.

The intermediate group, the weakly electric fish, represent an entirely different use of electric-field sensitivity. These fish produce weak electric pulses with which they both communicate and form pictures of the world around them. Walter Heiligenberg has shown that *Eigenmannia,* for example, a small electric fish common to the murky streams of the Amazon, generates a field between its mouth and the rest of its body. Tiny electric currents flow along the field lines from the head to the thousands of electrically transparent pores which the fish uses to measure current flow. A characteristic spatial pattern of current flow exists when the world around the fish is empty. Whenever an electrically opaque object—whether obstacle, predator, or prey—enters the field, however, the flow is disrupted, and a vague electric "shadow" appears on the part of the array of organs which normally receives current from the paths now blocked by the object (Fig. 10–3).

In fact, since the amount of current flow is constant, the displacement of the field lines by an object probably leads to an *increased* flow to the receptors around the shaded area, and hence an enhancement of the shadow. This electrical version of lateral inhibition means that electric fish map the elevation of any target directly on their relatively cylindrical bodies. For example, an object above the fish will cast an electric shadow on its back. The placement of the shadow along its length corresponds in most cases to the distance of the object from the fish—the farther away the target the farther back toward the tail the

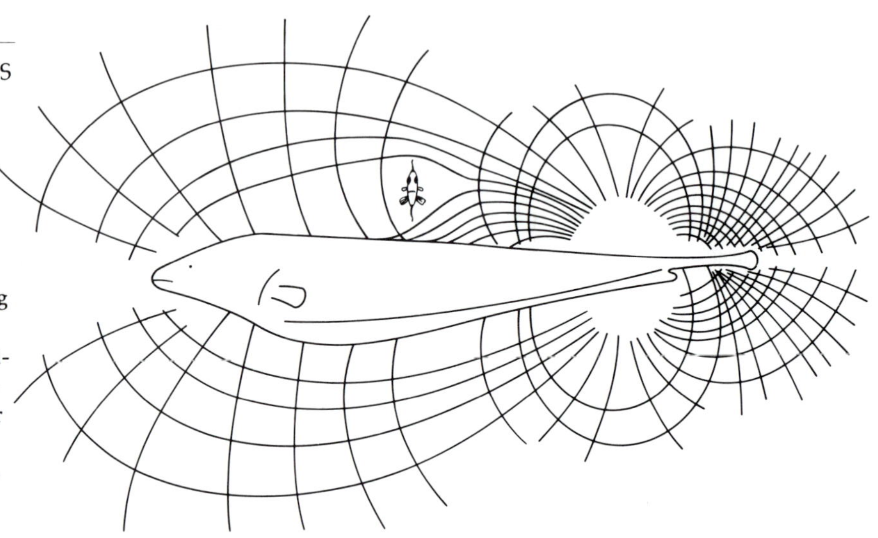

Fig. 10-3 The electric field around *Eigenmannia* runs from its mouth to the electroreceptors along its body. Any object whose electrical conductivity differs from the medium (murky fresh water in this case) distorts the field and casts an electrical shadow on the body of the fish.

shadow ought to fall. The size of the object, at least electrically, should correspond to the size of the shadow. Finally, by turning, *Eigenmannia* can also determine the horizontal direction (the azimuth) of the target.

Eigenmannia thus lives in a world of precise elevation and approximate distance, but a world which, without a special effort on the fish's part, lacks azimuth. The peculiar way these creatures sense distance even permits them to "see" behind obstacles to find what may be hiding on the other side. As yet we know nothing of the feature detectors which may exist to emphasize small-scale spatial or temporal patterns. Then, too, we know nothing of how this curious world is mapped on the brain, or whether it is (if it even could be) brought into register with the fish's visual world. That the fish must perceive a picture, however distorted from our perspective, suggests to us that its "electric cortex" may be composed of elements partially familiar to us already from the auditory and visual cortices of other, more conventional animals.

Dim as our sensitivity is to so much of the world, our progress toward a fuller understanding must be slow and halting. Who knows what senses we are missing altogether, how those we *do* know are mapped, or how evolution has formed the neural basis for the sign stimuli and memory each system undoubtedly has? In the absence of imagination or downright inspiration, in what direction will the experimental trail lead us?

SUMMARY

Each species concentrates its limited neural processing capacity on the small set of sensory modalities most important to its well-being, and ignores or shortchanges the others. Olfaction, for example, so central to the lives of most mammals, is rudimentary in our species, while electric and, so far as we know, magnetic fields are wholly outside our range of experience. Beyond discovering as yet unknown sensory "windows" of animals and exploring their functions, the greatest challenge to ethologically inclined neurobiologists today lies in understanding the specialized neural processing and the well-ordered representations of the information it generates that must be hidden in the brains of animals.

STUDY QUESTIONS

1. How could a hunting shark distinguish between the signals unintentionally broadcast by a hidden flounder (signals generated for the most part by the flow of its blood and the activity of its nervous system) and the current induced by the shark's own movement through the earth's magnetic field? There are several possibilities. How would you go about distinguishing among them?
2. Given the curious yet distinctive time course in the firing of odor receptors, how do you suppose the information might be processed for olfactory memory recall? Is there any test for this potential mechanism?

FURTHER READING

Bullock, T. H. "Seeing the World Through a New Sense: Electroreception in Fish." *American Scientist* 61 (1973): 316–25.

Dethier, V. G. "A Surfeit of Stimuli, a Paucity of Receptors." *American Scientist* 59 (1971): 706–15.

Gamow, R. I., and Harris, J. F. "The Infra-red Receptors of Snakes." *Scientific American* 228, no. 5 (1973): 94–101.

Haagen-Smit, A. J. "Smell and Taste." *Scientific American* 186, no. 3 (1952): 28–32.

Lissmann, H. W. "Electric Location by Fishes." *Scientific American* 208, no. 3 (1963): 50–59.

Schneider, Deitrich. "Sex Attractants of Moths." *Scientific American* 231, no. 1 (1974): 28–35.

Wilson, Edward. "Pheromones." *Scientific American* 208, no. 5 (1963): 100–14.

Part III

COMPLEX BEHAVIOR

CHAPTER 11

Motor Programs

The control of "motor output" in behavior has received relatively little attention from ethologists. Rather than how behavior is physically accomplished, their concern has been predominantly with describing it, finding out how behavior is triggered, and why it is adaptive. It is generally taken for granted that all the appropriate wiring must be in there somewhere, and we have assumed that neurobiologists will work out the details to the extent that it is possible sooner or later. As a result, the original ethological concepts of motor programs and especially "fixed-action patterns" have fallen into disrepute from general apathy and sloppy usage. Nevertheless, motor programs and fixed-action patterns are real and important phenomena, and the question of how they are organized has wide implications in behavior.

FIXED-ACTION PATTERNS

Motor programs are best thought of as self-contained units of behavior of which fixed-action patterns are a conceptually crucial special case. Characteristically, true fixed-action patterns

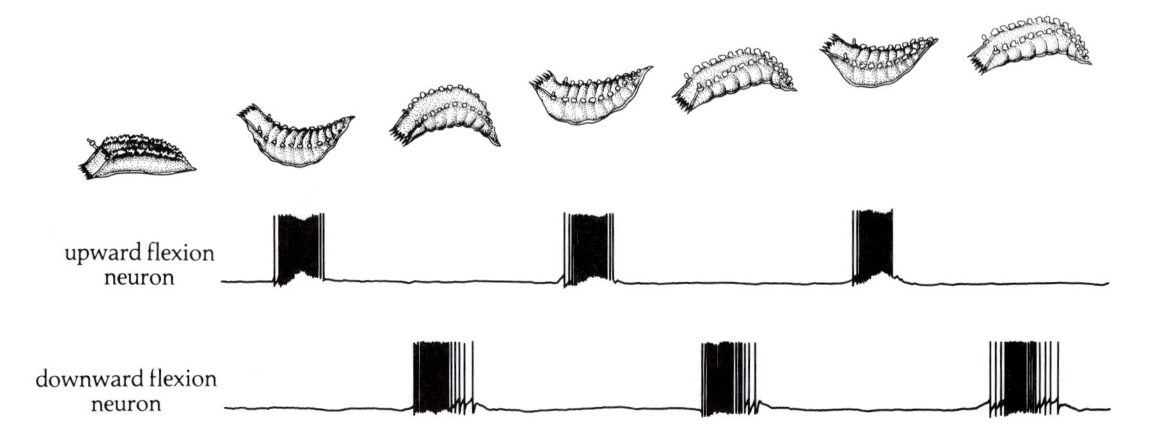

Fig. 11–1 The escape response of *Tritonia* consists of alternate bending of the dorsal and ventral surface. Electrophysiological recording shows that the two sets of muscles are activated alternately, suggesting that an oscillator (a circuit which alternates rhythmically between two states) and switch (the IRM) in the nervous system may be all that controls the behavior. [From "Giant Brain Cell in Mollusks, by A. O. D. Willows. Copyright © 1971. Scientific American, Inc. All rights reserved.]

have four things in common. First, they are under innate genetic control (and hence are "species typical") and are more or less stereotyped: all geese roll eggs in the same way. Second, feedback is often unnecessary or relatively unimportant: if the egg is removed the goose will finish the egg-rolling routine without it. Third, the behavior can usually be evoked in its complete form by brain stimulation, presumably by hitting the releaser or "trigger" cells. Finally, the coordination of several muscles is required, so the behavior must be more than a simple reflex.

A well-known class of simple FAPs are the escape responses of aquatic organisms such as scallops, crayfish, and sea slugs. *Tritonia*'s response (mentioned in Chapter 5) merely involves alternate flexing of the dorsal and ventral muscles (Fig. 11–1), a simple but crucial piece of coordination. *Tritonia* must have an oscillator which switches signals between the two muscle groups at an appropriate rate.

A similar alternating-muscle oscillator may be seen in the honey bee's sting (Fig. 11–2). The barb itself is formed from two interlocking pieces which can slide lengthwise relative to one another. When honey bee stinging is released, the barb is

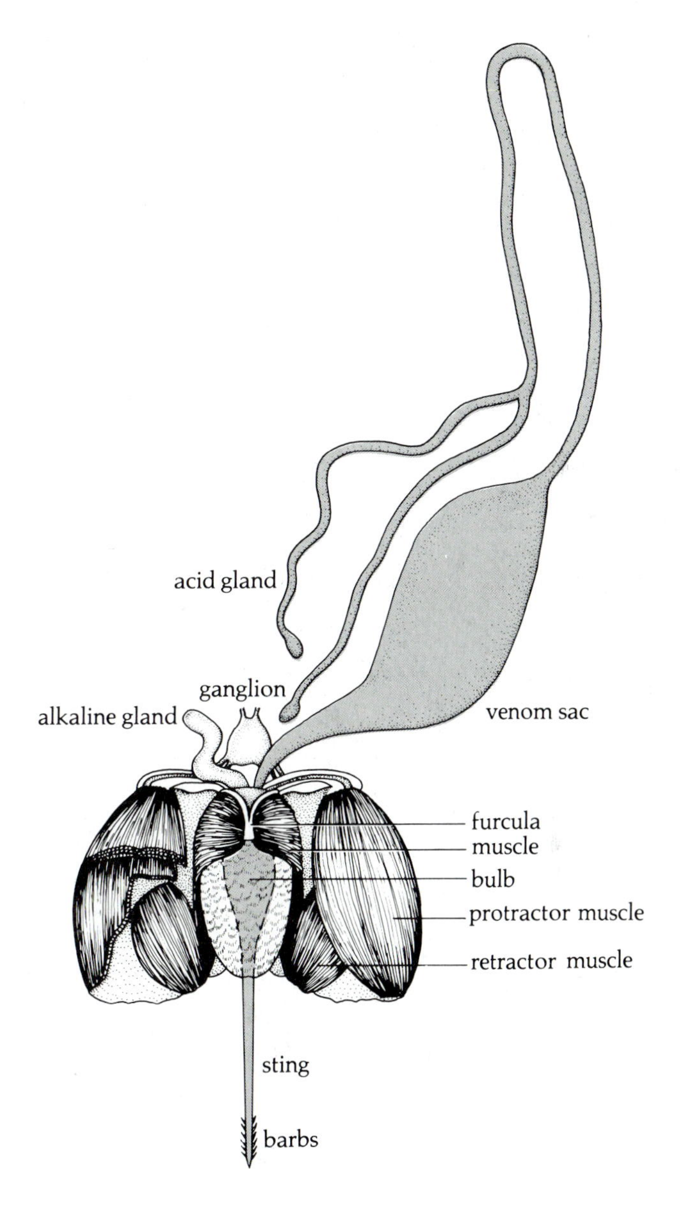

Fig. 11–2 The sting apparatus of the honey bee contains its own motor program which continues to function long after the bee has flown away. The sting itself consists of two barbed half cylinders which move up and down relative to one another. The hollow core they create provides a pathway for the venom. Once the stinging behavior has been released, the bee inserts the sting and then pulls away. The barbs keep the sting in the victim, and so the apparatus is torn out of the bee's abdomen. The retractor and protractor muscles contract alternately, driving the sting in. The furcula muscle also contracts rhythmically, forcing venom down the sting core. A one-way valve between the venom sac and bulb allows the bulb to be refilled between furcula contractions. The last abdominal ganglion, which always breaks away with the apparatus, contains the two motor programs.

driven into the offending object and the insect flies off leaving its sting, its venom sac, and two FAPs behind. The first oscillator alternatingly flexes the muscles attached to the two parts of the sting, pulling back on one barbed half at a time and, as a result, driving the other half farther in. Thus the sting slowly pulls itself in until it is entirely buried. Meanwhile, a second oscillator causes the venom sac muscles to contract rhythmically, forcing poison down the hollow center of the barb and into the victim.

Another similar sort of oscillator-driven motor program is involved in the neurogenic (i.e., slow) flight (Fig. 11–3) of butterflies and grasshoppers. (Fast "myogenic" fliers like the honey bee depend on the mechanical "resonance" of the thorax.) In-

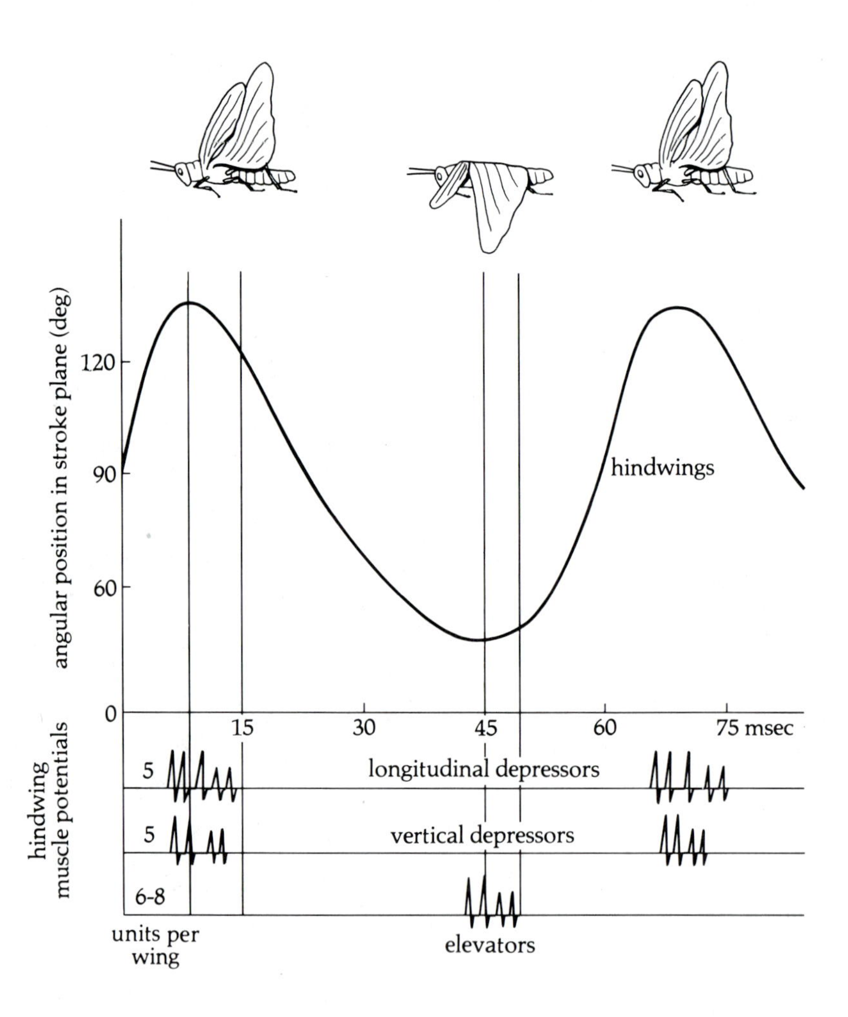

Fig. 11–3 Locomotion is often controlled by a relatively simple motor program whose basic function is to move limbs first one way and then the other. This is illustrated here by locust flight. The depressor muscles pull the wings down, while the elevator muscles raise them. [Redrawn from *Neural Theory and Modeling: Proceedings of the 1962 Ojai Symposium,* edited by Richard F. Reiss, with the permission of the publishers, Stanford University Press. © 1964 by the Board of Trustees of the Leland Stanford Junior University.]

terestingly, a single oscillator seems to be common to breathing, walking, and stridulation in Orthoptera. In the case of stridulation the patterns are sufficiently complicated that at least two oscillators must be interacting to produce them. The neural circuitry which produces such coordinated behavior is now being worked out with the emphasis at present on the basic designs of stable oscillator circuits (Fig. 11–4). In each case the oscillators are located in the insect's thorax, but the switch lies somewhere in the brain. Sometimes the brain is simply *inhibiting* the behavior which would otherwise be continuously ex-

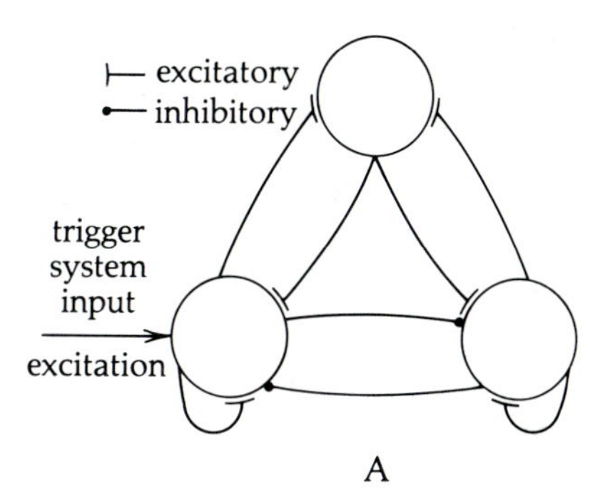

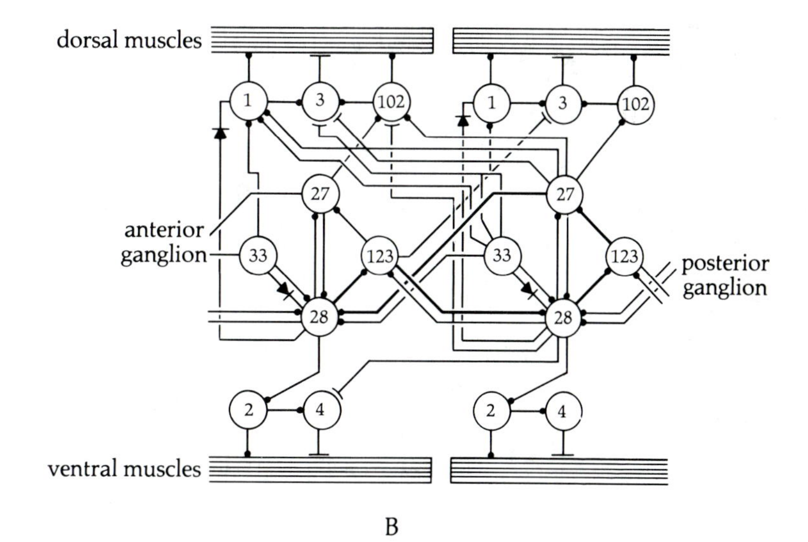

Fig. 11–4 The oscillators and switches which are thought to be responsible for the careful timing and coordination of motor output programs can be modeled. In A, the simplest three-cell circuit which could give rise to the *Tritonia* escape behavior is shown. A more direct but difficult approach is to construct a "road map" of cellular connections, and record the interactions of the cells. Part B shows the result of such an analysis on the cells which give rise to the repetitive swimming pattern of leeches. The basic element turns out in this case to be a four-cell oscillator (nos. 27, 28, 33, and 123) unit.

pressed. If an insect's head is cut off, the headless animal may begin to fly or perform some other behavior which its brain had been inhibiting. In fact, female praying mantises release male copulation in just this way.

Though easiest to study when they do, few FAPs depend on

free-running oscillators. Two examples familiar to each of us are swallowing and vomiting. Swallowing (Fig. 11–5) is triggered by a stimulation of the upper throat, and results in a stereotyped and coordinated firing of muscles which usually serves to move the stimulus (food) down the esophagus. But the pattern is the same if the stimulus does not move, or indeed if it is immediately withdrawn. Gagging and vomiting have their own, all too familiar releasers and stereotyped sequences of muscle movement.

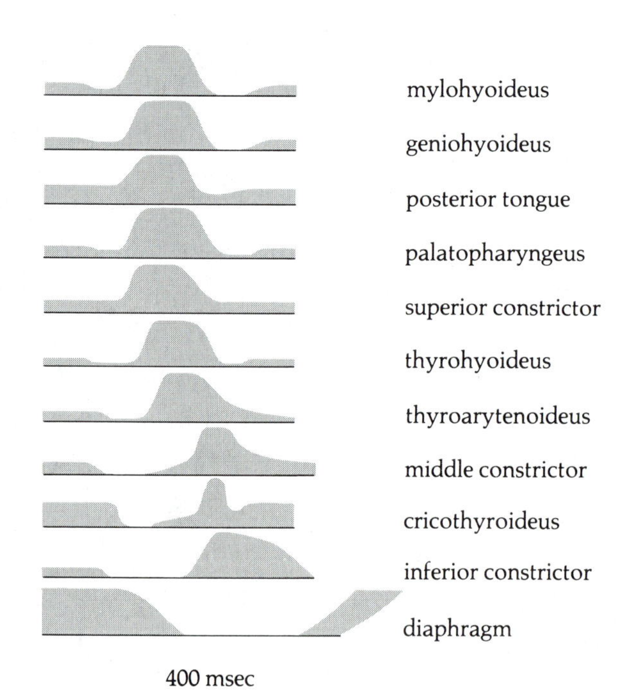

Fig. 11–5 Diagrammatic summary of the sequence, timing, and intensity of muscle contraction for swallowing in dogs. The height of the shaded area shows the strength of the contraction for each muscle involved. Swallowing works because the muscles in the throat and esophagus contract in a regular, well-timed order according to an innate motor program. [From R. W. Doty and J. F. Bosma, "An Electromyographic Analysis of Reflex Deglutination," *Journal of Neurophysiology* 19 (1956): 49. Courtesy of Charles C Thomas, Publisher, Springfield, Illinois.]

MATURATION

The notion that coordinated motor programs are innate has aroused a surprising amount of controversy. For example, as we have seen, Spalding maintained that the directed pecking of newly hatched chicks was innate. Later, behaviorists with another set of expectations observed that the accuracy of the pecking improved with time, and interpreted this as learning (Fig. 2–1). Ethologists were persuaded that the increase in accuracy was due rather to maturation, and this led to the ingenious experiment described in Chapter 2. One group of newly hatched chicks were fitted with prismatic goggles which deflected their line of sight slightly. As a result, their initial pecking was misdirected. With the passage of time the scatter in their pecking decreased, but they were never able to compensate their aim. Directed pecking must then be innate, although quibbles can still be heard about whether pecking might not be "practiced" during hatching.

Perhaps the ultimate example I know for how far one must go to rule out learning in what appears to the initiated to be a fixed-action pattern comes from John Fentress's study of the stereotyped grooming behavior of mice. This behavior, familiar to anyone who has ever kept a pet gerbil, hamster, or mouse, is released by tactile stimulation, but also appears ex nihilo as a "vacuum activity" or in conflict situations (see Chapter 12). The grooming consists of alternating rapid forward movements of the arms and paws beginning over the back of the head and ears and progressing finally to the nose and whiskers. Like the pecking of chicks, grooming behavior appears well after birth, and could be interpreted as the result of trial-and-error self-conditioning to the "good" feeling of some consequent sensation of cleanliness. In fact the behavior is wholly innate. Fentress amputated the front legs of mice at birth, long before any sign of gooming began, and the stumps nevertheless attempted to perform the entire species-typical grooming sequence in the normal contexts and in total absence of sensory feedback from the head. The behavior appears as part of normal maturation rather than as the result of learning. Similarly, surgically declawed cats rhythmically sharpen their nonexistent claws throughout their lives, showing no sign that the absence of feedback affects the behavior in any way.

The "maturation" we have been describing, however, is just a word, not an explanation. What is going on inside the chick, for example, to perfect its pecking motor program? Why should it need to improve? Couldn't evolution design a good, prewired program? One of the best-studied cases of maturation of a motor program is locust and cricket flight. These animals pass through several instars or moltings during which they are flightless and, initially, wingless. Nevertheless, if an early instar is lifted from the ground it begins to beat its tiny, stub-like wing pads. Loss of contact with the ground is the major releaser for the flight motor program. The wingbeat, however, is at only 10 Hz—half the adult rate—and the four wings are completely out of phase. When the day before adulthood is reached, however, the rate has increased to 20 Hz and the wings are perfectly coordinated. We know that the maturation is primarily due to the wiring-in of stretch receptors which provide feedback to fine-tune the oscillator (Chapter 19). Perhaps evolution cannot build all crickets so that they are mechanically identical (adult size, for instance, varies), and so had to put in a mechanism by which each cricket may calibrate itself.

Perhaps one of the easiest examples to misinterpret of the maturation of an innate program is the burying behavior of ground squirrels. The misleading features are that it seems too complicated to be innate and a certain amount of "plasticity" is apparent. The normal adult pattern is quite complicated. The squirrel first searches for an appropriate spot, digs a trench, rams the food in with its head, scrapes the dirt back over and tamps it, and then camouflages it. Although some of these behaviors seem stereotyped, the sequence does appear gradually over a period of two weeks when the squirrels are several months old. But only maturation is involved: individual elements are virtually perfect from the start, and the only task of the squirrel seems to be to get them all going, and then to put them in the proper sequence. Whether the squirrel "knows" the correct order or learns it, however, is not known.

But although the ground squirrel has a series of complex innate motor programs, they are not classical FAPs. The squirrel clearly adjusts its movements at least roughly to compensate for the contingencies of the situation—the hardness of the soil, the nature of the available camouflaging materials, and so on. This means that although the motor program is prewired, it is designed to accept specific sorts of external information or feedback. Indeed, wherever the details of a behavioral situation are not wholly predictable, evolution has built into motor programs mechanisms which adjust key elements to fit behavior to need.

COMPLEXITY OF INNATE PROGRAMMING

Indeed, the world is full of examples of motor behavior so complex and apparently flexible that the ethological notion that genes might encode such a performance seems wholly out of

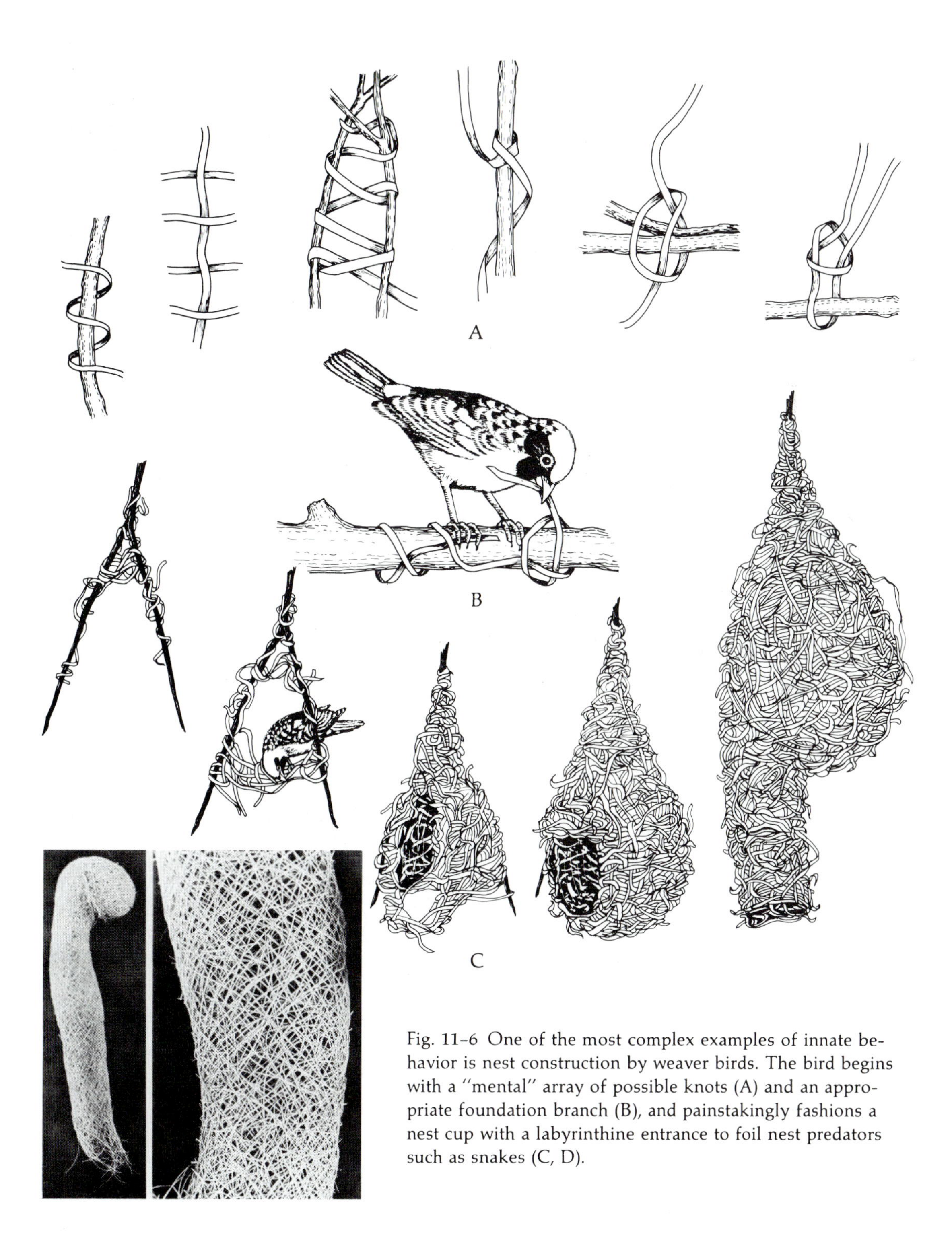

Fig. 11-6 One of the most complex examples of innate behavior is nest construction by weaver birds. The bird begins with a "mental" array of possible knots (A) and an appropriate foundation branch (B), and painstakingly fashions a nest cup with a labyrinthine entrance to foil nest predators such as snakes (C, D).

the question. Birds' nests, for instance, can be marvelously elaborate (Fig. 11-6), requiring accurate and painstaking weaving and "thoughtful" concern for location, balance, and design. Yet all this is innate. Birds build perfect nests in the absence of models or opportunities to learn, and only our limited imaginations prevent us from understanding the rules by which their genetic programming can accomplish these preordained feats. The key must lie in behavioral subroutines that can switch as construction proceeds and both predictable and unpredictable contingencies are encountered. A more convenient example for study than bird nests or beaver dams is that invertebrate architectural tour de force, the spider web.

Fig. 11-7 Orb webs require less silk per unit of catching area than any other design. This web consists of strong supports and radii over which the spiral of sticky catching threads is laid. The spider waits in the center, and determines the location of ensnared prey by detecting vibrations of the radii with her legs. The carefully reinforced opening in the center allows the spider access to both sides.

Each species of web-building spider, of course, builds the particular sort of web best suited to its hunting strategy. Some construct barrier webs which, like the wires running from the ground to the dirigibles over London in the Second World War, serve to knock down flying objects. The spider waits below to capture the hapless insects. Others construct platforms with trap doors out of which they spring to attack passersby. The aesthetic epitome, however, is the orb web (Fig. 11-7). Its construction is a classic example of rigidly programmed behavior which nevertheless has a flexibility—a carefully preordained flexibility—which automatically suits the operation of the program to the contingencies of the natural world. The spider begins her performance by producing a strand of silk whose

→

Fig. 11-8 Orb-web spiders start building their webs by (A) letting out a silken strand, which catches onto a branch, establishing a bridge (B). The spider may reinforce the bridge by traversing it several times, laying down more silk. On a thread attached to one of the bridge strands (C) the spider drops to a lower branch pulling the thread to form a Y (D), which creates the web's hub. A radius thread, attached to the hub, is carried up one leg of the Y (E) and down the branch, where it is tightened and fastened. Additional radii and frame-support threads are formed by the same procedure (F-I), and the spider may strengthen the hub with more threads once the radii are completed. Working from the center out (J), the spider lays down a temporary spiral—a scaffolding—spanning the radii. Then the spider, using viscid or hackled threads depending on the species, makes the permanent spiral (K-L), working from the edge of the web to the hub, while simultaneously rolling up (and later eating) the dry silk of the temporary scaffolding. As a final touch, some species may bite out the threads of the hub, creating easy access to both faces of the web, or add a "decorative" band of silk, called a stabilimentum, in the hub area.

end is carried out by the breeze (Fig. 11–8A). If it catches on something, the spider begins construction. Otherwise, she hauls in her fishing line and tries again.

Once she has an anchor, the spider runs back and forth across this silk bridge reinforcing it with additional strands (Fig. 11–

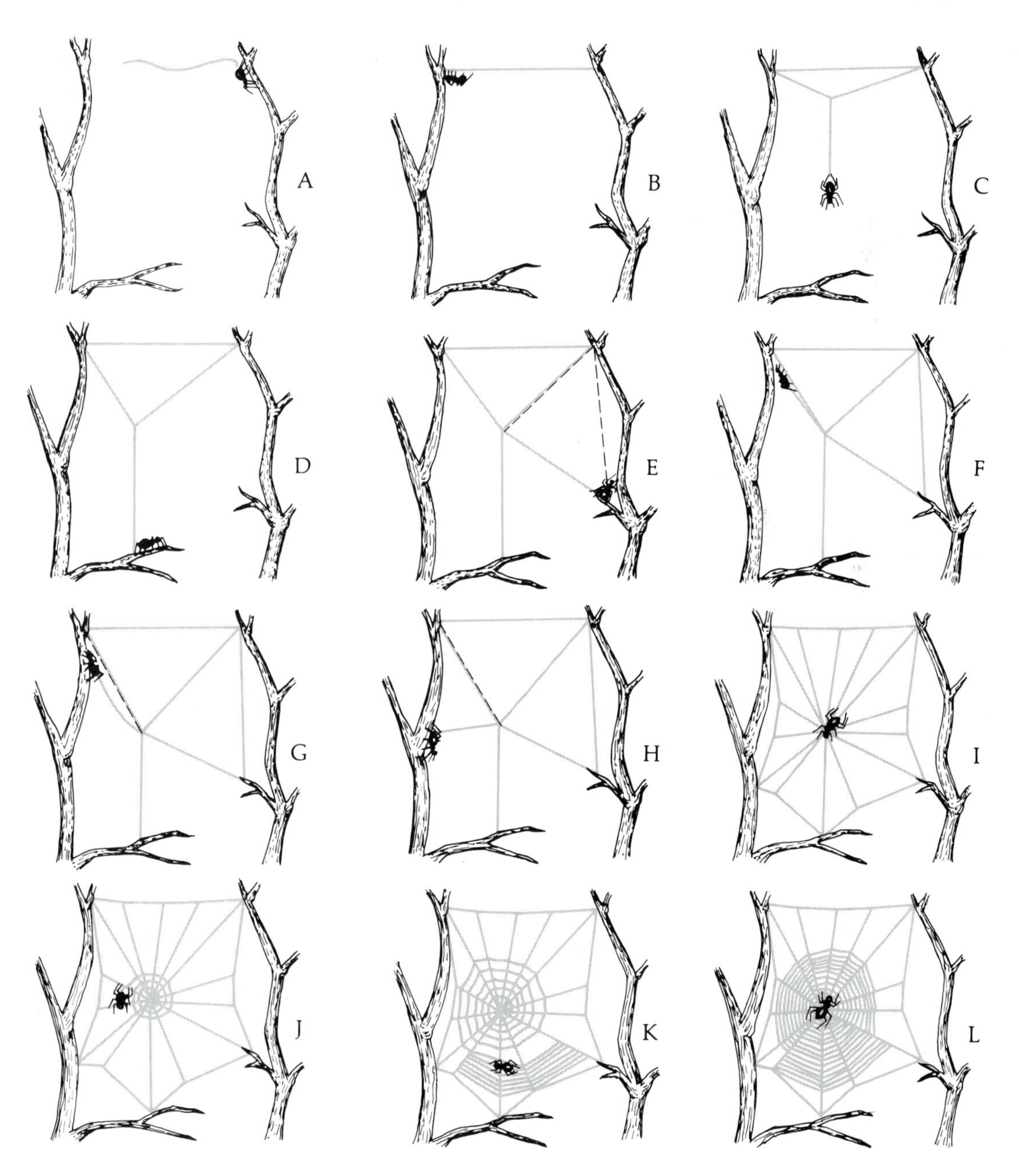

8B). Then she goes to the center, selects one strand (which stretches under the spider's weight), attaches another strand to it, spins out a new thread, and lowers herself to something solid (Fig. 11-8C). She then glues down this strand, one of the radii of the ultimate web (Fig. 11-8D), and back up it she goes to the center. At the center she again glues a thread, but now proceeds out one of the radii and carries the strand down to another support point (Fig. 11-8E). And so it goes, in to the center and back out until the framework is finished (Fig. 11-8F to 11-8I). Then the spider begins a spiral scaffolding from the center (Fig. 11-8J) to the edge, which she then uses for footing as she spins out a closely spaced, sticky spiral from the outside in (Fig. 11-8K to 11-8L), consuming the scaffolding as she goes. Once back in the center, she constructs a small reinforced circle connecting the radii, and then cuts the strands thus encircled to provide a hole, which permits her to move from one side of the web to the other. And now she waits for her prey. Armed with an innate processing system by which the output of her subgenual organs informs her of the location of prey, she can localize and remember the position of half a dozen dead *Drosophila* tossed into her net simultaneously. A battery of alternative motor programs for subduing (or, if necessary, freeing) the various insects she may catch provides her with the flexibility she needs to harvest her food. It is sobering when we look at the behavior of animals to realize that in all this motor complexity and adaptability there is no hint of learning.

MOTOR PROGRAMS AND LEARNING

Other examples of motor behavior really *do* look like learning. We talk, for example, of infants "learning" to crawl and walk because it is clear that they are learning from their seemingly endless series of mistakes. It is curious, however, that many other animals begin to walk at birth or as soon after as their muscles permit. Clearly, they come knowing how. Why not humans?

Observations of infants reveal that the process is not entirely without innate organization. From birth, children are programmed to move their limbs in alternation. A small baby can be "walked" from the beginning (Chapter 30). The first problem for the infant is somewhat more basic: getting into starting position for crawling. This takes literally weeks, but in most children the "drive" to crawl sustains their efforts until one of the movements tried, seemingly at random, seems to help. Suddenly this successful shot in the dark is seized upon and previous strategies, formerly tested over and over again, are abandoned as the new element is perfected. Slowly other successful elements are added and the process begins to accelerate. At length, and often in a matter of a few minutes, the last few elements fall into place and crawling begins (Fig. 11–9A).

The enormous self-congratulatory smile (Fig. 11–9B) on the occasion of the first success is one of several clues that the child innately *knows* how crawling should in some sense "feel." Apparently the child is provided with the answer but must find his own solution: an evolutionary strategy which will allow his "program" to work regardless of when during his first year's threefold change in weight he solves the problem. Once perfected, both crawling and walking programs must be recali-

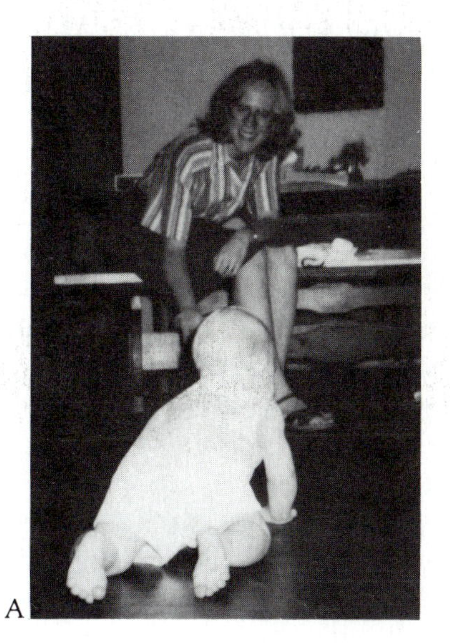

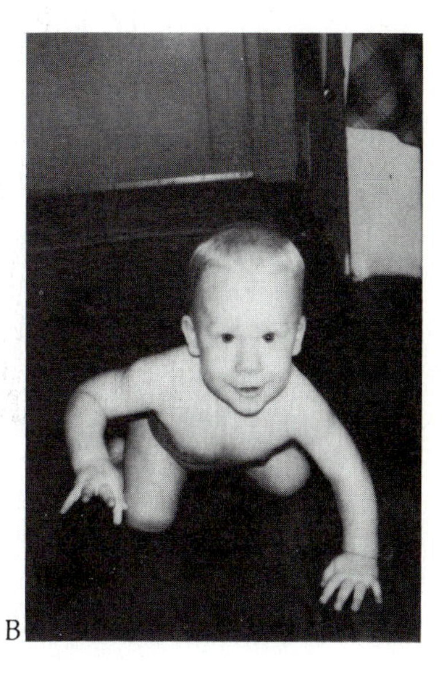

Fig. 11–9 After weeks of incompetent attempts, a human infant may perfect the crawling motion in a matter of minutes. A. The first successful attempt of this six-month-old infant began only seconds earlier, and he has already progressed about 3 m toward his goal (which proved to be not the proud parent in the background, but an electrical cord out of the picture to the right). B. The infant's smile throughout the whole process suggests that there is something inherently pleasing in his success.

brated with growth, a process whose painful inefficiency may be seen in the awkwardness of adolescence.

HARDWIRING LEARNING

What the programmed learning of crawling has in common with classical FAPs is the stereotyped and mindlessly automatic nature the behavior soon takes on. Who thinks about which muscles to move when during walking, running, or crawling? It is as though these coordinated muscle patterns have become "hardwired," incorporated into the brain just as firmly as any wholly innate motor program. Indeed, as we shall see in Chapter 17, this is the only interpretation which seems capable of explaining bird song learning. And it seems clear that even motor tasks learned without any apparent innate guidance also become hardwired with time and practice: most of us can, after the months of arduous learning as children, tie our shoes with our eyes closed or, as is so often necessary, scrawl notes in darkened lecture rooms.

Although chronically deaf children cannot learn to speak, those who become deaf after speech has begun are minimally affected. Apparently speech too becomes hardwired into the system, and with it the ability to generate whole units of speech, not just words, as essentially reflex actions. Short epigrams, verses, even considerable orations may be filed away and drawn upon without the intervention of conscious speech processing. An especially dramatic example of this hardwired aspect of speech involves the case of a longtime proctologist who suffered from aphasia, a brain lesion in the speech area which can destroy the ability to form or even to understand the simplest three-word sentences. At the mention of the patent medicine "Preparation H," however, the otherwise disabled physician would launch into a well-worn, grammatically perfect diatribe against the product which had apparently been stored in his brain as a motor tape.

Many of our familiar motor tasks follow this pattern: those happy few who type well, for instance, or play the piano can perform in the dark on imaginary keyboards without ever consciously choosing the keys. Beethoven's continued mastery throughout his encroaching deafness is a triumphant instance of how hardwired a motor program can become.

No matter how effectively we hardwire abilities like crawling, shoe tying, or piano playing, though, the systems must be flexible enough to accommodate themselves to changing circumstance—the size of the body, the material of the laces, the characteristics of the keyboard. That motor programs, whether innate or learned, may undergo calibration, may have something to tell us about play. Ethologists have always had difficulty with the concept of "play": we are in the position of the Supreme Court Justice who said of pornography, "I can't define it, but I know it when I see it." More than one ethologist has privately wondered whether animals may not play simply because they enjoy it. But "enjoyment," like "sweetness" and other pleasurable sensations, is defined by our genes, and is therefore beyond the pale of objective evaluation.

The notion that play is part of a calibration program with a temporary set of releasers which may bear no relation to later contexts is difficult to resist. Lion cubs must perfect their pounces before they are left to fend for themselves, and the distinction we like to make between their fetching practice on windblown leaves or their parents' tails, and actual attacks on small animals, is probably an artificial one. As we will see in Chapters 28 and 31, however, play in more social animals such as primates and humans seems to have taken on an increasingly large function in socialization, in learning the rules of the group into which we are born.

The emerging picture of motor behavior, then, whether it be the wholly innate performance of spiders, the goal-directed learning of infants, or the plastic learning we see in piano playing, is that all are routines stored, either from the outset or ultimately, as discrete neural programs. We can key these programs as needed, linking them into exceedingly complex chains of which all the details become wonderfully subconscious and machine-like. FAPs represent the purest case for study, unpolluted as they are by any substantial amount of learning or sensory feedback. As such, they are proving to be ideal candidates for neural mapping.

Although the challenge they represent is more formidable than that of a simple FAP, the more alloyed cases of motor programming, which depend partially on feedback, offer a greater opportunity for understanding behavior. They are clear examples of the complexity and flexibility evolution can derive from simple, wholly innate subroutines and neural switches. An understanding of how motor behaviors are orchestrated by the nervous system, genes, and evolution is crucial to the development of ethology.

Finally, even the cases of motor programs complicated by learning have a wealth of untapped potential, for here there is a generally unappreciated opportunity to study the neural basis of learning by finding and analyzing the sites of innately guided motor learning. Motor learning is not, as was previously thought, different in some basic way from the sort of learning which depends more exclusively on sensory input and "perception." Instead, it almost certainly involves the same cellular changes, takes place in relatively restricted areas of the brain, is easily elicited, and ought to produce firing patterns in the cells responsible which can be directly interpreted in terms of observable motor output. While the rhythmic, phase-locked firing of cells generating "learned" walking should be fairly obvious, who knows the neural "signature" a circle of pine cones engenders in a wasp? How do such circuits look and act before, during, and after the process of hardwiring? Out of such intellectual feats as motor program development may well come a better understanding of human and animal learning.

SUMMARY

The old notion that motor behavior is fashioned out of chains of reflexes choreographed by conditioning has succumbed to the realization that coordinated patterns of muscle movement exist which are entirely innate. Some of these prewired behavioral units or "motor programs" are relatively isolated from external feedback and so, once initiated, proceed inexorably to completion. Such "fixed-action patterns" are the most extreme class of motor programs. Many motor programs do not appear until the time in an animal's life when they are needed, and so were initially thought to involve learning rather than maturation.

The complexity and flexibility possible from motor programs continues to strain our comprehension; particularly difficult to grasp are the strategies by which the various innate elements of a complex behavior may be orchestrated without learning, and in the face of substantial environmental variability. Hand in hand with the realization of the power of innate motor programs has come an appreciation of learned motor routines which arise from an innately set goal (and feedback mechanism for recognizing progress), combined with the necessary motivational programming to direct an animal's attention and experimentation. Most often this self-programming approach makes liberal use of innate bits of motor units and, once success has been achieved, proceeds to so fix and hardwire the learned motor program that subsequent feedback is of little or no importance.

One of the most striking things about innately directed motor learning and the self-calibration of "play" is the total lack of any obvious external reinforcement. The absence of reward clearly distinguishes motor learning and play from the operant conditioning of behaviorism.

STUDY QUESTIONS

1. It is not clear whether the innate elements of burying behavior in squirrels is innately ordered or arranged by trial and error. How could you distinguish between the two?
2. Attempt to dissect the web building of spiders into an orderly set of subroutines organized by a master program with a set of recognition or contingency cues to signal when one or another of the component tasks has been accomplished. How could you now go about testing and refining your model?

FURTHER READING

Bentley, David, and Hoy, Ronald. "The Neurobiology of Cricket Song." *Scientific American* 231, no. 2 (1974): 34–44.

Feder, Howard. "Escape Responses in Marine Invertebrates." *Scientific American* 227, no. 1 (1971): 92–100.

Geschwind, Norman. "The Apraxias: Neural Mechanisms of Disorders of Learned Movement." *American Scientist* 63 (1975): 188–95.

Levi, Herbert W. "Orb-weaving Spiders and Their Webs." *American Scientist* 66 (1978): 734–42.

Pearson, Keir. "The Control of Walking." *Scientific American* 235, no. 2 (1976): 72–86.

Willows, A. O. D. "Giant Brain Cells in Molluscs." *Scientific American* 224, no. 2 (1971): 68–75.

CHAPTER 12

Motivation and Drive

"Motivation" and "drive" are terms which, for lack of anything more accurate, we use to refer to the apparent self-direction of an animal's attention and behavior. Animals may be little more than machines responding according to their genes to the stimuli around them, but it is clear that the level of their responsiveness and the precise nature of their responses can vary. A greylag goose, for example, will roll eggs (among other things) into its nest and remove broken eggs not only while incubating (when it makes sense to do so) but also during the week or so before the eggs are laid and for a week after they have hatched. Some species also inspect their eggs for cracks and discard those that are damaged, a behavior which disappears shortly before hatching for the very good reason that the otherwise mindless parents would discard eggs in the process of hatching. How is it that these birds know when to incubate, when to roll or to inspect and perhaps discard eggs, and when to ignore eggs altogether? Male redwing blackbirds, which feed peaceably with females and other males throughout the winter, are driven with the coming of spring to stake out territories, break into species-specific song, court every passing female, and attack anything which even vaguely resembles the red epaulets of a rival male. How do they come by this sudden change in personality?

CLOCKS AND DRIVE

Both of these examples hark back to the ring doves' "brooding clock" discussed in Chapter 4. A stimulus sets a timer which automatically alters the behavior after some preset interval. In the case of the doves, the sight of eggs in the nest began the sixteen-day countdown for the production of crop milk. And so in each case it is clear that an animal's innate programming is sufficiently elegant to switch behavioral "subroutines" on and off as appropriate. In many cases these relatively long-term changes in behavior are controlled by hormones. The territorial and courtship behavior of many (and probably all) birds, for example, is driven—switched on and modulated—by the androgen level in the bloodstream. Castrating the birds eliminates the behavior while implanting hormone pellets brings it on. Implantation will even cause females of some species to sing, court other females, and attack intruding males. (Females of other species seem to go through a hormonal critical period after which no reversal of sex-specific behavior is possible.) Of course, this simply moves the real question of behavioral switching back one square to what might be controlling hormone output. In the case of the blackbirds, the "unmoved mover" is daylight. Spring is marked by increasingly long days, and birds kept indoors can be brought into breeding condition if their lights are left on longer and longer, regardless of temperature and other potentially useful cues. Similar sign stimuli switch on the direction-specific migration programs or the hibernation preparations of many animals, while changing temperature is the cue which begins the physiological and behavioral changes which produce the courtship display of lizards. And who is to say that we humans are above this sort of influence? "Spring fever" may not be a scientifically documented phenomenon, but who doesn't know what we mean when we speak of it?

Other longer and shorter term behavioral changes depend on hormones. The most familiar example of the class of short-term behavioral changes are estrus cycles of mammals. Females of many species come into breeding condition once per cycle and for this brief period their behavior undergoes dramatic changes. In each case an endogenous free-running hormone cycle is determining behavior. Perhaps the shortest term examples are cases of rapid hormonal stimulation, the quick release of adrenalin as we prepare to be "angry," or the bursts of androgens in male redwings when they catch sight of other males.

The longest term hormonally mediated behavior change is associated with puberty, the changeover in an animal's life from juvenile to adult hormone levels. In songbirds this change switches on song production and causes sex-specific adult plumage to appear. In female mammals this same change initiates the estrus cycle and the development of species-specific sexual characteristics, while in human males, for example, it instigates the appearance of facial hair, a change in vocal pitch, and a wide variety of other very familiar changes in motivation and behavior well-known to man throughout the ages.

All these examples of relatively long-term alterations in behavior may be explained satisfactorily in terms of conventional programming. Neither estrus cycles nor the "drive" to fly south in the fall depend on learning, though the effects of early imprinting may now come to be expressed. Shorter term behavioral changes fall into much the same pattern. "Hunger," for example, is said to "motivate" an animal to look for food. From blowflies to humans, hunger is controlled by sensory devices monitoring either the levels of blood sugar or the amount of stretch on the walls of the stomach, or both, and it is these sensations which command an animal's attention. Again, many aspects of behavior are controlled by our internal clocks. There is a circadian rhythm to our desire for sleep, for example, over and above any real bodily demand, which seems to work much like an automatic timer throwing switches. Such examples fit neatly into the conventional view of the clockwork animal.

CIRCADIAN RHYTHMS

How animals decide *when* to do what they do remains one of behavior's central, most recalcitrant mysteries. Maintained under conditions of constant daylength, temperature, and humidity, European warblers will begin physiological and behavioral preparations for migration roughly a year after their last migratory preparations. Periodic cicadas will spend either thirteen or seventeen years in the dark isolation of the earth before emerging simultaneously (Chapter 21).

Minute intertidal midges will synchronize their times of emergence so they can find each other and mate before their two-hour life span elapses. The entire species elects 2 to 5 A.M. as the "correct" time of day, and they must coordinate this correct time with low tide in order to find a suitable place to deposit eggs. Both contingencies must coincide before the midges' behavioral ritual can begin.

Examples of this sort of temporal precision in even the simplest of animals are endless. In each case the "drive" to perform some behavior is clocked by a timer the period of which is innately specified, be it a tenth of a millisecond or ten years. But how do these timers, on which so much of behavior depends, actually work? Alas, we have many intriguing pieces of the puzzle, but no answers.

By far the commonest timers are *circadian,* or daily, and it is about them that we know the most. They are endogenous: they continue to run and the animal continues its daily patterns of behavior in the absence of external cues. Behaviors that show a circadian rhythmicity have a characteristic free-running period of about (but not exactly) twenty-four hours (Fig. 12–1), and can be reset by species-specific hierarchies of cues (*Zeitgebers* or timegivers) including light, temperature, humidity, sound, odor, pressure, and so on. A clock's susceptibil-

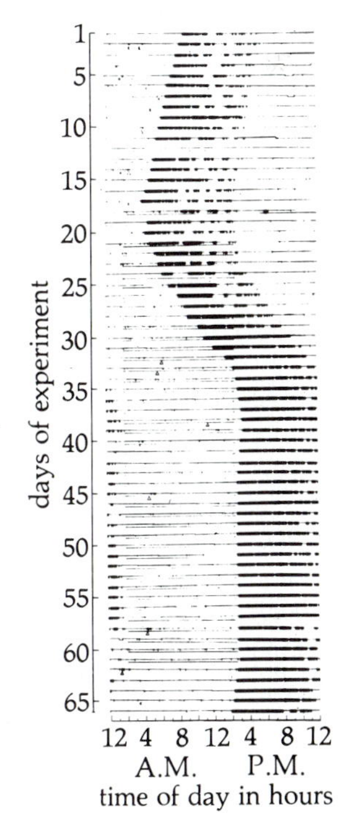

Fig. 12–1 The activity cycle of a flying squirrel, *Glaucomys,* drifts in constant darkness (days 1 to 17). When an artificial daylight schedule (double line, days 21 to 61) is imposed, the activity rhythm gradually comes into phase with it. When continuous darkness is resumed, the cycle again begins to drift.

ity to resetting has a time course of its own, peculiar to each sensory pathway. A flash of light, for example, may be taken as a sign stimulus for dawn, and if administered during the six hours before an animal's expected dawn it will advance the clock. In the dark *after* dawn *should* have occurred, a light flash delays the clock, and hence the creature's expectation of the next dawn will be delayed. These changes in the clock, however, in no way affect the myriad of trip switches which key each particular behavior at its appointed time.

In trying to get at the clock itself, scientists have found it remarkably well buffered against external conditions. A variety of toxic, life-threatening chemicals have no effect on the clock's function, while D_2O (heavy water), for reasons no one quite comprehends, slows it considerably. Temperature, too, has relatively little effect on biological clocks. In this case the system's compensation for the vagaries of the environment is accomplished actively: while most biochemical processes are dramatically slowed by a drop in temperature, most clocks actually speed up very slightly as the temperature drops.

To complicate things further, biological clocks seem rife with backups, not just for *Zeitgebers,* but for the clocks themselves. In *Acetabularia,* a large and beautiful single-celled alga, the nucleus holds the clock: transferring a nucleus on New York time into an enucleated cell on California time results in an East Coast cell. And yet the various cellular rhythms run perfectly well in the absence of a nucleus, and cells without nuclei can even be phase-shifted to a new dawn. Clearly the cytoplasm contains one or more backup clocks which are in nature chemically overridden or synchronized by the one in the nucleus.

These two patterns—multiple clocks and chemical communication of clock information—are ubiquitous. For example, the isolated brain of a moth pupa can be exposed to an artificial dawn and then implanted in the abdomen of a headless host pupa on a different schedule. The second moth will emerge at precisely the right time of day for the disconnected brain floating its abdomen. The complex behavioral programs that result in emergence must be generated outside the brain, but are released only by a well-timed surge of the appropriate hormone from the brain.

Honey bees have at least two clocks which can be experimentally disentangled: bees learn what time of day particular flowers are in bloom and visit them only during the appropriate hours (Chapter 17). This would appear to be the work of one simple timing mechanism, but if a group of 10 A.M. foragers are knocked out with CO_2 for four hours, the next day will see them at the feeder they have been trained to at both 10 A.M. and 2 P.M. The CO_2 treatment has stopped one clock, but not the other.

And yet, despite the undeniable importance of circadian rhythms in initiating and synchronizing behavior and in modulating drives, attempts to discover the nature of the clock at the whole-animal, cellular, and molecular levels have failed. In the end, it will probably be the determined efforts of behavioral geneticists (Chapter 20) which will disclose the springs and gears of the clockwork.

HOW DOES MOTIVATION WORK?

The reductionistic drive of scientists has resulted in many attempts to explain "motivation" as a unitary process. One early model, for example, postulated that motivation was like an appetite, causing an animal to satisfy its particular "hunger" by seeking out the appropriate stimuli and, presumably, satiating the appetite. Intuitively, however, it does not seem to be the rule that animals seek specific stimulations, or try to "fill up" on them. Geese do not seem to get "hungry" for eggs to roll. On the other hand, their response does resemble the waning of an appetite, as though a motivation specific to egg rolling has been affected. For example, after repeated encounters with prodigal eggs, a goose will begin to take longer and longer to roll them back in, and will finally ignore them altogether. Similarly, a male spider trying to approach a female on her web without being eaten—at least until he mates with her—will tap the web repeatedly until he exhausts her prey-capturing response before he ventures close enough to begin the courtship ritual.

Such examples are endless, and represent a sort of behavioral "habituation" in which some central processor has become dulled by repeated stimulation. (Habituation, as discussed in Chapter 5, is easily distinguished from sensory or motor fatigue.) So far, so good. But to the keen observer—Lorenz, for example—there is something more going on than mere appetite or simple habituation. Lorenz noticed three things in his observations of birds which characterized motivation. First, the "vigor" of a behavior may decrease after a few performances, but since full-blown, energetic motor programs using the same muscles may be elicited by other releasers, simple muscle fatigue cannot be held responsible. Second, the strength or quality of the releaser necessary to elicit a motor program may rise with repeated stimulation. Hence, although a beer can may excite the first egg-rolling response or two in a series, increasingly egg-like objects become necessary to trigger the response repeatedly. This could be interpreted as habituation of a high-level feature detector. Finally, and most bizarre of all, Lorenz noticed that for many behaviors the strength of the stimulus necessary to release a behavior in the continued absence of appropriate stimuli declined with the passage of time until almost anything would do. Eventually, for some behaviors literally *nothing* was necessary, and the behavior would be performed spontaneously. For example, a well-fed domestic chicken will peck at food and ignore, say, pencil marks on a piece of paper. Denied food, however, it will begin to peck at the marks and, denied them, will eventually perform the same stereotyped, well-directed pecks at blank paper.

Such examples of "vacuum activity" led Lorenz to postulate a conceptual mechanism for explaining how motivation works. This famous "hydraulic" model was inspired by its real-world analogue, the European flush toilet (Fig. 12–2).

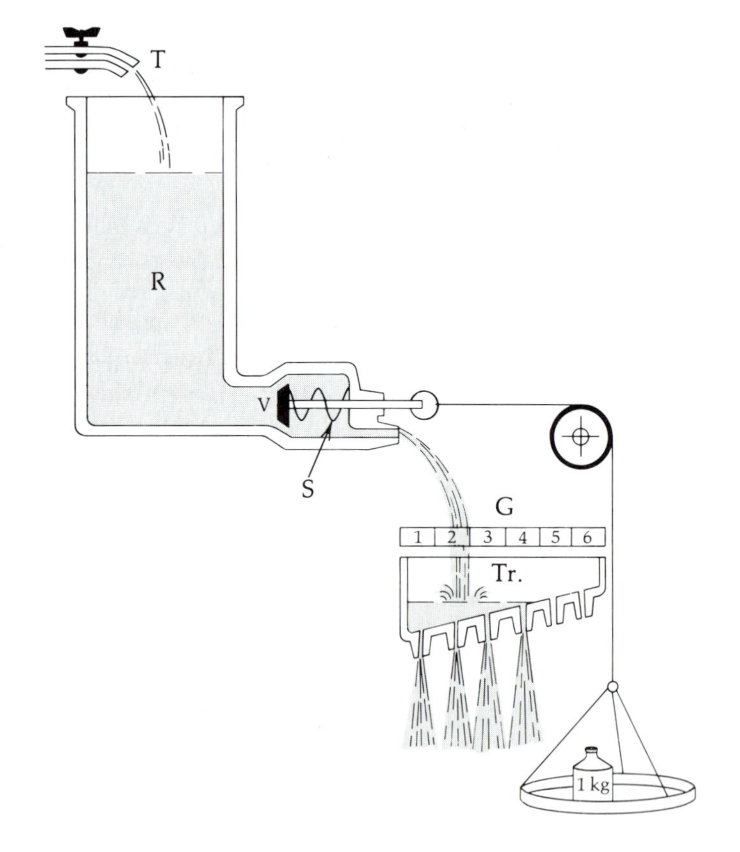

Fig. 12–2 In attempting to explain the periodic variations in behavioral motivation which he observed in his animals, Konrad Lorenz developed this "psycho-hydraulic" model. Hypothetical motivational waters flow in from a tap (T) into a reservoir (R) where they are stored. The water represents what Lorenz calls "action-specific energy." A releaser placed on the scale in the lower right pulls open the spring-equipped valve (V) so that the waters can "turn on" the behavior. The gradual nature of the behavior which makes the response stronger as the interval between performances increases is analogous to the graded water pressure which produces a stronger flow when the valve is opened. Lorenz illustrates this with the graded trough (Tr.). The lower threshold (smaller weight) necessary for releasing a behavior after prolonged abstinence is a consequence of increasing water pressure over the spring (S). "Vacuum" activities are a consequence of the water pressure's finally exceeding the spring tension, so that the system triggers itself internally.

Lorenz's intuitive model supposes that there exists a reservoir of motivation which is, at least when the program has that subroutine turned on (and a handle was added to the tap to indicate this), slowly filling, and as a consequence increasing the pressure on the outlet valve. Releasers (represented by weights placed temporarily on the pan) cause the valve to open, and some quantity of motivation is lost through the valve. The motivational "pressure" represented by the level of the reservoir has two consequences. First, the pressure determines the force on the spring-loaded valve, so that as the pressure builds, less energy—a weaker releaser—becomes necessary to trigger the mechanism. If the spring is weak enough or the reservoir large enough, the system will eventually release itself in the manner of a vacuum activity. The fullness of the reservoir also determines the force with which the motivational waters leave the valve and hence, according to Lorenz, the vigor of the behavioral response (represented by the graded trough).

Any model for such a fuzzy concept as motivation is going to be challenged, and the explicit similarity of the hydraulic model to toilets rendered it particularly inviting to attackers. The notion that satisfying a need was equivalent to flushing a toilet, however intuitively appealing, provoked the insulted scorn of behaviorists. The literature is now full of counterexamples devised to discredit Lorenz's unaesthetic hypothesis. And yet it is surprising how well the model fits much of the behavior we see around us. There is a clear buildup of "nervous energy" or "frustration" in many animals when a particular behavior remains untriggered for long periods. Most of us have seen bored domestic cats perform mock prey chases on bits of string and "disembowel" balls or toy mice. Similarly, nervous energy builds up when a behavior is actively repressed, as when an animal is caught between the desire to attack and fear of the consequences. The strength of the pent-up frustration is dramatically illustrated by the vigor with which the behavior, or some irrelevant "displacement" activity, is finally performed. Evenly matched male gulls facing off over a territorial boundary will "take their aggressions out" violently on nearby tufts of grass rather than risk injury from a fight.

But to make Lorenz's model work for a wider range of behavior, a series of increasingly complicated adumbrations become necessary. Some valves must leak, some input taps must be connected to floats in the reservoir which progressively shut off the flow of motivation as the reservoir fills, negative pressure on the other pan must be possible, and so on. Worse yet, feeding behavior, the archetypal exemplar of the appetitive model, just does not seem to fit Lorenz's hypothesis. Motivation—hunger in this case—is controlled internally, so that overt feeding behavior need never be performed. Something, perhaps the stretch receptors, must be imagined to be controlling another valve, releasing the accumulated motivational waters without any associated behavioral output. So, too, it is for "thirst." Cells in the hypothalamus monitor the ion level in the blood and make us feel thirsty when water is needed to thin the body fluids. Injection of salt near these cells makes an animal immediately and desperately thirsty, and injections of pure water just as quickly cause the thirst to vanish. No behavior, such as drinking, is required to satisfy the apparent need in this case.

The simplicity of the actual feeding and drinking systems stands in marked contrast to the elaborate jerryrigging needed for the hydraulic analogue, and argues that "motivation" must not be a simple single process. Nevertheless, the wide applicability of Lorenz's reservoir model to one class of behaviors not explained by any hunger hypothesis, and the successful application of a model based on a straightforward internal servocontrolled system with the well-known sensory monitors of hunger and thirst to others, offers hope that evolution may have developed only two major strategies for switching and modulating behavior within animals. The problem of motivation needs to be attacked with the same sort of elegant physiological analysis which revealed how blowfly wiring controls feeding, directed this time at behavior with Lorenzian characteristics. For all we know, the "reservoir" nature of motivation will turn out to be a relatively simple but previously undreamed-of elaboration of the "feedback" circuitry seen in feeding.

Any ethological analysis of motivation presupposes that an animal's behavior, its moods and predispositions included, is the ultimate result of its genetic programming. No one has to tell a hummingbird when to stop gathering the nectar she eats and begin the arduous task of catching the insects her newly hatched babies will need to grow. When we think about our own behavior, however, we are reluctant to apply this same supposition. We imagine ourselves to be different; we fancy that we control and even generate our own desires. However attractive this may be, the distinction is difficult to maintain when we look more objectively at the behavior either of our own species or that of other higher mammals. Our own species' blind, illogical behavior in areas as diverse as courtship rituals and feeding habits argues that subconscious, innate programming, once adaptive to a life vastly different from the one we live now, is at work; whereas the clear displays of some amount of conscious "intention" on the part of mere housepets suggests that even some nonhuman motivation may have components not entirely genetic. The evidence suggesting that other animals may be ever so slightly more than just machines will be taken up (with some trepidation) in Chapter 18.

SUMMARY

Motivation and drive are basically short- and long-term versions of the same general class of phenomena: the ability of animals to switch their behavioral priorities. In classic behavioristic terms, motivation is basically a recurring "hunger" to be satisfied, and as such is used as the reinforcement to shape behavior. Lorenz saw something more in motivation then mere hunger and thirst, however. To him, the building hunger seemed to be actually a lowering releaser threshold, and the behavior itself, rather than what it accomplished, was the reward—as though the *act of eating* rather than food in the stomach would satisfy hunger. Clearly the motivational process must come in at least two forms. As a longer term force, drive serves to turn on and off or modulate behavioral programs, and as a consequence an animal's responsiveness to particular stimuli. The switching generally involves releasers, and is controlled by clocks. Where physiological changes are involved, specific hormones often serve to initiate preparations for a behavior, and maintain the particular behavioral state. Motivation and drive, then, serve as the key elements in selecting and orchestrating the behavior of animals.

STUDY QUESTIONS

1. To what extent could habituation be used to explain Lorenz's observations, assuming virtually all behaviors to be partially habituated? What testable prediction would such a model generate?
2. What, if anything, does Lorenz's flush-toilet model have in common with the properties of a nerve cell? Can you specify a cell—an IRM if you like—with a set of inputs and outputs which would mimic Lorenz's model? Could you test for such a cell?
3. Shore birds exhibit both a circadian rhythm and a 25-hour tidal rhythm. Both can be demonstrated under laboratory conditions. Though the circadian rhythm is taken as obviously endogenous, there

is a widespread assumption that the tidal rhythm is learned. The evidence for this belief is that the phase of the rhythm with respect to the natural day is ever changing, and the phase also differs geographically. Is there a flaw in this logic? How could you test the notion that shore birds, unlike the marine species they feed upon, have to (and so are able to) learn a regular but noncircadian rhythm?

FURTHER READING

Beach, Frank A. "Behavioral Endocrinology: An Emerging Discipline." *American Scientist* 63 (1975): 178–87.

Crewes, David. "The Annotated Anole: Studies on the Control of Lizard Reproduction." *American Scientist* 65 (1977): 428–34.

———. "Hormonal Control of Behavior in a Lizard." *Scientific American* 241, no. 2 (1979): 180–87.

Davidson, Eric. "Hormones and Genes." *Scientific American* 212, no. 6 (1965): 36–45.

McEwen, Bruce. "Interactions Between Hormones and Nerve Tissue." *Scientific American* 235, no. 1 (1976): 48–58.

Meier, Albert H. "Daily Hormone Rhythms in the White-Throated Sparrow." *American Scientist* 61 (1973): 184–87.

Menaker, Michael. "Nonvisual Light Reception." *Scientific American* 226, no. 3 (1972): 22–29.

Nathanson, J. A., and Greengard, Paul. " 'Second Messengers' in the Brain." *Scientific American* 237, no. 2 (1977): 108–19.

Palmer, J. D. "Biological Clocks of the Tidal Zone." *Scientific American* 232, no. 2 (1975): 70–79.

Penegelly, E. T., and Asmundson, S. J. "Annual Biological Clocks." *Scientific American* 224, no. 4 (1971): 72–79.

Saunders, D. S. "The Biological Clock of Insects." *Scientific American* 234, no. 2 (1976): 114–21.

Truman, James W. "How Moths 'Turn On': A Study of the Action of Hormones on the Nervous System." *American Scientist* 61 (1973): 700–6.

von Holst, Eric, and von Saint Paul, Ursula. "Electrically Controlled Behavior." *Scientific American* 206, no. 3 (1962): 50–59.

CHAPTER 13

Invertebrate Orientation and Navigation

"SIMPLE" ORIENTATION

Virtually all animals have to orient themselves in the world to find food, or mates, or their homes. Until perhaps the late 1940s most scientists assumed that animals possessed only simple taxes and kineses—"reflex" orientations like those by which bacteria and planaria, for example, are programmed to move in response to light, or temperature, or whatever. A familiar example of such allegedly simple orientation is the system of olfactory "homing" in insects. If you are a bee looking for a flower, a moth looking for a mate, a wasp looking for a bee, you have only to fly around until you smell the appropriate odor and then fly upwind (a behavior known as "positive anemotaxis": *anemo,* wind, and *taxis,* response to; Fig. 13–1). The lesson of the last three decades of work on invertebrate orientation, however, is that insects are far more complicated than anyone had dared to imagine. Even such an obvious and simple example of orientation as anemotaxis, for example, can have its complications.

The unrecognized stumbling block here is that measuring the wind direction while standing on terra firma is not at all like

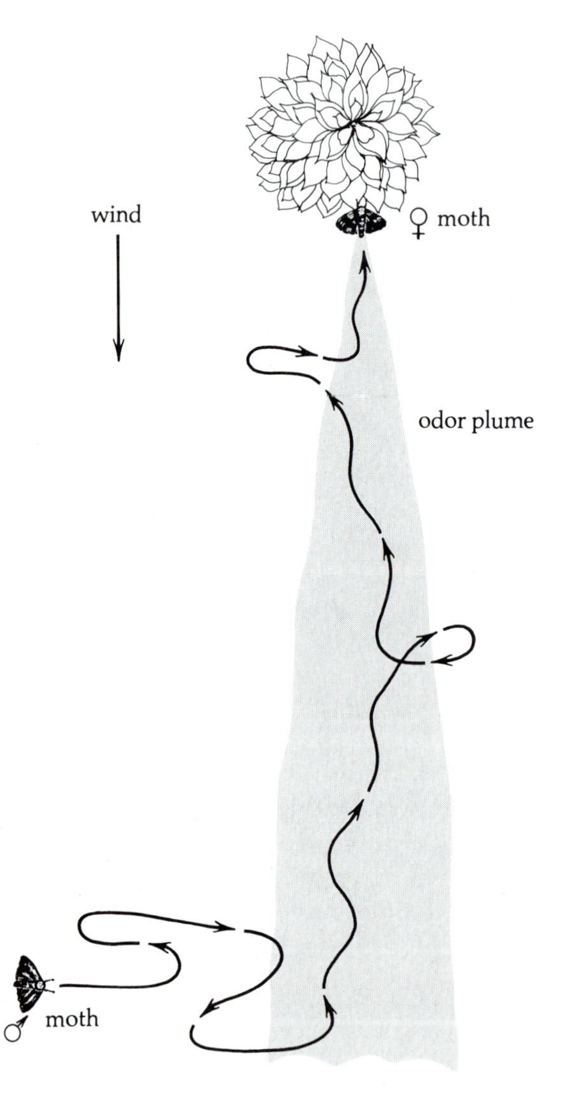

Fig. 13-1 Male moths find females by their specific odor, or pheromone. The pheromone is blown downwind in a plume, and males search for these trails by casting back and forth across the wind. When they detect the olfactory sign stimulus, the males fly upwind to the source.

making the same determination in flight. So strong are our terrestrial biases that many otherwise intelligent humans simply cannot imagine the difference, or even be convinced that one exists. Nevertheless, if you stand in a deep stream you feel the water rush by, but when you allow yourself to float any sense of movement is quickly lost as the water accelerates you to its speed. Only by looking at the shore can you tell which way you are moving. So it is for flying animals: only by looking at the ground are they able to determine wind direction and, naturally enough as we saw in Chapter 7, direction-specific movement detectors do exist to deal with this relationship. The trouble with this explanation, of course, is that many animals—moths, mosquitoes, and migrating birds, for example—regularly orient to the wind *at night,* when no point of reference should be easily accessible.

So although evidence is beginning to suggest that insects and birds may well be endowed with some mysterious technique for judging wind direction, they have until quite recently been thought of as capable only of relatively uncomplicated *orientation,* rather than the more intricate reckonings involved in *navigation.*

SOLAR ORIENTATION

Tinbergen's wasps pose the dilemma faced by such overly reductionistic theories elegantly. These solitary insects often inhabit a desolate and featureless environment, and are often off hunting for half an hour or more. Tinbergen thought they had memorized the sun's direction on the flight out and then performed some sort of constant-angle phototaxis to get back home. The difficulty here is that during the period of hunting the sun's azimuth can change significantly (Fig. 13-2). A wasp flying 1 km to hunt near noon in the early summer and flying back along some remembered angle to the sun would miss her nest by 100 m. And yet, as Tinbergen found (Chapter 3) but

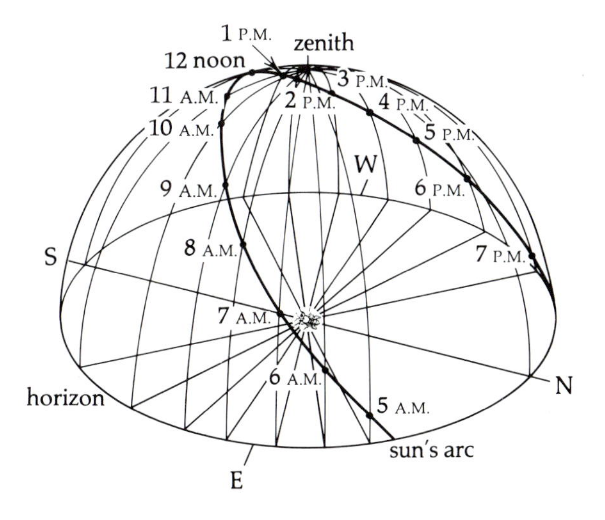

Fig. 13-2 To an animal on the earth, the sun appears to move east to west at 15° per hour along a great circle on the celestial hemisphere. The sun's azimuth direction, however, changes much less regularly, moving fastest near noon and slowest around dawn and dusk. In this example—summer solstice in Princeton, N.J.—the noon rate is 44° per hour while the dawn movement is only 9° per hour.

could not explain, the wasp is sure of her nest's position. On her return she will be able to locate an experimental array of pine cones placed around her nest only if, during her absence, it has been moved no more than 1–2 m.

Von Frisch found virtually the same phenomenon with his honey bees, but in their case the evidence is even more convincing. Here the path of a searching forager can be highly irregular, hardly keeping a constant angle to the sun; and yet a bee, once it finds a food source, departs directly toward home. Von Frisch's students showed that the bees were actually using the complexities of true sun navigation by training them to a feeder and then moving the feeder around the hive as some bees were feeding. The departing bees flew off in the direction which would have been appropriate to the old location, thereby ruling out the possibility that the bees were following some elementary, mysterious homing gradient (Fig. 13–3). Furthermore, if the bees were trapped at the feeder for a time, so that the sun's movement was appreciable, the bees flew off in the correct compass direction, adopting, in consequence, a significantly different solar angle from the one they had used on the flight out.

The most convincing evidence for this sort of sun-compensated orientation comes from von Frisch's studies of the re-

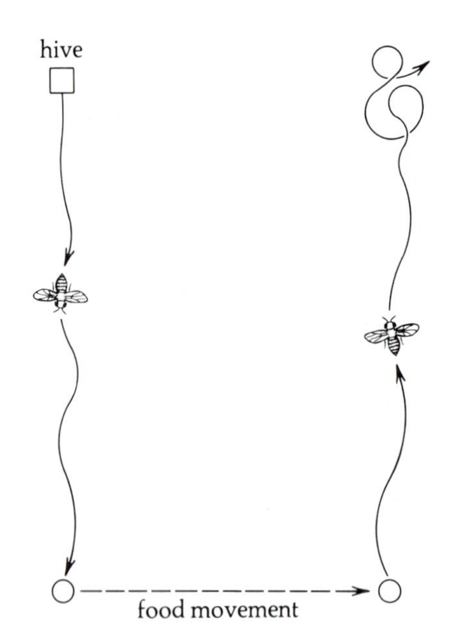

Fig. 13–3 When bees are displaced while feeding on an artificial food source, they head off in the direction which was appropriate toward home on the basis of their flight out. Clearly they are ignoring or not seeing their home. Animals with a true homing sense, on the other hand, would not be fooled by this maneuver, and would depart directly for home. This "map' sense which indicates true homing is discussed in Chapter 14.

SUN COMPENSATION

Given our limited sensory apparatus and our restricted, human-centered imaginations, one of the hardest problems in science is figuring out how it is that animals can do what they do. Once we have stretched our minds and imagined a plausible mechanism, we must reach still further for the means of verification: how can we design experiments which will tell us if what we have imagined is in fact true?

One of the most intriguing of these mental puzzles is deciphering how on earth bees manage to make room in their communication system for the inexorable, yet variable, movement of the sun. When reëmerging from the darkness of the hive the forager bee faces the task of recalling in which direction the food she has found lies. The sun, which acts as her compass, has moved since her last trip a few minutes earlier, necessitating some adjustment of whatever navigation system she uses. How does she manage to correct her picture of the world? Three possibilities come to mind: the bees could compensate exactly if they could somehow magically know the sun's arc and perform some spherical geometry; or they might calculate and use the sun's average speed of 15° an hour—a fair guess, since the error over the few minutes spent unloading and dancing in the hive would not be large; or they might simply extrapolate from the most recent rate of movement they have seen, a strategy of intermediate accuracy. (Actually, there is a fourth alternative which would yield nearly perfect accuracy: the forager bees could simply remember the azimuth between the sun and food from the same time on a previous day. In fact, foragers will use this trick with a very familiar source, but it is of no use under most natural circumstances.)

These three possibilities will diverge the most when the sun is moving its fastest: near noon on the summer solstice (Fig. 13–2). The sun's movement is plotted precisely in Figure 13–4, as well as the consequences of compensation by these three methods, calculated for 11 A.M. and noon. Our test involves training a group of foragers to a feeding station and then, after they have been foraging for a few hours, closing the hive. Recreating the hypothetical conditions of a sudden summer storm, we allow returning bees in but let none leave. During their incarceration we set out a broad array of feeding stations for them to choose among, and move the hive to another field so they cannot "cheat" and use landmarks for orientation (their usual strategy on these occasions). Then, two hours later, we open the hive and allow foragers to fly out to the food. The direction they now choose indicates their notion of how the sun has been moving. The results of an 11 A.M. hive closing are shown in Figure 13–5A. It is clear from their behavior that the bees have no precise idea of the sun's motion. Instead, they seem to have extrapolated its position from the 11 o'clock rate of movement. Looking more closely, however, their extrapolation seems to be about twenty minutes out of date. When the experiment is run with the hive closed at noon, the bees once more behave as if they have extrapolated, though

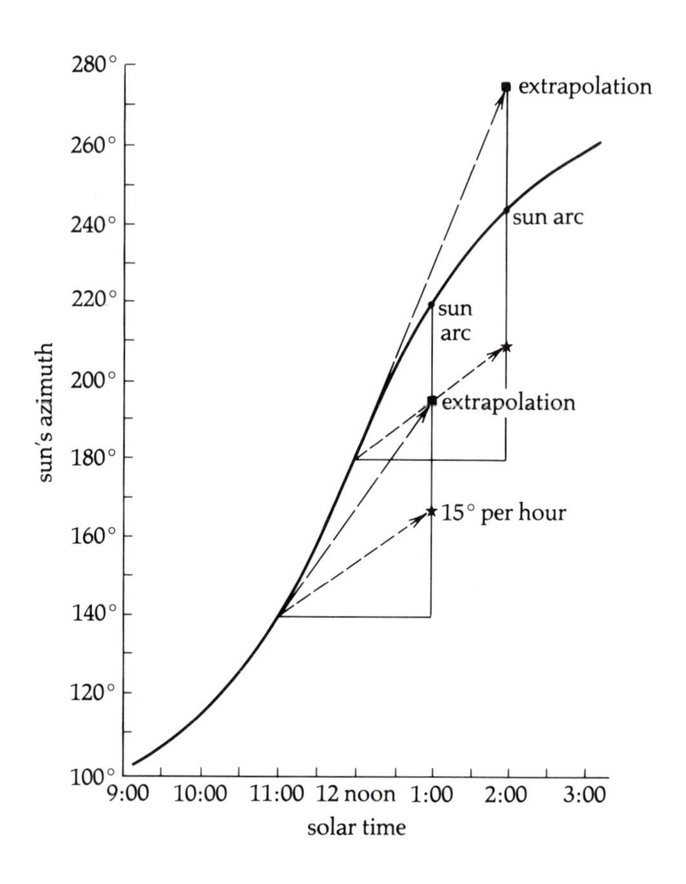

Fig. 13–4 The sun's azimuth direction changes most quickly around noon, as shown in this example from the summer solstice in Princeton, N.J. If bees are prevented from seeing the sun for two hours around noon they must compensate for its movement, and in an experimental situation any errors will be noticeable. Shown here are two closing times—11 A.M. and 12 noon—at which bees were trapped in their hives. For each closing time, predicted compensations according to three possible mechanisms are illustrated: a 15° per hour average, extrapolation of the rate of azimuth movement at closing, and true calculation by spherical geometry.

again it looks like they have used an out-of-date assessment of the sun's speed (Fig. 13–5B).

As so often happens, this systematic glitch in the data tells us more than we thought we were asking. It calls to mind some curious, long-forgotten results of Martin Lindauer's which have never been made to fit any previous hypothesis. Lindauer once tried moving a station every half hour by just the angle the sun had moved. His foragers found the station quickly after each move and began flying there directly, but in their communication dances in the hive (Chapter 24), they continued to indicate the former location. Only slowly, after several trips back and forth, did the dances begin to "catch up" with the correct location. When Lindauer moved a station concomitantly with the sun, the dances *never* caught up.

All three experiments—Lindauer's and ours—suggest that bees might be *averaging* navigation information—the sun's rate of movement, the relative azi-

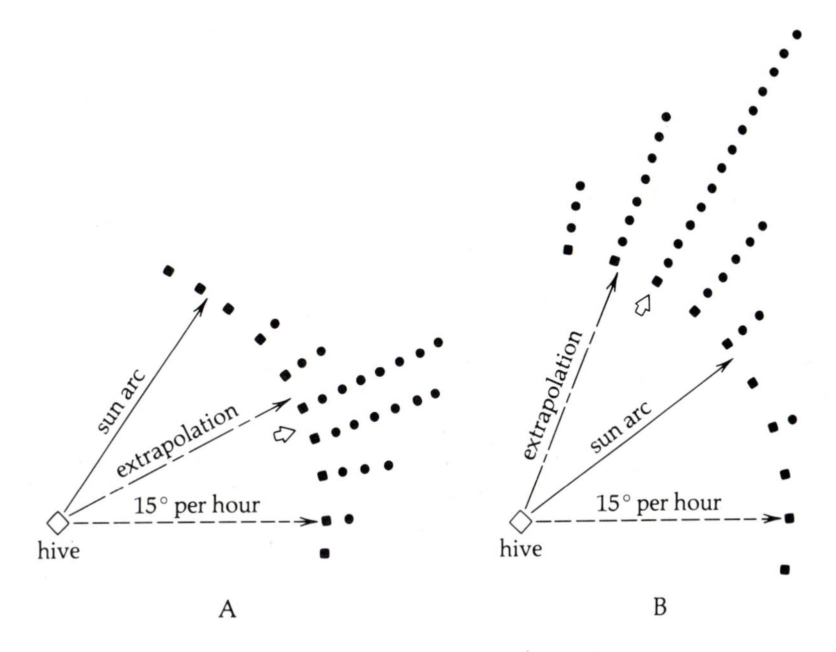

Fig. 13–5 When bees are forced to compensate for the sun's movement from 11 A.M. to 1 P.M. (A) or from 12 noon to 2 P.M. (B), they misjudge the sun's azimuth as if they had extrapolated its movement.

muths of sun and food—over about 40 minutes. Averaging all the information at their disposal would give the bees from the closed hives an extrapolation of the sun's position that would be about 20 minutes out of date, and their compensation for changes as abrupt as those in the experimental situations would be gradual, as they averaged in new data bit by bit.

That bees extrapolate the sun's motion makes sense: extrapolation is just as accurate as spherical geometry under normal circumstances, but far less intellectually demanding. But why should bees have a running-average strategy? Its effect could be to suppress mistakes and miscalculations by averaging an occasional wrong piece of data with many correct ones, and to smooth out the "noisy" or inherently inaccurate measurements that must be a necessary consequence of trying to use a compound eye. This running average would also substitute for missing data, as would be the case for our sun compensation experiments. The result is that bee navigation would be well buffered against the unpredictable vagaries of wind and sky that might otherwise destroy the clockwork precision of the bees' remarkable system of navigation and communication.

cruitment dances of bees (described in detail in Chapter 24). In these dances the bees signal the azimuth between the sun and the food. Without reemerging from the darkness of the hive, foragers which dance for prolonged periods shift their dance directions to compensate for the constantly changing angle between the sun and the food. Now we know that virtually all vertebrates and invertebrates that need to can perform the sun compensation necessary for true solar navigation.

Sun compensation may be a common ability, but it is by no means a simple intellectual task even for humans. Although the sun moves along its arc through the sky at 15° an hour, its azimuth can change much more slowly near dawn and sunset, and faster near noon (Fig. 13–2). The exact rate depends on the time, the exact latitude, and the date. The most dramatic rates of movement occur at the equinox along the equator when the sun passes directly overhead. The sun's azimuth is 90° (east) from dawn to noon as it rises straight up in the east, then suddenly is 270° (west) until sunset. Whether animals know about these variations, either by performing the necessary spherical geometry or by remembering the sun's behavior on the previous day, or merely guess, perhaps by extrapolating recent movement or by using the sun's 15° an hour average rate of movement, is not known for sure. Recent evidence suggests that bees, at least, compensate by extrapolation.

We also know that bees must *learn* that the sun moves. Learning is crucial here since the sun moves clockwise (left to right) or counterclockwise through the sky depending on the latitude (and in the tropics, the time of year). This learning appears not to be reversible in that older bees are unable to adjust to change, so it may have a "critical period." The ontogeny of this orientation ability remains largely unexplored.

POLARIZED-LIGHT ORIENTATION

So far so good. Insects and many other invertebrates use the sun as their compass, even to the point of being tricked by mirror images and lights, and are programmed to compensate for its preordained yet varying movements. As it happens, however, many insects go about their business perfectly well when the sun is hidden by clouds, landmarks, or experimentors. This puzzling ability was first discovered in certain tidal crustaceans which scurry toward safety—in this case the ocean—when disturbed. Their determination of the direction in which safety lay was clearly based on celestial cues, since animals transferred to beaches on the opposite sides of their island homes would flee in the same compass direction—a behavior which led them instead to land.

The experiments were difficult to understand. Blocking the animals' view of the sun and providing a mirror image of it in the opposite direction caused them to reverse direction abruptly, so it seemed clear that the sun must be the cue the animals found salient. However, when researchers blocked their view of the sun and provided no substitute at all, the animals were unaffected. The old notion that animals are simple and can do things in only one way made these results impossible to reconcile.

It was von Frisch and his bees, again, which helped unravel this mystery. He showed unambiguously that the navigation system of bees, like any crucial piece of equipment, had been built with backup systems to take over if the primary one—in their case the sun—should fail. The bees' backup system is the sun-centered pattern of polarized ultraviolet light in the sky (Fig. 13–6), a fact which was made clear by a useful manipulation of the ever-informative bee dances. When performed on a horizontal surface with a view of the sun or sky, these miniature reconstructions of the foraging journey point directly at the source of food. What the dancers see in this carefully controlled experimental situation, however, is determined by the researcher, and by varying and manipulating the parameters of possible cues, we can work out how bees (and, as it turns out, insects in general) perceive, process, and use navigational information. Von Frisch found, for example, that given a view of as little as 1 percent of the blue sky, the bees' dances point directly at the goal. Von Frisch began his search for the salient cues by blocking the dancers' view of the sky with various optical filters. To his surprise a UV-blocking filter which is perfectly transparent to human eyes destroyed dance orientation, while a UV-transmitting filter which cuts out light visible to us so that it looks like a sheet of black glass permits the dances to go on normally. At the suggestion of a colleague, von Frisch tried a polarizing filter: instantly the dances took up new orientations, and rotated as he rotated the filter. Clearly, bees were able to orient their dances by even a tiny patch of polar-

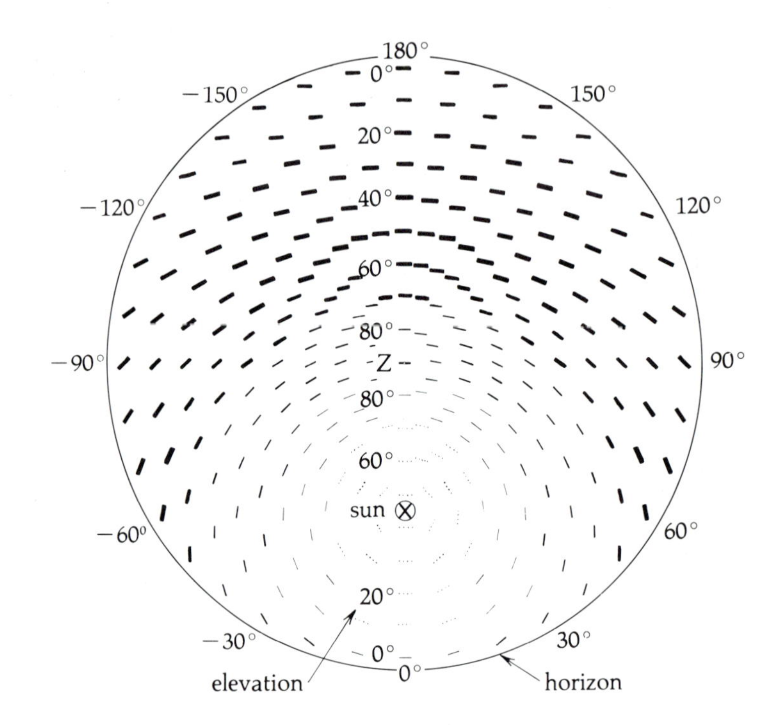

Fig. 13-6 The pattern of polarized light in the sky depends on the position of the sun, shown here in this "fish-eye" view at an elevation of 45°—halfway between the horizon and the zenith (Z). At any point in the sky, the polarization is perpendicular to the great circle connecting that point and the sun; and the degree of polarization (indicated here by the thickness of the lines) depends on the distance from the sun, being greatest 90° away.

ized UV light—two cues available to them to which we are blind.

The pattern of polarized light in the sky has an elegant geometry which is symmetrical about the sun. The pattern arises from the scattering of unpolarized light from the sun off the molecules in the atmosphere. The interaction (Fig. 13-7) is such that light scattered perpendicular to the direction of the incoming light is completely polarized, while light scattered at other angles is polarized less (such that the degree of polarization is roughly proportional to $\sin^2 \infty$). Outgoing polarization is perpendicular to the plane in which both the incoming and the outgoing light path lies (Fig. 13-7). As a result, the sun must lie on the great circle perpendicular to the polarization direction of any patch of sky (Fig. 13-8).

Although the spherical geometry of scattering creates a regular pattern, we have no idea whether bees and other invertebrates detect or use the perpendicularity rule. If they do, they still face the problem of determining at which of the two intersections with the sun's elevation the sun is actually to be found (Fig. 13-8). Alas, there is no evidence that any animal, verte-

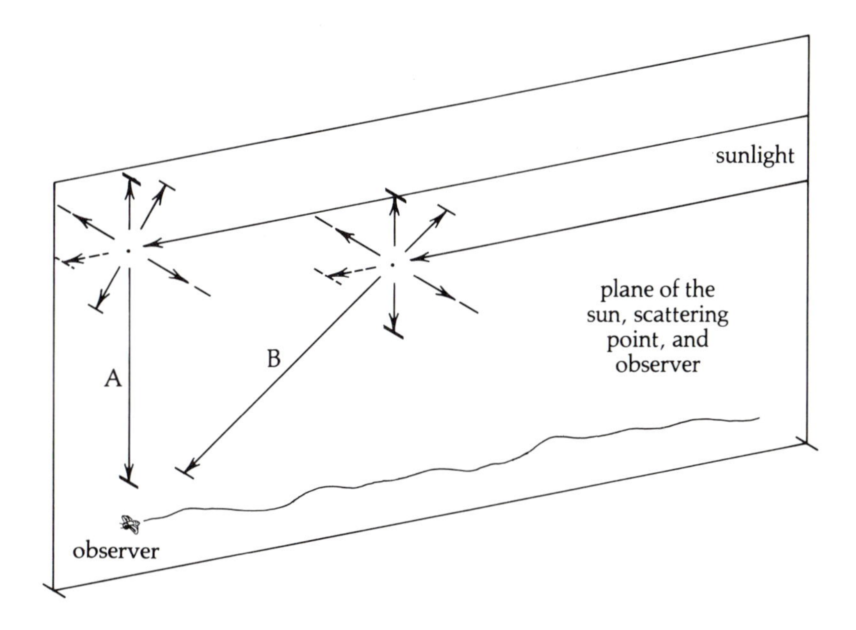

Fig. 13-7 The polarization patterns in the sky arise from the scattering of unpolarized sunlight off molecules in the atmosphere. The orientation of the polarization induced by the scattering is perpendicular to the plane of the sun, the scattering point, and the observer, as shown here. The degree of polarization depends on the angle between the incoming sunlight and the scattered light. In this illustration the sunlight is parallel to the ground, indicating that the sun is close to the horizon. Skylight reaching the bee from point A (overhead) was scattered 90° and so is highly polarized, while skylight from point B, having been scattered only 45°, is only mildly polarized.

brate or invertebrate, ever detects the sun's elevation, and even granting this ability, every satisfactory explanation of the remarkable ability of insects to navigate by polarized light is contradicted conclusively by at least one experiment. My own guess is that we are trapped in this dilemma by our assumption that there is only one strategy for analyzing polarized light, when the aesthetically unappealing possibility that there are two levels of polarized light analysis, one taking precedence over the other, would rescue any of the present theories.

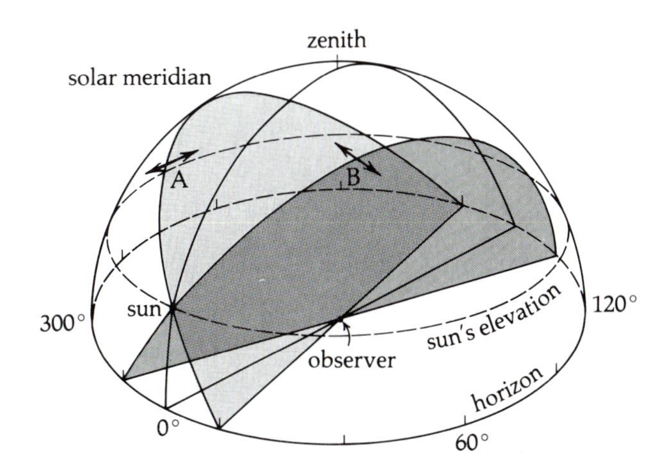

Fig. 13-8 Since the angle of polarization at any point in the sky is perpendicular to the plane of the sun, the point, and the observer, it defines the plane of the sun. In the example, the patterns at A and B each define a plane, and the planes intersect at the sun. If an animal can only see point B, for example, the sun's position is ambiguous. If the animal knows the sun's elevation, however, only two locations are possible. Which of the two is correct is another problem. [From "Polarized Light Navigation by Insects," by Rudiger Wehner. Copyright © 1976. Scientific American, Inc. All rights reserved.]

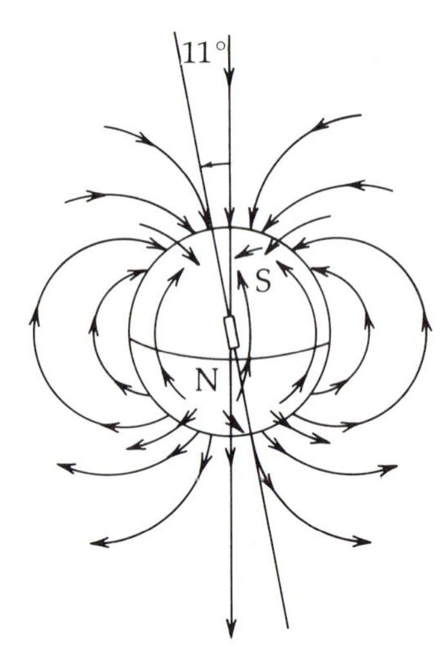

Fig. 13-9 The magnetic field of the earth is equivalent to the field which would be produced by a thousand-mile-long bar magnet oriented 11° from the earth's axis of rotation.

MAGNETIC FIELD ORIENTATION

If the ability of bees to make use of polarized light is perplexing, consider their behavior on cloudy days. Even when the bees are given a full view of the cloudy sky, their dances are disoriented. This makes sense: the clouds block both a view of their primary cue, the sun, and their backup system(s), the patterns of polarized light. And yet, as they must, bees and other insects navigate perfectly well to and from the food. In fact, if offered a light as an artificial sun, the bees perform perfectly oriented horizontal dances. In other words, *bees know where the sun is* even though they cannot see it. How can this be? What cues could their cloudy-day backup system use?

One possible backup compass could be the earth's magnetic field (Fig. 13-9). Although this idea is admittedly farfetched, there is increasingly strong evidence of magnetic field sensitivity in honey bees.

The experiment which has yielded the most convincing results involves forcing bees to perform their recruitment dances on a horizontal surface with no view of the sky. We would expect no consistent orientation to be possible. Nevertheless,

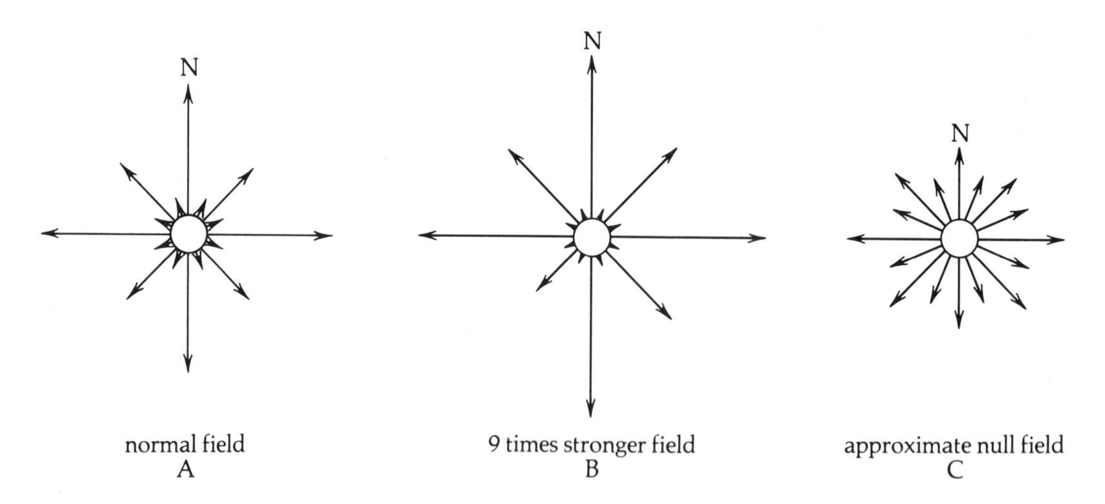

Fig. 13-10 When bees are forced to dance on a horizontal surface without visual cues, the dances become oriented to the eight points of the magnetic compass (A). When the magnetic field around the bees is artifically strengthened this magnetic orientation is enhanced (B), while when the field is cancelled the orientation vanishes (C).

after a few days of being forced to dance without celestial cues, the bees begin to orient their dances north, northeast, east, south, southwest, west, and northwest (Fig. 13–10). Cancelling the earth's field causes this curious reorientation to disappear, while rotating it causes the orientation to shift appropriately.

Almost all animals have daily rhythms of eating, sleeping, and so on, and there is increasing evidence that bees at least may use the minute daily patterns of change in the earth's field (Fig. 13–11) to calibrate their internal clocks in the absence of other cues. For bees the ability to know what time it is must be especially crucial, since many flowers produce nectar at particular times of day. But animal clocks, like cheap watches, inevitably run fast or slow (see Chapter 12). Animals overcome this difficulty by resetting their watches each day, generally to the time of dawn or dusk. If maintained in isolation from such temporal cues most animals begin drifting systematically out of phase with "true" time.

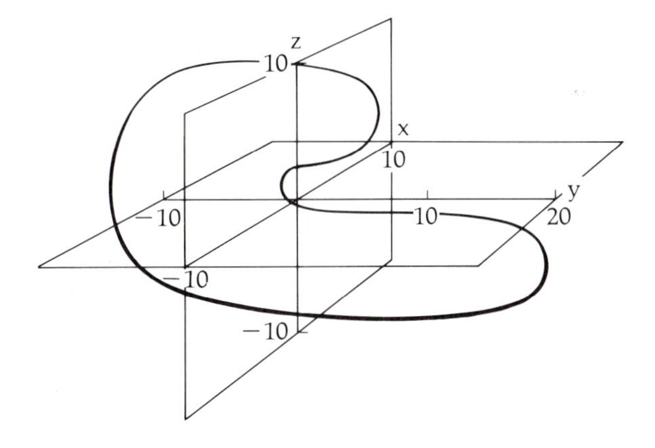

Fig. 13–11 The exact strength of the earth's magnetic field changes slightly each day in a regular way. Here, against a background of about 50,000 gammas, is the pattern at Würzburg, Germany. This minute variation is postulated to be the source of bees' exhibited ability to keep accurate time in the absence of other cues. During magnetic field storms the bees' time sense is disrupted.

Bees, however, are different. Living as they do in the perpetual darkness of their hives, their opportunity to observe dawn or dusk must be limited and undependable. When we place a hive in an indoor flight room under constant illumination, bees trained to visit a feeder at particular times will continue to check the feeder at the right time with no sign of the drift which characterizes most other animals deprived of normal time-setting cues. We might deduce from this that bees simply have very good clocks, but curious things begin to happen when we manipulate the magnetic field felt in the hive. In a field which is either abnormally strong or randomly varied the bees are no longer able to keep time. Either their "watches" are not anti-magnetic, so that some generalized disruption results, or there is circadian-rhythm information in the magnetic field strength.

The strength of the earth's magnetic field has a daily rhythm (Fig. 13–11) unique to each latitude and season. The field arises from two sources: the interior of the earth and the ionosphere. The interior of the earth has by far the stronger effect, yet it is the ionosphere which generates the daily variation. The jet streams carry ions around the earth and this flow of ions, like any charge movement, generates a magnetic field. The daily variation comes as the atmosphere is heated by the sun during the day and cools at night. This expansion and contraction of the atmosphere pushes the jet streams alternately north and south.

And yet the magnitude of these daily changes is very small—about one part in 500—and any system which detected them would have to be outrageously sensitive. Then, too, these minute changes must be completely obscured by the occasional and unpredictable solar flares which dump unusual numbers of ions into the atmosphere. One of the strongest points of evidence suggesting that bees do in fact sense the earth's field is that during such "magnetic storms" the time sense of bees in flight rooms seems to be abolished.

Now that the effects of magnetic fields on bees (as well as on some birds, bacteria, and fish) have been convincingly demonstrated, we must begin to ask the ethological "how" questions. This intellectual exercise has become the subject of lively controversy, and until recently opinion favored an induced-current model. This hypothetical system is based on Faraday's Law: moving a conductor through a magnetic field induces a current flow and, conversely, a current flow generates a magnetic field. So if animals came equipped with low-resistance conductors and highly sensitive ammeters, they could measure at least the direction of the field. Induction has indeed proved to be the basis of the magnetic field sensitivity of elasmobranchs such as sharks and rays. Their electric-field-detector organs act as conductors, and the animals' sensitivity is so great that when swimming no faster than 2 cm/sec they induce a detectable current flow. However, as we shall see in the next chapter, the discovery that bar magnets disrupt pigeon homing rules out this

theory in birds: the static field of a permanent magnet should not affect an induced-current detector which responds only to motion *through* a field.

Another alternative is permanent magnetism, which arises in special crystals, like the familiar lodestone (magnetite, Fe_3O_4), which are naturally magnetic. The permanent magnet of the "lodestone" alternative requires the animals to house little magnets which might attempt to line themselves up with the field. The torque generated in doing this might then be measured by the nervous system.

Several recent discoveries have allowed this theory, long in disrepute, to stage a comeback. For example, several species of bacteria are now known to synthesize a chain of magnetite crystals which steer them down the earth's field lines into the mud in which they live (Fig. 13–12). In the northern hemisphere, where the field lines point north and down, these chains have their north-seeking end toward the front, while in the southern hemisphere it is the south-seeking end which leads the way. Honey bees and pigeons, too, have about a hundred million magnets concentrated in specific tissues, but no conclusive evidence yet links them to the elusive detector.

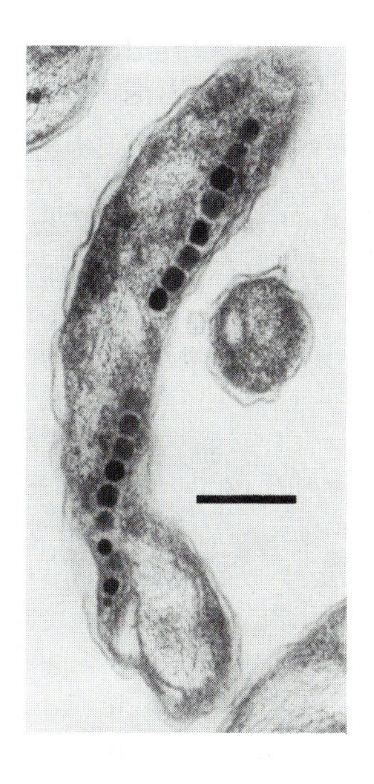

Fig. 13–12 Many species of bacteria regularly swim down. In some, the basis of this orientation is one or more long chains of magnetic crystals (dark in this electron micrograph) which passively rotate the bacterium into alignment with the earth's field lines. Since the field lines are nearly vertical (see Fig. 13–9), this takes north-seeking bacteria downward in the northern hemisphere.

MAGIC?

Regardless of what the detector(s) turns out to be when it is finally isolated, magnetic field sensitivity serves as another reminder of von Frisch's lesson that animals live in their own sensory worlds: worlds to which evolution has denied us admission. The history of research on magnetic sensitivity, the electric senses of fish, and the echolocation abilities of bats illustrates that, after our species-specific blindnesses, our greatest limitation has been our inability even to *imagine* the bizarre things animals might be doing to subsist in their unique environments. There is, in fact, every reason to suppose the existence of unknown sensory modalities, or at least unknown uses of known ones.

Honey bees and certain ants, for instance, make use of mysterious "traditional" mating areas. In the case of the bees, these special places are located 20–30 m above the ground, and occur at a density of about one per square kilometer. These "drone congregation areas" are quite restricted in size—perhaps 20 m in diameter—and are in the same places year after year. This is curious, since queens fly to these places to mate only once in their lives and no drones survive the winter, yet each year new drones and queens find the exact same spot. How do they do it? To locate these distinct but evasive areas we "fish" for drones with a queen who emits her alluring pheromone suspended from a helium-filled balloon. When she reaches the confines of a congregation area, she is suddenly mobbed by drones. The ground beneath may be an open field, or a wood, or even a pond. The terrain may be flat or steep, but drones from a hive brought in from hundreds of miles away find these places on the first or second day in their new home, and queens find them, mate several times, and return to the hive in 15–30 min. Are we missing an obvious cue, or are the bees using a part of the sensory environment to which we—and the imaginations of a generation of experimentors—are blind? Perhaps they are even using an exclusive yet-to-be-discovered sense, while we fish laboriously to discover a place they find so easily.

Through evolution, animals seem to have seized upon virtually every imaginable cue—and almost certainly several as yet unimagined ones—to construct prewired, self-calibrating strategies for orienting in their own species' sensory world. Since our evolution seems to have left us with little more than the ability to memorize landmarks, only modern technology in many cases can put us in touch with the real world. And where technology fails, only our imaginations can light the way.

SUMMARY

The ability to find a goal—home, or food, or a mate—is crucial to animals, and even the simplest strategy often involves precise information processing and sensory cues outside our own experience. In returning home after a circuitous search for food, bees and wasps must know the direction and compensate for the ever-changing rate of the sun's movement, allow for the changing effects of the wind, and integrate both distance and direction over the prior legs of the journey. Like any crucial system, the navigation programs of insects usually provide one or more backup routines to take over in the event the first one fails. Hence when the sun disappears most insects switch their navigation systems immediately to the patterns of polarized light in the sky, and when these indicators are hidden by clouds, to some third system which may use the earth's magnetic field or other cues of which we are not as yet aware. For ethologists there is a dual challenge in orientation and navigation: first, to find the sensory cues and discover their detection mechanisms; second, to decipher the programming, the strategies of information analysis which make it possible for animals to interpret and use the sensory information in the world about them in this well-defined context.

STUDY QUESTIONS

1. One of the major anomalies in the behavioral data on polarized-light orientation is the observation that although the horizontal dances of bees are poorly oriented on hazy days to most artificial patterns of polarized light, they are well oriented to horizontal patterns—as well, in fact, as to an artificial sun. What might be happening?

2. If insects really can orient on cloudy days, but like bees cannot see the sun directly, there must be a backup system. Three possibilities come to mind. First, insects might be extracting the sun's position or polarization out of the atmospheric "noise" by averaging: the quality of the signal would increase as the square root of the time. In the dance, there is simply not enough time to perceive celestial information. Second, insects might remember the sun's position on previous days relative to landmarks in the vicinity of home. Third, insects might use the earth's magnetic field as a compass, and remember the sun's position in magnetic coordinates. How could you use the bee dance to distinguish these three possibilities? Are there others?

3. Contrary to popular opinion, bees do not fly in bee-lines, but rather in a slow zigzag pattern, aiming themselves alternately left and right of their home course. Why might this be better than flying in a straight line?

FURTHER READING

Gould, James. "The Case for Magnetic Field Sensitivity in Birds and Bees." *American Scientist* 68 (1980): 256–67.

Kalmus, Hans. "Sun Navigation of Animals." *Scientific American* 191, no. 4 (1954): 74–78.

Wehner, Rudiger. "Polarized-Light Navigation by Insects." *Scientific American* 235, no. 1 (1976): 106–15.

CHAPTER 14

Vertebrate Navigation

With the discovery that mere insects can actually navigate came the realization that birds and other vertebrates ought to be capable of the same feats. After all, many birds and mammals migrate vast distances—vast at least compared to the typical 1-km home range of bees. Yet many long-distance migrants turn out to have systems for navigation which are, if anything, simpler than that of the bee. European warblers, for example, manage their 500-km odyssey from Europe southwest through Spain, across Gibraltar, and then southeast into their winter range in central and southern Africa (Fig. 14–1) by a prewired program which merely directs them to fly southwest for 40 days, and then southeast for 20–30 days. This migratory version of the triggered-clock strategy we saw in the ring dove feeding cycle (Chapter 4) may seem crude, but then Africa is a large target. So, too, many species of small songbirds are programmed to begin flying southeast from the North American coast out into the trackless Atlantic when a cold front passes by. If luck is with these millions of tiny birds which set off with each front, their prize is a winter in South America. If the winds fail them, they perish.

Fig. 14–1 In the fall, European warblers fly southwest through Spain and then southeast into Algeria. The directions, and the time at which to change from southwest to southeast, appear to be innate since birds isolated in the laboratory will orient in the proper directions for the appropriate lengths of time during the fall.

MAPS AND COMPASSES

Other vertebrates, however, perform more impressive feats and appear to use something beyond simple endurance to get from one place to a distant other. Many, in fact, are able to travel long distances to relatively small targets. Sea turtles reach tiny Ascension Island from 3000 km away seven years after having hatched there. Nesting birds displaced twice that distance return to their nests in only slightly more time than it would take them to fly straight home. To explain this precision there must be something more to the navigational systems of at least some vertebrates. That something extra is a "map" sense, an ability on the animal's part to judge its position relative to home.

A map sense might take either of two basic forms. One possibility, by far the simpler, is that an animal might use a "getting warmer" scheme, determining which way to go by measuring and comparing graded cues successively over time. This may be the strategy employed by salmon as they display their astounding homing talents. The other possibility is that an animal might be able to measure its position relative to home by means of some coordinate system. We commonly use both a Cartesian system of orthogonal axes of latitude and longitude—Princeton is 1° 55′ south and 3° 40′ west of Boston—and a "polar" system—Princeton is 225 miles southwest of Boston. Animals could conceivably form a grid (whose axes need not be perpendicular) from any of a number of hypothetical cues.

Whatever their system, it seems clear that at least some birds do perform bicoordinate navigation. The evidence is quite simple. Imagine yourself taken from home blindfolded and released in some unknown spot. A compass alone would do you no good since you would have no idea in what direction you had been taken. Nor would a map by itself be of any use. If you had a map sense so that you *knew* where you'd been taken relative to home and therefore which compass direction you should adopt, you still would have to have a compass to find that direction. (Neither are necessary for the "getting warmer" method.) If someone had tinkered with your compass so that it was systematically wrong by, say, 90° to the right, your setting off in a direction 90° to the right of home—east when you had been taken south, south when you had been taken west—would demonstrate at once that you had both a compass and a map sense: you must have known where home *should have been.* This is precisely what happens with birds.

CELESTIAL ORIENTATION

Studies of caged migratory birds prove that their usual compass is celestial: the sun during the day, the stars at night. Given a view of the sky, these birds will show a strong preference for the direction of migration. When experimentors shift the position of the sun or stars abruptly, the birds shift direction accordingly (Fig. 14-2A). Of course, both sun and stars move naturally as time passes, so like the insects birds must not only be able to learn the direction of movement, but they must be endowed as well with some as yet unknown system for compensation. Using a planetarium at Cornell, Steven Emlen showed that young birds are programmed to locate a pole star, identify the direction of celestial rotation, and learn the constellations.

Migrants, however, leave something to be desired as experimental subjects. They are nearly all nocturnal, and in addition provide data during only two brief periods a year. By far the most useful animal for studying vertebrate navigation, the honey bee of the bird kingdom, is the homing pigeon. This

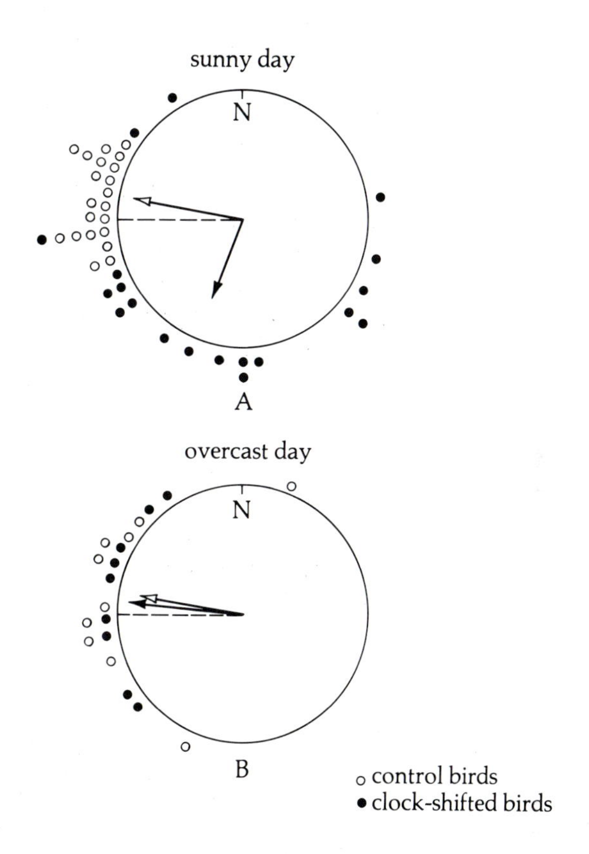

Fig. 14–2 Birds use the sun as their primary compass. However, since the sun moves, they must consult their internal clocks to determine the sun's azimuth. When birds are clock-shifted so that their sense of time is running six hours fast, they misinterpret the sun's azimuth and set off 90° left of home (A). On cloudy days, however, clock-shifted birds are not fooled, having resorted to a backup compass for directions (B). The dots at the periphery of the circles each represent one bird's "vanishing bearing," or the direction it headed off when released, the dotted line is the true homeward direction, while the arrow is the mean for the group. The length of the arrow is a measure of the quality of the orientation, reaching to the edge if the birds all agree but very short if there is virtual randomness.

diurnal laboratory version of the wild migrant will return home quickly and reliably after displacements from home of hundreds or even thousands of kilometers (Fig. 14–3B). When released, pigeons circle briefly (Fig. 14–3A) and then depart in the direction of home.

Demonstrating the existence of a map/compass system in pigeons is remarkably simple. First we must rotate their compasses. This is done by "clock-shifting" a group of birds—keeping them in rooms out of sight of the sun and sky, with an artificial light cycle out of phase with the real day. The pigeons quickly reset their circadian rhythms to this new "dawn." If, for example, the lights are set to come on and go off six hours early, the internal clocks of shifted birds will indicate noon when it is actually dawn outside. Hence, a pigeon using the sun as its compass should interpret a dawn sun in the east as a noon sun in the south, so that its compass, its entire frame of reference, should be rotated 90° to the left. In fact, regardless of the direc-

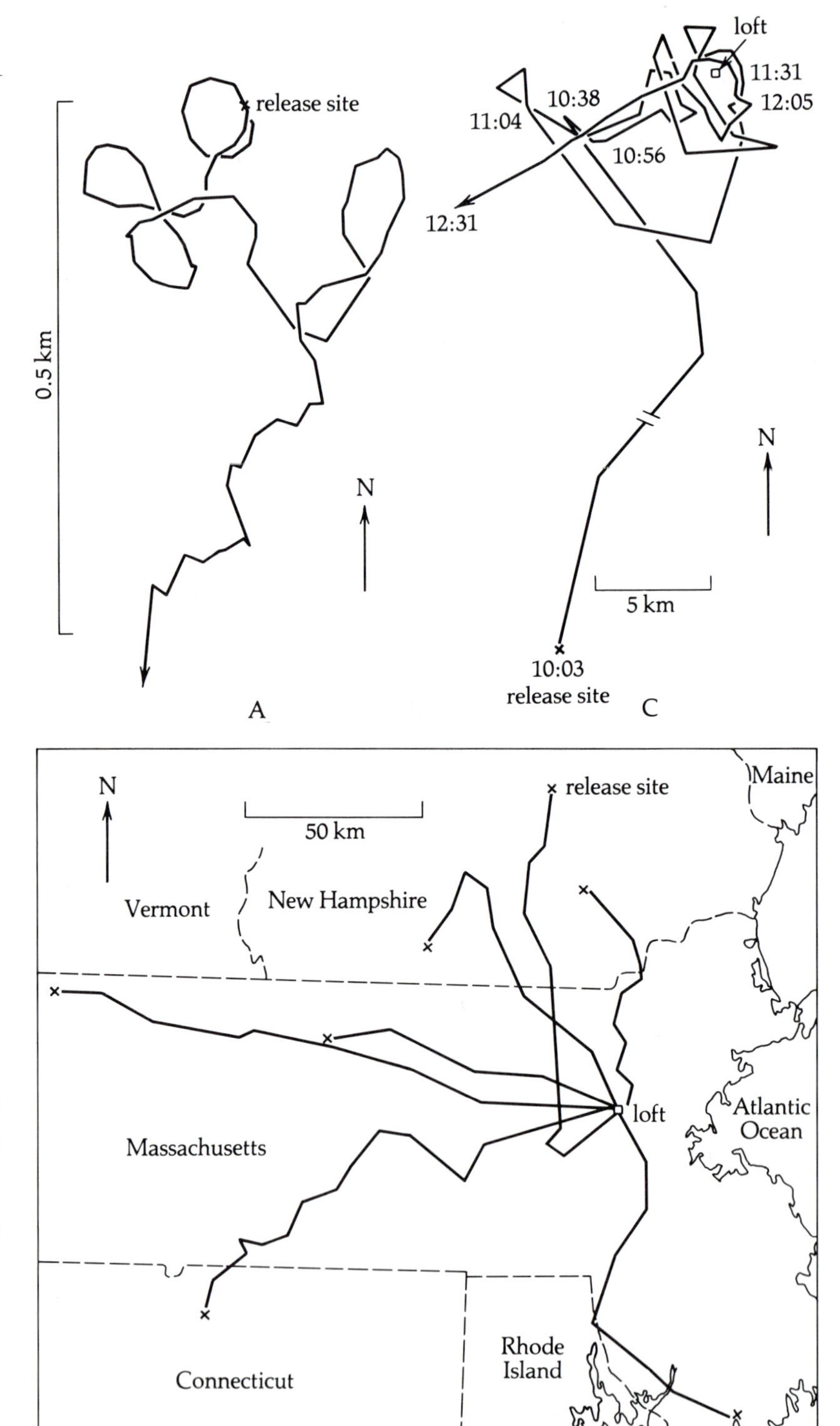

Fig. 14-3 A. Pigeons typically begin their journey home by circling the release site, but quickly show a vague preference for the correct direction. After wandering around for reasons yet to be understood, they set a rough and irregular course for home. B. The actual routes are rarely straight and direct, and indicate that new map measurements must be taken from time to time to make mid-course corrections. At most sites, pigeons depart systematically left or right of home, a phenomenon known as release-site bias. C. When pigeons arrive in the vicinity of home, they use visual landmarks to locate the loft. Pigeons wearing frosted contact lenses have no form vision, but nevertheless know when they are in the vicinity of home and fly wide circles nearby. Such tracks indicate that the accuracy of the pigeon's map sense is a few kilometers.

tion of displacement, pigeons whose clock has been shifted six hours early depart approximately 90° to the left of home (Fig. 14–2A).

MAGNETIC COMPASS ORIENTATION

Birds, like bees, often have to navigate on cloudy days, and just like bees they do so perfectly well. What is their backup compass? Like the bees', it is not celestial: clock-shifted pigeons depart directly for home on cloudy days (Fig. 14–2B). Old evidence, which almost no one took seriously at the time, suggested a magnetic sense. Wolfgang Wiltschko, a German orientation specialist, claimed that robins captured during migration would orient in the proper direction—north or south, depending on the season—in the absence of visual cues. When an artificial magnetic field was substituted, though, the birds reoriented in the new direction, while when the earth's natural field was cancelled the orientation became random.

But Wiltschko's evidence was somehow unsatisfying. For one thing, the experiment worked only in a peculiar circular cage with radial perches (Fig. 14–4). Using tangential perches so that the birds could actually face the correct direction caused the phenomenon to disappear. Then, too, the data had to be averaged in three stages in just one special way or the effect could not be seen. Unlike other such experiments, however, when his particular setup was used, Wiltschko's results refused to go away with repetition.

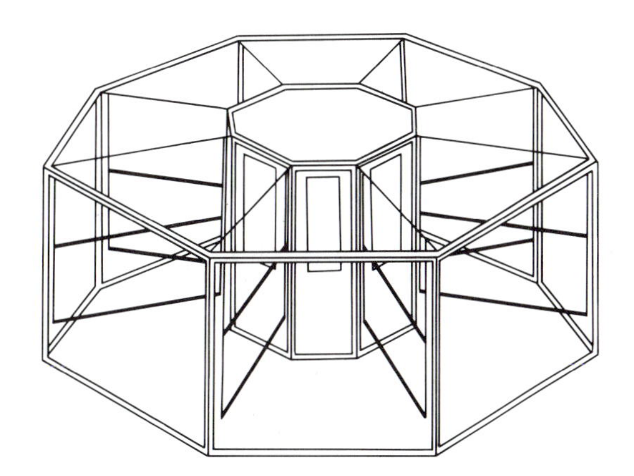

Fig. 14–4 One version of the Gustav Kramer cage. The experimental bird is able to hop around the octagonal doughnut from one radially directed perch to the next. Microswitches under the perches register the hops. From these measurements a "preferred" direction is calculated which generally matches the appropriate migration bearing for that season.

In fact the backup compass *is* magnetic. In a classic series of experiments William Keeton of Cornell University released pigeons carrying either magnets or brass bars of the same weight. This had been done before, but by scientists who, having failed to learn the lesson of the bees, only tried releases on sunny days. Keeton, like them, found no obvious effects under these conditions. On *cloudy* days, however, differences between magnet- and brass-equipped birds were striking: brass-bar pigeons headed toward home, while those with magnets flew off in all directions (Fig. 14–5).

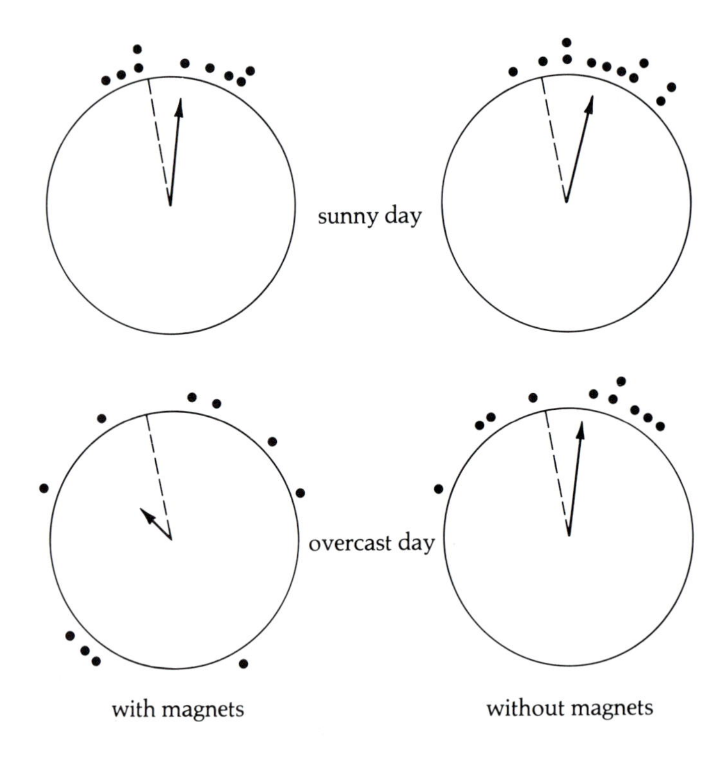

Fig. 14–5 When pigeons are released with magnets glued near their heads on sunny days, the birds, relying on the sun, are not much affected. On cloudy days, however, magnets have a dramatic effect.

The key to Keeton's success in this sort of study seems to be the infamous climate of Ithaca, N.Y. Only birds which experience cloudy weather during some of their early practice flights are later able to home well under clouds. Apparently there is a critical period in the development of pigeons' magnetic compass sense, and Cornell pigeons are guaranteed the opportunity to perfect it. And there may be some sort of automatic solar/magnetic calibration to be performed since, in contrast to more experienced birds, young pigeons taken from the loft for their first flights cannot orient even on sunny days when they are carrying magnets.

How, then, do pigeons sense the earth's field? A set of elegant experiments by Charles Walcott of Cornell University ruled out the possibility of an induction detector. Walcott fitted his birds with miniature Helmholtz coils (Fig. 14–6) run off batteries, and released controls with their coils not hooked up and experimental birds with their induced fields pointing either up or down, as determined by the polarity of the battery. On cloudy days birds with coils whose north ends pointed *up* flew away from home (Fig. 14–7). This means both that a static field affects orientation, and that "north" to a pigeon is the direction in which the field lines dip into the earth regardless of the polarity of the field. Experiments with caged European robins had already indicated that the robin compass was blind to polarity but sensitive to dip direction. These criteria decisively rule out induction as the detector system. In fact the magnetic sense of pigeons, like that of bees and bacteria, is probably based on a permanent magnet detector. Preliminary findings

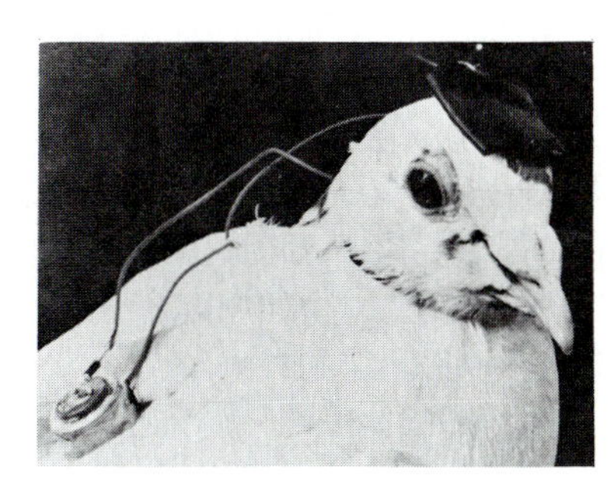

Fig. 14–6 A pigeon outfitted with a pair of coils, one on the head and the other around the neck, to create an artificial magnetic field. The polarity of the battery determines whether the north of the artificial field points down ("S up") thereby adding to the earth's field, or up ("N up") which largely reverses the field felt by the bird.

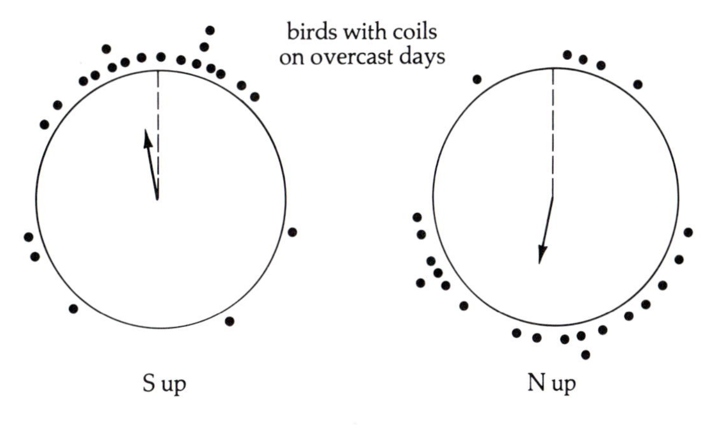

Fig. 14–7 "S up" birds home normally on cloudy days, but "N up" pigeons seem to misinterpret their magnetic compass and fly away from home.

indicate that at least many homing pigeons carry dense arrays of magnetite crystals in their heads, and what these sensitive magnets might be doing in return for room and board is currently the subject of intensive experimentation.

THE MAP SENSE

So far, so good. Pigeons are programmed in ways not yet understood to use the sun as their compass, and to substitute a magnetic sense when the sun is invisible. But bees do all this and more, and are far more willing to share the secrets of their navigation system than birds. The incredible mystery of pigeon homing, however, is the map sense, a gift bees lack. We can estimate the accuracy of the homing pigeons' map sense from some experiments by Klaus Schmidt-Koenig and Walcott. Walcott's airplane tracking data implied that pigeons navigate to within sight of familiar landmarks, then make their final approach visually. Schmidt-Koenig and Walcott equipped pigeons with frosted contact lenses so that they could not form clear images of anything more than a few meters away. For the most part these birds homed normally until they came within a few kilometers of the loft. After that, they either wandered around the vicinity or performed hesitant and awkward landings in the neighborhood (Fig. 14–3C). It seems clear that these birds knew the position of home to within 5 km.

The first theory of how the pigeon map might work was put forth by Britain's G. B. T. Matthews in the 1950s. He proposed that birds might be miniature versions of human navigators, equipped by instinct rather than training with an ability to infer where they are on the basis of the sun's position. Consider how our species navigated before the days of radio, satellites, and computers. A very patient human navigator can find his way on a sunny day with a clock and a sextant (a device to measure the elevation of the sun). By comparing the time of the sun's highest elevation—local noon—with his watch, the navigator would be able to calculate his longitude (his east/west position on the globe). For example, noon in Princeton comes three hours before noon in Pasadena, and each four minutes of difference represents 1° of longitude. From the sun's noontime elevation, a navigator could calculate his latitude (his north/south position): at any given time the sun is always lower in the sky in Princeton than in Atlanta, and each degree of sun elevation corresponds to 1° of latitude. With a sextant and a watch, then, an exasperated human can construct a map, determine how far he has been displaced on the earth's surface, and then calculate in what direction he has been moved. (Of course the clock must be very good, especially if he has not been home to reset it recently.) He still needs a compass, however, to find that direction. Since north of the tropics the sun is always in the south at noon, it serves as a natural compass. Waiting for noon and hoping that the sun will be in view at that unique instant, however, is a luxury that neither men nor animals can enjoy. With a clock, a real compass, a sextant, a proper set of charts, and a good pocket computer, though, the same calculation may be done for any local time of day.

Our cumbersome solar coordinate system is not, however, the source of the pigeon's map. A variety of experiments exclude it, the most convincing being clock shifts. Consider our group of birds clock-shifted six hours early and taken 100 km north for release at dawn (Fig. 14-8). Since their watches said noon and the sun was low on the eastern horizon, a solar coordinate system would have convinced them that they were west of home—a quarter of the way around the world, in fact, somewhere in the Pacific Ocean north of Hawaii. Under those circumstances they should have departed basically east, regardless of whether they had been transported 100 km north, south, east,

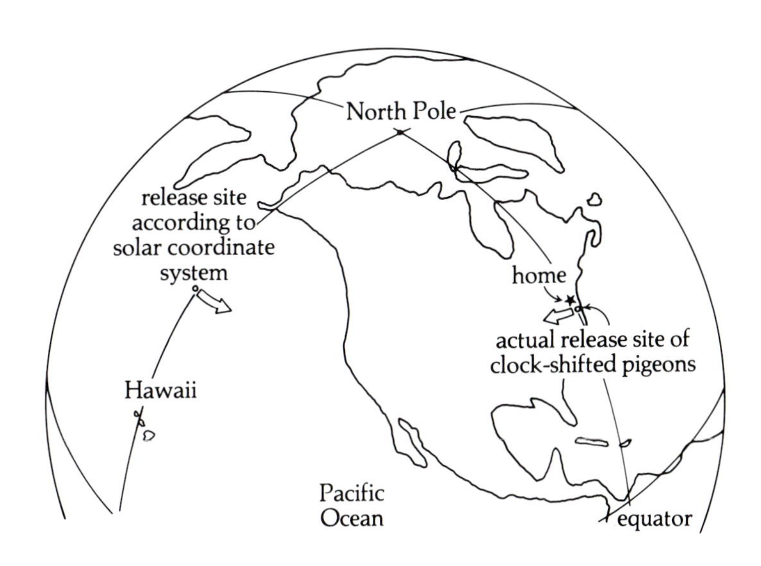

Fig. 14-8 When birds are clock-shifted six hours fast and released at dawn 100 km south of home, at least three predictions of their behavior are possible. If the birds have a sun-coordinate map, they will interpret the low sun and their noon clocks as indicating a displacement into the Pacific, and depart east regardless of their compass system. If, instead, the pigeons have a true map sense (and so know they are 100 km south of home) and use the sun as a compass, they will misinterpret the dawn sun as a noon sun in the south, and so fly away from it to the west. If birds had a true map and a nonsolar compass, of course, they would depart to the north. In fact, on sunny days these pigeons would go west, while on cloudy days they would fly north.

or west. In fact, the shifted pigeons departed 90° to the right of home—west when taken south, for example. They knew where they had been taken without recourse to the sun, and their ability to home on cloudy days reinforces this conclusion.

AN INFRASONIC OR OLFACTORY MAP?

Again, we are in the position of searching for a sense to which we are blind. Where should we begin? At present there are three leads, all difficult to accept. Take, for example, the perplexing discovery that pigeons can hear low-frequency sounds of 0.1 Hz. This immediately led ethologists to wonder why. As it happens, meteorologists have discovered that low frequencies produced by storm systems travel enormous distances—hundreds or even thousands of miles—and they now routinely track storms acoustically. They also report that the world is full of acoustic landmarks, mountains and forests which murmur and sing at great distances as the wind passes over them, each in its own characteristic way. Could pigeons be listening in, and using what they hear as a map or a compass? If so, the map would have polar coordinates: that is, the distance and direction of sound sources would be learned for the home loft, and compared with the infrasonic landscape at a release site.

A second hypothesis supposes that pigeons might have an olfactory map. According to this theory, birds learn to associate odors with wind direction. An onshore wind may always smell salty, while a breeze from another direction may carry the odors of a pine forest. Birds displaced from home might recognize the local odor as one which is always on winds from, say, the northeast at home, and so fly southwest.

The "olfactory hypothesis," as developed in Italy by Emilio Papi, grew out of a curious series of "palisade" experiments begun by Kramer and carried further by H. G. Walraff. The object was to determine what, if anything, pigeons must see from their lofts in order to home. Blocking their view of the ground up to 3° above the horizon destroyed the ability of inexperienced birds to home, while blocking out the sky—but not the horizon—permitted normal homing. This was true even if the horizon was "palisaded" near sunrise and sunset, so the birds could not see the sun. Glass walls had the same effect, which convinced researchers then that visual information was irrelevant. (However, glass blocks certain wavelengths of light, such as UV, and distorts polarization patterns, both of which, though invisible to us, we now know are available to pigeons, so there may be information involved which has not yet been considered.) One interpretation of these results is that there is some map information moving in the horizontal plane, which might lend credence to the olfaction hypothesis.

Papi has also observed that pigeons whose nostrils have been plugged or whose olfactory nerves have been cut, or who have been overwhelmed with odorous substances applied to their nostrils before release, orient poorly and home badly. Such negative evidence is quite indirect—any of these treatments might affect homing ability in other ways—and Keeton's group has been unable to repeat any of Papi's results here in the States. But two other experiments which Papi has performed are more suggestive. In one, birds were raised in lofts whose air supplies were controlled by fans. One odor, olive oil, was

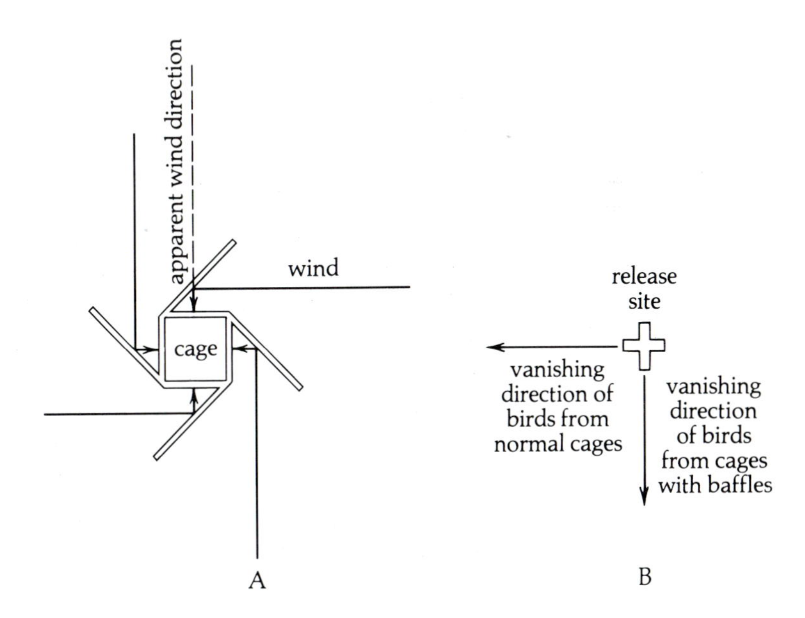

Fig. 14-9 The olfactory theory of pigeon homing postulates that pigeons memorize the odors which come from each compass direction, and when taken to a new release site simply match the local odor with their memory of windborne odors to determine the direction of displacement. Hence, birds raised in a cage with wind baffles which rotated wind direction (A) were observed to make systematic mistakes in judging home direction (B). The odor system would provide a crude map, but the birds would still need a compass to determine direction.

always blown from one direction, while the odor of turpentine came from another. When these pigeons were taken out and released, experimentors applied one of the two odors to their birds' beaks. Their departure bearings correlated with the directions from which the respective odorants had been blown in the loft, rather than the direction of home from the release site.

In the second experiment birds were raised in a "deflector loft," which cleverly rotated the apparent wind direction (Fig. 14–9). The birds' subsequent vanishing bearings were similarly rotated. The latter experiment has been repeated by Keeton, but the Plexiglas deflectors also provide visual reflections, block UV light, and distort polarization patterns so that interpretations of the results which do not implicate olfaction abound. In fact, recent work at Cornell with a complicated deflector loft which rotates the wind in one direction but reflects the visual cues in the opposite direction indicates that the pigeons are actually being deflected by the visual information.

A MAGNETIC MAP?

Finally, we come to the arcane possibility that birds might be making use of a magnetic map in orientation. The first hint of this bizarre alternative came from Keeton's study of "release-site bias"—the tendency of homing pigeons (and Ithaca's bank swallows, for that matter) to depart systematically right or left of the true homeward direction, depending on the place at which they are released. At some release sites this bias is extreme: 13°–50° at Campbell, N.Y., for instance (Fig. 14–10A). When Keeton tried to discover why the bias should be large some days and small on others he discovered a rough but remarkable correlation with the intensity of concurrent magnetic "storms." Since these data were for *sunny*-day releases, and the storm intensities were small compared to the background field of the earth (a large 500-gamma storm is 2 percent of the total field), it seems unlikely that any compass effect could be involved.

Still more striking is Walcott's discovery that birds released at magnetic anomalies—places at which the earth's field strength is slightly irregular—are disoriented in proportion to the strength of the anomaly, even on sunny days (Fig. 14–10B). When released from a place at which the earth's field strength is only 6 percent higher than normal, for example, pigeons are

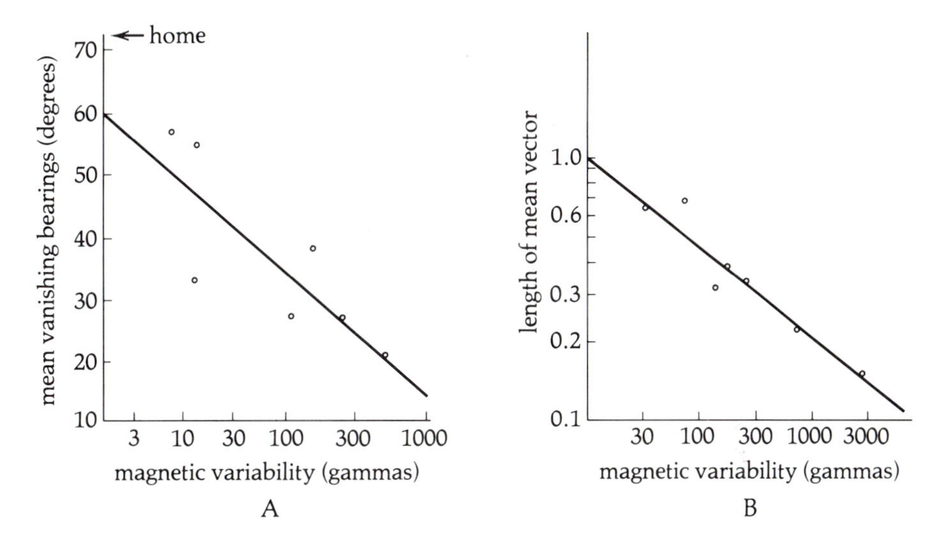

Fig. 14–10 Evidence for a magnetic component to the pigeon map sense comes from a variety of sources, of which these are the two most easily quantified. A. The mean vectors of pigeons are shifted during magnetic field storms—counterclockwise in this example from Campbell, N.Y.—in a roughly gradual fashion (1000 gammas represents a 2 percent change in field strength). B. At magnetic anomalies, the quality of orientation is reduced in proportion to the strength of the anomaly. The point at the lower right corresponds to the anomaly in Figure 14–11A.

almost totally disoriented (Fig. 14–11A). Then, too, birds transported to release sites in iron (vs. aluminum) boxes or in artificial magnetic fields are clearly affected on sunny days. The most reasonable interpretation at the moment seems to be that Walcott's birds, for example, their perfectly good primary compass shining in the sky, simply had no idea where they were in relation to home.

In fact, there *is* map information to be gleaned from the earth's magnetic field. From the equator to the magnetic poles, for example, the intensity of the field doubles. Birds with the sensitivity we must infer from the magnetic storm and anomaly curves would be able to determine their magnetic latitude to within a few kilometers—about the accuracy of birds wearing Schmidt-Koenig's frosted contact lenses.

There are two major difficulties with this magnetic map notion. First, how can Keeton's birds which carry magnets on sunny days manage to home? Somehow, they must factor out

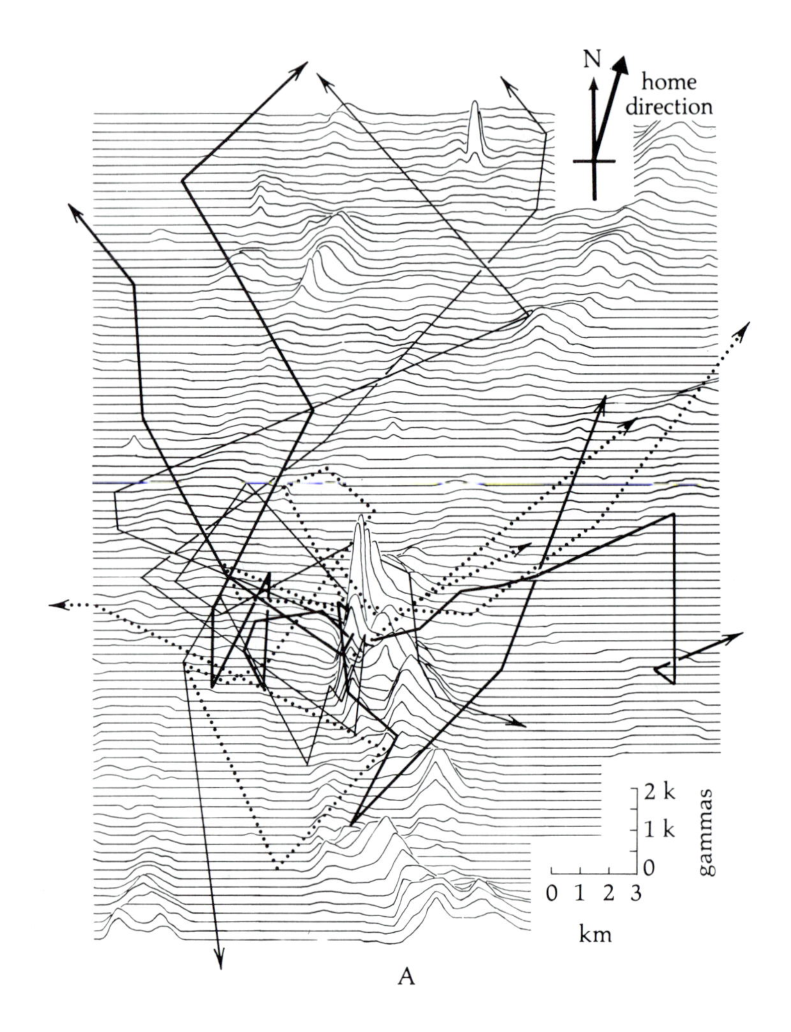

Fig. 14-11 The earth's magnetic field strength increases about 6 gamma/km in the northeastern U.S. Shown here in a three-dimensional plot is the anomaly in field strength. A. When birds are released at Iron Mountain, R.I., they are almost completely disoriented even on sunny days. Their confusion persists until they have been out of the anomaly for some time. B. At normal sites—Worcester, Mass., in this example—pigeons depart in one general direction. Note the slight clockwise bias typical of Worcester.

the static field of the bar magnet. Second, how could birds possibly measure magnetic field strength to an accuracy of one part in 5000 or better—a performance unrivaled by most other known sensory systems and only rarely (and expensively) surpassed by modern techonlogy?

Actually, there are three ways they might increase resolution. One would be to ignore the total strength and only measure differences around some mean, the way pigeons extract the minute changes in air pressure generated by changes in altitude of a few feet. The other two methods would require increasing either the number of detectors or the time interval over which the nervous system is averaging incoming information, or both.

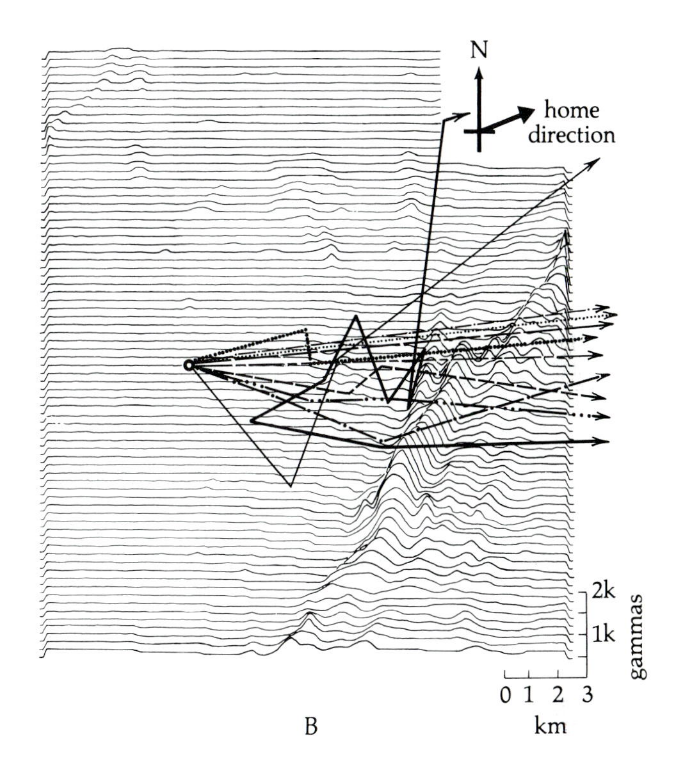

In either of these cases, resolution goes up as the square root of the number of samplings. With a hundred million permanent magnet detectors and several seconds in which to make a decision, the pigeon may not face so very difficult a task.

In the end, however, the map sense of birds and wide-ranging vertebrates (and perhaps even that surprising long-distance migrant, the monarch butterfly) in general remains a mystery central to the study of animal behavior. The existence of yet another elusive information-processing system ties together a variety of ethological themes: we are forced to recognize one more previously discounted sensory world to which the exigencies of our niche have denied us admission; and our ignorance of its development, from whether it might involve critical periods to how any required learning takes place, challenges us to look all the more diligently for a solution. The unraveling of this most incredible feat of programming awaits that optimal combination of imagination, incisiveness, and experimental finesse which is ethology at its best.

HOMING IN SALMON

From the top to the bottom of the phylogenetic scale there are creatures which must travel great distances to very specific destinations. This "homing" behavior requires the animals to have two separate navigational systems. To get within "shooting distance" of home, which might be anywhere from centimeters to thousands of kilometers away, they must employ some method of long-distance navigation. There are a myriad of cues in the sky and in the geomagnetic field of the earth which may be used for this purpose, cues which can give an animal sufficient information to reach the general vicinity of home. Then a second, much more precise system must take over, a system which will enable the animals to locate the sometimes amazingly small area they seek. For a bee whose food search has taken her immense distances in relation to her body size, odors, colors, and visual landmarks are the cues which allow her to arrive at her own hive door, which might be in one of hundreds of identical (to us) trees in a forest or hives in a beekeeper's yard. For both homing pigeons and digger wasps on the other hand, the burden is carried purely by visual cues: odor plays no part in this final discrimination.

Only recently has one of the natural world's most startling homing phenomena been worked out: the mystery of salmon homing. Adult salmon battle their way back to the very tributaries they hatched in two to three years before, mate, deposit eggs in gravel nests called "redds," and die. How do they get home? Something about their natal home stream remains in their memories, but what?

The young salmon hatch in their redds and remain there safe in the gravel for two months, then set up individual feeding territories for the next fourteen months. At this time, as with juvenile birds, increasing day length sets off a physiological change, called the "smolt." Surges of hormones cause the young fish to change color, adapt their systems to salt water, and swim downstream. When they reach the ocean they feed and grow, until at three years they stop eating and begin the tortuous journey home.

How the fish traverse miles of ocean or lake to reach their native rivers remains a mystery; but a lengthy series of ingenious and critical experiments by Arthur Hasler and his colleagues has proven conclusively that odor is the cue which leads the adults to the streams and tributaries in which they hatched. Previous work suggested that the "smolt" stage was the time at which the fish form a "memory" of the home stream, and Hasler hypothesized that the memory was one of odor. To test this, he subjected hatchlings to uniform conditions in holding tanks throughout the smolting period. They were raised in water piped in from Lake Michigan; no water from any tributary was allowed to reach them. A third of the fish, however, were exposed to a minute concentration of morpholine, an organic odor not naturally present in streams, another third to phenethyl alcohol (PEA), and a control third remained unexposed to any artificial odor.

The fish from all three groups were marked with fin clips and released at a point in Lake Michigan midway between two tributary streams. At spawning season Hasler and his colleagues "baited" one of the two tributaries with morpholine and the other with PEA in an effort to create artificial "home" streams for the now-adult salmon. When the baited streams (and sixteen unbaited ones) were surveyed, PEA-raised salmon were found in the PEA-scented stream, morpholine-raised fish in the morpholine-scented stream, and control fish scattered equally among all eighteen streams surveyed (Table 14–1).

These results reveal a clear-cut instance of classical imprinting, this time to an olfactory cue. As mentioned in Chapter 4 and developed in Chapter 17, imprinting involves rapid, long-lasting, irreversible learning during a brief critical period. Four hours of exposure to a scent during their critical smolting period is sufficient to ensure the salmon's return to the odor a year and a half later. No amount of exposure to any chemical or water from any other stream either before or after the brief smolt stage has any effect. Hence, the homing of salmon as revealed by this remarkable combination of field observation and carefully thought-out behavioral experiments is a classic example of genetically programmed flexibility, a rigidly determined mechanism which uses the unpredictable but constant olfactory landscape of each stream to make sure that fish return to their tried-and-true hatching grounds to spawn.

TABLE 14–1 *Effects of Olfactory Imprinting on Salmon Homing*

	Fish Recovered		
Stream type	*Morpholine-imprinted*	*PEA-imprinted*	*Control*
Morpholine-baited	659	20	76
PEA-baited	8	343	55
Unbaited	14	9	154

Source: Arthur D. Hasler, A. T. Scholz, and R. M. Horrall, "Olfactory Imprinting and Homing in Salmon," *American Scientist* 66 (1978): 347–55.

SUMMARY

Many vertebrates migrate long distances to change habitats with the seasons. All must have compasses, and the general pattern is to use a celestial compass—the sun, the zenith polarization pattern (used by some amphibians at least), or the stars—when they can, but to substitute a magnetic compass when celestial cues are unavailable. All of the possibly celestial compasses must be time-compensated. Some vertebrates migrate simply by taking up a learned or innate direction, either for a preordained interval or until their goal is reached. Some even use a two-stage version of this strategy. The most outstanding feature of some vertebrate navigation systems (and perhaps too of monarch butterflies which fly thousands of kilometers to isolated wintering grounds) is a map sense. The basis of this ability to judge where they are in respect to their goal is unknown, though evidence suggests that at least one component in the pigeon map depends on small but systematic variations in the earth's magnetic field.

STUDY QUESTIONS

1. Darwin thought that pigeons and other animals might determine where they had been taken not by a map sense, but by keeping track of the approximate distance and direction they had traveled by integrating the compass directions and times on the various legs of the journey. This route-based navigation strategy, very unlikely for animals with much distance to cover but thought to be the system used by insects, was ruled out for pigeons and several other species of birds about forty years ago before anyone worried much about the sensory cues involved. How do you suppose this experiment was done?

2. The most obvious objection to the hypothesis of a partially magnetic map is that the map sense is not affected by a large, static field applied to the pigeon's head. Surely the map measurement, depending on subtle variation in field strength, ought to be swamped out by this large field.

   a. When Walcott released magnet-equipped pigeons on sunny days from anomalies, the birds were still disoriented. What does this mean?

   b. How, then, must any magnetic map processing be organized to account for the ability of pigeons to home with magnets? How could you test your idea?

3. If birds are checking their map coordinates from time to time, what path would you expect a clock-shifted bird to fly? How could you

distinguish between a strategy of occasional, discrete map checks and a running-average system such as bees use in sun compensation?

4. Pigeons, like bees, must allow for the sun's movement. How could you discover their system?

FURTHER READING

Carr, Archie. "Navigation of the Green Turtle." *Scientific American* 212, no. 5 (1965): 79–86.

Emlen, Steven. "The Stellar-Orientation System of Migratory Birds." *Scientific American* 233, no. 2 (1975): 102–11.

Gould, James L. "The Case for Magnetic-Field Sensitivity in Birds and Bees (Such as It Is)." *American Scientist* 68 (1980): 256–67.

Hasler, Arthur D.; Scholz, A. T.; and Horrall, R. M. "Olfactory Imprinting and Homing in Salmon." *American Scientist* 66 (1978): 347–55.

Keeton, William T. "The Mystery of Pigeon Homing." *Scientific American* 231, no. 6 (1974): 96–107.

Sauer, E. G. F. "Celestial Navigation by Birds." *Scientific American* 199, no. 2 (1958): 42–47.

Walcott, Charles. "The Homing of Pigeons." *American Scientist* 62 (1974): 542–52.

Williams, T. C., and Williams, J. M. "Oceanic Mass Migration of Land Birds." *Scientific American* 239, no. 4 (1978): 166–76.

CHAPTER 15

Animal Communication

The goal of most animal communication, like the ultimate aim of animal behavior in general, is reproduction: genes must ensure their survival into the next generation. Most animals are solitary, and so must actively seek out their own kind for mating. And since most animals have no opportunity to learn where to look for the opposite sex of their species or how to recognize a suitable mate once there, much of the behavior associated with mating is innate. In this chapter we will look at a sample of the vast repertoire of signals that animals have developed for identifying themselves as potential mates.

Sexual communication accounts for the greatest part of information transfer in the animal world, and because of the crucial importance of mating in an animal's life nature has invested enormously in perfecting the necessary communication systems. The evolutionary strategy most often observed employs a chain of species- and sex-specific releasers to enable the sexes to locate each other with minimal error and maximum efficiency. The real interest animal communication holds for most ethologists is the potential it offers for research into how the nervous system works: how it manages to encode information into one sex and decode it in the other.

CHEMICAL MESSAGES

The simplest and evolutionarily the oldest communication system is chemical. Some animals carry or release their sex- and species-specific compounds to the world at large, indicating their sexual readiness. When members of the opposite sex equipped with the appropriate highly specific binding proteins encounter the chemical, their orientation programs cause them to locate and mate with the source of the molecular siren song. Examples of this strategy range from bacteria which recognize their special chemical when they bump into each other, on up. Slime molds, for instance, are sexual aggregations of amoebae drawn together when "hunger" has made them receptive to their specific pheromone: cyclic AMP. Some one amoeba begins the process when it senses that times are getting tough by releasing cAMP, and nearby amoebae move up concentration gradients until they find the signaller. Surface molecules on the signaller must then convey a species-specific message, since only cells of the same species "stick together." Then the congregated amoebae unite to form a sluglike unit which moves upward to a high point, forms a stalk, and releases spores (Fig. 15–1).

Only slightly more elaborate than this is the mating system of moths. As described in Chapter 13, the female releases a species-specific pheromone which causes males to turn and fly upwind (Fig. 13–1). Since moths generally mate at night, however, other messages must be communicated by pheromones and tactile cues. Once the male lands, for example, he must search on foot in the dark for the female and distinguish her from, say, a sleepy caterpillar or a twig. The female, too, must determine whether she has been discovered by a male of her own species.

Because they mate in the daylight hours, a full repertoire of signals is known for many butterflies. As described in Chapter 3 (Fig. 3–9), male queen butterflies pursue anything that produces a high-contrast flapping pattern—moths, tumbling leaves, other males, etc. When the male overtakes a flapping object, he extrudes a set of pheromone-soaked "hair pencils" from his abdomen and waves them in front of the prospective mate. If the target of this olfactory display settles he hovers overhead, hair

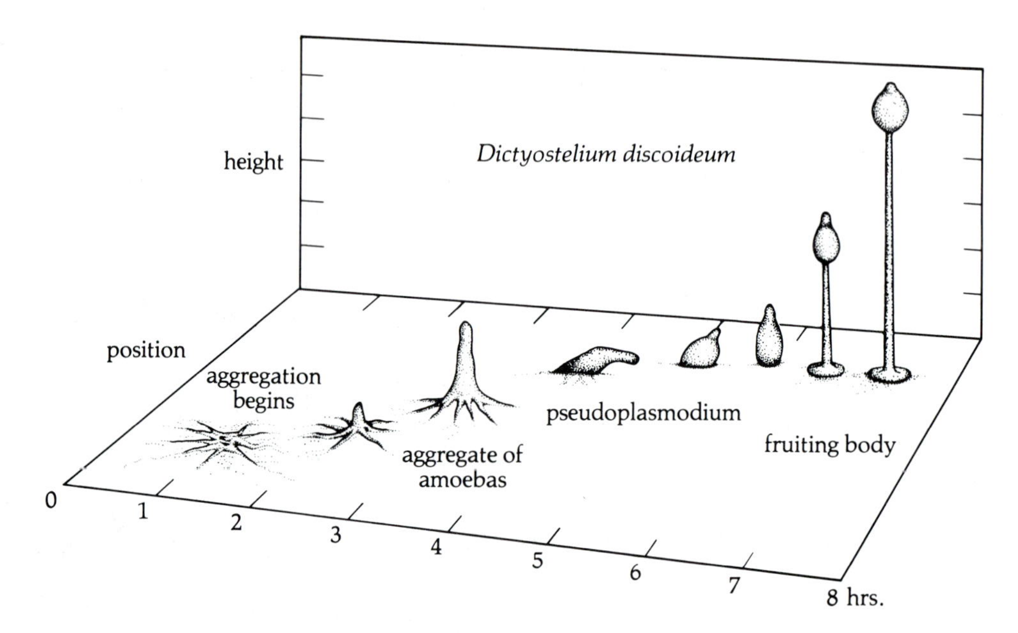

Fig. 15–1 When times get hard, one amoeba of the soil-living slime mold *Dictyostelium* begins releasing a phenomone, cyclic AMP, which attracts others in the vicinity. The resulting aggregation forms a unitary slug which sets off toward the surface. There, the slug differentiates into a base, a stalk, and a fruiting body with spores. As the wind-blown spores carry their genes to new, possibly more favorable habitats, the organism disintegrates into its many component cells, once more independent, which continue their briefly interrupted struggle for survival.

penciling until she closes her wings. This is his signal to land. The three questions of species, sex, and readiness have been answered to the satisfaction of each, and so since their genes can be quite sure to be recombining with a complementary set, the two are permitted to mate. Each of these examples falls into the established, necessarily innate and automatic pattern in which simple sign stimuli trigger interlinked motor programs.

EVOLUTION OF PHEROMONES

A few hints about how the evolution of the encoding and decoding of chemical messages might have come about have turned up recently. In many cases, insects find or recognize their host plants on the basis of odor, and relatively specific receptors for one or more of the scents characteristic of the food are found in the insects' olfactory arsenals. Surprisingly often, the pheromones of an insect are related to naturally occurring chemicals in its diet. Remove a queen butterfly larva from its natural source of food and raise it instead on some nutritious but bland laboratory fare, and its ability to produce its species-specific pheromone as an adult is gone. The presumptive evolutionary sequence is straightforward: the animal's ability to recognize its range of host plants came to be based on a releaser chemical (often the plant's own toxic defensive secretion, a tolerance of which often leads to an insect's food-plant specialization). The insect then would have come to have a special receptor class for that compound; individuals feeding on the host plant metabolized (and detoxified in the case of defensive substances) this olfactory sign stimulus into a related chemical; a simple alteration in the coding of the binding protein for the releaser could have given rise to a selective sensitivity to the altered metabolic product exuded by conspecifics. From here, the possible evolution of the behavioral responses and the morphology of specialized scent glands is obvious.

A variety of cases consistent with this scenario exist, some of which appear to be in that elusive and unstable class of partially evolved systems whose rarity Darwin feared would rob his theory of its most suggestive proofs. Female Douglas fir beetles, for example, locate their host plant by its odors—alpha-pinene, camphene, limonene, and others—substances which are toxic to virtually all other animals. The females attack and begin boring into the tree, eating the trunk and its fluids, and producing feces ("frass") containing the metabolites frontalin, a tiny quantity of 3-methylcyclohex-2-en-1-one (3-2-MCH), a large amount of MCH-ol, transverbenol, and other components. Of these compounds, MCH-ol seems to be the pheromone most enticing to males. The males arrive, locate a pheromone-producing hole, and signal the female by means of a species-specific stridulatory sound. Now both the male and the female begin producing large amounts of 3-2-MCH, a compound which appears to block the receptors tuned to MCH-ol as well as the fir's own odor. Thus the female (and when enough pairs are at work, the tree itself) becomes olfactorily invisible to other fir beetles, male and female alike. As a result, further attacks by females and the continued attraction of males is inhibited.

VISUAL MESSAGES

Fireflies use a more efficient and considerably more elegant system to convey their critical message. Males fly about flashing their luminescent semaphores according to a species-specific code (Fig. 15–2). They have the visual channel all to themselves at night and, at least before man's arrival on the scene, all they could expect to see after dark would have been other fireflies and a few luminescent fungi. Because the usual nocturnal predators, birds and bats, find fireflies distasteful, they ignore their flashing invitations. Fireflies, like monarch butterflies, contain cardiac glycosides; but unlike monarchs their diets are palatable to their predators, so they must manufacture these noxious

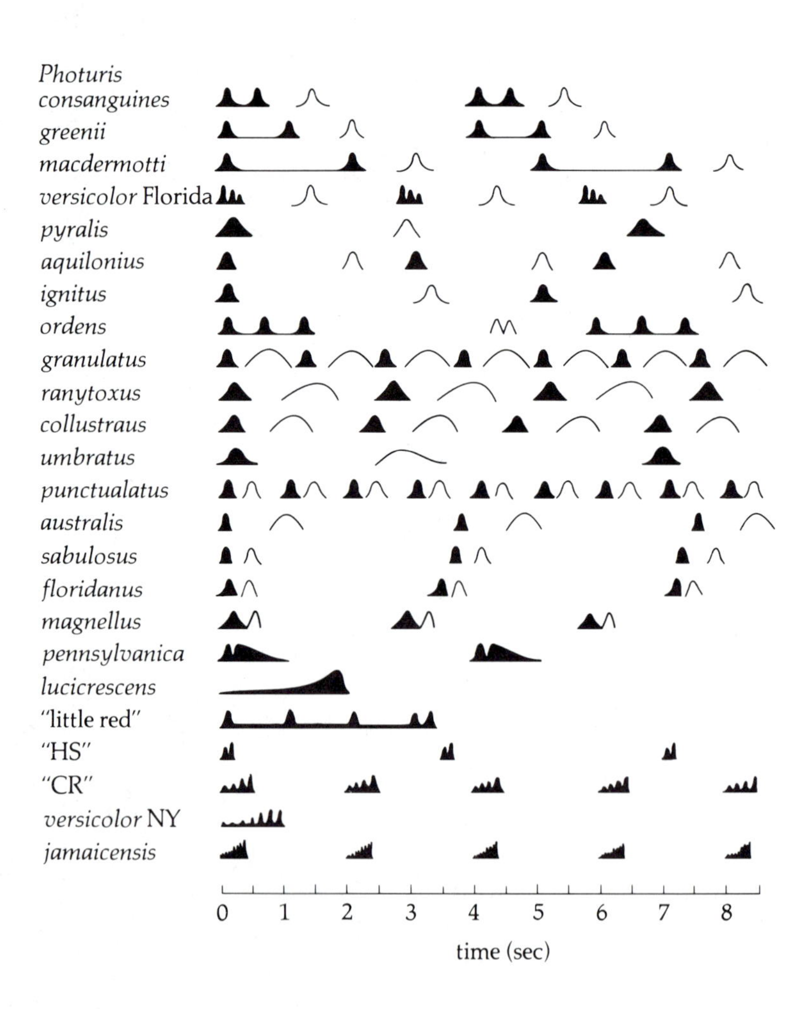

Fig. 15–2 Each variety of firefly has its own code. The males of most species fly about at a well-defined time at night, at a special elevation in a particular habitat, advertising themselves with a characteristic flash pattern (shown here in black). Occasionally (perhaps once every several thousand displays) a female will respond (shown in white) with a flash at the species-specific delay time. The male will then begin to home in, land, and mate.

chemicals themselves. That fireflies must use their lights to warn predators of their toxicity seems clear: the larvae glow continuously with structures which are not used to make the adult lanterns, and adults which are not courting glow or flash irregularly nevertheless.

Females wait on the ground or in bushes until, seeing the right signal, they respond with an appropriate answer—often just a simple flash. The key feature in the display for the female firefly is the temporal pattern of the male's signal. For the male, it is the latency of the female's response—the time from the end of his signal to the beginning of hers—that sets his mating pro-

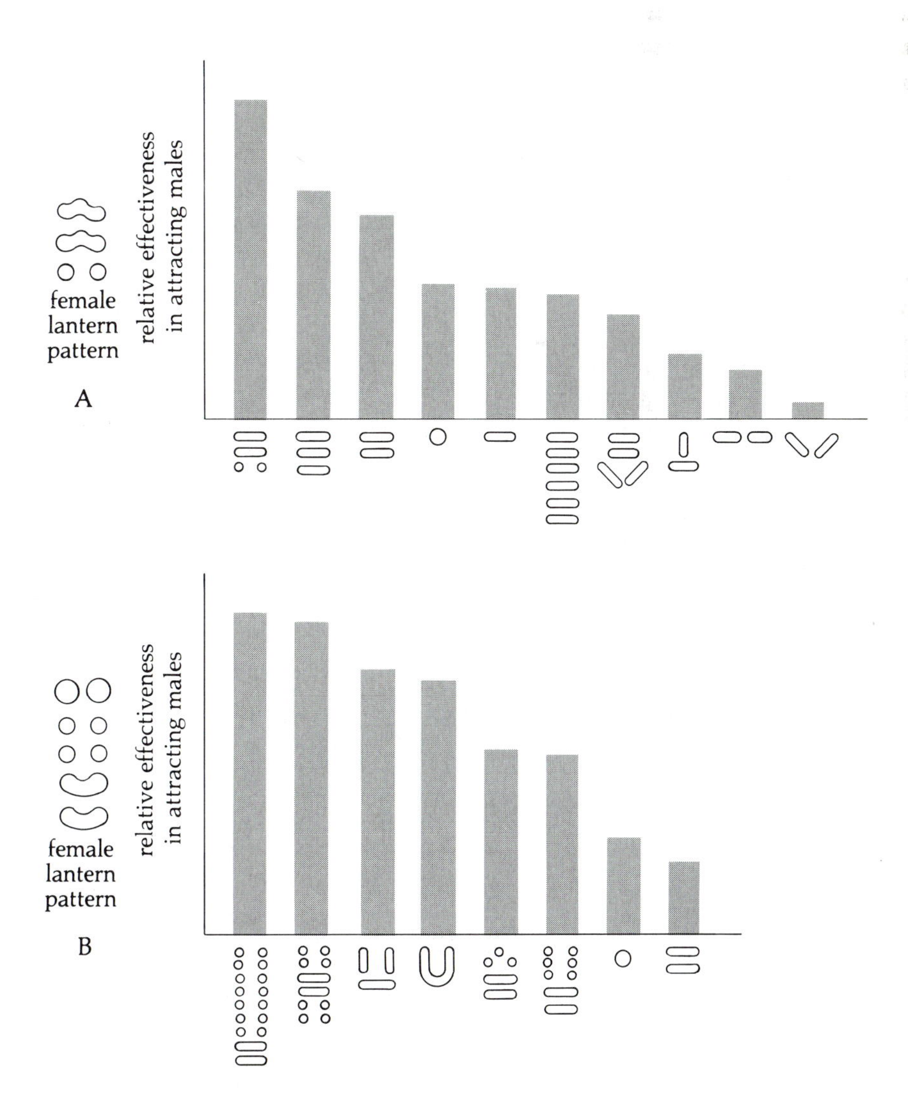

Fig. 15-3 In some species, the light organs of the females have a clear spatial pattern. From model experiments it is clear that males use the pattern as a final cue in species identification. In these two examples the actual female pattern is illustrated at left (A is *Lampyris noctiluca;* B is *Phausis splendidula*) while the models tested are at right. Among the models, those on the left were the most preferred, those in the middle were less attractive, and those on the right were least stimulating to the males.

gram in motion. The timing of the female's response varies among species, and the males use this code to zero in until they get near enough to land. You can, in fact, by trial and error "crack" the codes of the various species of firefly with a penlight. Once you get the delay period right you can easily lure any given male to your hand or to the light itself. Still, the male will not actually attempt to mate with hand or penlight. In at least some species some ill-defined aspect of the pattern of lights on the female's lantern becomes involved in the next stage of courtship (Fig. 15–3), and it may be that specific olfactory or tactile cues permit the male to be certain that his final search on foot in the dark has turned up a conspecific female. Some species, in fact, have secondarily lost their lanterns and rely on pheromones entirely.

It is relatively easy to hypothesize a mechanism which might drive firefly communication. Males need oscillator circuits to generate the flashing and to recognize those responses with the species-specific latency; females need complementary filter circuits to recognize the male signal and to release their replies at the proper moment. Unlike the continuous-message systems already described, it is possible to imagine that the pulse-code-generating circuit of the one sex could serve as the recognition circuit of the other. Attractive as such a notion is, the females of at least one species must be too complex to fit. Female *Photuris versicolor* are predatory, and seem to have several pulse-code circuits. After they have mated with a male of their own species they begin to recognize and respond appropriately to males of several other species. Unlike most other predators, these femmes fatales find the males appetizing, and a fine source of just the right proteins to make a new generation of fireflies: they devour the hapless suitors of other species when they land (Fig. 15–4).

Fig. 15–4 Females of certain species of predatory fireflies will, after having mated, attempt to "seduce" males of other species by mimicking the responses of their females. The unlucky male (seen here on the right) provides extra protein for the offspring of the female.

The use of rhythmic species-specific visual systems is widespread. Fiddler crab males "beckon" to passing females, and encoded in the timing of their waving claw is the species identity of the male (Fig. 15-5). They also make use of a nocturnal

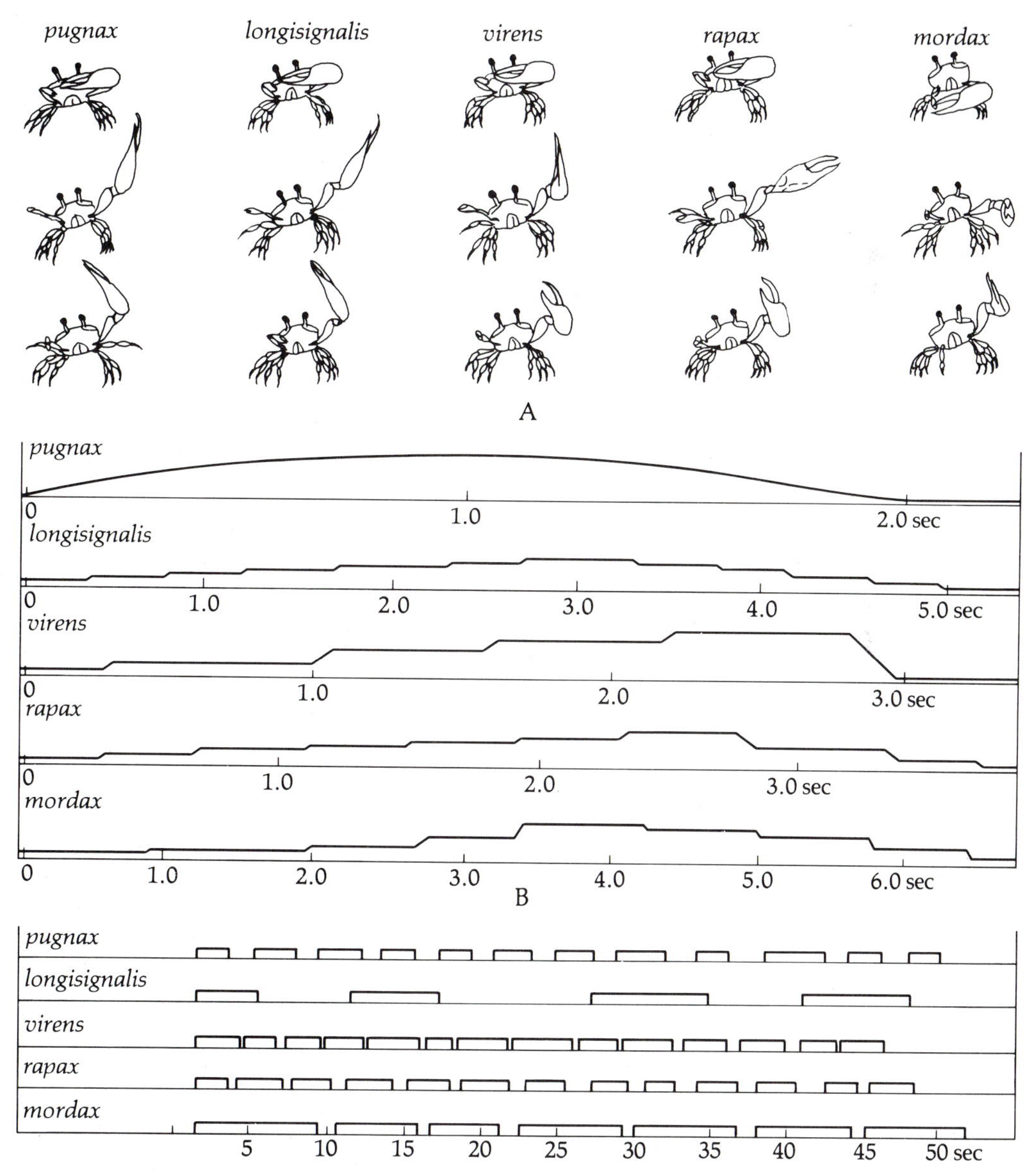

Fig. 15-5 Male fiddler crabs beckon to females with a species-specific wave. Although no model experiments have been performed, the time course of the wave and the wave interval are probably the critical cues in this exchange.

backup, an auditory code generated by the rapping of their giant claws on the ground. Nor is this strategy limited to invertebrates. Male lizards, for instance, attract females with a conspicuous head-bobbing display whose temporal pattern varies from species to species (Fig. 15-6).

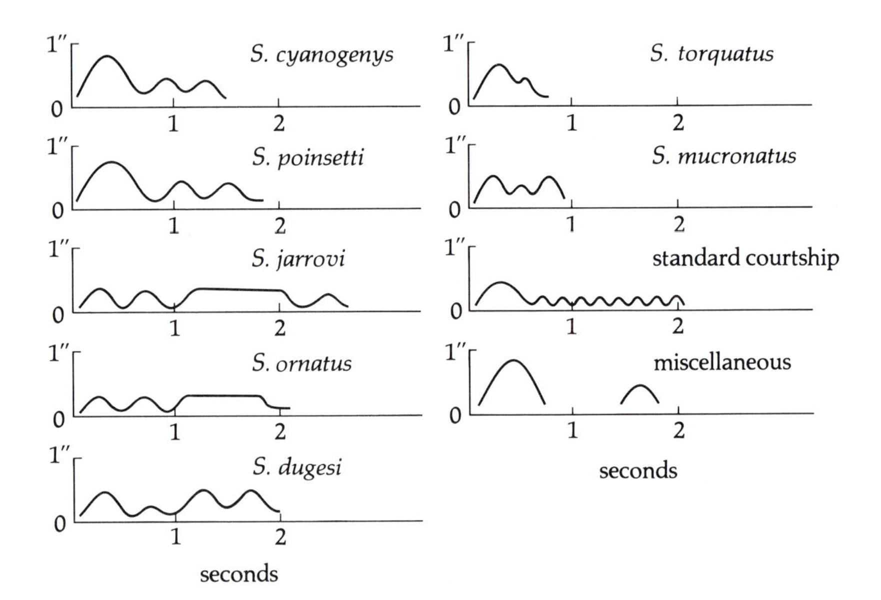

Fig. 15-6 Male lizards indicate their species by a series of stereotyped head bobs. As in the case of fiddler crabs, there have been no model experiments so the exact species-specific parameter females use to distinguish among species remains to be isolated.

AUDITORY MESSAGES

Signal-generating oscillators almost certainly underlie each of these displays, and the suggestion that oscillators are being used by females to decode the signal is difficult to resist. Formidable technical problems have prevented recordings from the cells themselves, however, so the matter remains in doubt. Neurobiologists have been more fortunate in the analysis of cricket song. Crickets are basically acoustic fireflies, and use the same sorts of species-specific, pulse-coded songs (Fig. 15-7). The sexually ready male broadcasts a song of invitation, and receptive females respond by moving toward the sound source. Once a female arrives, the male switches over to a courtship song and, if all goes well, mating follows.

Evidence suggesting that oscillators are the underlying mechanism in cricket song are convincing. For one thing, the chirping

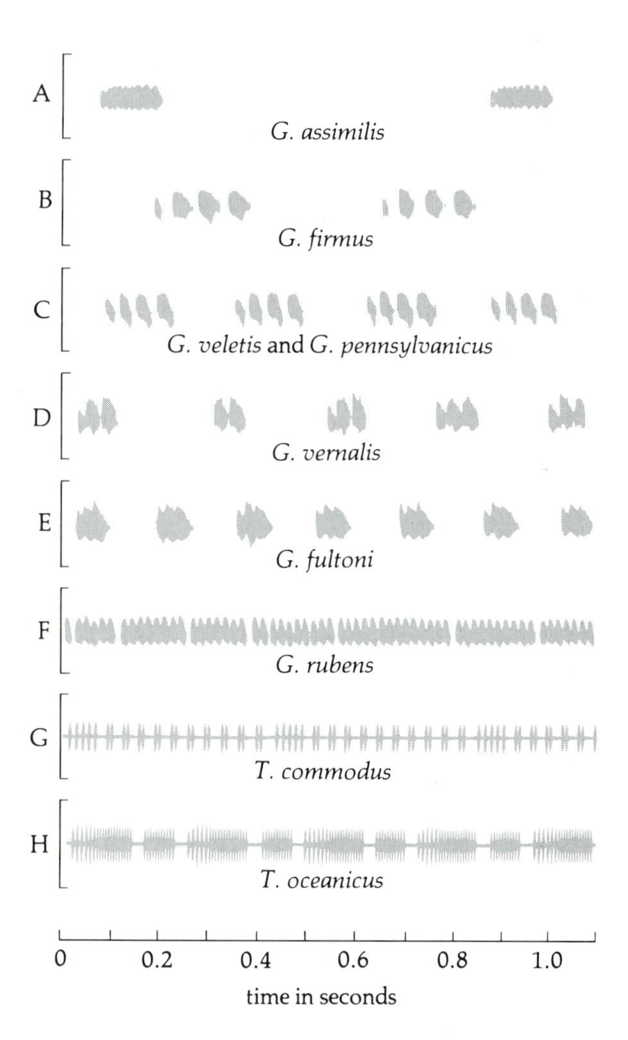

Fig. 15-7 Male crickets broadcast a pulse-coded calling song which females recognize and approach (A–F). The last two songs illustrated here are the most complex, probably requiring three separate oscillator circuits to create (G, H). When a female approaches, males switch into a courtship and then a mating song which is more similar among species. Pulse intervals appear to be the critical parameter for the communication of crickets.

of males is phase-locked with breathing, and breathing is phase-locked with flying and walking. This suggests a master oscillator with secondary oscillator circuits to run each separate behavior. The rate at which crickets sing (and walk and breathe) increases with temperature, and a female responds only to songs produced at the rate which is appropriate for her body temperature. This means that her filter—the releaser for phonotaxis—has the same temperature-dependent characteristics as the males' oscillator circuits, and again it is tempting to suppose that it may be the same circuit. The same genes may be used to build each since, as we shall see in Chapter 19, hybrid females prefer the songs of hybrid males to all others.

If the instructions for how to encode a signal could be used to decode it as well, the evolutionary possibilities would be limitless.

Alas, nature appears not to have been quite as reductionistic as we who study her. It is not at all clear that female crickets pay much attention to the complex patterns of the males' songs. The carrier frequency is sorted for by the resonance of their ears, but beyond simple frequency the songs present a wealth of potentially useful information, which may or may not be picked up by female crickets. The relatively elaborate song of *oceanicus*, for example (Fig. 15–7G), contains five closely spaced rasps in its chirp, followed by about ten widely spaced trills, each composed of a pair of very tightly spaced rasps. It is seductively simple to suppose that the females might use an oscillator filter to sort for the interrasp interval or for the length of the chirp, which would then synchronize or trigger a second oscillator filter to test the trill. But let us not become lost in the labyrinth of complex pattern recognition which led the early ethologists astray (Chapter 4). Like the herring gulls' red-spotted beaks, which turned out to have two wholly independent simple releasers for pecking in chicks, the actual pattern, the acoustic whole, might be utterly irrelevant, and only the interrasp intervals—those in the chirp, the trills, and between the trills—have salience. If this is so, and the song elements therefore constitute three independent releasers, then we ought to be able to scramble their pattern (preserving the proper intervals) and get perfectly normal levels of female responsiveness nonetheless. In fact, females find such garbled songs as acceptable as normal ones. This by no means implies that the same genes may not be building oscillators in both sexes (a possibility which is explored more critically in Chapter 20); it simply means that the relative configuration of the three temporal elements of the males' song pattern is arbitrary.

It has always been a bit puzzling that several species of crickets are virtually mute. A possible explanation is that predation has silenced them. The presumed culprit is a recently discovered parasitic fly which homes in on singing males and deposits newly hatched larvae which feed on the victim. In species thus parasitized, certain males show a curious behavior. Silent "satellite" males will gather around a singing male and, presumably, try to intercept the females he attracts, thus both avoiding the parasites and propagating their genes. (The ways in which stable behavioral polymorphisms such as this could be established and maintained will be explored in Chapter 23.)

The use of auditory pulse coding is widespread: the male Douglas fir beetle mentioned earlier has two auditory codes, and the female has one of her own as well. *Drosophila* males pulse the output of their flight oscillators in a species-specific pattern as part of their courtship ritual (Fig. 15-8). Water striders, which normally use their legs to listen for and localize prey on the water's surface, also broadcast a signal to attract mates.

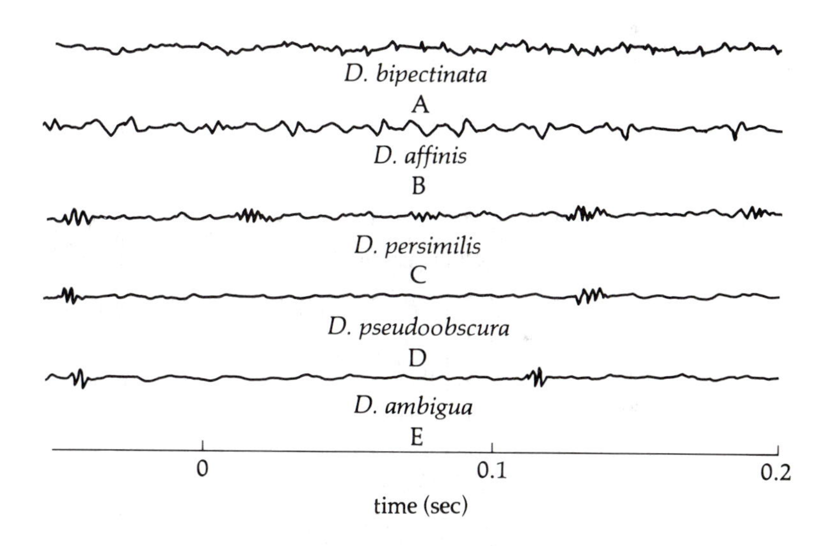

Fig. 15-8 *Drosophila* also have species-specific songs, though they are produced so inefficiently by the flies' wings that they are inaudible to us. The songs consist of single (A and B) or multiple (C–E) wing flappings, separated by species-specific intervals of silence. Odors allow males to locate females, and diurnal species also employ visual cues. The female is provided with a small but effective repertoire of signals to indicate an unwillingness to be courted.

While most insects employ ears tuned to a single frequency, frogs have at least two and often three separately tuned pairs of ears. Bullfrogs, for example, are listening to their world at 200, 500, and 1400 Hz. Again, males broadcast and females respond, but now, because of the differential absorption of high frequencies by the air, females can compare the strength of the high and low frequencies and judge the male's distance. The middle frequency, which corresponds to the low-frequency component in the higher pitched voice of juvenile males, inhibits the approach of females. In a different medium, the weakly electric fish of Chapter 10 use an electronically pulse-coded set of courtship signals.

BREAKING THE CODE

The common theme of communication throughout the world of nonsocial animals is that within the constraints imposed by the sensory abilities of each species and of their keen-eyed or sharp-eared predators, evolution has managed species discrimination by means of one or a short series of simple releasers. It is encouraging to note just how simple the information processing in communication, and for that matter in other tasks, is. Whether it's crickets attracting mates, bats extracting faint echoes from their noisy world, or bees navigating by polarized light patterns, there is no persuasive evidence that the nervous systems of animals do anything more complicated than measure and match: measure the information coming in and attempt to match it against innate or stored information. But the more social animals—animals which have more than single species-specific messages to exchange—are increasingly complex, and resist simple explanations. Even the lowest level of sociality, in which the offspring's need for parental care causes generations to overlap, necessitates a manifold increase in communication. A parent gull, for instance, must recognize the scrawny, bedraggled, and thoroughly unbirdlike creatures which appear in its nest as its own, realize that their pathetic peeps are requests for food, and know that the food is to be placed in the gaping red throats which greet it on its return to the nest. In birds, it is doubtless the case that all of this intricate codebreaking is innate, accomplished by the conventional array of releasers and programmed responses, with the adumbration that in many species parents and offspring come programmed to learn to recognize each other as individuals. However, even such a simple social system as pair-bonding in gulls demands more than spot-detector and movement-recognition circuits. Gulls use a formidable array of mood-specific signals based on posturings of the body which permit each member of the pair to understand and coexist with the other. How can gulls be wired, as we know they are, to recognize the combination of head, neck, and wing positions we have learned to associate with each of their social signals? More complicated social systems often involve dominance, and require an even more complex system of communicating status and "intention."

Complex social communication, though to a great extent innate (dogs are born knowing how to judge when other dogs are scared, for example), is so intimately involved with learning that we may be able to sweep the chaos of complex releasers under the rug. In any case, such signals are so entangled with the workings of the social systems that they need to be taken up as these topics are treated in the next few chapters. Although most social communication is accomplished by the standard, well-worn array of releasers, we simply do not yet understand just how dogs, for instance, are (or could be) prewired to make and recognize the delicate and graded but unambiguously innate distinctions of ear angle and mouth shape that let them decipher the "moods" and intentions of their associates (Fig. 15–9). This ability to recognize relatively subtle species-specific signals, however, is widespread in animals with good visual resolution, and our species certainly falls into this category. Many human

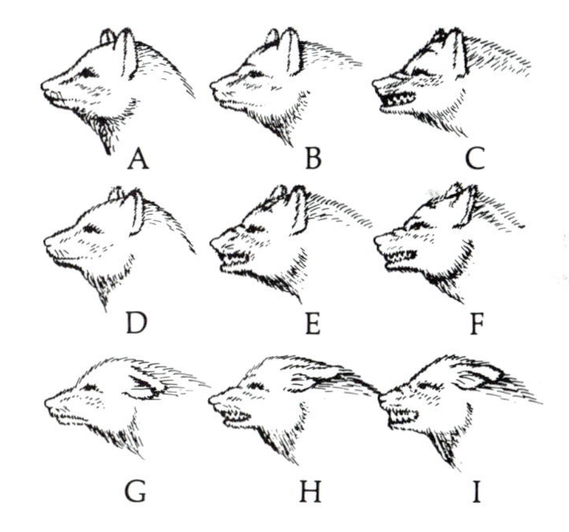

Fig. 15–9 Many birds and mammals employ body postures to communicate mood. Among the higher mammals this "body language" can be relatively subtle. This example is a matrix of anger and fear, with anger increasing from left to right and fear from top to bottom. Hence, dog A is calm and unafraid while dog I is both angry and scared; dog C is purely mad, and dog G is simply frightened. In addition to these posturings of the ears and snout, tail position is also important.

facial gestures—smiles, for instance, or eyebrow flashes—are largely innate. So, too, is their recognition and interpretation. One striking illustration is the uncanny ability of young infants to mimic the expressions of adults around them—that is, to see a smile, recognize it as such, and without degrees in anatomy and physiology to tell them which muscles to use, generate one in return (Chapter 30). It is tempting once again to invoke the notion of preordained neural pictures to account for something we are at a loss otherwise to explain. It is perhaps inevitable that we should sometimes be lost amid a plethora of potential information as we were with the crickets' songs or the gulls' heads.

The animals, of course, are using only a tiny fraction of the detail we see or hear. Inevitable, also, is our failure to notice or deem relevant many cues which are crucial to the animal's world. In the end, however, it seems likely that among the billions of yet-to-be-analyzed feature detectors beyond layer 4c are those releasers which decode the more subtle elements in animal communication.

ELECTRIC FISH COMMUNICATION

Weakly electric fish use their electrical pulses primarily to "see" in their dark, murky habitats, but they can also use their electrical pulses in social communication. Males and females have different pulses and pulse-repetition rates, and during the breeding season unmated males respond to female signals with a special electrical courtship "rasp." This sort of social communication, although physiologically uncomplicated, is no mean task. In Africa, for example, one species of electric fish may have to coexist with twenty other species, each saturating the water with similar signals. How does an individual sort out the messages of the opposite sex of its own kind from the social static?

Through a series of elegant electrical model experiments, Carl Hopkins has isolated the cue which allows the fish to "hear" only what they need to in the electrical cacophony. The sign stimulus for species identification in electric fish, unlike that of crickets, involves neither the frequency spectrum of the pulse nor the interval between pulses. Hopkins placed ordinary Ag/AgCl electrodes in the stream near unmated males at night, and played them a variety of synthetic pulse patterns. The only feature males of *Brienomyrus brachyistisus* required was a pulse 0.4 msec long (Fig. 15–10), a classic example of a sign stimulus.

From neurophysiological recordings in the laboratory Hopkins was able to show how the electroreceptive organs filter the incoming information. The sensors, like diodes, respond to only one direction of current flow. When the normal signal arrives, current flows in one side of the fish and out the other, activating the organs of one side, and then reverses 0.4 msec later, firing the other side (Fig. 15–10). Apparently all the CNS of such a fish does is to time the interval between the discharges of the sensors on its two sides. Signals that satisfy this crude criterion elicit a stereotyped rasp from males, which must itself have some elusive significance for females.

The ostensible simplicity of the electric-fish system is probably a consequence of their communication channel. Except during thunderstorms there is little distracting electrical "noise," and certainly no high-frequency attenuation of echoes to deal with. The only handicap they have had to contend with is the multilingual babble of their neighbors, and the system they have evolved succeeds effortlessly.

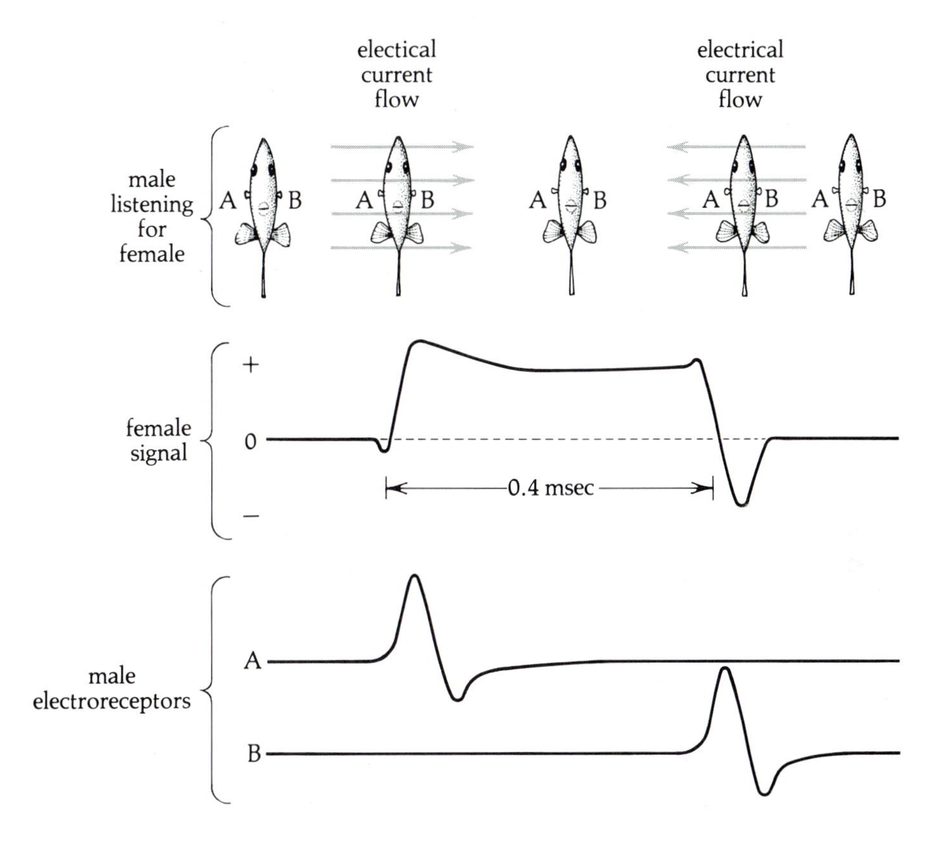

Fig. 15-10 Sexual recognition in an electric fish. The male *(top)* "listens" for the pulses of the female *(middle)* with electroreceptors on both sides of his body. These receptors respond only to current flow into the male's body, and so the cells on one side (A) signal the onset of the female pulse, while those on the other (B) are triggered by the end of the pulse. The male's receptor system recognizes the diagnostic characteristic of the female pulse, its duration, by sorting for pulses 0.4 msec apart between its two sides *(bottom)*.

SUMMARY

Communication is most often used to locate and court reproductively ready members of the opposite sex of the correct species. The general strategy involves one or a chain of releasers. The most ancient systems (and many which are more recently evolved) use pheromones, often extracted and barely modified from host plants or animals.

Pulse-coded messages are also widespread, as exemplified by insects, lizards, and electric fish. The primary cue in pulse-coded communication seems to be the time intervals in the message, a characteristic for which well-known feature detectors exist. One or (most often) several motor-program oscillators underlie the timing of the output and may be responsible for its decoding in the receiver.

More complex communication typical of higher animals involves innate recognition of features for which specific detectors are not yet known and, very often, learning.

STUDY QUESTIONS

1. Recently it has been suggested that the ultimate purpose of communication is deceit and manipulation of other animals. How well does this popular new perspective fit the examples in this and previous chapters? What insights, if any, does it bring?

2. In studying firefly communication, James Lloyd has lately made four observations of considerable interest: (a) A male which has already landed and is searching for a female which has been answering his call will often interject a flash during the signaling of another nearby male. This frequently has the effect of temporarily silencing the female. (b) Predatory females often interject flashes when males of prey species are being lured in, and then return the appropriate signal for females of his species. (c) Males of predatory species often mimic the calls of males of the set of prey species his females are tuned to. (d) Two or more predatory species may coexist in the same habitat. What sense, if any, can you make of these observations, singly and as a group?

FURTHER READING

Bennet-Clark, H. C., and Ewing, A. "Love Songs of Fruit Flies." *Scientific American* 222, no. 1 (1970): 84–92.

Bonner, J. T. "How Slime Molds Communicate." *Scientific American* 209, no. 2 (1963): 84–93.

———. "Hormones in Social Amoebae and Mammals." *Scientific American* 220, no. 6 (1969): 78–91.

Carlson, A. D., and Copeland, J. "Behavioral Plasticity in the Flash Communication System of Fireflies." *American Scientist* 66 (1978): 340–46.

Hopkins, Carl D. "Electric Communication in Fish." *American Scientist* 62 (1974): 426–37.

Tinbergen, Niko. "The Curious Behavior of Sticklebacks." *Scientific American* 186, no. 6 (1952): 22–26.

Wilson, E. O. "Pheromones." *Scientific American* 208, no. 5 (1963): 100–14.

Part IV

LEARNING

CHAPTER 16

Instinct and Learning

INSTINCT

In *The Selfish Gene,* Richard Dawkins envisages animals as robots programmed by their DNA to survive: programmed, that is, to perform specific behaviors that ensure the survival of their own despotic genes. His assertion that many animals seem to be creatures of blind instinct is not new. Inquiring minds since Aristotle have been intrigued by the ability of animals to perform *de novo* behaviors that we, as intelligent observers, are at a loss to interpret or explain, and which we have therefore lumped together under the catch-all term "instinctive."

As early as 1873, Douglas Spalding marveled at the seemingly prescient behavior of a female wasp which would labor untiringly to gather "food . . . she never tasted [for] larvae she would never see" (Appendix A). Viewed through the magnifying lens of modern methodology and experimentation, the digger wasp's behavior appears still more astounding in its clockwork complexity and precision. We have learned that with no previous experience the wasp will excavate a tunnel with elaborate side chambers, and then fly out in search of her only prey: honey bees. She will fly back and forth across the wind until she detects the odor of a bee, and then fly upwind until she sees a small dark object. She will hover 15 cm from the object until it stops moving, and then pounce. Having caught the object, the wasp will smell it to be sure that it is a honey bee, paralyze it by stinging, and take it home. There she puts it into a chamber in her tunnel and lays an egg beside it which will hatch and feed on the immobilized prey.

The wasp's behavior is of course entirely innate, the inexorable result of the workings of instinct. The aim of ethology has been not, as its critics have repeatedly suggested, simply and uncritically to label all behavior as instinctive, but rather to strip away the mystery surrounding instinct by laying bare its underlying mechanisms. We have seen already how the conceptual framework of sensory worlds, sign stimuli, innate releasing mechanisms, motor programs, and fixed-action patterns enables us to comprehend much of animal behavior. Recall that the hunting behavior of the wasp is a rigidly programmed series of releasers and motor programs, and as such is an archetypal model for insect behavior. Even the most determined advocates of the importance of environment ("nurture") over instinct ("nature") in the growth and development of an animal are willing to concede that the behavior of these primitive insects and many other "lower" animals is the product of instinct rather than learning, which enables higher animals to adapt to the contingencies of their environments.

We are now beginning to understand the more detailed workings of instinct, both its mechanisms (releasers and motor programs) and its potential for generating complex behavior. As we have seen, the work of neurobiologists has implicated specific classes of neurons—feature detectors—as releasers, and identified and mapped some of the simpler motor-program circuits. At the same time the study of more elaborate but predictable behavior such as navigation or the construction of nests and webs has made it clear that the innate programming for complexity and for an adaptability to the unpredictable contingencies of nature can be sophisticated in the extreme.

Instinct, then, has taken on a more substantial meaning. We can see it now as the programming of behavior for life in the natural world: programming which specifies sensory channels and sign stimuli, and orchestrates responses which may involve a battery of subroutines and backup programs to accommodate a tremendous range of possible situations. Rather than encompassing merely the rigid and impoverished behavioral repertoire of primitive organisms, instinct has been shown to possess a stunning flexibility and overwhelming richness. As a result, we no longer need invoke the barren behavioristic tenet of learning as an "explanation" of complexity.

And yet the digger wasp, our original example of instinct in action, presents us with an apparent contradiction: how, by instinct alone, can the wasp find her way home? In fact, we know that as she set out to hunt she memorized the location and appearance of her burrow entrance, and then remembered enough about her flight and search directions to get back. Learning, the traditional antithesis of instinct, is indeed essential to her survival.

How much and how well do insects learn, how does insect learning relate to learning in higher animals such as birds and mammals, and how does learning fit into the ethological view of behavior? After decades of thinking of insects as mere creatures of instinct, we are just beginning to appreciate not only how astonishingly complex their behavior really is, but how important learning is to them. Nowhere is this blend of learning and instinct more strikingly illustrated than by that supremely well-programmed animal, the honey bee.

As we learned in Chapter 13, honey bee navigation is a remarkable feat. When a forager bee leaves the hive in search of food, for example, she may fly hundreds of meters this way and that before finding a suitable patch of flowers. During the flight she will probably have to aim herself into the wind—perhaps considerably off her true course—to maintain a consistent flight direction. The sun, which is her usual compass, may have been hidden behind a cloud or a line of trees for all or part of her flight, and its position in the sky will change markedly just over the course of her search. And yet when the forager arrives at a food source she is able to calculate a true compass direction and distance back to the hive. She has allowed for how the ever-changing wind has affected her path. She has calculated the sun's position when it was not visible, and compensated for its predetermined but variable movement to the west. What is more, back at the hive she will translate all these computations by means of specific linguistic-like conventions into a dance which will "tell" other bees the distance and direction of the newly discovered food. Willing recruits attend these dances, and then use the information gained from them to find the food for themselves.

At every stage in this behavior, learning has been essential. The forager has first memorized and then calculated and substituted and averaged information like a small computer. And bees are not unique in this regard: as we have seen, learning is crucial to most insects, even if it seems at first glance a peculiar sort of learning. It is curiously constrained and machine-like, and despite the traditional dichotomy between learning and instinct, it is tempting to call it *instinctive* or *programmed learning.* The phenomenon is most obvious in the flower-learning behavior of honey bees.

HOW ANIMALS LEARN

More than two thousand years ago Aristotle noticed that on any particular foraging flight a bee will gather food from a particular kind of flower and bypass others. Clearly the bee must have learned to distinguish the one species of flower from all others. It does not seem to have occurred to Aristotle to wonder how these diminutive creatures could be doing this. Early in this century, though, Karl von Frisch raised the question and, in answering it, upset the prevailing view that animals in general, and insects in particular, were deaf, dumb, and virtually blind. Von Frisch reasoned that since flowers have so much to gain from bees that remain faithful to their one species and thus cross-pollinate them on each stop, and so much to lose if bees are fickle, that the colors and odors of flowers probably evolved as alluring signals for the bees to learn. Bees, he theorized, must therefore have color vision and excellent "noses." Indeed, he was able to show that bees are able to remember and distinguish one floral odor from among some 700 others, and can learn any color from yellow through the greens and blues and on into the ultraviolet.

So far, so good. Bees are smart enough to learn the particular colors and odors of flowers. Around 1930, however, Elizabeth Opfinger, one of von Frisch's students, made the disturbing discovery that bees can learn a flower's color *only as they approach it.* This is striking to watch. We often train bees to a feeder by catching them as they fly out from the hive and then carrying them to the food; when the bees leave, they circle the feeder again and again as if "studying" it before flying back to the hive. Yet when they return they clearly have no idea what the feeding station looks like. The bees search everywhere, often to the great discomfort of those of us involved in the experiment. After they have once landed on the food *of their own accord*, however, they display no hesitation in the future. Indeed, Tinbergen's digger wasps must operate under similar constraints since they learn their burrow landmarks only on the flight *away* from home.

By changing the colors of an artificial flower during the approach, landing, feeding, and departing flights of foragers, Randolf Menzel, a German neuroethologist, has investigated how bees learn food-source color in more detail. He found that bees remember only the color they have seen during the *final two seconds before landing* (Fig. 16–1). On the other hand, the bee learns the landmarks around the food only as she flies away. Indeed, if we remove the landmarks during feeding, or carry a feeding bee back to the hive, she will have no recollection of the landmarks she must have seen as she approached the feeder. It is as if bees possess a set of switches which turn color and landmark learning on and off at certain times and in certain

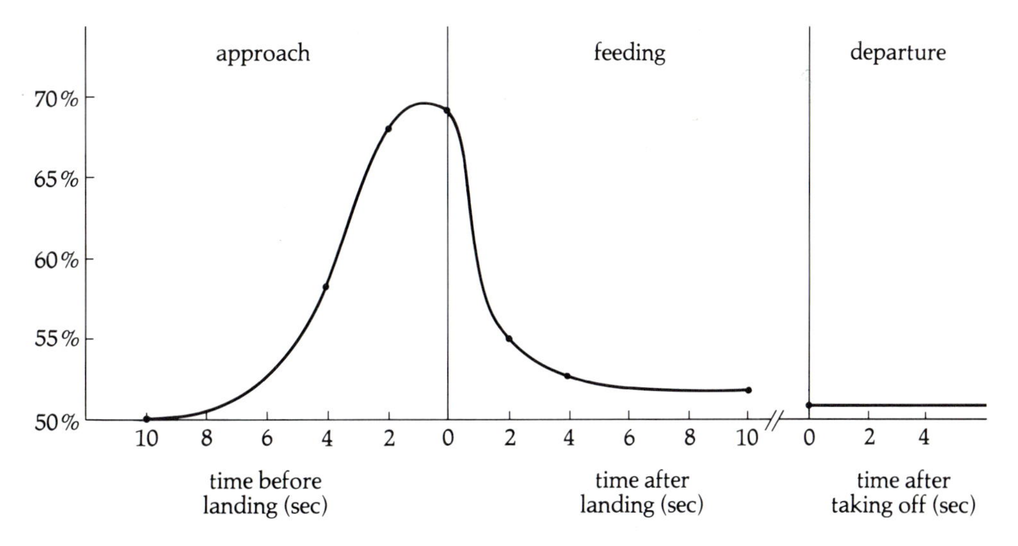

Fig. 16–1 Bees learn the color of a flower in the final seconds before landing. In a two-color choice test, bees which saw a color up to three seconds before landing or any time after landing or departing showed little or no preference for that color when they next returned to the food. Bees which saw a color during the final two seconds before landing showed a strong preference for that color—just the preference predicted from the one-trial point on the color-learning curve of Figure 16–2.

situations. In fact, there must be still another switch for odor, since the bee learns it only while she is on the flower.

The same pattern of constraints appears when foragers try to find home. On their return trip they depend on their elegant navigation system to get close to the hive, but locating the tiny opening in perhaps one of many similar trees from, say, 10 m away poses an enormous problem for an insect with such fuzzy vision. As a result, evolution has built in a program by which bees learn what the hive entrance and its surrounding landmarks look like. But since the appearance of the hive and the nearby vegetation necessarily changes with time or with the seasons, bees are programmed to relearn what the hive and its environs look like on the first flight out each day. Thus a colony can be moved several miles overnight without causing any difficulty for the bees. Moving it even a few feet at midday, however, inevitably produces swarms of confused and disoriented foragers searching frantically in the old location, even though they had flown out only a few minutes earlier from the new position.

For bees, then, learning has become specialized to the extent that specific cues can be learned only at specific times, and then only in specific contexts. In fact, their learning programs are even more specialized than this: although they acquire each bit of knowledge required for locating a food source separately and at different rates, this knowledge, once acquired, forms a part of a coherent and holistic set.

The first hint of this possibility came around 1930 when von Frisch followed up an observation by Auguste Forel that the bees which routinely shared his tea only showed up at table fifteen minutes or so before the scheduled hour. Von Frisch quickly confirmed his guess that bees learn to associate a time of day with a particular food source. Since many flowers produce nectar only during specific parts of the day, bees would waste a great deal of time and energy if they did not learn this. The strategy which underlies this ability, however, can lead to some odd behavior. For example, Martin Lindauer, von Frisch's most gifted student, trained bees to a feeding station that offered unscented food, but during one specific half hour he added scent. Given the choice on subsequent days, the bees steadfastly chose unscented over scented food except for just before and during that half hour. The bees, the feeding station, and the food remained the same, but the preference or expectation of the foragers switched. It is as though the bees had come equipped with appointment books in which they had noted down the food odor to be expected at each time of day. Similar experiments have shown that time, odor, color, shape, landmarks, and location are all learned as a "set," and that bees can learn and remember at least a dozen such sets; but if one component in a set is changed, the bee must learn that whole set all over again as if it were completely new.

This rather surprising revelation becomes obvious when we discover that bees learn the items in a set at different rates and with different degrees of precision. Odor is learned very quickly: after one exposure a bee will choose that food odor over another 90 percent of the time. Color learning is much slower. A bee with one exposure will choose the correct color about two to one over any other color, and reaches the 90 percent level only after six trials (Fig. 16–2). Bees never get much better than 90 percent to color, even when the choice is between such obviously different hues as yellow and blue, but to odor they become essentially perfect. So machine-like is this learning, though, that we can push the bee through her series of color-learning lessons in only one visit, letting her land on the feeding station and waving her away again and again the requisite six times. On her next test the forager will choose the correct color 90 percent of the time. Landmarks are learned more slowly still, and despite the bee's excellent clock sense, time of day is the last factor of all to become accurately fixed in her mind.

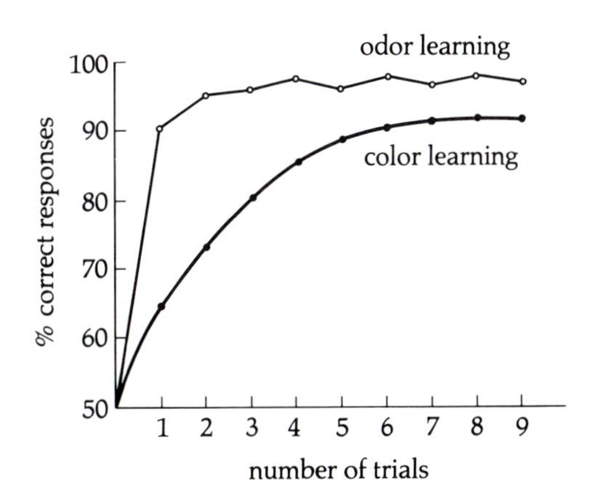

Fig. 16–2 Bee learning, like that of humans, improves with repeated presentation of the stimulus and reinforcement. The curves above represent the percentage of correct responses to a two-stimulus choice test after various numbers of training trials. Odor learning is very fast—bees are olfactorily smart—while color learning takes longer.

The holistic organization of bee memory stands out in sharp relief when we change one of the quickly learned components and observe the effects of that manipulation on the rest. For example, a forager may be thoroughly familiar with a blue, triangular, peppermint-scented food source which is available from 10 to 11 A.M., visiting the site day after day only during that period and choosing the color, odor, or shape over any other we might provide with a reliability of 90–100 percent. If instead of offering a choice we switch the odor to orange scent, which the forager immediately learns to choose over other odors with nearly perfect precision, her previous mid-morning preference for the color blue and triangular shape now vanishes. Instead the bee must laboriously master color, shape, landmarks, and time all over again.

In a very real sense, then, honey bees and other insects are carefully tuned "learning machines." They learn just *what* they are programmed to learn, just *when* they are programmed to learn it. This strict learning protocol makes them especially suitable for study since they will learn a well-defined cue reliably when presented with an equally well-defined releaser—no more and no less. And because of the organization of the insect nervous system, which is a strung-out series of ganglia, it often happens that a single ganglion can be separated from the rest and taught a simple task like lifting or lowering a leg to avoid shock. Hence, a particular learned response can be isolated to a few hundred nerve cells, one or more of which has changed in some crucial, yet-to-be-discovered way.

Despite its awesome anatomical complexity, the insect brain is still less forbidding than those of vertebrates because of the almost mechanical reliability of the learning program. This convenience has, for example, allowed Randolf Menzel to trace the anatomy of honey bee odor learning. To see what parts of the brain had to be active for the bee to learn, he used a liquid-nitrogen–cooled needle to chill specific tiny spots in the brain. Chilling stops neural activity; any cells essential to learning and memory must be warm for things to proceed along their normal course. When he performed this elegant experiment Menzel found that over the six minutes after he presented odor and food to the bee the sensitivity to disruption (i.e., the site of memory processing) moved systematically from the antennal lobes to the alpha lobe of the mushroom bodies, and then to the calyx (Fig. 16–3). When only one antenna was trained, Menzel

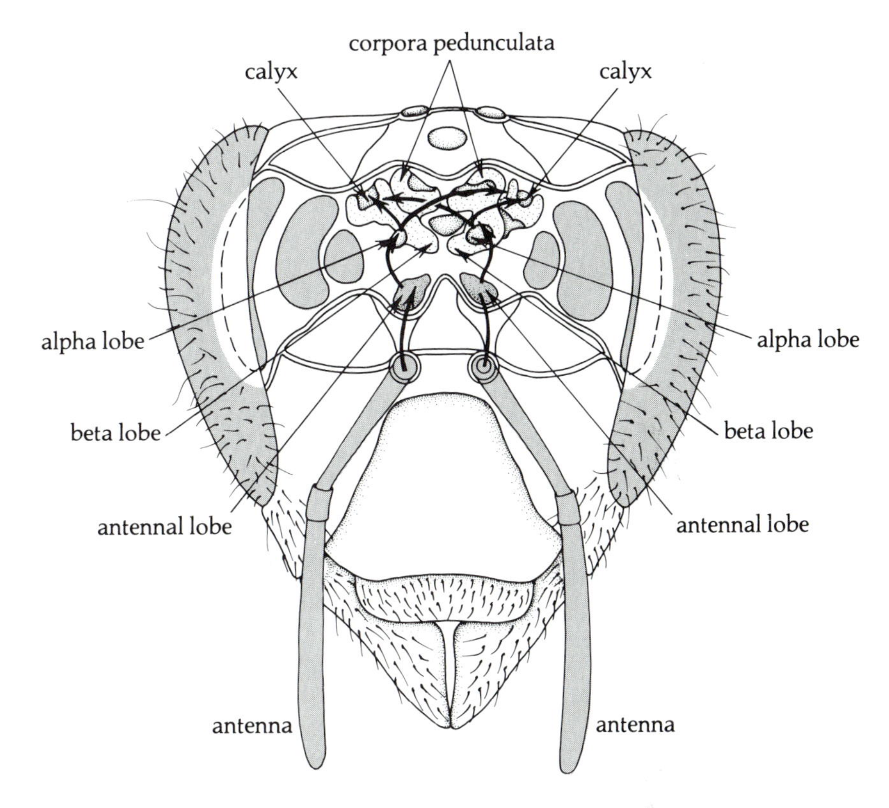

Fig. 16–3 Information flow during odor learning in honey bees can be traced by chilling various parts of the brain at different times after training with a needle dipped in liquid nitrogen. The first sensitive area is the antennal lobe, followed a few minutes later by the alpha lobe, and finally by the calyx. Toward the end of this sequence, the information from one antenna reaches the other side of the brain. [From "Learning and Memory in Bees," by Randolf Menzel and Jochen Erber. Copyright © 1978. Scientific American, Inc. All rights reserved.]

found that the processing remained unilateral until the last step, when a copy of the learned information was transferred to the other side.

Clearly the insect brain is a powerful and convenient place to study learning, and the next logical step is to listen with electrodes as the cells in the pathway "learn." There remains the nagging doubt, however, that insect learning is somehow different in *kind* from learning in birds and mammals, and hence might be irrelevant to understanding "real" learning. As we shall see in the next chapter, there is every reason to believe that learning involves basic processes shared across the animal kingdom.

SUMMARY

Instinct and learning are usually thought of as a pair of mutually exclusive behavioral opposites. Studies of insect behavior, however, suggest that learning in many invertebrates is "instinctive": genetically programmed with regard to context, time course, and cues. Indeed, it is difficult on the one hand to imagine how many insects could exploit their environments without being able to learn, and yet to see how in their brief lives they can afford the time to work out by trial and error the characteristics of the world around them that they ought to learn.

STUDY QUESTIONS

1. Assuming no intellectual limitations, why should bees be programmed to learn odor so quickly and so well, color less quickly and always with some margin for error, and shape less accurately and less quickly still?

2. Although bees can learn to distinguish different patterns of the same color, it is not obvious that they remember "shape" in the sense we conceive. Indeed, certain strikingly different patterns seem indistinguishable to the bees, and one suggestion is that the bit of information actually stored in their mental notebooks is the ratio of circumference to surface area. How could this idea be tested rigorously? What advantages can you imagine for such a strategy, and how would you test them?

3. If bees can learn color to a 90 percent criterion in six landings, why has evolution not programmed them to make six landings during their first visit to a single new flower?

FURTHER READING

Menzel, Randolf, and Erber, J. "Learning and Memory in Honey Bees." *Scientific American* 239, no. 1 (1978): 102–11.

CHAPTER 17

Programmed Learning

Ethology began as a search for the sources of "knowledge" which account for the highly varied and well-adapted behavior of animals. Unlike behaviorists, who theorized that animals must be conditioned into their species-specific behaviors by a combination of morphological and environmental contingencies, ethologists concluded that much of this knowledge must be inborn. The overwhelming role that instinct plays in the lives of animals is especially obvious in insects, and yet from the outset ethologists recognized the importance of learning even for these lowly creatures. Behavior, however, the way an animal reacts in response to various natural stimuli, is the *only* real measure we have of what and how well most animals learn.

MECHANICS OF LEARNING

As we saw in Chapter 16, bees and other insects learn surprisingly well, but in a curiously constrained and almost machine-like way. This seems fundamentally different from our usual picture of learning in higher animals (and particularly humans), in which we see learning as "plastic," self-directed, and unconstrained as to what, when, or how it is accomplished. It is tempting to suppose that the difference which we feel sets us apart from the insects is one of *kind* rather than just degree, that we are so different that the basic process of bird and mammalian learning is unique. We have always trusted (or perhaps hoped) that there were insurmountable barriers between vertebrate and invertebrate learning, but where may these lines of defense be drawn?

One of the processes we have previously considered characteristic of mammalian learning is a phenomenon known as short-term memory. "Learning" is our word for what happens between the perception of information by our sense organs and the ultimate storage of some part of that information in our brains. Memory, on the other hand, is the recollection and use of that information later. There is no direct line between the sense organs and the memory banks: our minds first sort and

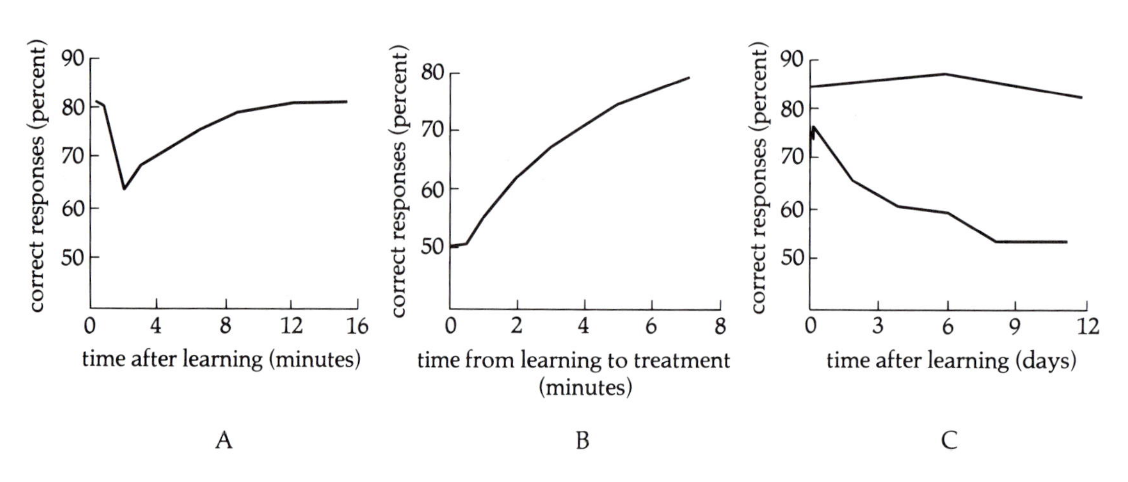

Fig. 17–1 Bees rewarded once on a certain color, and then offered a choice between that and another color, land preferentially on the training color. Their preference, however, depends on how long after training they are tested. A. "Memory" (the ability to recall the information) appears to decline sharply after training and then recovers, a result which suggests that some sort of active processing may make the information temporarily inaccessible. B. When bees are distracted or subjected to physiological stress after training, their ability to remember is impaired. In this example, when bees were shocked 30 sec after training they were unable to choose the correct color any better than is predicted by chance, while shock 7 min after training has no effect. The shock disrupts the active information processing (known as short-term memory) which precedes storage of data in long-term memory. The process which transfers information into long-term storage is called "consolidation." C. The stability of honey bee color memory depends on how often a forager is rewarded. In this case the lower curve, representing a bee trained only once, shows complete forgetting in about a week. In contrast, a bee rewarded three times (upper curve) will never forget what she has learned.

filter the input before it can be stored (Fig. 17–1A). This sorting is essential lest we clutter our memories with everything we sense, trivial and important alike.

During this "short-term" phase of active neural processing any shock or disturbance is likely to cause us to forget what we have just experienced. A disruption just a few minutes later, however, will have no effect. This is the source of the amnesia an accident victim experiences when he or she has no recollection of the accident or the events immediately preceding it. Menzel has shown that bees possess the same short-term memory phase. If we shock or chill a forager within five minutes of her finding a new source of food, she will forget all about it. If we wait instead until ten minutes after the discovery, the "knowledge" has passed into a long-term form and cannot easily be eradicated (Fig. 17–1B). We can see precisely the same processing at work in the lowly *Drosophila*, although the time constants are different (Chapter 20).

The experiments of behavioral geneticists have turned up other similarities (Chapter 20), and it is increasingly logical to suppose that learning involves a single set of basic processes which are shared throughout the animal kingdom from molluscs to man. If this is the case, then the difference in kind which we sense separating insects and higher animals must lie instead in the way learning is organized. Perhaps it is that insect learning is rigidly constrained, while in birds and mammals it is plastic: that bees lead such simple lives that evolution has thought of and provided for all possible contingencies and preordained the necessary learning, while higher animals live in such a rich and complicated world that evolution has had to provide them with brute, unfettered intelligence to enable them to figure out what to do.

Again, however, the facts fail to support such a basic difference. For one thing, many of the "higher" mental abilities that psychologists have discovered in their rodents turn up in insects whenever someone asks the question. For example, higher vertebrates have the ability to form what is known among aficionados as "cognitive maps," which enable animals to perform an ordered sequence of behavioral sets. Rats, for instance, can run mazes whose solutions depend on the animals' turning one way at one cue (perhaps a color or shape) and another direction at another cue. Further, a triangle might direct them right at one point in the maze and left at another: the rats will have formed

PROGRAMMED MEMORIZATION

A female digger wasp typically digs and provisions one tunnel at a time with paralyzed prey—bees, caterpillars, or whatever her species specializes on. When she has collected enough to feed a larva until it pupates, the wasp lays an egg and seals the chamber. This "mass provisioning," as it is called, leaves the larvae to fend for themselves. The first tentative step toward sociality is taken by species which must feed their young on a day-to-day basis. Since developing larvae have relatively modest daily needs a wasp can support several at once. One species of digger wasp, *Ammophila pubescens,* has developed this "progressive provisioning" to an art, maintaining a dozen or more burrows at a time. This must be an enormous intellectual task, since not only is each larva in a different stage of development with necessarily varying needs, but the burrows themselves may be far apart, and well camouflaged to avoid parasitic predators.

In a thorough study of the behavior of *A. pubescens,* the Dutch ethologist G. P. Baerends observed that before setting out each day to gather provisions for her young, the wasp would first visit each burrow in turn. Then, when she returned, she would stock each cell with just the amount of food each larva needed: small larvae received one, two, or three caterpillars, large larvae between four and seven, eggs none, and larvae which were beginning to pupate were not fed, but had their cells sealed. In an effort to determine what was going on in the mind of the wasp, Baerends began to experiment with removing and interchanging the larvae. He discovered that if he made his exchanges *before* the wasp made her preliminary morning round, each larva would receive the treatment appropriate to its age and stage of development. If, however, Baerends switched or removed larvae *after* the wasp's first visit, she would provision each cell as if the morning's occupant were still there. Cells which Baerends had emptied were stocked as though that morning's hungry larva were inside; cells which had contained pupae but now held eggs or young larvae were sealed off and abandoned.

During the morning inspection, Baerends reasoned, the wasps must make some sort of "shopping list," in which the detailed needs of each burrow are inscribed in association with its location and relevant landmark information. Working from her original shopping list the wasp, having learned more in a few minutes than many humans could—perhaps as many as forty-five items in sets of three—proceeds to organize the rest of her day as mindlessly as a machine. This sort of learning, then, phenomenal as it is, is as stereotyped as any other piece of insect behavior.

an ordered mental "map" which tells them how to interpret each cue in order.

Anyone familiar with the way bees use landmarks for navigation (especially when their using them will ruin an experiment) might wonder if bees must not need such sequential memory. Indeed, Euglossine bees in the tropics will develop and memorize a foraging sequence between a dozen or more widely separated flowers, traversing perhaps ten miles before completing their circuit (Fig. 17-2). Surely this tour de force demands an elaborate sequential memory. Honey bees, placed in the same sort of maze-learning situation as the rats, can master a maze with at least five choice points.

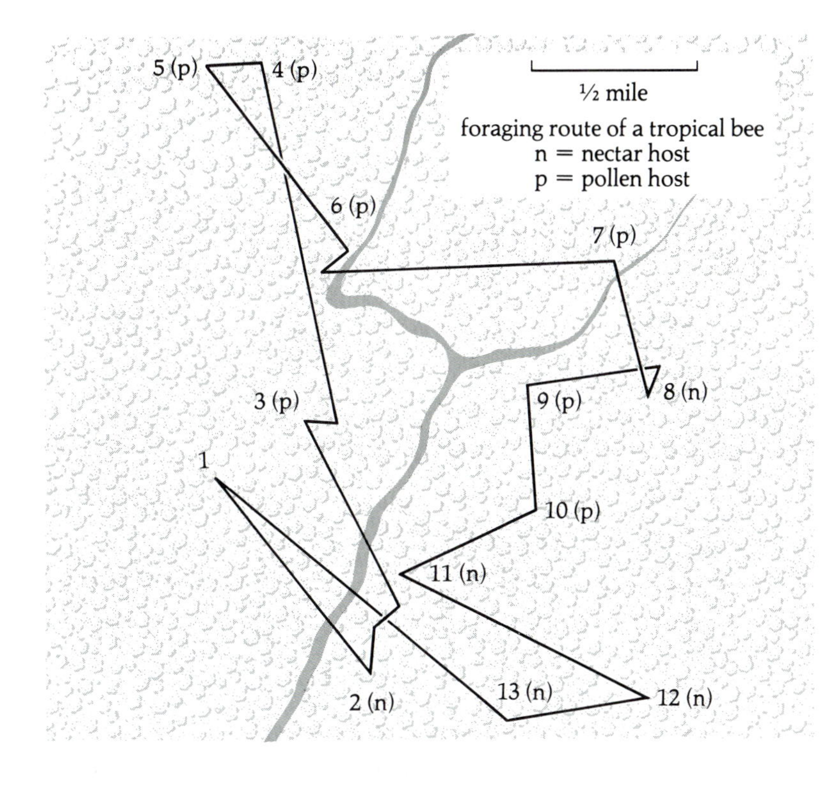

Fig. 17-2 A representative foraging route for a tropical bee in a Costa Rican forest. The nest is 130 m above the ground, under some loose tree bark (1). Each day the bee flies a stereotyped "trapline" route to vines, ground plants, bushes, and trees collecting nectar (n) and pollen (p) from the one or two new flowers produced by each plant every day.

FOOD-AVOIDANCE LEARNING

On the other side of the coin, examples of tightly programmed learning in birds and mammals are inexorably becoming the rule rather than the exception. One simple and fascinating example is "rapid food-avoidance conditioning." Many animals from blue jays to garden slugs come programmed to wait a species-specific length of time after eating a new food to see if they become ill. If they do—even if the sickness arose from a completely independent cause—they will never eat the food again. Even more curious, each species is programmed to identify the forbidden food in the future by its own set of cues. For example, rats will remember the suspect food's *odor* while quail recall its *color* (Fig. 17–3). This is the consequence of specific

Fig. 17–3 Perhaps the most familiar example of rapid food-avoidance conditioning is the reaction of a bird that has just eaten a monarch butterfly. The butterfly's distinctive, bright wing pattern provides the cue which birds memorize after their first mistake. The distasteful substances of the monarch are actually stolen from milkweed plants which manufacture the compounds for their own defense. Several other species of butterflies have copied the monarch's distinctive patterning, and so share a measure of protection from well-educated birds.

programming rather than any sensory bias: the animals can sense, and in other contexts remember, a far wider range of stimuli. Pigeons, for instance, will learn to associate food with a visual cue dramatically faster and forget the association far more slowly than with any other cue, but they associate danger in the same way with an auditory cue. Obviously there is something strangely *un*plastic about these cases of vertebrate learning. The animal is prewired to remember only the one or two cues deemed salient by evolution in such situations, and to ignore any others.

Our species is not immune to this sort of mindless mental bias. Researchers served an unfamiliar flavor of ice cream to children a few hours before they experienced the nausea produced by chemotherapeutic drug treatment, and the children refused the same ice cream several months later. This marked aversion developed even though the patients knew from long experience that the nausea was caused by the treatment rather than by any food. Children receiving the same ice cream *after* their treatment accepted it again eagerly.

IMPRINTING

A more complicated example of the programmed learning common to higher animals is "imprinting," a term which calls to mind a picture of Konrad Lorenz leading a long line of devoted baby geese across a Bavarian meadow (Fig. 2-6). Young animals which have to keep up with their ever-moving parents—antelope and sheep, for example, as well as geese—must learn to recognize those parents as individuals rapidly if they are to survive. To accomplish this, evolution has built into these creatures an elegant learning routine. The first task for a newborn duckling or calf is simply to stay with its mother as long as it needs to to learn her individual appearance. To do this, ducklings, for instance, are programmed to follow the first object producing the species-specific call they see moving away from them. When it stops moving away, the duckling stops.

While the sight or even the sound of an object moving away triggers the "following response," it is the *physical act of following* which starts the learning. We know this because a chick carried passively behind its mother, like a bee ferried to and from a feeder, fails to learn. The strength of the imprinting is directly proportional to the distance the chick has followed the parent. In fact, like honey bees judging how far away a food source is, distance and effort are interchangable. Obstacles and barriers in a chick's path strengthen imprinting just as effectively as further following. The learning involved in imprinting is further constrained to a brief, species-specific "critical period" of about a day (Fig. 17-4). Whatever the chick sees (or hears) later is virtually irrelevant, so that hand-reared ducklings, for example, steadfastly ignore real ducks. More recently, however, it has become clear that animals with imprinting programs are not completely naïve about what their parents should look and sound like. They are born with one or two general clues which they use if a decision between two moving objects becomes necessary—a contingency which must be very common in our crowded world. Hence, given a choice between, say, geese and people during its critical period, a duckling is programmed to pay attention to the more duck-like object.

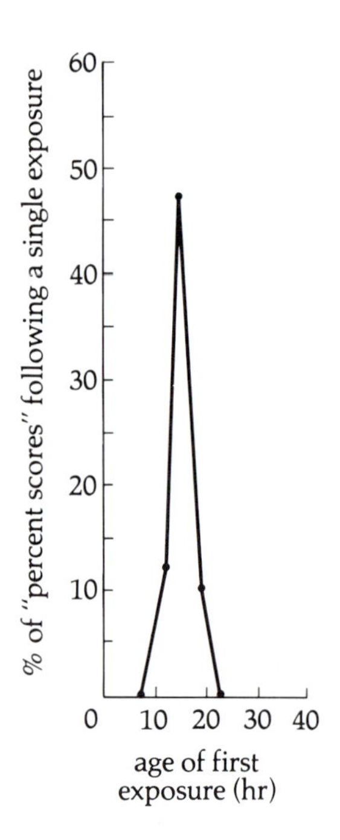

Fig. 17-4 The critical period for parental imprinting comes during a few hours in the first day of life. The experiment summarized here tests how well chicks memorized the features of the parental model after a single presentation at a particular time after hatching.

Like selective learning in general, imprinting is specific to cue. Wood ducks, which nest in holes near the water's edge, are a case in point. The parent must utter a species-specific call to get her chicks out of the nest. The downward frequency sweep embedded in the call is the sign stimulus. Since the wood duck's shore-nesting habit requires that the following response be elicited acoustically, the ducklings probably imprint on the individual acoustic characteristics or "accents" of their parents over any but the most obvious morphological features. Other species such as mallards attend instead to the morphology of the parents, although in mallards, too, the call is species specific. (Repetition rate may be the sign stimulus for mallards.) Mallard chicks, then, ignore intraspecific acoustic differences and focus on learning to recognize their parents individually on the basis of sight.

When Lorenz first worked on imprinting in geese, he noticed that hand-reared ganders would later try to mate with humans. He concluded that sexual preferences were set while chicks were learning to recognize their parents. Lorenz's keen intuition combined with his disdain for experimentation to lead him astray on this point. We know now that "parental" and "sexual" imprinting are two completely separate phenomena whose critical periods happen, as fate would have it, to overlap only in the case of geese. More frequently, sexual imprinting has a critical period of several weeks which begins when the animal is several months old. Like parental imprinting, if the available stimuli are controlled throughout the full length of the critical period and no choice is offered until it is over, sexual imprinting is strikingly irreversible. German ethologist Klaus Immelman raised male zebra finches with closely related Bengalese finches throughout their critical period for sexual imprinting. Before this exposure the males had been kept with their own species and afterward were offered only females of their own species. The zebra males mated reluctantly with the zebra females and raised broods at the last possible moment each year; but when offered the choice eight years after last seeing one, they unambiguously chose female Bengalese finches over their conspecific partners of the previous seven years. They had not forgotten the lesson they were wired to learn as juveniles.

As Chapter 4 suggested, an equally important class of imprinting is the process by which parents learn to recognize their young as individuals. For the first two days after its eggs hatch, an adult gull will accept any chick into the nest and defend it. After that, the parents will attack, kill, and eat an outsider. It seems likely that this imprinting on offspring is triggered by some specific cue—a sign stimulus—such as the begging mews or the colored throat patches of the young.

In such cases of programmed memorization, learning is simply a subroutine called into play by timers and releasers to increase the fitness of the animal. Species which have an evolutionary need to learn something particular have the necessary clockwork to do it, while closely related species with no such need lack this piece of harmless programming. Thus it is that birds parasitized by cuckoos and cowbirds imprint on their own eggs, while other species do not. Individuals of a species which nests so close together that a possibility of accidentally incubating a neighbor's egg exists also have the egg-learning routine, while birds whose nests are farther apart—herring gulls, for example—fail to notice a change in number, size, or even color of eggs. Herring gulls are not simply stupid. Although they cannot remember what their eggs look like, they memorize the exact appearance of their chicks and can distinguish them from all others. Herring gulls have to be "smart" about their chicks since they run the risk of having strange chicks wander into their nests. The kittiwake gull, on the other hand, whose precarious cliffside nest is in no danger of being invaded by another pair's young, fails to notice if either the eggs *or* chicks of an entirely different species—different in both size and color—are substituted.

SONG LEARNING

Another example of rigidly programmed plasticity is the song learning of many birds. Early ethologists noticed that for at least some species—doves, for example—normal species-specific songs developed in individuals raised in isolation. This was by then a familiar pattern characteristic of innate behavior, and so it was tempting to suppose that bird song was instinctive. And yet what about parrots and mockingbirds, which clearly learn much of their repertoire? Were there two patterns: the wholly innate and the wholly learned?

In the 1950s, first W.H. Thorpe and then Peter Marler began to suspect that the usual case for songbirds was, oddly enough, that although the song *was* learned, only the species-specific song *could be* learned. Consider, for example, the dilemma of swamp sparrows. This species lives in "islands" of its preferred habitat in the midst of the much larger range of its close relative, the song sparrow. How do swamp sparrow chicks, whether in the lab or the field, know to reject the songs of the ubiquitous song sparrows—and the dozens of other species living within earshot—and learn only the swamp sparrow song? The sequence, worked out by Marler and his students, is similar in many ways to classical sexual imprinting. It begins with an early critical period during which the birds memorize their species' song. Evolution has provided them with filters—IRMs—by which they distinguish their own song from those of other species and from the general noises of the environment. When they hear the appropriate cues, they turn on their mental tape recorders. Acceptable songs—those that trigger learning—can vary widely, so long as the one or two diagnostic clues which serve as releasers are present.

In the case of the swamp sparrows, the range of acceptable songs is at once both large and narrow. The songs are all different—they must be so that the birds can recognize mates and neighbors as individuals—and yet are all of a type: monotonous repetitions of a single "syllable" (Figs. 17–5A and 17–5B). Song sparrows, on the other hand, have several different syllables of

which some are delivered at increasing or decreasing rates (Figs. 17–5C and 17–5D). The marked difference in temporal pattern is so pronounced that at first sight it seems it must be the diagnostic cue by which the two species sort things out. But as we have so often seen, what is obvious to our species is as often as not subtle to another, and what is subtle (or invisible) to us is frequently glaringly plain to them. And so it is with sparrows. Single song sparrow syllables, well within the range of variation to be seen in swamp sparrows (to *our* eyes and ears, that is) but arranged according to the monotonous scoring of swamp sparrow song, are unambiguously rejected by young swamp sparrows (Figs. 17–5H to 17–5J). Yet the chicks *will* learn swamp sparrow syllables orchestrated to song sparrow tempo, although they later deliver them at an even rate (Figs. 17–5K and 17–5L). And so, just as in so many other cases, it is the *element*—the syllable in this case—rather than the fine structure of the patterning which is the key; and somewhere, hidden within the many apparently different swamp sparrow syllables but common to them all, is the specific sign stimulus which accounts for the species' ability to organize their world.

The process of song learning in birds is a wonderful illustration of how programmed learning works. For one thing, the song is normally learned long before the birds begin to vocalize. The lengthening days of the following spring trigger the release of specific hormones in the males, hormones which in turn initiate attempts to match first the elements and later the sequence of the stored song. By a trial-and-error process analogous to that by which infants learn to crawl, the birds learn slowly to manipulate their vocal musculature to produce a match between their vocal output and the "recording" in their brains. Once learned, the muscle sequence becomes a hard-wired motor program, so fixed that the bird may be deafened without affecting his song production (Fig. 17–6).

The fidelity of this learning is indicated by the tendency of the birds to preserve even the idiosyncrasies of the song they heard in their critical period, a fact which has probably led to the regional "dialects" found in the songs of many species (Fig. 17–7). Recent evidence indicates that females too remember well enough when they come to mate to choose the best facsimile of the song they heard months or years previously during their critical periods. For females, then, song learning is a kind

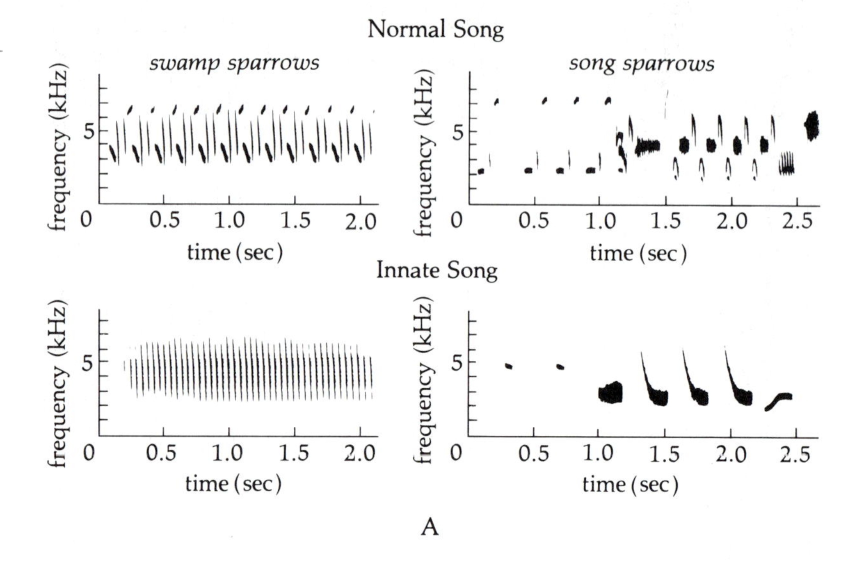

Fig. 17-5 A. Though their ranges overlap considerably, swamp sparrows and song sparrows remain faithful to their species' song type. Within each species, considerable variation exists between different birds, and serves as the basis for individual recognition. The normal swamp sparrow song consists of a trill of nearly identical syllables delivered at a constant rate. Birds raised with no opportunity to hear their species' song produce a trill of very simple syllables ("innate song"). Song sparrows, on the other hand, usually produce a song consisting of a trill of identical syllables delivered at a varying rate, followed by a single, complex syllable—a "note phrase"—followed by another trill and another phrase. Song sparrows with no opportunity to learn a song nevertheless generate an unmistakable facsimile of the normal song with shortened and simplified trills and note phrases. B. At first glance, the basic difference between swamp sparrow songs (a and b) and those of song sparrows (c and d) seems to be in the temporal patterning rather than in the syllables. Nevertheless, when baby swamp sparrows are offered a choice of various synthetic songs as potential models composed of syllables from normal songs of the two species scored in either the repetitious swamp sparrow manner (e–g) or in the complex and more interesting song sparrow fashion (h–j), the young accept only songs containing swamp sparrow syllables (k–m). Since young swamp sparrows are able to recognize the elements of their song even when the patterning is wrong (as in example m), the recognition must be based on the syllable per se—probably something to be found in the innate syllable type. Notice that though the pattern of the model was wrong, the swamp sparrow delivered its syllables at a constant rate (m).

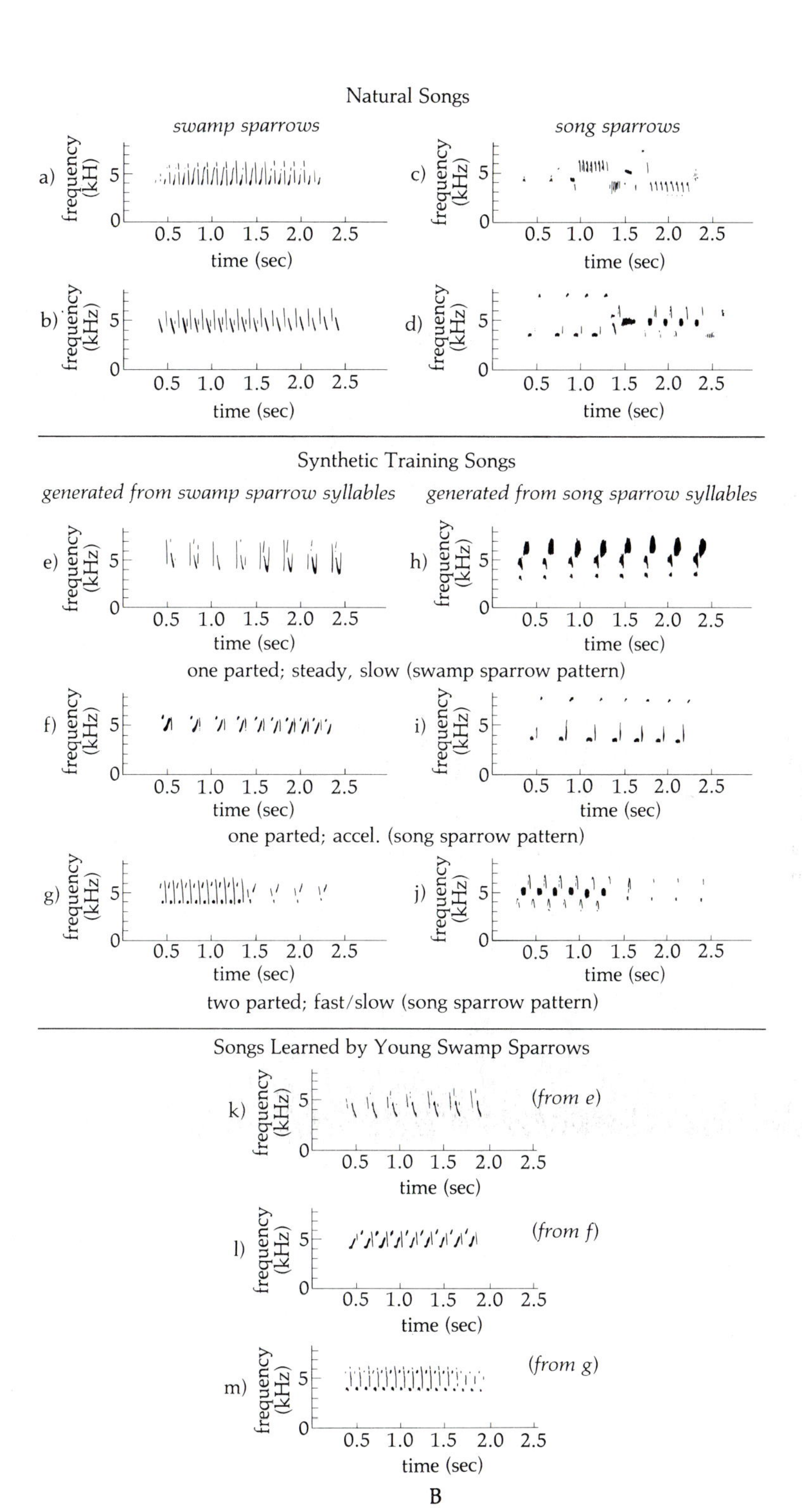

Natural Songs
swamp sparrows
song sparrows
a)
b)
c)
d)
frequency (kH)
frequency (kHz)
time (sec)
Synthetic Training Songs
generated from swamp sparrow syllables
generated from song sparrow syllables
e)
f)
g)
h)
i)
j)
one parted; steady, slow (swamp sparrow pattern)
one parted; accel. (song sparrow pattern)
two parted; fast/slow (song sparrow pattern)
Songs Learned by Young Swamp Sparrows
k)
l)
m)
(from e)
(from f)
(from g)
0.5 1.0 1.5 2.0 2.5

B

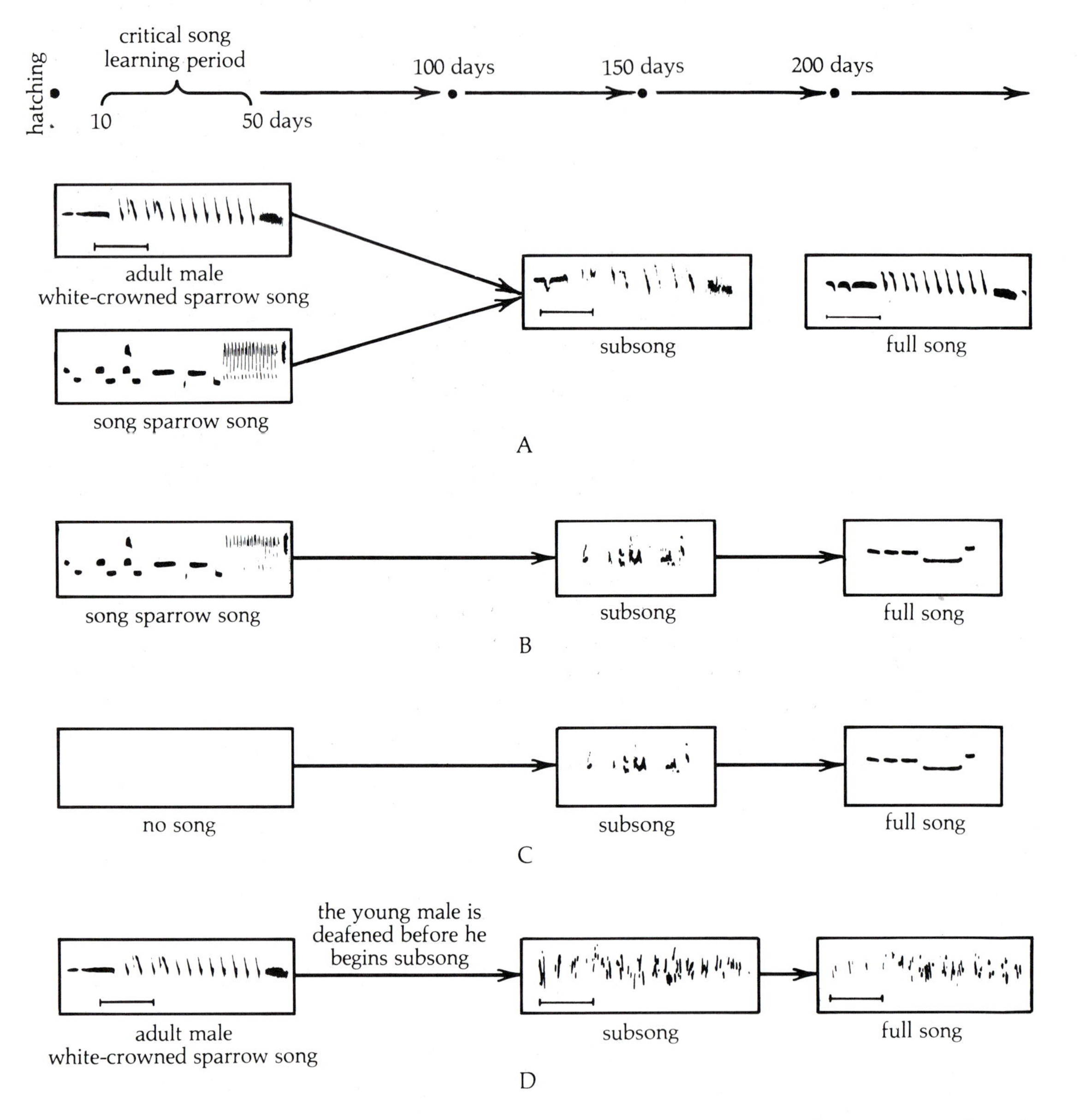

Fig. 17–6 White-crowned sparrows learn their species' song during a critical period from 10 to 50 days of age. At about 150 days they begin to vocalize and practice making syllable sounds. By 200 days they have developed a stable song which closely matches the song heard during the critical period. The song learning is selective, so that if offered a choice, birds will learn only their own species' song (A); and if offered the wrong song (B) or no song at all (C), the bird will learn nothing and sing only a simple tune. If deafened before he begins to practice (D) the bird will never sing anything melodic while if deafened after the song "crystalizes," subsequent singing is perfect. These visual displays of the songs were made with a sound spectrograph, a device which represents frequency on the vertical axis, time on the horizontal axis, and indicates intensity by the darkness of the trace.

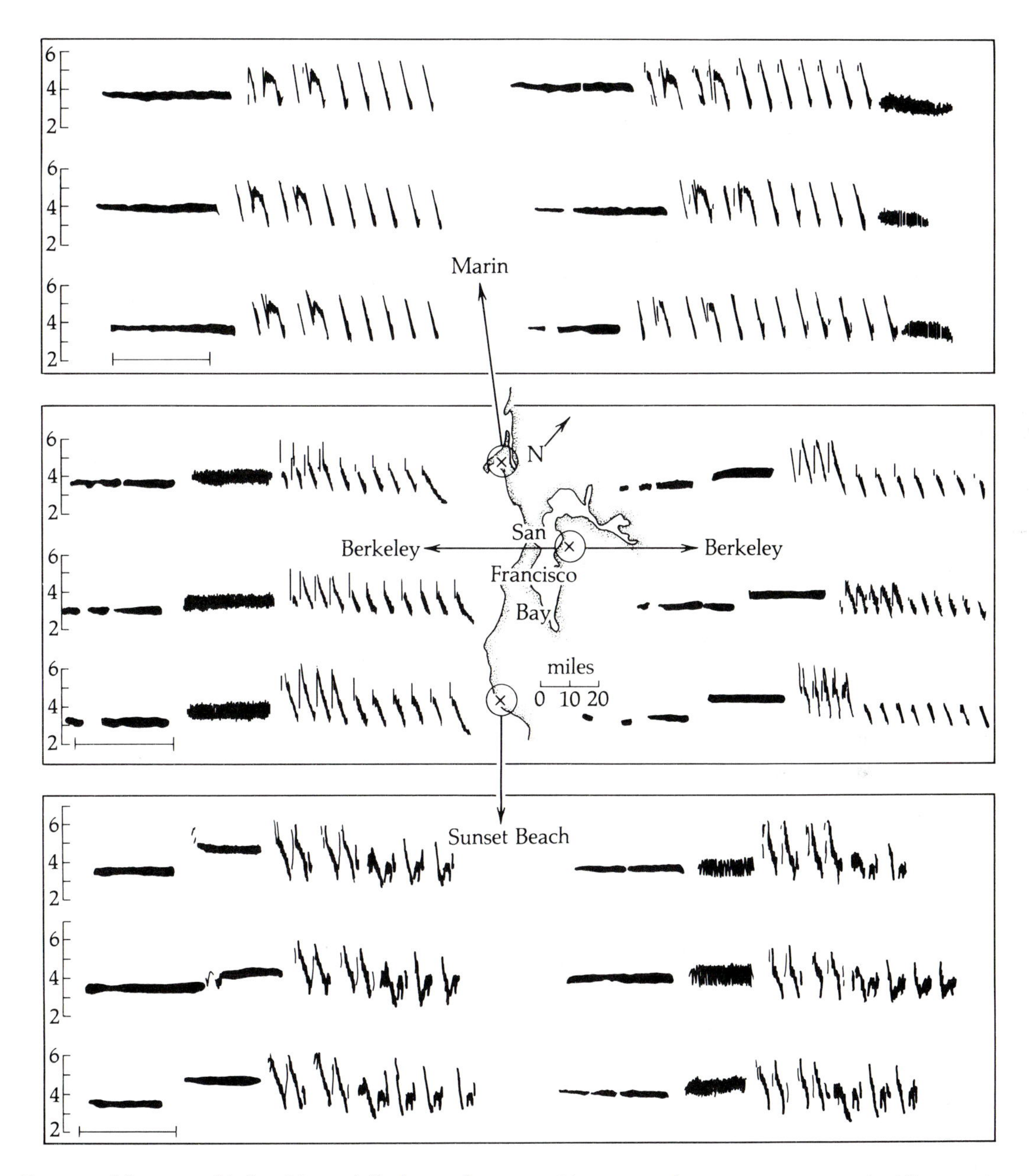

Fig. 17-7 Like many birds with partially learned songs, white-crowned sparrows preserve the idiosyncrasies of their region, a practice which leads to the formation of distinct dialects. These representative examples from around San Francisco illustrate three regional dialects.

of sexual imprinting, and should lead to their breeding preferentially within their own culture. Like imprinting, song acquisition is specific to context, critical period, and cues.

INSTINCTIVE LEARNING

For those of us who like to believe that nature is simple at heart, there is a common theme running through all of these stories: learning is adaptively programmed so that specific contexts, recognized by an animal's neural circuitry on the basis of one or more special cues, trigger specific learning programs. The programs themselves are constrained to a particular critical period—the two seconds before landing in the case of color learning in the bee—and to a particular subset of possible cues. Nothing is left to chance, yet all the behavioral flexibility which learning makes possible is preserved.

It may be that these and other examples of selective learning reflect the inability of genes to code for either the production or recognition of complex signals. Compare the examples of the learning-enhanced songs of this chapter with the simple, innate song produced by a bird isolated from auditory experience (Fig. 17-6) or the calls of species for which no amount of experience has any effect (Fig. 17-8). Only through the tactic of sign stimuli can genes be sure that the feckless animals that bear them will learn the right things in a world full of potentially learnable stimuli. In this way, the genes increase their species specificity during courtship, enable the animals they program to recognize

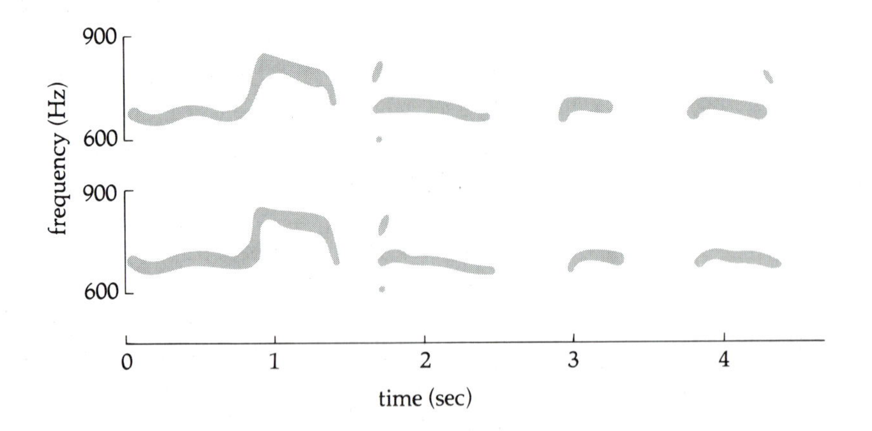

Fig. 17-8 Two examples of the familiar, wholly innate call of mourning doves are shown for comparison. Although the songs are from different individuals, they are essentially identical.

and protect their young, spot food, comprehend danger, and so on. But although releasers (picked out by the sorts of unitary feature detectors already well known from the auditory and visual systems of vertebrates) rather than complex neural pictures are responsible for *directing* learning, there seems little doubt that what is learned *is* stored as a gestalt pattern. It is the high resolution of these mental pictures (whether visual or auditory) which must make possible the individual recognition which has come to play such an essential role in social behavior.

To a large extent this modified picture of instinctive or programmed learning is helping to bring a constructive end to the centuries-old debate about whether "nature" or "nurture" is the source of the adaptive behavior of animals. In general animals probably do embody most of the characteristics of the preprogrammed robots envisioned by Dawkins and most early ethologists, but we now know that their programming is far more elaborate than had been imagined. Flexibilities hitherto unacknowledged enable them to sample their environments in particular contexts and to modify their behavior accordingly. Animals are perhaps in some sense molded and shaped by their environments and experiences, but within a set of complex, species-specific rules that do not at all fit the model that classical behaviorists maintained. Learning, as these far-flung examples suggest, is one of the standard, off-the-shelf programming tricks available to evolution—and despite the usual dichotomy, this kind of learning is the epitome of instinct.

And yet we must wonder whether the "self-programming robot" theory can adequately account for *all* the behavior taking place around us. What about the *really* plastic learning we see going on, the mastering of chance variations of environment, the solving of problems evolution probably could not have anticipated? Could it be that this apparent intellectual "free will" on the part of higher animals is simply an illusion created by our ignorance of how the programming could be arranged to deal with more complicated situations? Just as the wholly innate performance of spiders building their webs appears at first glance to be far too complex and "plastic" to be programmed, so too might other seemingly intricate and adaptable accomplishments. In the next chapter we will look into cases of cultural and plastic learning in birds, bees, and primates with this question in mind.

SUMMARY

Whether vertebrate or invertebrate, learning seems to involve the same basic physiological processes, and several similarities in programming are apparent. Learning, even in higher vertebrates, seems less a general quality of intelligence and more a specific, goal-oriented tool of instinct. Bouts of learning such as food-avoidance conditioning, imprinting, song learning, and so on, are specialized so as to focus on specific cues—releasers—during well-defined critical periods in particular contexts. Releasers trigger and direct the learning, and in general the learned material is thereafter used to replace the releaser in directing behavior. As a result animals know what in their busy and confusing world to learn and when, and what to do with the information once it has been acquired. Most learning, then, is as innate and preordained as the most rigid piece of instinctive behavior.

STUDY QUESTIONS

1. Birds deafened before they begin to practice produce what sound like scratches and buzzes, while birds simply prevented from hearing their species' song produce simple, melodic vocalizations (Fig. 17-6C). Why do you suppose the two are different, and what do those differences tell us?
2. Mockingbirds seem to violate the standard songbird pattern by constructing a medley out of phrases borrowed from other species. Beyond the question of why this species should be programmed to build its repertoire in this way, there is the very serious problem of how female mockingbirds are to recognize males of their species. What do you imagine the possibilities are (there are several)? How would you test your alternatives?
3. Many species of cuckoo lay their eggs in the nests of other species. Hence, baby cuckoos are exposed to the natal environment typical of another species. How, then, do you suppose cuckoos go about learning their songs? How, if at all, is imprinting likely to fit into their lives?
4. There are species of birds in which females do not undergo sexual imprinting. In these cases, the sexes are highly dimorphic. What might be going on here, and how would you test your guesses?
5. Although in most contexts animals habituate to (become bored with) repetitions of the same stimulus, such is not the case with chicks during imprinting: chicks offered a choice of things to look at choose to spend their time viewing photographs of their parents. Given a choice,

however, between familiar photos of the parents and new ones taken from different angles, the chicks prefer the new ones. What might be going on, and how would you investigate it?

6. What role might song learning in male and female sparrows have to do with the vast number of species of this group?

FURTHER READING

Brower, Lincoln. "Ecological Chemistry (How Blue Jays Learn Not to Eat Monarch Butterflies)." *Scientific American* 220, no. 2 (1968): 22–29.

Hess, Eckhard. "'Imprinting' in Animals." *Scientific American* 198, no. 3 (1958): 81–90.

———. " 'Imprinting' in a Natural Laboratory." *Scientific American* 227, no. 2 (1972): 24–31.

Immelmann, Klaus. "Sexual Imprinting in Birds." *Advances in the Study of Behavior* 4 (1972): 147–74.

Marler, Peter. "Song Development in Sparrows." *Journal of Comparative Physiology and of Psychology* 71 (1970): 1–25.

Smith, N. G. "Visual Isolation in Gulls." *Scientific American* 217, no. 4 (1967): 94–102.

CHAPTER 18

Cultural and Plastic Learning

It is clear that some species, our own especially, make their livings in environments which do not seem sufficiently predictable to allow their genes to plan out their lives in great detail. The potential subtlety of such insidious programming as we have seen in cases like birdsong learning is so vast that we must be open to the possibility—indeed, the probability—that underlying the many examples of apparently flexible behavior and learning is, in reality, the deft orchestration of instinct. To all appearances, though, at least some species *do* seem to sort out certain problems for themselves. How do they manage this? Are these animals being really clever, or is the programming too well hidden for us to see? Where can we draw the line we know must exist between instinct and plastic, unconstrained, self-initiated and directed learning?

ANIMAL "AWARENESS"

In examining cases of apparently plastic learning, what diagnostic features can we look for, and at what point can animals be said to "think," have ideas, comprehend what is happening to them, break out of their programming, and otherwise enter that exalted mental Valhalla of "self-awareness" in which we so confidently place our own species? Complexity we have discovered to be a treacherous guide. If complexity were the primary characteristic of cleverness, even honey bees would require us to invoke an almost self-conscious intelligence to explain some of their subtle behavior. Every spring, for instance, colonies divide and half the bees and the old queen leave to form a swarm in a nearby tree. Scouts fly out in search of suitable cavities in which to establish a new hive. The cavity must be chosen with great care, since its ability to keep its occupants warm and dry during the long, cold winter is crucial to the colony's survival. In New York State a swarm has about a 35 percent chance of surviving the winter, while the established colonly has a 90 percent chance. After their searches, scouts return and advertise what they have found by means of the same dance used to communicate food location (Chapter 24). At first several potential sites may be reported, but soon, usually within a day or two, most of the scouts will indicate just one location. Shortly thereafter the swarm will depart to the chosen site to set up housekeeping. How is this important decision reached? By marking individual bees and monitoring empty hive boxes set out as "bait," Lindauer and others have shown that each scout makes her own evaluation based on the cavity's size, exposure to sun, dryness, freedom from drafts, and so on. Each scout advertises her find with more or less of the "enthusiasm" which reflects its quality as a potential dwelling: a bit of water poured on the floor of a popular prospective nest can so dampen the enthusiasm of its discoverer that she will stop dancing for it altogether. So far this may sound like a forager bee reporting a flower patch, modulating her enthusiasm according to the sweetness of the nectar, but now the scout will stop and attend dances to *other* sites. She will then fly out and visit them, perhaps also reinspecting her own discovery, and then return to the swarm and dance for the best one. She has sampled the real estate market, compared the available possibilities, and has made a decision. When virtually all the scouts agree—that is, when all of the dancing indicates the same spot—the swarm will be roused and led to the chosen cavity (Fig. 18–1).

Are these bees being uncharacteristically clever and "plastic," or is their programming simply too subtle for us to decipher easily? The small size, too numerous legs, and phylo-

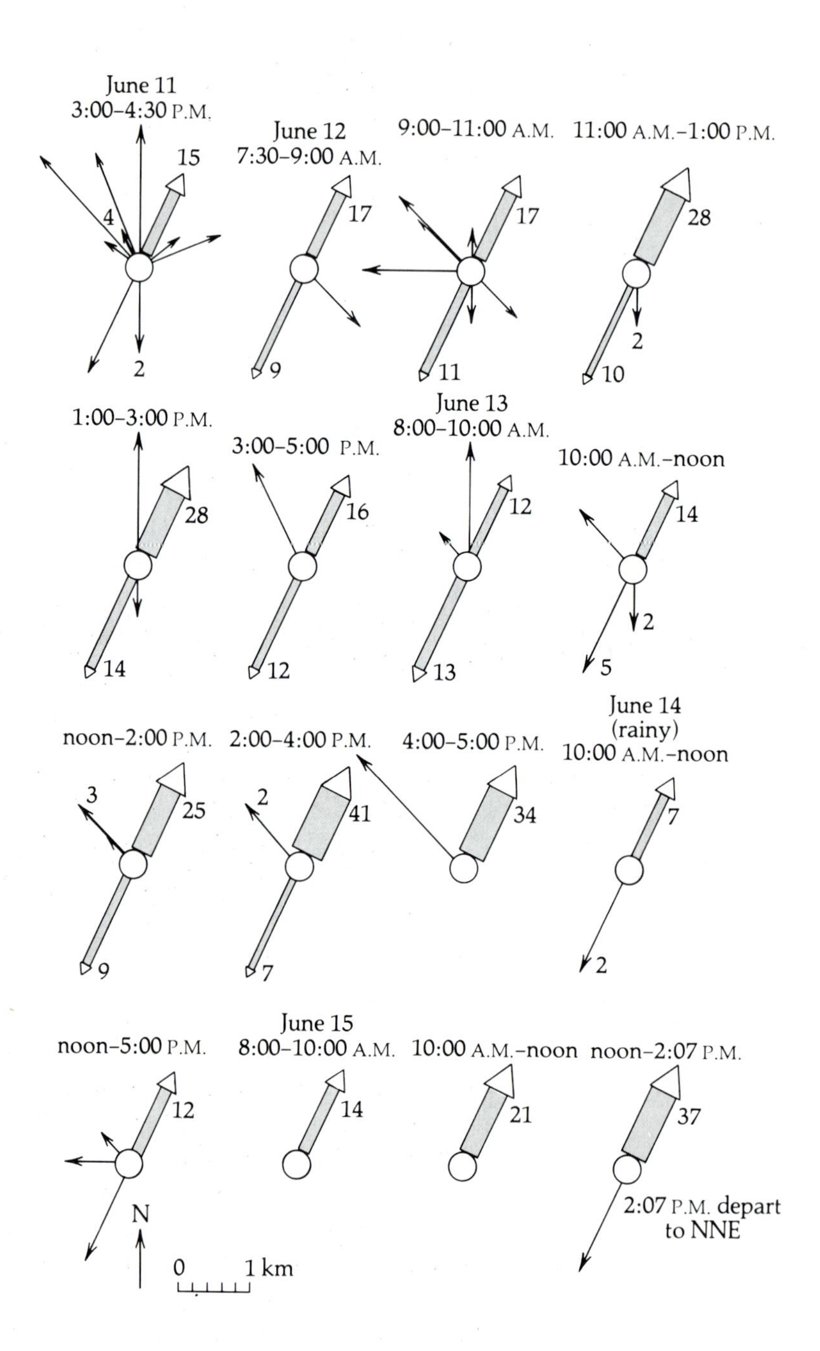

Fig. 18–1 Scout bees return to their swarm and signal the location of potential nest sites. At first many sites are advertised, but as scouts fly out to compare their own discovery with those of others, the choices narrow until relative unanimity is reached and the swarm departs for the chosen cavity. In this example, the drama begins on June 11 at 3 P.M. The first 29 dances indicate 11 different sites in all compass directions at distances from 350 to 1900 m away (the vector angles indicate direction, the lengths distance, and the widths number of dances). By the next day, the debate has been reduced to involve only two major alternatives, although new reports continue to come in. Unanimity is close late on June 13. After a day of rain, the swarm finally departs just after 2 P.M. on June 15.

genetic remoteness of bees from humans strongly inclines most people to the view that the apparently flexible decision-making behavior of swarm scouts is an illusion which *could* be explained as programming if we were only imaginative enough to find the clues. If this story had been about humans or chimpanzees, however, I doubt that computer-like preprogramming would be seriously considered. But after all, bees have swarmed nearly every year for millennia, and must face this general set of problems each time (although each year it is an entirely new generation which swarms).

Other examples place bees in situations which evolution could not easily have anticipated, and are less easily explained. Honey bees hate alfalfa because its flowers come equipped with a spring-loaded anther which, when tripped, gives them a nasty blow. Bumble bees, which evolved along with this flower, do not seem to mind, but once so treated honey bees will avoid alfalfa assiduously. When forced to it by being put into the middle of acres of solid alfalfa—a situation created only by modern agricultural practices—individual honey bees solve the problem in the face of potential starvation by adopting one of two strategies. They either learn to recognize tripped from untripped flowers and frequent only the former (Fig. 18–2), or they learn to chew through from the side of the flower to take the nectar without ever venturing inside.

Has evolution provided bees with two contingency plans for dealing with American agriculture, or have these diminutive creatures reasoned the problem out on their own? Perhaps the eeriest argument against bees' being nothing more than elegant pieces of clockwork comes when we try to train them for an

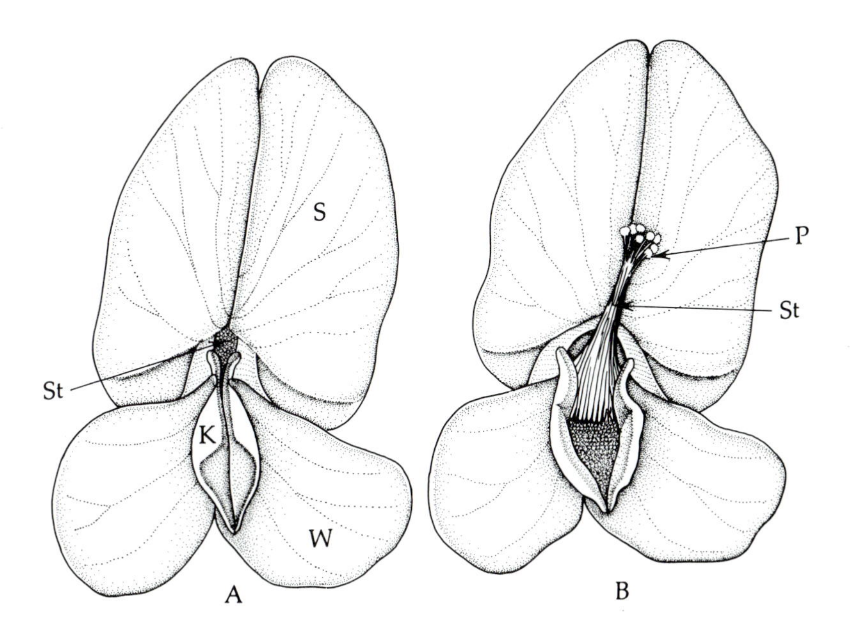

Fig. 18–2 Alfalfa actively deposits pollen *(P)* on the first bee to visit it. The stamen *(St)*, fixed tenuously to the keel petal *(K)*, sweeps quickly upward toward the standard petal *(S)* when a bee begins probing for nectar. Although bumble bees are undeterred by this assault, honey bees may be knocked out of the flower, and quickly learn to avoid alfalfa. Some honey bees, however, learn to distinguish untripped (and therefore dangerous) flowers (A) from tripped blossoms (B), and visit only the latter.

experiment. We begin by letting them find some drops of sugar solution in a dish near the hive entrance, and then move the dish about 25 percent farther away from the hive every few minutes. Hence, early on we may move the food only an inch or so, but later the food will be transported a hundred feet or more in a single jump. Virtually every student of bees from von Frisch on has noticed that there comes a time when at least some bees begin to "catch on," and anticipate the movement of the food by flying on to the next distance and waiting. What on earth could there be in the behavior of flowers which could provide a reason for bees' having evolved such a behavioral program?

Are bees really so disturbingly ingenious, or has evolution programmed them with such exquisite finesse that we are left in doubt as to the source of their abilities? Donald Griffin suggests in his charming book *The Question of Animal Awareness* that much of bee behavior is most easily explained if we assume that they are conscious beings like ourselves, though infinitely less intelligent. Experience and an anthropocentric viewpoint rightly prejudice most of us against the idea that creatures so small could possibly "know" what they are doing. And yet if we concede that even the programming of a 1-mg honey bee brain is too intricate to be distinguished easily from some sort of insect "free will," where does this leave us when we look at bird or mammalian learning, or attempt to analyze the sources of our own incredibly subtle and complex species-specific behavior?

Clearly bees are a problematical case to choose since we must begin with the assumption that most if not all of their behavior is innate. Likewise, our species is a poor example since we imagine (incorrectly, as Chapters 29–31 will illustrate) that all human behavior is plastic. Only among birds and mammals will our prejudices admit of both possibilities. Since the business of life among these animals is the propagation of genes, we can assume that through the course of species evolution those genes would have done all they could to ensure a successful outcome. In early behavior crucial to life—avoiding predators, recognizing food, getting together for mating, and raising the all-important young—evolution would have determined how much to wire in for security and how much to leave plastic and adaptable. While as human observers we are predisposed to see unconstrained learning as an unmixed blessing, the natural world is fraught with gene-threatening ambiguity. Wherever we look for pure constrained plasticity in learning, curious telltale constraints turn up which betray the existence of that genetic insurance policy, behavioral programming. Of all the behaviors necessary for the perpetuation of a species, the most basic component, species recognition, seems almost always to depend on releasers, either at the time of courtship or earlier during sexual imprinting when a more precise picture of conspecifics is formed. Parental care, too, displays all the hallmarks of instinct—sign stimuli, motor programs, drives, and programmed learning. The features of a species and the needs of its offspring are relatively predictable, and so ideally suited for the workings of instinctive behavior. But what about those often less predictable contingencies which confront all animals, food and danger?

CULTURAL TRANSMISSION OF KNOWLEDGE

In recognizing who and where their enemies are, many animals display learning. But what kind of learning? A recent study has demonstrated the existence of an unsuspected and intriguing case of what we might consider "programmed culture." Many birds attack ("mob") potential nest predators, but it has been difficult to imagine how the knowledge of which birds to mob could possibly be innate. The German ethologist Eberhard Curio placed two cages of blackbirds on opposite sides of a hallway in sight of each other. Between the two cages he installed a four-compartment box which allowed the occupants of each cage to see an object on their side, but not the object on the other. Curio then presented a stuffed owl to one cage and an innocuous and unfamiliar bird, an Australian honey guide, to the other. The birds seeing the owl went berserk with rage at the sudden appearance of this familiar predator, and attempted to mob it through the cage. The birds on the other side, seeing only the honey guide (which they had ignored on previous occasions) and the mobbing birds, began mobbing the honey guide. These birds would now pass on the practice of mobbing stuffed honey guides to other birds, and those pupils pass it on to still others through six generations. Needless to say, none of these birds had ever suffered at the hands of this peaceable Australian species. Curio was even able to repeat this mindless enculturation with bottles instead of stuffed birds as the objects of official hatred. The very special attention paid to even a single incidence of mobbing and the faithful transmission of the identity of the offending species (or object) in the absence of any obvious reinforcement argues strongly for a kind of old-fashioned programmed learning: imprinting. As in imprinting, the cue which instigates naïve animals to pay attention to and learn this crucial piece of information—the appearance of a nest predator—is almost certainly a sign stimulus: the mobbing call (Fig. 18–3A). The call is innate—it is recognized by and can be elicited from birds raised in isolation. Moreover the call, broadcast by birds during an attack on the threatening animal, is so similar between species (as is the danger call; see Fig. 18–3B), that birds of one species learn the (probably mutual) enemies of others, and so expand their range of cultural knowledge. Something as crucial to an animal's survival as the recognition of enemies, then, though transmitted culturally, can be seen to depend on programmed learning. Put quite simply, this sort of culture is innately guided.

Fascinating examples of cultural transmission of enemy identity abound, all probably based on the same sort of general mechanism. "Island tameness" refers to the propensity of animals living in areas not previously inhabited by people to show

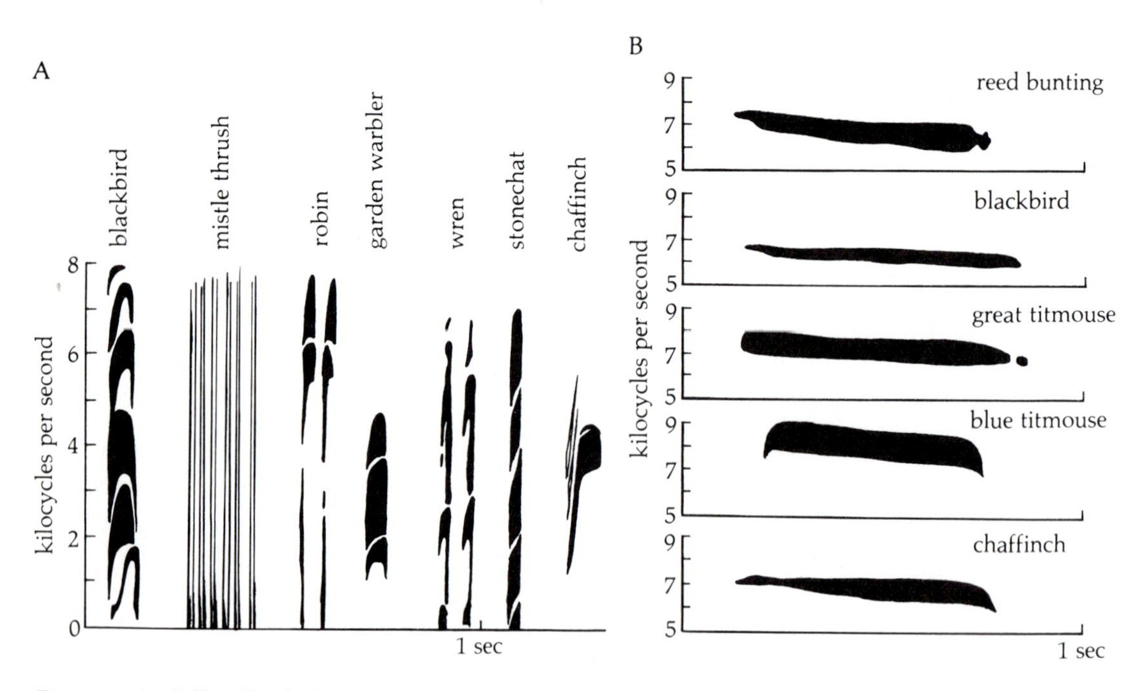

Fig. 18–3 A. Calls of birds from several families given while mobbing an owl. B. Calls of five different species given when a hawk flies over.

absolutely no fear of our fell species, at least until they learn their mistake. Animals now on preserves which presumably were tame in the long-distant past retain strong cultural prejudices against us from more recent times when they were hunted, though that may have been dozens of generations ago. Elephants are tame where they have never been hunted, while those from other areas are aggressive. A well-recorded but incidental case of cultural learning occurred in 1914 in what is now Addo Park in South Africa. A hunter by the name of Pretorius was asked to exterminate a herd of 140 elephants. After a year Pretorius had their numbers down to about 20 or so, but the survivors proved very reluctant to put themselves in the way of the hunter. Pretorius gave up the attempt, and the area became a preserve in 1930. Although they have not been shot at since, and all the elephants hunted by Pretorius have doubtless long since died of old age, this fourth-generation herd remains shy and strangely nocturnal, and is said to contain the most dangerous elephants in Africa. Presumably young elephants observe their elders directing their species-specific alarm trumpetings at tourists, enter our species on their enemies list, and react toward humans in the future with the same alarm behavior. This, of course, will impress the lesson (too fatal to be learned by experience) on their own offspring. Only if the park were cleared of people for two decades or so might the elephants' hatred for people be extinguished.

Cultural learning (which despite its high-class nomenclature is in these cases nothing but mindless copying, probably based, as we have seen, on releasers) occurs in the equally necessary job of food gathering and the related, equally critical customs of food taboos. Although many species are specialists and so can be programmed to recognize food innately, others regularly depend on their parents to teach them how to make their living. Oyster catchers, for instance, eat a variety of foods from along the shore, but have developed a flair for opening mussels. Ethologists observing these birds have noticed two distinct strategies for opening shellfish: stabbing and hammering. There are birds which use one method, and birds which use the other: no one practices both. "Stabbing" oyster catchers deftly insert their long, thin beaks into the tiny opening between a mussel's shells through which the mussel circulates water for feeding and respiration. The birds then cut the adductor muscle and the shell falls open. "Hammerers," on the other hand, remove the mussel to dry ground and rely on brute force to hammer a hole in the weak point of the shell. This sounds like a case of plasticity. If mussel-opening were innate, why shouldn't all birds do it the same way?

In fact, oyster catchers are not born programmed to harvest mussels in these ways at all. Many of the required movements are innate, but the FAP building blocks may go unused: many colonies exist which eke out a meager living poking in the sand for worms. The oyster catchers' overwhelming success as a species, however, depends on their learning to master their peculiar trade, which enables them to exploit the shellfish "market" or niche. This learning, despite its use of the preexisting motor programs which these birds' less sophisticated cousins use to delve for worms, takes place over two long years as the young apprentice themselves to their parents and slowly pick up and perfect their specialized technique. As it happens, both of a bird's parents, and hence both of its teachers, use the same approach—stabbers mate with stabbers, hammerers with hammerers. This makes sense because otherwise the offspring, which seem barely able to learn *one* strategy even working at it full time, would receive conflicting instructions. It suggests, too, some sort of class bias in mating, since the white-collar stabbers do not breed with the blue-collar hammerers.

What has evolution programmed a young oyster catcher to do? Do the genes instruct each young bird to pay particular attention to how its parents feed, to ignore a world full of other things to watch and learn, and to copy what its parents do until it can feed at a high enough rate to raise and feed its own offspring? What evolution seems to have done is to instill in each bird a drive to learn in detail the tricks and strategems its "culture" has worked out. But how do the young birds know *what* to watch, *what* to learn? Most likely, the long bright-red beak of the parents, the instrument whose deft manipulation must be copied, is a sign stimulus which focuses and directs their attention. As with most other examples of programmed learning this case is not entirely flexible with regard to time either. Even when old birds have been "culturally deprived" and have had no opportunity to learn either strategy and so must make their living digging worms, they show no interest in learning the new vastly more productive techniques. Could it be that their "critical period" for learning how to harvest food has passed?

The faithful but mindless transmission of information about what and how to eat and not to eat is not restricted to "lower" animals. Our own species provides ample illustrations of similar sorts of cultural inertia. Consider, for example, one of the cases of food taboos from James Trager's fascinating *Food Book.* In the 1600s some Jewish planters settled in what is now Surinam. They kept slaves, some of whom escaped and established villages up the rivers in the jungle. Even today, 350 years and fifteen to twenty generations later, the descendants of the early slaves maintain two sets of dishes and utensils—one for meat and one for dairy products—in good kosher tradition.

INNOVATION

In discussing cultural learning, the question of innovation always arises. How does a culturally transmitted practice originate? Did some prehistoric oyster catcher "invent" hammering and another stabbing? Birds are often said to be "creative" or "innovative" in their song learning since they never produce perfect copies (though a cynic will say these idiosyncrasies reflect sloppiness and careless listening rather than imagination). There are, however, at least two well-documented cases of apparent innovation of a more suggestive sort in recent years. In England there was an outbreak of cream robbing by blue tits. The birds would remove the cardboard caps from bottles of milk left on the doorsteps of British homes, and drink the cream which floated on top (Fig. 18–4). The practice spread from one town to a whole county, and later to other species of birds, before a new cap design put a temporary halt to the theft. The birds subsequently solved the new problem as well, and only the advent of special boxes on the doorsteps (followed by the demise of home delivery altogether) has put an end to this criminal practice.

Fig. 18–4 Blue tits discovered the cream under the caps of milk bottles, and how to get at it. The practice was transmitted culturally all over England and into Europe, and spread to some two dozen species of birds.

It is reasonable to suppose that some bright or lucky blue tit—perhaps one that had been hand-reared—first discovered the secrets (1) that milk bottles contained cream, and that cream was desirable, and (2) that the tops could be opened. The remaining birds probably learned the secrets culturally. Was this learning a consequence of some specific watch-what-others-are-eating program? The apparent intellectual abilities of the blue tits might gain impetus from the harvesting motor programs of the species as well. The tits opened the original bottle tops, whether foil or cardboard, with a stripping motion. A prominent part of their natural behavioral repertoire consists of stripping bark off trees to search for hidden insects. So deep-seated and mindless is their propensity to "peel" surfaces that tame blue tits have been observed stripping wallpaper off houses. Moreover, even the most carefully managed packaging system inevitably turns out caps which leak or with small perforations, which would have been all it took to focus the attention of some lucky bird on this novel prey item it came prewired to harvest.

In fact, motor-program preadaptations which can facilitate certain sorts of learning are widespread. In a set of experiments designed to teach birds to pull a string with their feet to get food, for instance, species which normally manipulate food with their claws—tits, for example—master the task quickly with little variation in technique among individuals, whereas species such as the finches, which normally employ their beaks to manipulate food, have great difficulty in cracking the problem and eventually solve it in many different ways. Although we might conclude from this that tits are simply smarter than finches, it seems likely that a slight alteration in the task could invert any such imagined intellectual hierarchy.

The unromantic suggestion that cultural transmission of food preference may in many cases involve little more than mindless copying has strong experimental support. In one memorable study, for example, a hen was taught to eat only a particular color of seed—red, say. Given a choice of identical seeds dyed a variety of colors, the hen would predictably select only the red seeds. The remarkable thing was that without prior training, her chicks also developed the same strong preference for red. Even though all the other colors were perfectly edible, the young chicks' subtle observation and imitation of their parent and their unwillingness to experiment shaped their subsequent diet. Obviously, a powerful and adaptive urge to imitate in this critical context is programmed in by the genes.

Similar cases of cultural transmission of knowledge have occurred in Japanese macaques. A particular troop, confined to an island and fed by humans, has been observed by scientists for years. Their food is typically dumped on the sandy shore, and the macaques normally ingest a good dose of grit with their meals. One day—one specific day—a young female named Imo took one of their favorite foods, a sweet potato, to a nearby stream and rinsed it (Fig. 18–5). Whether she did this by accident or by design is not known, but the practice was soon adopted by her peers—other low-ranking female macaques—and slowly spread up the hierarchy. The dominant males were the very last to try something new. Subsequently the same female discovered that wheat floated, and so could be separated from the sand which otherwise inevitably clung to it, and the discovery spread in precisely the same way.

Fig. 18–5 A Japanese macaque washes a sweet potato in a stream to remove the sand clinging to its skin. The practice was discovered by a single monkey, and transmitted culturally through the troop, first to playmates, then to sisters, then mothers, and so on from the bottom of the hierarchy to the top.

How has evolution provided animals with this capacity for innovation? Animals, especially young ones, seem driven to be ever alert to what others are eating and how, and are quick to copy. One chimpanzee discovers how to drink from a water fountain and soon everyone "knows" the secret. The illogical power of the cultural tradition is suggested by the apparent lack of comprehenion on the part of the copiers of what they are doing, indicated particularly by their willingness to copy irrelevant actions along with the ones that matter, or to maintain a cultural tradition long after its basis in reality has vanished. But the act of discovery itself is another matter. Was Imo a prophet, an experimentor, or just lucky? The proclivity of young animals, particulary human children, to "play" with food, combined with their distressing willingness to put virtually anything into their mouths, may suggest some innate urge to experiment. Perhaps in higher animals it is the young, too naïve to know any better, who are designed to be the primary source of cultural innovation. The more mature are the equally indispensable vehicles of cultural transmission. Only by bringing such phenomena into the realm of careful ethological experimentation, however, can the roots and mechanisms of cultural innovation and transmission be examined.

LEARNING AND DRIVE

In trying to come to grips with plastic learning we must reenter the quagmire of drives and motivation. For example, we might imagine that it is the lack of any obvious reinforcement which is the key to distinguishing at least some cases of programmed learning from the sort of plastic, instinct-free learning behaviorists seek to deal with. (What, after all, is the tangible reward for learning what to mob or memorizing the appearance of a clutch of eggs?) Behaviorists traditionally have studied learning by means of operant conditioning. The behavior of the experimental animal is "shaped" by rewarding the subjects with food for ever more exact versions of the behavior ultimately sought. This procedure differs from classical conditioning in which the stimulus in an innate stimulus-response behavior is paired with, and ultimately through learning replaced by, an irrelevant stimulus. Operant conditioning, however, cannot be entirely free from the programmed prejudices of animals. Reward and reinforcement are defined by the innate sense of what is good and bad for the species. Pigeons will learn for seed but not for meat, while dogs have the opposite prejudice. If animals come programmed to learn how to harvest food, the tricks they can be taught in such circumstances and the breakdowns reported by Keller and Marian Breland take on a slightly different color. In "The Misbehavior of Animals" the Brelands report that the animals they tried to train to perform tricks for a food reward would revert to natural food-gathering behaviors even if starved and left unrewarded. In one of many instances, a raccoon which they had painstakingly trained to take a coin and deposit it in a

bank, was unable to achieve the next step in his training. He was to have taken one coin at a time from a pile of coins and put it in the bank. Instead, he would seize two coins and begin rubbing them together, dipping them from time to time in an imaginary stream, again and again until the trainers lost patience.

The Brelands interpet their observations to indicate that only natural elements of a food-gathering behavior can be conditioned easily. Indeed, when a scientist actually observed pigeons in a Skinner box pecking keys as part of an elaborate criterion for food, he observed that the birds opened their beaks in the species-typical feeding pattern as they hit the keys. Apparently the pigeons were treating the keys as food, perhaps as something to be hammered until they yielded their edible contents. When the reward was to be a liquid, the pigeons pecked with an entirely different flourish to their beaks, one typical of drinking.

Again, we are caught in the tangle of innate motor programs, drives, cue biases (recall that the pigeons look for visual cues in feeding), and learning. Patterns are beginning to emerge from the chaos, however, in cultural and plastic learning. Virtually every case which has been analyzed in detail depends on sign stimuli to direct attention and to turn on learning. The learning normally focuses on specific sensory cues and, when it involves muscle movement, makes liberal use of motor programs or if need be wires its own. A surprising number of cases of even the most patently plastic learning seem to be guided initially by the now-familiar pieces of behavioral clockwork available to the genes. In the most plastic cases, situations in which animals need to have the maximum flexibility to master the unpredictable, the programming of successful species seems to consist of setting an inborn goal—crawling, feeding, or what have you—and providing a powerful drive to achieve it. But with the goals to be achieved thus defined and the internal motivation to solve them turned on, individuals of such adaptable species may be left to find the solutions through observation, trial and error, or even insight, a sort of cognitive trial and error.

We see, then, that it may not be unreasonable to think of even plastic, malleable learning as a programming ploy, perhaps a backup strategy to be called into play when all else fails. How much of this unromantic view of creativity might be correct, minimizing as it does the role of self-awareness, is difficult as yet to determine. As we shall see in the last chapters of this book, evidence from observations of our own species' behavior tends to support the notion that plastic learning and conscious thought are not as far removed from instinctive control as we might like to think.

SUMMARY

Learning serves to provide animals with information either too complex or too unpredictable to be innately encoded. Far from being the antithesis of instinct, modern evidence and reappraisals indicate that most learning is innately directed, adaptively specified by the genes with regard to time, place, cues, and goals. In this light, many cases of even cultural learning now reveal themselves as simply more elaborate versions of programmed learning, depending most often on releasers to initiate and direct cultural transmission. Even insight learning is suspect since it turns up in insects, and seems usually to depend on chance even in primates. Indeed, the basic message of evolution is that without genetic variability there cannot be progress. Since behavioral variability is essential for just the same reasons, the role of mental mistakes and lapses in the process of creative thought and recombinations of data, particularly in science, cannot be overestimated. Although it is clear from introspection that our specific thoughts and actions are no longer under the explicit control of our genes, unbiased observation of animals and the recent deciphering of their programming strategies should make us cautious about assigning the source, motivation, and direction of even the most insightful behavioral performances to creative intelligence or self-awareness.

STUDY QUESTIONS

1. How would you distinguish between aesthetic creativity and sloppiness as explanations for the variability in learned bird songs? What selective advantage might there be for imperfect learning of song?

2. The basic difficulty in dealing with "self-awareness" lies in distinguishing between the cleverness of evolution and that of the animals it has shaped. How would you go about isolating the source of the insightful cleverness bees show while being trained or when faced with alfalfa? What of birds confronted with milk bottles?

3. Rats and ants are excellent subjects for maze-learning tasks. Generations of learning theorists have considered maze mastery as

involving a generalized form of learning. As an ethologist, might you have other expectations? How might you test them both with other species, and with rats offered a superficially simpler maze task?

4. Rats are regularly trained to negotiate mazes beginning and ending at specific locations. To master the maze it was generally thought by researchers that the rat memorizes the series of correct turns at choice points—"left, left, right, left, right, right" for instance—and various elaborate hypotheses were constructed to explain the rapidity and accuracy of this kind of learning in various sorts of mazes. Later someone observed that if such a maze were simply moved to a different location in the room, experienced rats had to relearn the task. What do you suppose the rats were really doing? How would you test your idea? In terms of intellect, do you think the rats are doing more or less than had been thought? How would you investigate the possibility of a true left/right-series memorization ability in rats?

5. Rats, pigeons, and certain other species are very adept at learning a sequence of behaviors to obtain food. For example, a pigeon can be taught to peck first a red key, then a yellow one, then green, and finally a blue key in that sequence regardless of the position of the keys. This is usually interpreted in terms of generalized intellectual capacity. Indeed, if during testing we omit, say, the green key, the pigeon pecks red, yellow, blue without the slightest hesitation, a behavior which is often said to imply an ability to "generalize." If we replace the word "obtain" with the word "harvest," what ethological hypothesis comes to mind with regard to these abilities? Are they as intellectually impressive as they look? If your skepticism were well founded, what conceptually simpler tasks along the same lines ought pigeons to *fail?*

6. Songbirds have a repertoire of about twenty distinctive, innate calls to signal distress, alarm, hunger, threat, and so on. Many of these calls are associated with distinctive body postures—tail up, head feathers ruffled, and wings quivering, for example—which convey the same meaning, at least to experienced birds. How could you determine whether the posturings are innately recognized (or, as seems likely, innately produced)? If you were to find that posture recognition is not innate, what implications might this have for the relatively complex posture-recognition signals in gulls and dogs?

FURTHER READING

Curio, E., et al. "Cultural Transmission of Enemy Recognition." *Science* 202 (1978): 899–901.

Griffin, Donald R. "A Possible Window on the Minds of Animals." *American Scientist* 64 (1976): 530–35.

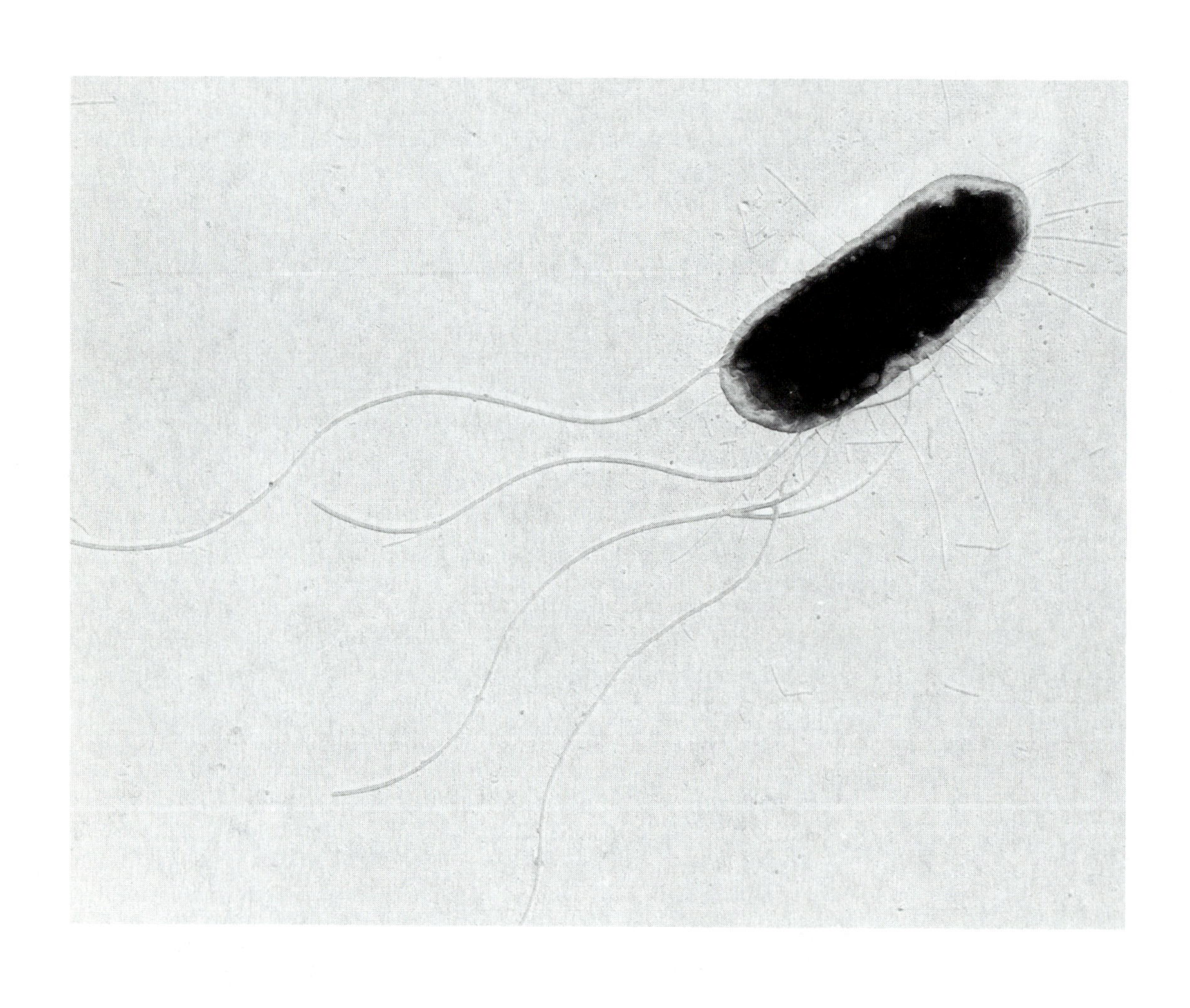

Part V

BEHAVIORAL GENETICS

CHAPTER 19

Classical Behavioral Genetics

As ethologists we seek to understand behavior with the expectation that much of what we see happening around us is the consequence of innate programming. Sign stimuli, motor programs, drives, and programmed learning are all largely preordained by an animal's genes, requiring only the environment to which the species evolved to guide the creature through its normal development and reproduction. Although we take it for granted that genes manage the incredible feat of building an organism and then telling it what to do, ethologists rarely mention genetics or use genetic techniques—a strangely illogical phenomenon since the logical extension of Lorenz's "dissect the parts" argument must lead to that ultimate unmoved mover, the gene itself. This is not to say that it is easy to imagine just how the long chains of bases in DNA with their modest four-letter alphabet, transcribed into messenger RNA from which the ribosomes print out structural proteins and enzymes, could account for the egg-rolling response in geese, flower learning in bees, or our own erratic behavior. Indeed, a substantial group of scientists view behavior as an "emergent process," and believe that to understand the workings of the genes and the nerve cells they construct and wire is not to understand behavior. Behavior, they maintain, transcends its underlying mechanisms, and is thus beyond the purview of reductionistic science. The inescapable corollary to this attitude is "why bother?"

Unfortunately, in addition to finding this metaphysical view needlessly defeatist, most ethologists have little taste for the laboratory environment or the highly quantitative techniques necessary for research in behavioral genetics. As with motor programs, there is the sense that something important is probably there to be learned, but surely someone else will work out the tedious details. Indeed, progress until lately has been painfully slow, and the results have thrown relatively little light on behavior. The situation is rapidly changing now, thanks to recent developments, and there is every reason to suppose that behavioral genetics will ultimately reduce the wonderful behavior of at least simple organisms to the raw biochemistry that must be its lowest common denominator.

The reason for optimism is the development of an ingenious new approach to behavioral genetics. In general, behavioral genetics can ask two sorts of questions and, as we shall see, employs two basic ways of attempting to answer them. The questions are the genetic versions of the standard ethological ones: *what* behavior is controlled or affected by the genes, and *how* do the genes manage to control it? The techniques are classical Mendelian genetics—hybridization and backcrossing—which exploit naturally occurring genetic variation, and the more recent approach known popularly as molecular biology which intervenes directly in the genetic machinery through mutagenesis. Mendelian strategies will be the subject of this chapter, while the next will take up the molecular tactic.

HYBRIDIZATION

One way to study the genetic basis of behavior is to cross two species which are related closely enough to interbreed, but which nevertheless display species-specific behavioral differences. Then we observe the behavior of the hybrid offspring for clues about what the genes are doing and how. Ideally this first generation is backcrossed to the parental species. If only two or three genes are involved, the behavioral trait(s) will "segregate" in a characteristic and predictable way which tells us not only how many genes are responsible, but also what they do. However, this is rarely possible in practice because hybrids are al-

Evolution and behavior share one crucial common element: the genes. An understanding of how genes work and interact with each other is crucial to our appeciation of how behavior is programmed and under what constraints it has evolved. The first important breakthrough in the study of genetics was made by Gregor Mendel in the mid-1800s. Mendel is remembered for his discovery of a simple, quantitative model to explain inheritance. When Mendel crossed red-flowered peas with white ones the resulting peas produced plants with red flowers. Mendel guessed that each parent supplied the offspring with one gene for flower color, and that the red gene was "dominant" over the white one. When he crossed this presumed first-generation hybrid with white peas, half the resulting offspring in the next generation—the "backcross"—had white flowers and half had red. Mendel guessed that each hybrid was "heterozygous"—that is, although the "phenotype" was red, the "genotype" included one red and one white gene. These heterozygotes, then, had each contributed either a recessive white or a dominant red gene to each offspring while all of the white peas, being homozygous for white genes, had passed on a white gene. Hence, half the new generation had two white genes and were white, while half had one red and one white and so were red (Fig. 19-1). Mendel's model could also predict the results of crosses of two or three genes at once.

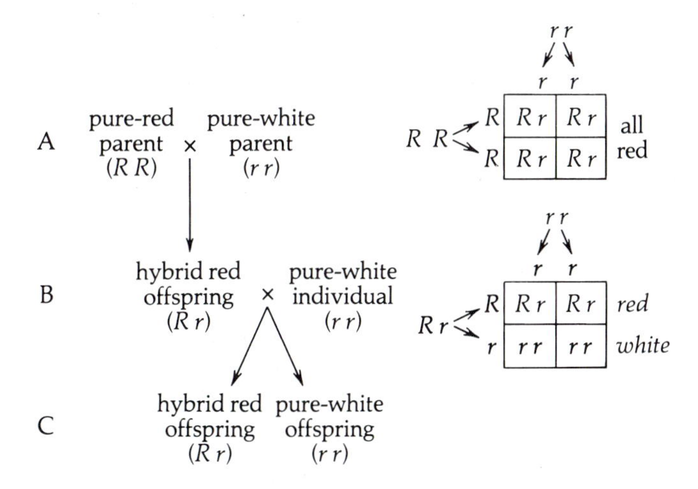

Fig. 19-1 In the parental generation (A), pure-red and pure-white peas are mated. Pure reds have only red genes *(R)*, and pure whites have only white genes *(r)*. The matrix on the right shows the possible combinations as each parent contributes one gene to its offspring. All the offspring are heterozygous red. In the second generation (B), the heterozygote is crossed to a white pea. As the matrix indicates, two outcomes are possible: heterozygous red again, and homozygous white (C).

Although a variety of anomalies and complications plague the model, it is basically correct. But it is only a model. Underlying it, and accounting for its puzzling curiosities, is the biochemical clockwork which builds and runs animals and plants—DNA, RNA, ribosomes, and all the rest. We know now that the red gene is dominant because it codes for a red pigment protein which hides the colorless product of the white version (allele) of the gene. We know too that such differences in the exact structure of gene products are the source of the variation upon which natural selection works. A specific change in the sequence of bases in the DNA generates the difference. And those changes, whose unpredictable results are submitted each generation to the arbitration of evolution, arise most often from mutation.

Although none of the many mutant genes with obvious behavioral consequences has been examined in sufficient detail as yet, the precise change responsible is known in a variety of cases. For example, sickle-cell anemia, a disease which results from decreased ability of blood cells to absorb oxygen, is the result of a single change in the 1800-base-long DNA message for hemoglobin. The seventeenth base coding for the beta chain is uracil in normal cells, but adenine in sickle cells. This change results in the sixth amino acid in the hemoglobin chain being valine rather than glutamic acid. Since blood cells with one normal and one sickle gene produce a mixture of normal and sickle hemoglobin, and thus absorb enough oxygen for normal functioning, the sickle gene is considered recessive. In the homozygous organism, however, the sickle-cell gene leads to a variety of disabilities and often to early death, symptoms caused by the proclivity of the red blood cells to collapse into jagged crescent-like shapes after giving up their oxygen in the capillaries and then to clog these crucial channels. Given the apparent "unfitness" of the gene, it is surprising that the sickle allele comprises about 40 percent of the genes in certain parts of Africa. Evolution, however, has its own logic. Individuals heterozygous for the sickle-cell gene are more resistant to malaria, a protozoan disease whose area of infestation almost perfectly maps the natural distribution of sickle-cell anemia. Of course, if malaria is ever eradicated from central Africa, the frequency of the sickle allele will slowly drop as evolution works its inexorable selection for "fitness," and no longer has to weigh the benefits of greater resistance to disease against the chance of anemia.

most always sterile, and behavioral differences inevitably involve several genes. Moreover, hybridization experiments are of necessity opportunistic, depending on those rare cases in which the defenses erected by the genes of each species to prevent matings with other species can be circumvented. Hence, we can almost never cross the animals that it would be best to hybridize. Despite these problems, though, the technique has been widely used. McGrath, for example, crossed chickens and pheasants and analyzed the distress calls of the offspring (Fig. 19–2). The distress call was already known to be innate, as indeed it needs to be for a chick to communicate to its parent. The characteristics of the pheasant's shorter, FM call dominated that of the hybrids, although the frequencies were intermediate. The frequency proved to be a secondary consequence of size and the consequent vocal resonances, as the difference between large and small hybrids indicates. What can we conclude from this experiment? It is clear that the calls have a genetic basis (though isolation studies had demonstrated that already), and it shows that the genes for pheasant calls are dominant. But the sad truth is that studies of this sort give us little hard information. They are interesting, but powerless to tell us what we so vitally need to know: how genes code for innate behavior.

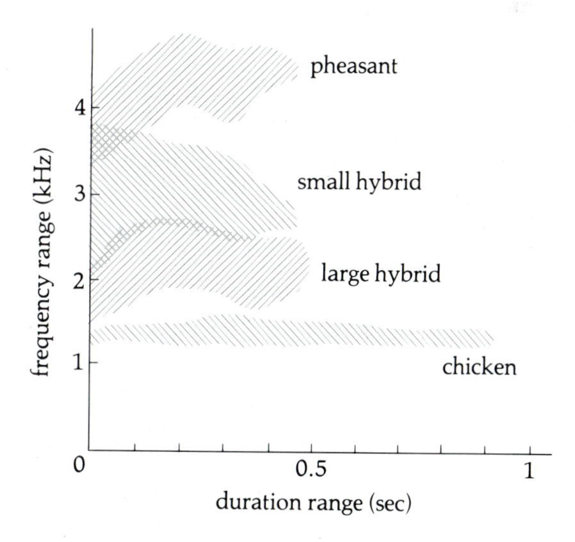

Fig. 19–2 The distress calls of chickens and pheasants are innate. When these two species are crossed, the offspring produce calls of intermediate frequency which fall into two classes, those of larger birds (resulting from matings of pheasant males with chickens) which are lower in frequency than those of small hybrids (roosters with pheasant hens). The call length and frequency modulation of pheasants are preserved in both classes of hybrids.

Lade and Thorpe performed a more elaborate set of crosses of various doves, and examined the songs of hybrids. Again, raised in isolation as eggs and chicks, doves develop perfectly normal songs: they require no outside information to perform this behavior. In this case the hybrid songs showed no clear relationship to either parental song (Fig. 19-3). Again, it is difficult to know what to make of these results. Another interesting if fruitless case is a hybrid duck that Lorenz derived by crossing Chiloe teals and Bahama pintails. In his early days Lorenz had done a wide-ranging comparative study of duck courtship displays, and created a behavioral phylogeny based on which elements were shared among species and which were unique. This ethological family tree agrees closely with ones drawn along more traditional morphological lines. Elements of the courtship

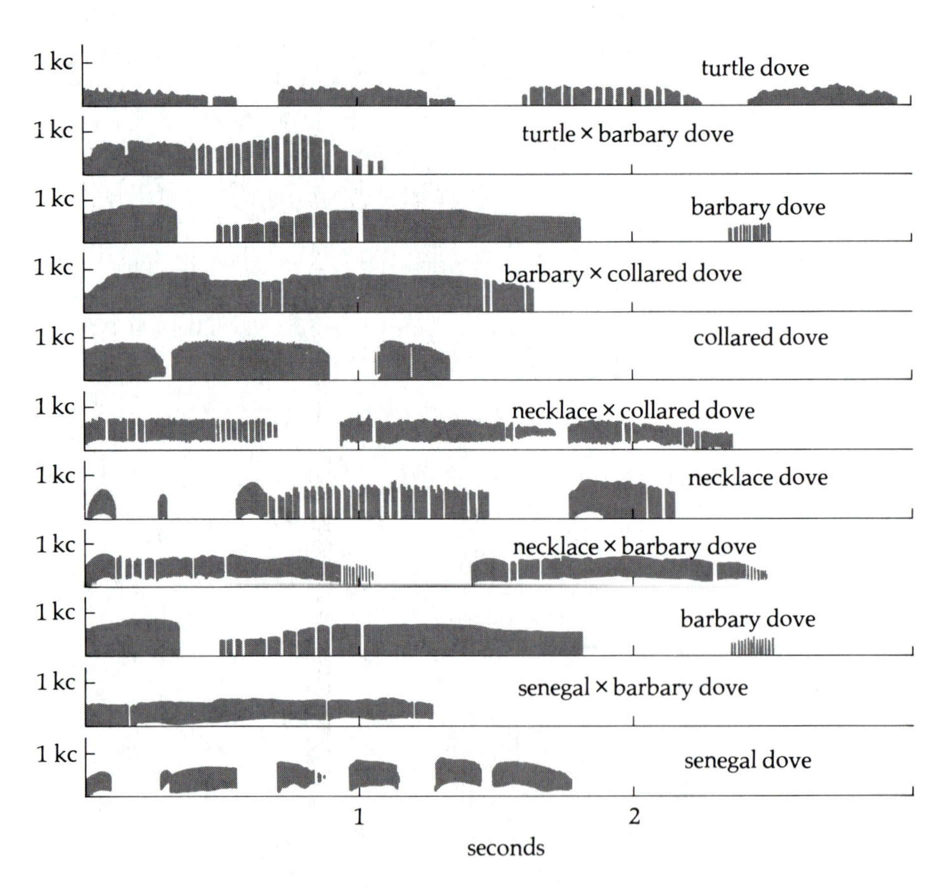

Fig. 19-3 The calling songs ("coos") of the many species of doves are wholly innate. When species are crossed the hybrids produce new calls, which result from the interaction between the two very different sets of song genes. If only a few genes were involved in song production, these hybrid songs could provide clues about how this behavior might be genetically encoded.

display of Lorenz's hybrid resembled those of neither of the two parental species. They matched instead the display of another species which Lorenz had classified as a primitive precursor of the parental species. The primitive behavior had not been replaced, but simply repressed. A similar phenomenon occurs in a particular mutant common in *Drosophila.* Fruit flies evolved from insects with four wings, disposing of their second pair of wings by turning them into halteres—tiny, rapidly vibrating structures which act like gyroscopes to aid in orientation. The mutants have the second pair of wings restored, although they are abnormally swollen. Apparently evolution made halteres by repressing but not expunging the wing information. Cases like this in which characteristics are repressed by evolution offer tantalizing hints about how behavioral programs may evolve.

Another interesting outcome of crossbreeding techniques involves lovebirds, that group of parrots which are distinguished behaviorally by early, lifelong pair bonds and real nests. William Dilger successfully crossed two species of African lovebirds, *Agapornis roseicollis* and *A. fischeri.* Each species has its own species-specific method of courtship and technique for transporting nest material. Both species cut long ribbons of vegetation to use in nest building. Fischer's lovebirds, *fischeri,* carry this material to the nest site in their bills, one piece at a time. The peach-faced lovebirds, *roseicollis,* on the other hand, carry several strips on each trip tucked into the feathers of the lower back and rump (Fig. 19-4). Hybrids *attempt* to tuck the strips, but are

Fig. 19-4 Peach-faced lovebirds cut strips of vegetation (or in this case, paper) and transport it to the nest tucked into the back and tail feathers. This pattern is entirely innate. Hybrids from a cross with a species which carries the strips in its beak attempt to tuck the strips, but are utterly incompetent at this task.

never successful in transporting any material in this manner. They hold the strips in the wrong place, fail to let go, begin to preen (a displacement activity?) and drop the strips, or fail to entrap the strips properly by smoothing their feathers. As time goes on the hybrids begin to carry material to their nests in their bills, but only after vain attempts to tuck it in their plumage.

It is curious that on the one hand the hybrids greatly improve their carrying with experience, but on the other hand are never able to dispense entirely with the fruitless tucking movements. We might suppose that the improvement is the result of the presence of the *fischeri* behavior pattern in the hybrid; alternatively, the improvement might result from some innate plasticity in the peach-face behavior. To isolate the source of this improvement it would be worthwhile to devise an experiment in which Fischer's birds would be made as incapable of tucking as the hybrids.

Dilger's hybridization experiments produced another interesting set of results. The two parental species were observed to devote different percentages of courting behavior to a "switch-sidling" display (Table 19–1). From field observation alone this difference would almost certainly be attributed to heritable differences in male behavior. Indeed, frequency of switch-sidling by male hybrids displaying to female hybrids was intermediate, falling roughly halfway between those of the males of the parental species. However, the percentages observed when the hybrid males courted females of either parental species were

TABLE 19–1 *Comparison of Behavior of Two Species of African Parrot and Their Hybrid Offspring*

	Transport of nest material		Precopulatory activity of ♂♂
Species	Carrying in bill	Tucking in rump	Time spent switch-sidling
A. roseicollis	3%	97%	32% (to *roseicollis* ♀)
F_1 hybrid, initially:	6%	94%	33% (to *roseicollis* ♀)
F_1 hybrid, with experience:	41%	59%	40% (to hybrid ♀)
			50% (to *fischeri* ♀)
A. fischeri	100%	0%	51% (to *fischeri* ♀)

Source: William C. Dilger, "Behavior of Lovebirds," *Scientific American* 206, no. 1 (1962): 88–98.

identical with those shown by males of that species toward the females. This implicates some heritable trait or behavior on the part of the female. It is clear from this discovery that the hitherto-mysterious switch-sidling is a kind of displacement behavior. The male is caught between desire for the female and fear of reprisal if she takes his advances amiss.

In the end, however, what do the lovebird hybrids tell us? The female's control of switch-sidling could have been discovered by observing a mixed pair during courtship. Since these hybrids are sterile, backcrossing to work out the number and specific action of the genes is impossible, and so our pressing questions remain unanswered.

BACKCROSSING HYBRIDS

This rather bleak picture of the species hybridization technique brightens slightly when we come to crickets. As we saw in Chapter 15, crickets have distinctive species-specific calling, courtship, and aggressive songs. More important, they produce fertile hybrids. David Bentley and Ronald Hoy exploited this fortuitous combination by crossing two Australian crickets, *Teleogryllus oceanicus* and *T. commodus.* They focused their analysis on the calling song, the initial signal by which females distinguish species and guide themselves toward potential mates. The songs (Fig. 19–5) begin with a chirp of several loud pulses followed by a series of trills of several pulses each. Each pulse

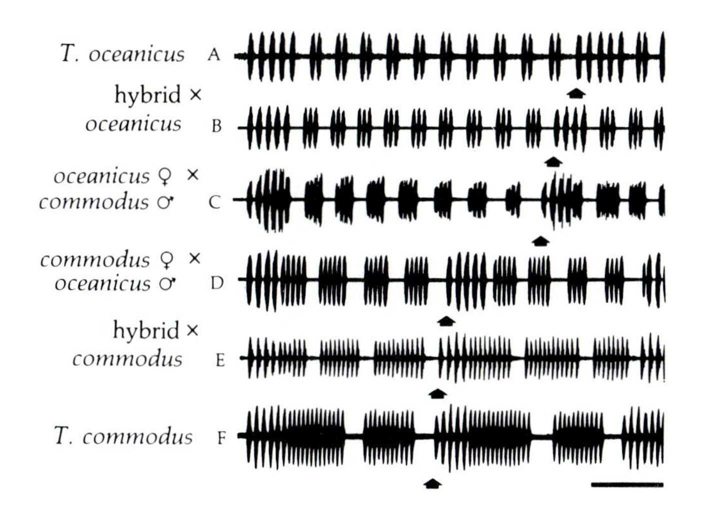

Fig. 19–5 Songs produced by two species of crickets, their hybrids, and the backcrosses. Each song phrase consists of a chirp and several trills. Chirps and trills are composed of pulses. New phrases begin at the arrows. The time marker is 100 msec.

reflects one stridulation. Sound intensity is controlled by the number of nerve spikes going to the muscles for each sound pulse. The signals for the song are generated by a group of neurons in the thoracic ganglion and when the cricket is not singing are inhibited by the mushroom bodies in the brain. Figure 19-5A represents the relatively stable calling song of *oceanicus* which begins with a five-pulse chirp followed by about nine trills of two pulses each. Figure 19-5F shows the more variable song of *commodus,* which is composed of a chirp of about six pulses followed by a few trills of approximately eleven pulses each.

If the song parameters were controlled by just a few genes, something might be learned by examining the songs of hybrids (Figs. 19-5C and 19-5D) and backcrosses (Figs. 19-5B and 19-5E). Unfortunately, hybrid songs are intermediate in all respects, although the trill number seems to be partially sex-linked or maternal (compare Figs 19-5C and 19-5D). Moreover, the backcross songs are intermediate between hybrid and parental songs.

It is unfortunate that in a case in which the behavior is so quantifiably different and the hybrids, in this case at least, are fertile, none of the song traits has sorted out during the crosses. The lack of individual variation among the backcrosses implies that many genes are involved. Whether this will be a general property of species-specific song remains to be seen.

The behavior of hybrid females is intriguing. When presented with a choice between a hybrid male's song and the song of a parental male, the females choose the hybrid song. At first glance it seems likely that the male's singing might be controlled by the same genes which specify the female input filter, the male motor-program circuit and the female IRM circuit being perhaps the same. If nature were really simple at heart it would use the same group of nerves for generating the output in males and recognizing the proper input in females. Hence, the hybrid output and input filters would match perfectly.

There is some support for this notion from other data. As everyone who has lived in the country knows, the rate of cricket chirping varies in a dramatic way with temperature. Choice tests show that females respond only to songs produced at rates appropriate for the temperature they feel. Hence, the output and input filters adjust for temperature in the same way, a result most easily explained if the same cells are involved. Tree frogs do the same thing.

There are problems with this pretty story. Female *Drosophila* hybrids, for example, prefer one of the parental songs to those of their hybrid brothers, as do hybrids of grasshoppers, which are close relatives of crickets. Perhaps this should have been expected. As mentioned in earlier chapters, song output in insects appears to be controlled by oscillators, two or more of which must interact in crickets. As such, the output "filter" is not the passive element of electronic circuits. And if the females were equipped with the same oscillator to generate an internal reference pattern to compare with the sounds of courting males, how could she make any sense of the auditory input—unless, by chance, the tune she is humming to herself happens to be in phase with her suitor's? If crickets depend on sign stimuli, we would expect the female to be looking for *elements* rather than patterns. When Hoy offered female crickets a choice between natural songs and scrambled versions which preserved the time intervals but destroyed the complex pattern, the females were equally attracted to each. Apparently it is the simultaneous presence of the three appropriate intervals in the song which is the releaser for female phonotaxis. Perhaps, instead of the assumption that the same filter is operating in each sex, the genes code for certain common cells—the timers, for instance—but sex-specific genes arrange these elements in different ways. A closer look at the songs of hybrid *Drosophila* and grasshoppers with an eye to potential acoustic sign stimuli might be rewarding.

What has classical behavioral genetics then taught us about crickets? That songs are genetically specified we already knew. Song recognition is based on temporal intervals rather than patterns, but that work was done without the need of hybrids. And finally, in the rare case which involves fertile hybrids, too many genes are involved in the behavior to allow further analysis.

CROSSING RACES

One way to avoid all the problems of hybrid infertility is to use just one species and cross animals with naturally occurring behavioral differences. Of course the differences will be smaller, but at least the breeding is easier. Such behaviorally different races are fairly common in animals which experience a range of environments and a certain degree of reproductive isolation. One early experiment along this line was peformed in the mid-1800s by an Austrian monk. He wanted to combine the superior gentleness of Italian honey bees with the greater industriousness of the German strain. What he got, however, were colonies of stormily aggressive hybrid bees. The monk, Gregor Mendel, decided to switch to peas.

More recently the behavioral peculiarities of the central African race of honey bees, known popularly as the Brazilian or "killer" bees, has once more drawn attention to the genetic basis of aggressiveness. Of course, the first thing behavioral geneticists must do is to *quantify* the behavior in question in order to detect and evaluate differences resulting from crosses. Apiculturists have an elaborate and amusing set of measures for aggressiveness in bees. A 2-cm black leather ball is suspended 5 cm in front of a colony. The hive is then "stimulated" (i.e., kicked) and the ball bobbed up and down for 60 sec. Then the ball is carried slowly away. The measured parameters are latency to first sting, number of stings in the ball (easily counted since bees leave their stings and venom sacs behind), number of stings in the gloves of the person conducting the test, and distance from the hive at which the last sting is delivered.

African honey bees, *Apis mellifera adansonii,* are often highly aggressive. The mean latency is 4 sec while for the gentler Italians it is 12 sec. The measure of number of stings in the ball presents problems when analyzing the Africans since it seems clear that more bees would like to sting the ball than can find room. Then, too, the beekeeper conducting the test rarely holds out for the full 60 sec, and it almost never happens that the observers can manage the measured retreat required by the protocol. And whereas Italian colonies rarely pursue an intruder more than 20 m, African colonies regularly keep up the chase for 1000 m.

Crossing yields roughly what we would expect intuitively: aggressiveness is mostly dominant, the hybrids exhibiting a latency of 6 sec. Backcrosses tentatively suggest a four-gene system—that is, the *differences* between the two races can be accounted for by four genes, although aggression per se doubtless involves many more which the two groups share. The enormous practical difficulties in breeding, maintaining, and testing bees make more detailed analysis extraordinarily difficult, even in this unusual case of a very prominent intraspecific behavioral difference combined with an apparently finite number of genes. Then, too, these two races have a variety of other behavioral and physiological differences—"genetic backgrounds" for the aggressiveness genes to work in—which complicates everything. For example, the African race prefers small nests, and will even build comb in a hole in the ground, whereas Italian bees require large cavities in trees; African bees swarm frequently, while the Italian ones are considerably more stable; African bees forage at low light levels, even in moonlight, and avoid flying out during the middle of the day, while Italian bees prefer midday and avoid dawn and dusk; and so on.

INBREEDING

One way to eliminate the complications that interindividual and interracial differences create is to inbreed. If animals are inbred rigorously, all the original genetic variation will ultimately be extinguished. Any heterozygosity will eventually vanish. Of course, whether a gene pair which was heterozygous to begin with will wind up as a homozygous recessive or dominant is a matter of chance. As a result, two strains of mice begun from a single pair might be expected eventually to differ in half the genes which were heterozygous to begin with. Separate breeding populations of wild mice, for example, differ genetically by about 30 percent. When inbreeding is severe, as when siblings are mated with each other, about 19 percent of the total heterozygosity is lost with each generation. If 30 percent of the mouse genes were heterozygous at the outset, by the seventieth generation the probability that even one gene would remain heterozygous is only one in ten. Since these conditions of inbreeding lead to the expression of many recessives normally hidden by dominant genes, the phenotype of an inbred strain is likely to be quite different from that of the founder group, and unlike anything to be seen in nature.

Certain practical difficulties make inbreeding more complicated. Many of the recessives which come to be expressed are deleterious, and inbred strains tend to become weak. As a result, inbred animals must be outbred periodically to related strains to restore vigor (i.e., heterozygosity). Inbred strains will thus be less different than theory predicts. Nevertheless, inbred strains—even of mice and rats—have significant potential in the study of heritable behavior, so long as we keep in mind the very artificial behavioral machine we have created. The individuals in a strain must have almost identical genotypes, and differences between strains, although random, are large. The environment can be varied to demonstrate how the expression of the genotype is affected, or it can be held constant and various strains compared to assess what sort of things are inherited.

A good example of this approach is Bovet's analysis of differences in avoidance learning for mice in shuttle boxes. Putting aside for the moment what this test measures or means, it is clear that the learning curves for mice from a heterogeneous population vary greatly from one individual to another (Fig. 19-6A) while within an inbred strain individuals differ only slightly from one another (Figs. 19-6B to 19-6D). On the other hand, differences in the learning curves between strains loom large—as large, in fact, as in the heterozygous population. Avoidance learning, therefore, displays the characteristics of a heritable trait.

It is intriguing that when the schedule of training was altered in a seemingly trivial way the learning curves were greatly af-

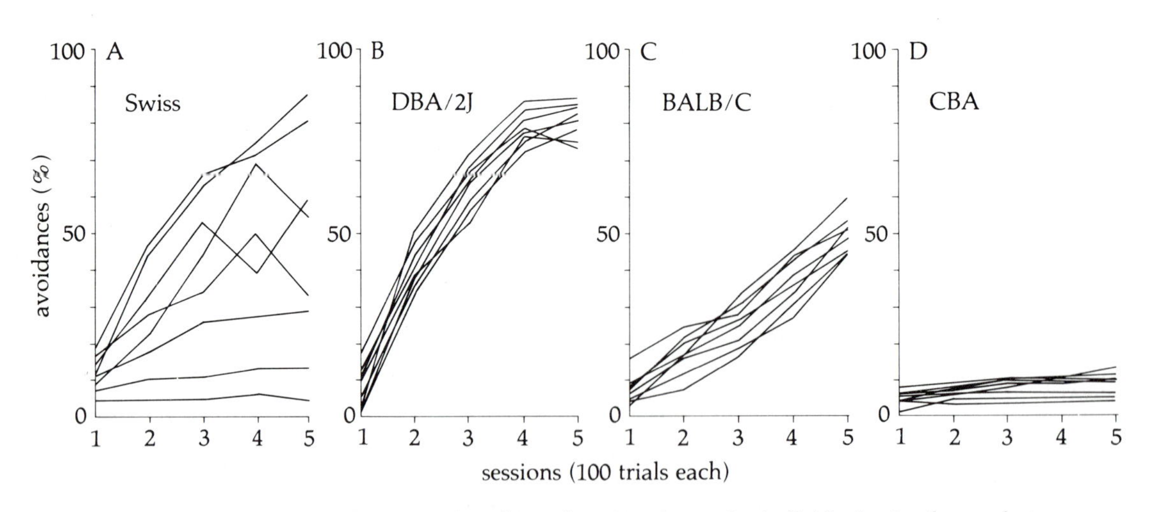

Fig. 19-6 Avoidance-learning curves for individual mice from a heterogenous population (A) and three inbred strains (B-D). Each curve represents the results of five sessions of 100 trials each.

fected, even reversed. Mice which had done badly before suddenly did well, while the quick strain now appeared to be quite slow. It certainly comes as no surprise to ethologists that there is an optimum combination of environment and genotype. However, this sort of approach by itself tells us nothing about the *mechanisms* which underlie learning except that they are genetically controlled, which we had already guessed, and that the relative timing of cues can be crucial to the workings of the learning program, which the phenomena of critical periods and short-term memory phases have also taught us to expect.

SELECTIVE BREEDING

Behavioral geneticists also use selective breeding to achieve a favorable recombination of genes with respect to a particular trait or group of traits, just as pigeons for ages have been bred to home well, chickens to lay more eggs, roosters to fight, or cows to produce more milk. The success of commercial applications of this technique to select for heritable morphological features in plants is a familiar story. The same approach has worked to select for heritable behavior, as beekeepers, for example, have bred for even-tempered bees. In the scientific world *Drosophila* have been bred for mating speed and for longer and shorter circadian rhythms, and in the original mouse-intelligence experiment mice have been bred for "brightness" and "dullness" (as measured by maze-running speed).

Tryon's classic selective breeding experiment illustrates the usual outcome of such genetic "shaping." Tryon took the mice from the two ends of the "intelligence" distribution and bred them as separate lines. After only six generations of such selection he had populations almost completely lacking in overlap (Fig. 19–7). When he crossed these two lines, the offspring were again "average" by his measure. In general, though, selective breeding suffers from the same limitations as the rest of classical behavioral genetics: if the behavioral difference is caused by more than a handful of genes, explicit analysis of what the individual genes are doing is impossible.

There is one instance, however, of a two-gene behavioral difference's being uncovered by this technique, though quite by accident. American foulbrood (AFB) is a bacillus which infects honey bee larvae and kills the pupae after their cells have been

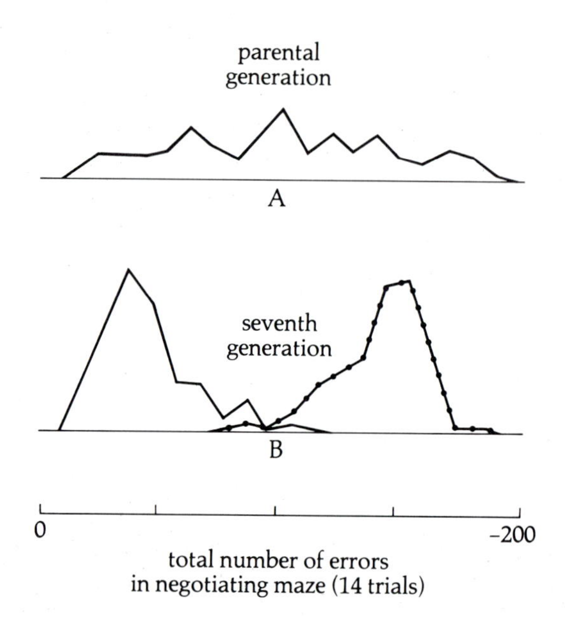

Fig. 19–7 When tested in a maze, individual rats make varying numbers of wrong turns (A). When rats which do well on this task are interbred, and those which do poorly are interbred in a separate group, the scores of the two progeny groups are skewed somewhat toward the performance of their parents. As the best and worst are again interbred as separate groups, the offspring in the next generation are dramatically different, and by the seventh generation the two groups show almost no overlap (B). Clearly maze-running ability has a strong genetic component, although whether it involves intelligence (as the original classification into "dull" and "bright" implied) is unknown.

capped. It is thought that the bacteria make their way through the wax into adjacent cells, spreading the infection. This disease can sweep like an epidemic through a hive, killing off many of the future bees and seriously weakening the colony, and will spread from one hive to another in a commercial bee yard. Beekeepers treat this disease by a modified scorched-earth policy: infected hives are typically burned or buried. An enterprising but foolish beekeeper once began volunteering to dispose of such colonies, and instead of burning the diseased comb, simply left it out so that his own bees could rob it of its honey. Of course his bees became infected, and only a few colonies survived. It happened, however, that after several years this strong if inadvertent selection pressure for an ability to survive AFB led to colonies specifically resistant to the disease. Surmising that this ability might result from a difference in the physiology or in the behavior of the bees, Walter Rothenbuhler set out to discover which was responsible.

Observation revealed that bees from resistant colonies would uncap and remove a dead pupa soon after it died. Susceptible colonies, also called "nonhygienic," did not. When the two strains were crossed, all of the first-generation colonies were nonhygienic, suggesting that *recessive* genes control hygienic behavior. When the hybrids were backcrossed to the hygienic strain, about a quarter of the resulting colonies were hygienic (Fig. 19–8), which tells us that only two genes account for the behavioral difference. Observation of the other colonies resulting from this backcross revealed that about a quarter uncapped the dead pupae but failed to remove them promptly. Another quarter of the colonies would remove dead pupae if they were already uncapped, but would not themselves perform the uncapping. The last quarter of the colonies were hopelessly nonhygienic and refused to remove even uncapped pupae. Rothenbuhler's refreshingly simple model accounts for these results.

It is curious and perhaps significant that hygienic behavior is controlled by *recessive* genes. It is very unlikely that *one* gene could specify, by itself, such complicated behaviors as uncapping or removal. Rather, a combination of many genes is doubtless involved. It seems reasonable to suppose that nonhygienic behavior is due to a blockage or defect in some link of the chain which generates the behavior pattern, one of which leads to uncapping and the other which ends in removal. In this regard it would be interesting to discover the mechanism of dominant blockage in these two forms of honey bee behavior. At present, however, the research trail is cold. Even with this rare, two-gene, genetically "clean" difference, it is unclear what further can be done.

And so in case after case the classical approach to behavioral genetics, although interesting, has told us relatively little that is of genuine ethological interest. In the next chapter we will see how the ability to *create* single-gene mutations in most or all the genes affecting a behavior provides the real key to dissecting the genetic control of behavior.

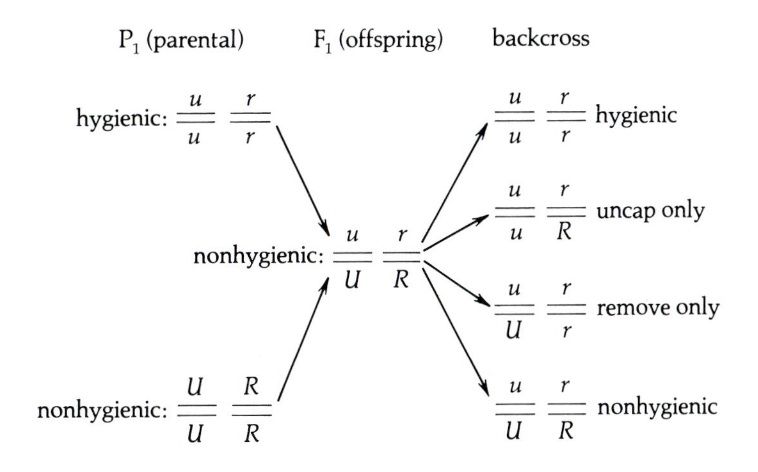

Fig. 19–8 Rothenbuhler's model of the proposed genetic mechanism of nest-cleaning behavior in honey bees. The two parental strains are crossed to produce the F_1 hybrids. The hybrids are backcrossed to the hygienic strain to yield the combinations at the right. *U* and *u* are the dominant and recessive genes for uncapping, respectively. *R* and *r* are the dominant and recessive genes for removal, respectively. The dominant genes block the behavior.

SUMMARY

Genes build and program animals, and it is on genes that evolution operates. Any real understanding of behavioral evolution requires that we understand what single genes do, and thus which evolutionary steps are open and which are closed to animals. Certain obviously adaptive changes in receptors or programming may be utterly impossible in that step-by-step random walk which mutation and recombination generate. The failure of classical behavioral genetics to forge that crucial link between our understanding of the mechanisms of behavior and of its evolution arises from its inability to deal with single genes. Behavioral genetics, as it has previously been practiced, depends instead on crossing species, races, or inbred lines, and observing what must be the effect of many genes interacting with each other.

STUDY QUESTIONS

1. It is typical for a gene which blocks a cellular process to be recessive. That is, a defective gene on one chromosome is "hidden" or compensated for by the good (and hence dominant) copy on the other chromosome. In the case of AFB resistance, however, the "bad" copy blocks the "good" one. How can this be?
2. In the experiment on chicken/pheasant hybrids, what behavioral experiments of interest went undone? Imagine the possible outcomes and how you would interpret them.
3. Interpret the nest-material behavior of the hybrid lovebirds and its gradual change in terms of genetically coded behavioral programs. Does your model suggest any experiments?

FURTHER READING

Bentley, David, and Hoy, Ronald. "Neurobiology of Cricket Song." *Scientific American* 231, no. 2 (1974): 34–44.

Dilger, William. "The Behavior of Lovebirds." *Scientific American* 206, no. 1 (1962): 88–102.

CHAPTER 20

Molecular Ethology

Classical behavioral genetics, that seemingly ultimate tool for dissecting the Lorenzian black box of behavior, has yet to answer any pressing ethological questions. As we have seen, a naturally occurring behavior which distinguishes one group of animals from another almost inevitably arises from differences in many genes at once, so that what single genes do and how they do it remains shrouded in mystery. The key to understanding the circuitous path from genome to behavior, therefore, probably lies in creating and sorting for single-gene mutations which block or alter a behavior pattern. Then and only then can the specific actions of each gene be analyzed one by one.

This single-gene approach sounds much simpler than it is. For one thing, most mutations are recessive: if we create a defect in one copy of a gene in a diploid organism (except for bacteria, almost all creatures have two sets of genes, one from each parent), the copy in the other chromosome normally continues to produce a functional protein or enzyme, and little or no effect will be evident. Then, too, there are a host of other problems to confuse and confound things: blockage of the behavior may be lethal; the gene may be involved in other processes as well; the effects of knocking out two quite different genes may be the same; and so on. Nevertheless, the success molecular biologists have achieved in working out intricate, multigene biochemical pathways controlling physiological functions suggests that all these difficulties may, with diligence and imagination, be overcome, allowing us to get a glimpse of the ultimate biochemical basis of behavior. The focus of this new technique of behavioral genetics (whose practitioners may well be called "molecular ethologists" since they are interested in the molecular basis of behavior, and use the single-gene mutation technique of molecular biology) is at present on relatively simple organisms: those in which the causal chain from the gene to behavior may be relatively short. The emphasis is understandably on behaviors for which the oddball, one mutant in a group of thousands, stands out, thus making the choice of both species and behaviors opportunistic.

UNICELLULAR BEHAVIOR

The most extreme case of an organism chosen for behavioral-genetic analysis strictly on the basis of convenience is the familiar bacterium *E. coli*. *E. coli* is haploid—has only one copy of each gene—so any mutations must be directly expressed. It has only three thousand genes, of which the functions of perhaps a third or so are known. *E. coli*'s behavioral repertoire, on the other hand, is quite modest. It moves toward some substances and away from others. Elegant genetic analyses have shown that it does this by sampling the environment with some two dozen

→

Fig. 20-1 *E. coli* has two dozen different classes of receptors which contribute to behavioral decisions. Each receptor class may have tens of thousands of actual receptors. Each receptor (so far as we know) is a binding protein. Some receptors such as arginine simply hand the chemical they sense over to a transport mechanism which takes the compound into the cell (top group of five receptor types). In a second type of receptor (next eight types down) the binding protein signals the cell that it has encountered something noxious; and needless to say, the compound is left outside. The third and fourth classes of receptors (next twelve types down) both hand over the substance to a transport system and signal the cell that something good has been found. The responses of the classes which are transduced are averaged in three groups (the integration system) through an enzyme, MCP, specific to each group. The three MCPs then undergo separate adaptation/habituation chemistry and are averaged by the tumble generator where the decision about whether things are getting better or worse is made. The tumble generator then passes its message to the flagellar motors. There are about twenty single-gene mutations known in this pathway so far. Two receptor classes, glucitol and galactitol, have yet to be assigned to an MCP group, and doubtless other receptor classes are yet to be discovered.

kinds of receptor molecules (one for each of six different sugars, for example), integrating the input from hundreds of thousands of these receptors, comparing the resultant sensory sum with its memory of the situation a few seconds in the past, and then

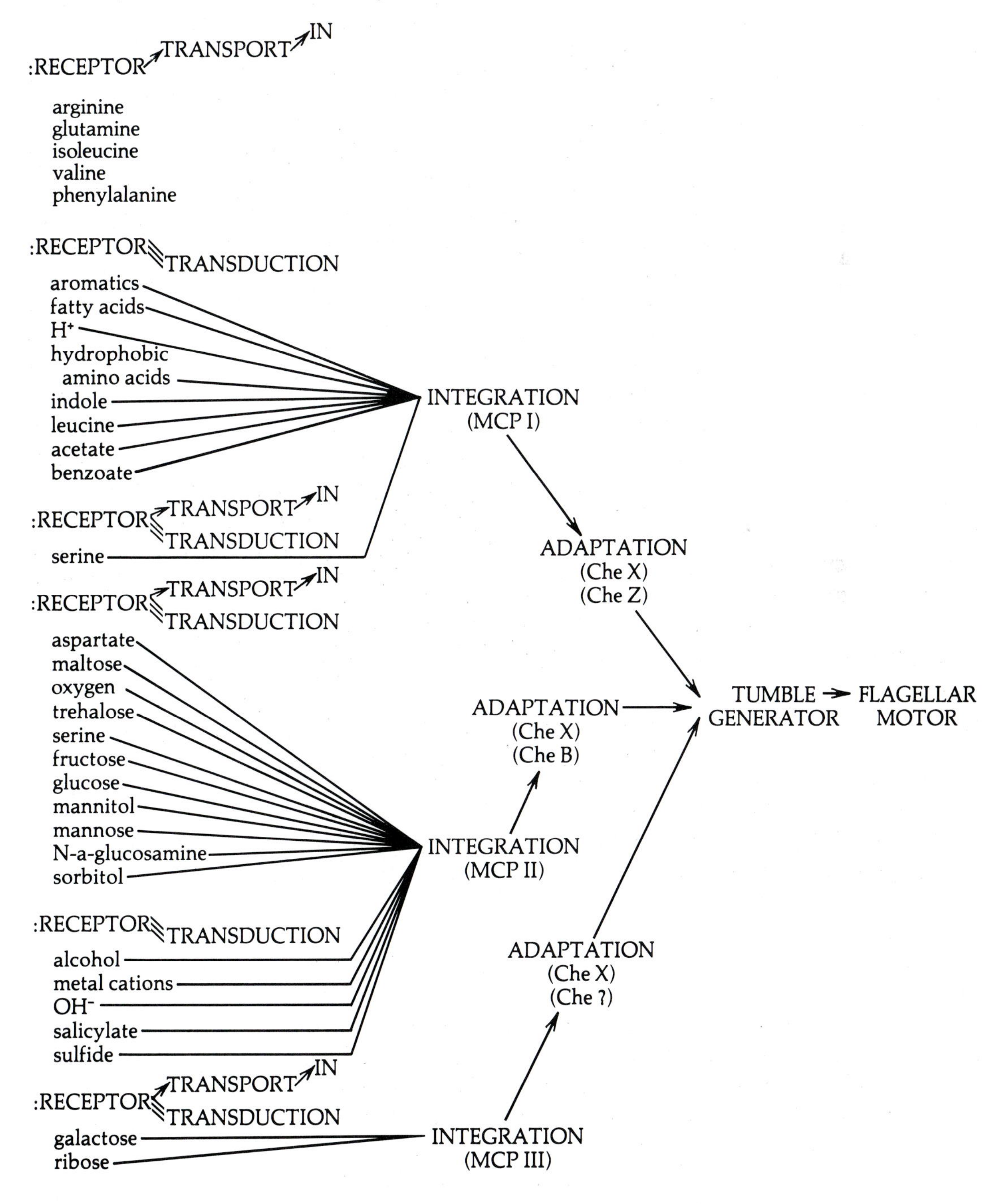

"deciding" whether to continue swimming in the same direction (if things are getting better) or to try another direction. The separate components of this behavior which have been identified to date are shown in Figure 20–1. For example, the galactose system depends on a galactose-binding protein as a receptor. A bacterium both senses galactose—that is, moves up concentration gradients of this sugar—and metabolizes it, but mutants with defective binding proteins do neither. A second protein transports the galactose into the cell to be "eaten." A mutation in this second protein leads the cell into concentrations of a food that it cannot eat. Still another protein is responsible for transduction—letting the cell know that the binding protein has found a galactose molecule. A defect in the gene coding for the transduction protein will result in a cell which, although blind to galactose, nevertheless consumes the sugar as greedily as ever when it happens upon it. Thus the relationships between at least these genetic steps have become clear from studies of chemotaxis-defective mutants. The genetic makeup of *E. coli* is so convenient to analyze that a relatively complete understanding of its nuts and bolts must soon emerge.

The difficulty with studying *E. coli* is that it is not obvious how its behavior is relevant to animals. For example, the obvious intermediary between the gene and behavior is the nerve cell. Since ethologists are interested in the mechanisms which underlie behavior, much of their concern is directed toward nerves. How do nerves work, and how are the very different characteristics of various classes of cells generated?

Although *E. coli* is an electrically active chemosensory receptor, it may not be at all typical. For one thing, receptor cells with specific binding proteins—taste or pheromone cells, for example—normally have only one kind of binding molecule, and so are sensitive to only one compound or group of compounds. The elaborate twenty-four-channel input and three-channel integration system of *E. coli* is probably not to be found in multicellular organisms. And although it is electrically active, and probably uses the electrical potential of the cell surface to integrate receptor information and transfer the "decision" to the flagella where directional changes are accomplished, *E. coli* is a bizarre nerve cell. It shows a depolarization and repolarization reminiscent of the way signals are produced in real nerves, but the process is a thousandfold slower and the electrical response is superficially the same whether the bacterium decides the world is getting better or worse. On the other hand, *E. coli* is not wholly without familiar bits of clockwork. Particularly in its habituation or adaptation the chemical processes resemble those of *Drosophila.*

A more promising nerve cell analogue is *Paramecium.* Its genetic structure is at once complex and trivial. *Paramecium* is polyploid—it has many copies of each gene—but by adjusting conditions properly we can cause all but one set of genes to be thrown out of their vast library. Hence, we can mutagenize *Paramecium* with chemicals which change the DNA bases, cause them to become haploid, and look for strange behavior. The normal behavior, dependable if not spectacular, consists of backing up when things get worse (Fig. 20–2), and then setting out in a new direction. These animals seem to have all the complexity of real nerve cells and more. We know that the direction the cilia (which propel the animal) beat is controlled by the voltage across the cell membrane, which is controlled in turn by sensory receptors. Mutants have been found which go only forward (*pawn*), or perform spontaneous avoidances (*paranoiac*), or swim too fast (*fast*). Each has a problem with its membrane potential, and hence a defect in some crucial part of the elaborate machinery which allows nerve cells to control behavior is expressed: *pawns* lack the voltage-dependent ion gates which provide the positive feedback necessary for action potentials, *paranoiacs* have a deficit in membrane repolarization, while *fasts* have an elevated membrane potential. With specific unitary mutations like these, single-gene analysis is ferreting out, one by one, each of the pieces of the elaborate molecular machinery responsible for making nerve cells work.

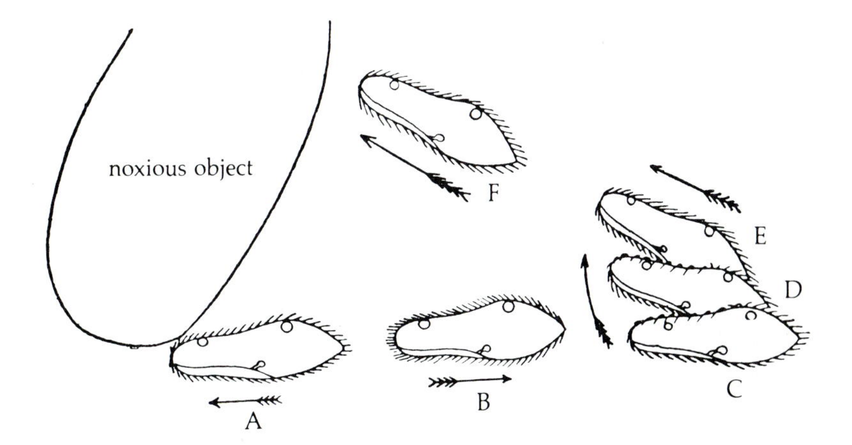

Fig. 20–2 Avoiding reaction of *Paramecium* in schematic form. While moving left, the organism encounters a noxious change in the environment (A). A short-lasting ciliary reversal follows (B) which moves the animal back to the right (C). The *Paramecium* then reorients by pivoting about 30° clockwise (C–E). Finally normal ciliary activity is restored and the animal moves off to the left (F).

NEURAL DEVELOPMENT

How nerve cells interact with each other is the next level for behavioral analysis, and unicellular organisms are obviously inappropriate for solving the puzzle of how nerves interact to generate behavior. For behavioral geneticists these concerns can be reduced to three seminal questions: (1) How do the genes build specific sorts of nerves? (2) How do the genes tell particular cells which sort they are to become? and (3) How do the genes tell the cells which nerves they must connect with in order to form circuits that will both process and integrate sensory information, and direct the behavior?

The most promising candidate for an analysis of how nerves interact to process sensory output and create behavior is the lowly nematode. It has about the smallest number of nerve cells to be found in an organism with more than one—approximately 260—and many nematodes are hermaphroditic: that is, they normally fertilize themselves. Their proclivity to inbreed with themselves under a well-known set of environmental conditions means that induced recessive mutations are expressed relatively soon as they drive themselves to homozygosity.

One tempting approach to studying nematodes is to map all the nerves and their connections, to try to reason out their functions from the anatomy or from neurophysiological recording, and then to create mutations affecting behavior and to look at the consequences. As usual, this procedure sounds much simpler than it has turned out to be. For one thing, 260 is a much larger number than it seems: the complexity is staggering. Clear-cut analysis of what is going on in the nematode brain is currently beyond our ability.

Another problem has been that no one has been able to record from the nerve cells because they are so small. When a behavioral mutant is isolated, then, only its anatomical defects can be investigated. Mutations which affect physiology rather than anatomy simply cannot be analyzed. As a result, only two lines of research have proven useful. One approach has been the mapping of self-contained subsystems, a feat which has been performed most successfully with the group of cells controlling taste and feeding. Unfortunately, mutations in the feeding system are lethal. The other approach has been to study how the nerves are made and wired during development, and then to make mutants in which this orderly process is disrupted.

The rules of neural development—how genes build a nervous system—and how outside input during critical periods affects cellular connections are crucial questions for behavior. The genes somehow tell each cell whether it is to become a visual nerve or a liver cell, and if a nerve cell how to find and recognize the particular cells, perhaps several centimeters away, to which it is to pass on its information.

There seem to be two general strategies by which its genes tell a nerve cell what to be and where to go. Sidney Brenner, one of the growing army of molecular-biologists-turned-neurobiologists, calls them the European and American plans. Under the European plan what a cell does is determined by who its ancestors were, while in the American plan what a cell does is specified by what its neighbors are doing. Nematodes subscribe to the European plan. For example, the long ventral cord of nematodes consists of twelve nearly identical ganglia spaced out along the nerve track. These nerves control swimming and mating, and all arise during development from one of twelve precursor cells. The subsequent pattern of cell division is the same in all twelve ganglia even though the eventual organization of the ganglia may be very different. As a result, cells that become nerves on only some (or even just one) ganglia are nevertheless made in all twelve and then killed when not needed (Fig. 20-3). The fate of a cell is determined at its birth, at

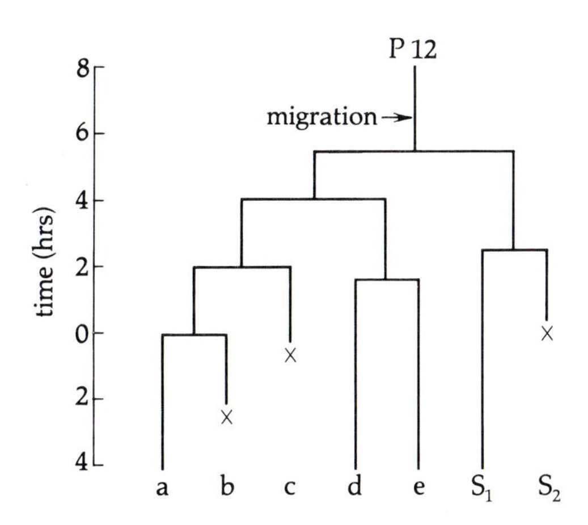

Fig. 20-3 Each of the twelve clumps of nerve cells in the ventral cord of the nematode *C. elegans* arises from a precursor cell which follows a fixed pattern of division. The last group, which began with precursor P12, does not need cells b or S_2 because it is at the tip of the tail, and has no clump further back to communicate with. Nor does it need cell c, a neuron which is used only in the middle clumps of males. Nevertheless, the entire stereotyped division process is gone through, and the superfluous cells killed.

which time the genes provide it with a specific set of instructions which depend on who its parent cells were.

By way of contrast, young cells from one part of the body in an American-plan organism can be placed in another part and will develop so as to be indistinguishable from their fellow cells. The genome of these cells takes its cues from those around it, and uncovers the appropriate specialized genes which issue instructions specific to each cell. These instructions, however, once issued, seem to be irreversible, since after a certain critical time cells become developmentally "imprinted." If transplanted *after* they have received their instructions, they will develop as ordered, regardless of the neighborhood into which they are moved. In these cases, American-plan cells inevitably switch over to the European system after their critical time is reached.

The nematode strategy of building extra cells and then killing them appears to be widespread, and may provide an insight into the universality of nervous system phylogeny. In the spinal cord of vertebrates, for example, the two vertebral segments serving the front and rear limbs have many more motor neurons than the other two dozen or so segments. Nature builds the spinal cord, however, by generating the full complement of cells in every segment, and then editing out the unnecessary ones. Hence, in these two vertebrae which control all the complex motions of the limbs, motor neurons have the best chance of survival.

Although wasteful in terms of cells, the genetic instructions for the construction of nervous systems from worms to people seem to work best by delineating not which specific cells should be *created,* but which ought to be allowed to *survive.* There must be a powerful logic, a sweeping informational efficiency about this system which we have yet to grasp.

The "targeting" instructions which nerves receive from their genes tell them where to go and whom to talk to, and single-gene mutations offer fascinating clues about the nature of these instructions as well. Oddly enough, some of the most interesting mutants have turned up in that hopelessly diploid organism, the mouse. The reason, of course, is that mice have been inbred for hundreds of generations, and this has resulted in the expression of deleterious recessives with bizarre behavioral consequences. To date, more than 250 such mutants have turned up, and their strange effects are described by their names: *reeler, weaver, staggerer, waltzer, shaker, quaker, spinner, tumbler, tipsy, zig-zag, gyro,* and so on. Many involve massive neural deficits: lack of otoliths, the organs for gravity reception in the inner ear; no myelin, the substance used to insulate axons and speed the conduction of action potentials; no cerebellum; no corpus callosum, the enormous nerve track which connects the two halves of the cerebral cortex.

Other mutants, however, have more subtle and instructive problems. The *reeler* mutation in mice, for instance, causes the layers of the mutant cortex to wind up inverted—the bottom one on top and the top one on the bottom. Normally the mouse brain builds a layer at the bottom which then migrates through the other layers to the outside. Hence, the six layers of the visual cortex develop in the order 6, 5, 4, . . . 1, but are found in the opposite order in the mature cortex (Fig. 20–4A). The normal visual pathway is from the eye to the thalamus to the visual cortex (Fig. 7–6). As we have seen in Chapter 7, the fibers climb through the bottom layers (6, 5, 4) and form synapses in layer 4c. In *reeler,* the visual fibers from the thalamus first encounter layers 1, 2, and 3 (Fig. 20–4B), pass straight through layer 4, and continue up to layer 6. There the fibers abruptly turn and proceed back down through the layers in the normal order, and only then form synapses in 4c. In this case the behavior of developing nerves is just as mindless and machine-like as the behavior of the whole animal which they come later to control. And, of course, the ultimate source of both sets of rote controls is the genome.

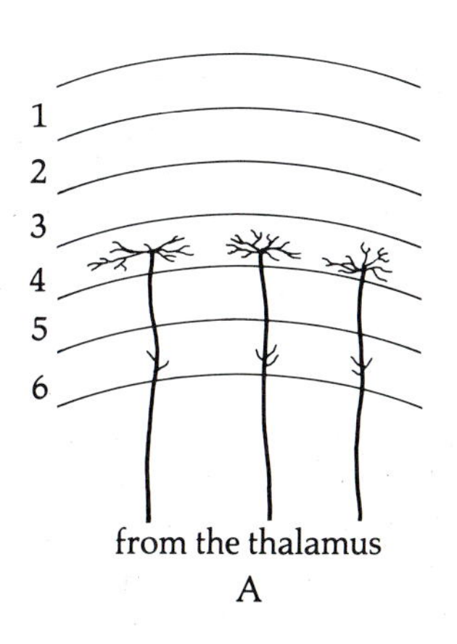

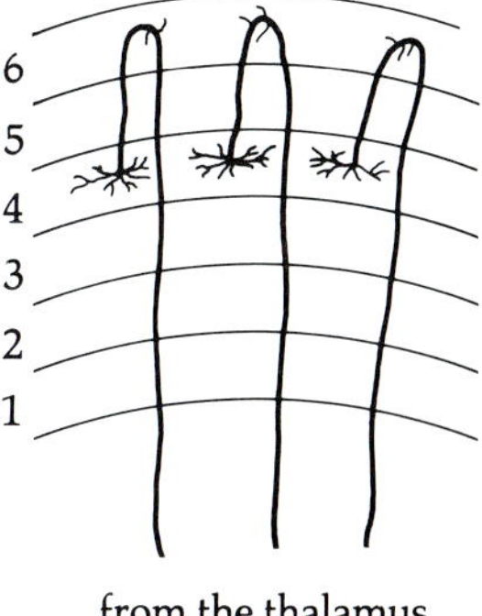

Fig. 20–4 A. In normal mice, the primary visual input to the cortex comes up from the LGN of the thalamus through layers 6 and 5 to terminate in the layer 4c of the visual cortex. B. In *reeler* mice, whose single-gene defect gives rise to an inversion of the cortical layers, the fibers from the LGN pass up through all the layers to the surface of the cortex where they reverse direction and return back down to layer 4c. The instructions by which visual fibers find their targets must specify finding layer 6 first, and only then going on to layer 4c.

The one-step-at-a-time system of programming leads to some strange results. For example, mammals have a nerve which invariably enters the neck from below, runs up the head, and then, without making any synapses, travels back down through the neck to its actual target. Apart from such anatomical curiosities, this developmental strategy also provides some valuable flexibility which, along with the self-calibration provided by various critical periods, allows the nervous system to correct for or accommodate major anatomical and physiological disruptions.

The most familiar example of the nervous system's ability to cope with massive foul-ups is the Siamese cat. This breed is distinguished by its unique coloring, which is the consequence of a single-gene, temperature-sensitive mutation of a pigment gene. The gene produces dark fur on the extremities (which are cooler than the body), and light fur everywhere else. Beauty has its price, and this pigment gene badly disrupts the organization of visual fibers as they progress from the retina to the LGN of the thalamus. Nevertheless the visual system absorbs the defect, in part by reorganizing the lateral geniculate and then by making adjustments in the cortex itself. The result is a cat which, although slightly cross-eyed (the neural compensation is not quite perfect), appears otherwise to have normal vision (see Guillery's "Visual Pathways in Albinos," listed under "Further Reading").

The possible universality of the patterns seen in the visual cortex is suggested by work on crickets. David Bentley mutagenized crickets and then sorted for animals blocked in their escape response. This was done by blowing compressed air on the crickets and capturing the jumpers with a vacuum cleaner. Among the survivors was a single-gene mutant lacking sensory hairs on its vibration-sensitive cerci. Electrical recordings revealed that the now-hairless sensory cells were perfectly normal, but the interneuron, which integrates the information from all the cells and sends the "escape" message up the cricket to the brain, was abnormal. But did the mutation affect the interneuron directly or secondarily? Rod Murphy, reasoning from analogy from the deprivation experiments on cats described in Chapter 7, waxed the cerci during development. Just as they did in the cats, the receptor cells remained perfectly functional, but the higher level integrators stopped listening to them.

DROSOPHILA

However, neither mice nor crickets are especially suitable genetic tools. They have long generation times, take up too much valuable laboratory space, and are 95 percent diploid at best. (They are not completely diploid because maleness in these species involves the absence of one of the pair of sex ["X"] chromosomes. Hence, the genes on the X chromosome of males are haploid.)

Perhaps the most accessible and exciting system for behavioral genetics is *Drosophila.* Here, fully one-quarter of the male fruit fly's genes, those on his single X chromosome, are haploid, so any mutation created there will be directly expressed. A vast arsenal of genetic tricks has been developed in the decades since T. H. Morgan first selected this fly for analysis. More important, the behavior of *Drosophila* is complex enough to be interesting. As we saw in Chapter 5, flies locate food by following odor gradients—the odor of fermenting fruit for *Drosophila*—and their mating ritual (Chapter 15) makes use of various olfactory, tactile, and visual codes, and includes a species-specific song. But besides eating and courting, flies are interesting because they learn. To date, Seymour Benzer and others have been able to create mutations in one or more genes controlling phototaxis, chemoreception, visual development, mating, circadian rhythms, sight, learning, memory, and life span. Three examples will illustrate the strategy of the single-gene behavioral genetics of *Drosophila.*

Little is understood about the mechanisms which underlie circadian rhythms, the daily cycles which prompt much of an animal's behavior. For instance, *Drosophila* in the wild emerge from their pupal cases shortly after dawn ("Drosophila" means dew-lover in Greek). This gives them time to stretch, harden their cuticles, and search for food before it starts getting hot. In the laboratory it has been observed that the flies tend to emerge some multiple of 24 hours after experiencing an artificial "dawn"—any flash of light after an extended period of darkness. No natural dawn is necessary after that, and *Drosophila* pupae maintained in the dark will "pace" their development so as to emerge like clockwork 24, 48, or 72 hours later (Fig. 20–5A). Clearly the pupae have their own clocks which they set one to

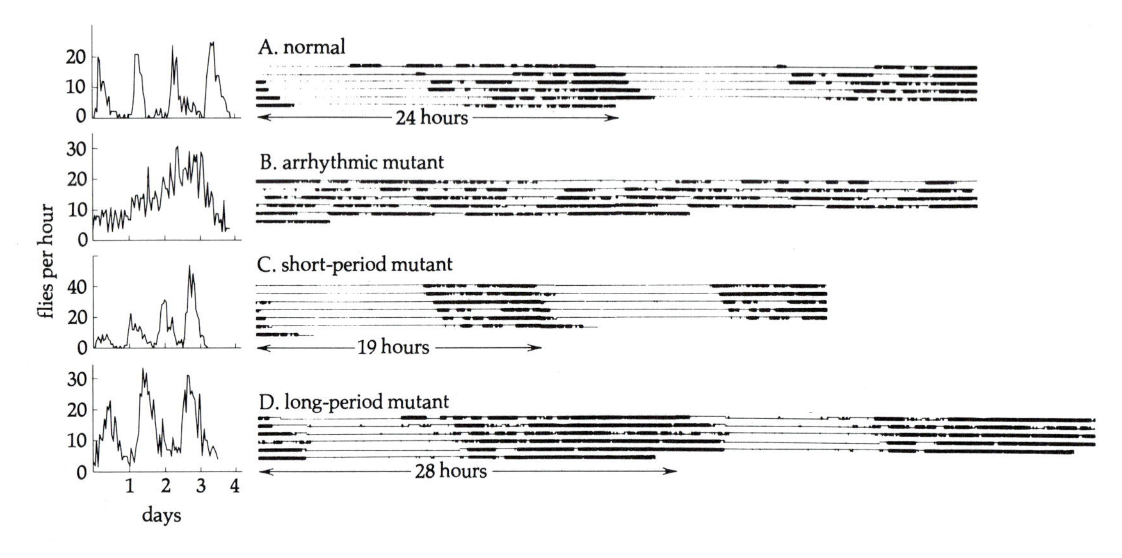

Fig. 20–5 Normal *Drosophila* adults emerge from their pupal cases just before dawn, an interval which is timed as a multiple of twenty-four hours from some previous onset of light. The eclosion rhythm of four strains is shown at the left: normal flies (24-hour rhythm), arrhythmic flies (no period), short-period flies (19-hour rhythm), and long flies (28-hour rhythm). The activity rhythms of these strains *(right)* show corresponding periodicities.

three days before they have finished developing. This is important in the life of the fly since the process of emergence actually begins several hours *before* dawn, so the pupae must anticipate the correct time. As adults, the flies continue to use their internal clocks to regulate their behavior. Since circadian rhythms are innate—animals refuse to be taught that days are thirty hours long—genes must hold the key to the clock.

The first problem in any behavioral genetic analysis is to sort out the desired mutant from a population of normal individuals. The second problem is to analyze what the search turns up. To select for circadian rhythm mutants in fruit flies, Seymour Benzer and Ronald Konopka allow the larvae to attach their pupal cases to a plate which, after exposure to an artificial dawn, is put into a device which automatically shakes any emerged flies into a bottle. Benzer and Konopka analyze the daily activity rhythms of flies collected in the early and late bottles to find abnormal individuals with reproducible aberrations in their circadian rhythms. After extensive sorting in this manner, three important mutants have been found: one with a 19-hour day, one with a 28-hour cycle, and one with no consistent period at all (Figs. 20–5B to 20–5D). All three mutations map to the same place on the X chromosome, so it is reasonable to suppose that they represent three different versions (alleles) of one of the crucial enzymes or control elements in the circadian clock.

FATE MAPPING

In *Drosophila,* the trail of research into single-gene behavioral mutation runs slightly further. In any cell most genes are inactive. Genes specific to kidney, liver, or bone cells will be shut off in a muscle cell. Hence the presence of a specific mutation in the genome is irrelevant and unexpressed in most cells. In *Drosophila* it is possible to create sexual mosaics (gynandromorphs), animals half of whose cells are male and half female. The gynandromorph is created by the loss of the maternal X chromosome in one of the two cells produced by the egg's first division (Fig. 20–6A), and the result is an organism divided more or less sharply into male and female groups of cells (clones). The division occurs randomly, so male and female tissue will differ from fly to fly. Since in the female half any X chromosome mutation is hidden by the second X, the mutation will only be expressed if the tissue in which it is active happens to be male. If the anatomical "focus" of the gene happens to fall in female tissue, the gynadromorph will be normal. Since the line dividing the tissue belonging to each of the two sexes can fall anywhere in the fly, and since only one piece of tissue will always be male in all the various gynandromorphs displaying the mutant behavior, a comparison of a series of such gynandromorphs quickly isolates the one patch common to them all. This area, then, must logically be the site of the mutation.

The gynandromorph trick greatly reduces the amount of work necessary to analyze how a mutation affects the animal. For example, a mutation called "wings up" generates flies which cannot fly and, curiously, do not run to light. Is the mutation affecting the wing muscles, the wings themselves, the nerves in the thoracic ganglion which direct the muscles, the brain, or something else? This enormously powerful technique places the locus of the "wings up" mutation in the wing muscles, and pinpoints the circadian rhythm defect in a particular part of the

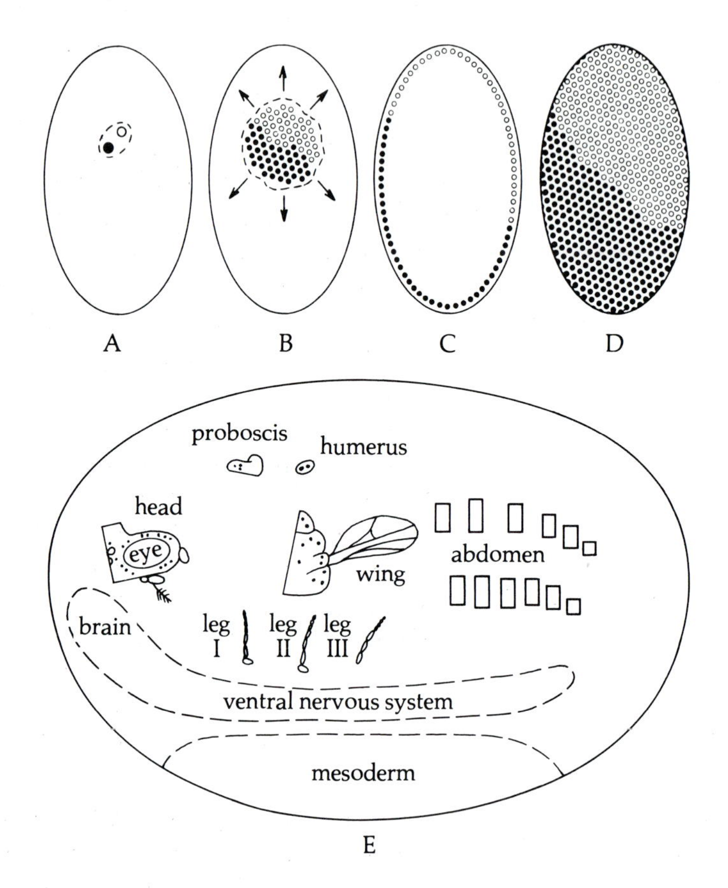

Fig. 20-6 Schematic diagram illustrating the formation of a gynandromorph. A. During the first division an X chromosome is lost yielding a clone of male nuclei (open circles) and one of female nuclei (filled circles). B. The nuclei continue to divide without mixing, and then migrate radially to the surface of the egg. C. Later cross section of an egg after cell walls have formed D. The egg is divided along an approximate line into male and female halves. E. Fate map of one half of an egg illustrating locations of cells which will ultimately form the indicated structures in the adult fly.

brain where, needless to say, a great deal of attention is now focused.

The gynandromorphic technique, called "fate mapping," has permitted genetic confirmation of the hierarchy of neural control of courtship and patterned motor output which we had inferred from the behavioral experiments of Chapters 13 and 15. For example, whether a gynandromorph courts females depends on whether the appropriate thoracic ganglion controlling the wing muscles *and* the brain are male. Apparently the brain contains the "will," since male-brained flies will approach females and extend their wings, but unless the thorax is male as well so that the song wiring is intact, the fly is essentially mute and his drive to court will come to nothing.

Selection of mating mutants has turned up a bizarre variety of sexual deviants, each apparently blocked in one of the specific releasers of motor programs in *Drosophila*'s courtship ritual. One, *fruitless,* illustrates the power of the single-gene approach.

Males carrying this mutation court females and other males indiscriminately. How could the whole sexual recognition system have gone so wrong? One hint comes from the observation that normal males court *fruitless* males. The answer to the puzzle is that *fruitless* males produce the female pheromone; hence, males court them. The reason *fruitless* males court other flies indiscriminately is that they smell *themselves.*

LEARNING

Perhaps the greatest ultimate potential of *Drosophila* behavioral genetics arises from the recent isolation of mutants with deficits in learning. The ability of flies to learn in the first place is modest: their IQs come in only slightly higher than blocks and stones. After having been trained to associate a particular odor with an electric shock, one fly in three at best will have learned to avoid the odor. Still, though we cannot teach all the flies all the time, a group of particularly dull flies can be distinguished from the normal, moderately dumb population.

A half-dozen different learning mutations have been found on the X chromosome. *Drosophila* displays many of the features of vertebrate learning including short- and long-term memory (Fig. 20–7), habituation, sensitization, and so on. Beyond this, larvae and conveniently isolated parts of the adult nervous system can learn: a leg (with its thoracic ganglion) may be taught to hold itself up or down and, if anything, learns better when the brain has been removed.

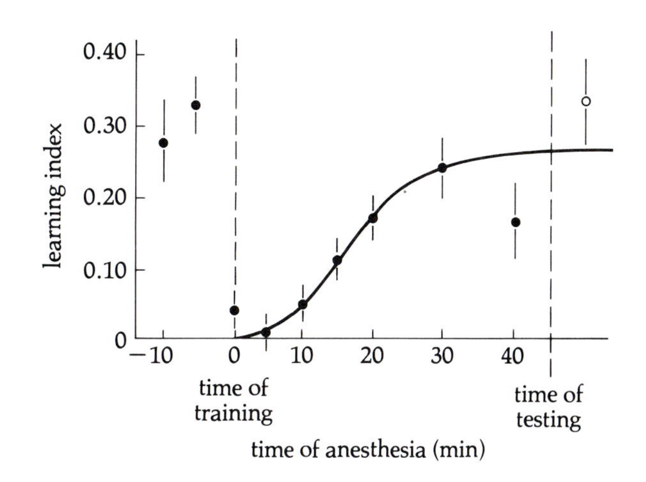

Fig. 20–7 Effect of anesthesia on the memory of *Drosophila* trained to an odor-avoidance task. Flies chilled soon after training remember very little, while those treated before testing or at least 30 min afterward perform normally. This experiment indicates that flies have an active processing phase (short-term memory) and consolidation phase very much like those of bees and mammals, though the time course is somewhat slower.

Analysis of learning mutations has turned up intriguing clues, though few solutions, to the questions surrounding learning. One mutant, *amnesiac,* seems blocked in retrieval from long-term memory. It remembers what odor to avoid for a few minutes, but then "forgets." Another, *turnip,* seems blocked in short-term memory since it normally never learns, but when a heterozygote is created—a fly with one normal and one *turnip* gene—it remembers for a very short while. Finally, one learning mutant, *dunce,* cannot learn and also fails to habituate. Habituation, as described in Chapter 5, is the phenomenon by which repeated presentations of a stimulus cause a normal behavior response to wane. Repeatedly presented with sugar on one foot but never allowed to feed, a fly will eventually stop extending its proboscis in search of reward. This process takes place centrally (in the brain, we imagine), and may be distinguished from simple sensory adaptation—the receptors getting tired, say—by the simple trick of stimulating a new unadapted set of the same class of receptors and observing the result. Present the other leg of a merely *adapted* fly with sugar and its proboscis would extend. If the animal were actually *habituated,* on the other hand, nothing would happen. In fact blowflies *do* habituate, and refuse to respond to the messages from their unadapted receptors. Previously, habituation was not considered part of learning, but now it seems clear that the two processes share at least one gene. In fact, *dunce*'s biochemical defect helps confirm Kandel's model for *Aplysia* (Chapter 5 readings). Kandel postulated a central role for cyclic AMP in habituation, and *dunce* lacks one of the two enzymes which metabolize cylic AMP.

The lesson from these beginnings is that, given animals with the right combination of behavior and genetics, the single-gene approach to behavior has an almost breathtaking potential for unlocking Lorenz's ultimate black box. Armed with the knowledge of how the biochemical underpinnings of behavior are organized and orchestrated both during development and during overt behavior, we can begin to understand how behavior can evolve—what changes can take place, where they lead, and why some possibilities are open and others closed. This ultimate ethological endeavor holds the key to many behavioral abnormalities previously relegated to the realms of phenomenology. Ironically, increasing numbers of mental problems are being recognized as chemical imbalances, and so can be treated pharmacologically rather than through psychoanalysis. In this way, medicine and ethology may be seen to be treading paths which will ultimately converge at the genome.

SUMMARY

The strategy of single-gene behavioral genetics is to generate mutations in genetically convenient animals, sort for defects in a behavior, and analyze their source. From a set of such defects affecting the same behavior, behavioral geneticists reconstruct the interactions and pathways which underlie and together give rise to normal behavior. Through such an understanding of the workings of these genetic building blocks, we may hope to comprehend how behavior is organized and might evolve. The single-gene technique, though still in its infancy, has already provided insights into the physiological properties of nerve cells, the rules by which they wire themselves together, and how they alter their properties on the basis of sensory input. Single-gene techniques have also provided hints about how such systems as circadian rhythms, the series of releasers and motor programs in fly mating, and learning might be constructed. Despite the delicate balance between genetic simplicity, behavioral complexity, and physiological convenience which must be maintained in the choice of organism, this ultimate mechanistic approach promises to provide models for realizing ethology's ultimate goal, which is understanding simultaneously the how and why of behavior.

STUDY QUESTIONS

1. How would you decide whether *E. coli* and *Paramecium* adapt or habituate?
2. All of the *Drosophila* work depends on punishment learning. Reward learning is difficult because (a) flies interrupt their behavior to feed, and (b) once they are full, they no longer respond to the stimulus and so look dumb. Can you devise a reward technique which might work?

FURTHER READING

Adler, Julius. "Chemotaxis Behavior of Bacteria." *Scientific American* 234, no. 4 (1976): 40–47.

Benzer, Seymour. "Genetic Dissection of Behavior." *Scientific American* 229, no. 6 (1973): 24–37.

Guillery, R. W. "Visual Pathways in Albinos." *Scientific American* 230, no. 5 (1975): 44–54.

Part VI

Evolutionary Mechanisms

Part VI

EVOLUTIONARY MECHANISMS

CHAPTER 21

Behavioral Ecology

The most significant intellectual challenge to the 1800s was the problem of the species: why should there be so many different kinds of plants and animals around the world? A young naturalist, Charles Darwin, guessed the answer, and in so doing changed our image of the world for all time. His discovery of the basic principle which generates the overwhelming diversity of life around us was based on two observations—first, that there is morphological, physiological, and behavioral variation among animals in any one species; and second, that more young are born than can possible survive on the food available. This latter observation was derived from a gloomy essay on population by Malthus which pointed out that most families live in poverty because they have as many children as they can feed and more, so that life for the masses goes on at the edge of starvation. Any increase in the food supply, Malthus saw, will be promptly absorbed by additional offspring who would otherwise have died of starvation or disease. Darwin reasoned that if the size of a population is a food-limited constant, if variation is heritable, and if a particular variation could improve even marginally the chances of those individuals carrying it to survive and produce healthy offspring, then the frequency of that variant in subsequent generations must rise. The process, of course, is natural selection, and the underlying mechanism is genetic.

Darwin, who knew nothing about genes, thought in terms of the survival of the individual animal, but we believe now that evolution acts directly on *single genes,* and is best thought of from that point of view. Put bluntly, the one goal of genes is *their own* survival. They contrive to build elaborate fortresses, commonly known as plants and animals, to protect themselves from the harsh world outside, and see to it that these living fortresses feed and sustain themselves as long as they are useful. Like all good playwrights, the genes tell their one-man shows where to go and what to do, struggling against rival genes in other animals be they prey, predators, or direct competitors, all for the purpose of transmitting a millionth of a billionth of a gram of coded instructions into the next generation.

The variety of species in the world is certain evidence that survival is a problem with at least a million tried-and-true solutions. *Homo sapiens* is one such answer, "correct" at least for the present, whereas *Australopithecus boisei, A. robustus,* and *Homo erectus* are not any more. What are the strategies of that unique combination of genes that build and pilot each animal through youth, courtship, reproduction, and parenting? What other strategies were available but rejected, and why? With some idea of the sensory mechanisms and processing strategies available—information whose critical relevance forges ethology's link with neurobiology—we turn now to ethology's evolutionary interest, behavioral ecology: the pursuit of the design principles by which evolution has fashioned the special combination of habitat choice, diet and foraging strategy, social organization and reproductive strategy, mating system and parental investment, which makes each species different.

HABITATS AND NICHES

As it happens, packing the world with a set of a million ever-so-slowly changing animal species has been accomplished with a relatively brief list of rules. The most important features of this stability require that different species must (1) avoid mating with each other, (2) avoid competing for exactly the same food since the systematic victory of one means the inevitable extinction of the other, and (3) avoid completely consuming their prey, be it animal or vegetable, and thereby eliminating both species.

We discussed the usual way animals go about observing the first of these requirements in Chapter 15: communication. The second regulation leads directly to the major generalization of behavioral ecology, the Niche Rule: to a first approximation, no two species can long be eating exactly the same thing at the same time at the same place. An animal's niche is its profession, its strategy for making a living. In this sense the world is organized like an enormous Rotary Club, with only one member (species) for each profession (niche) allowed per area. Hence, of the two species of woodpecker found on the islands of New Guinea, some islands have one species, some the other, but none has both. On the Galapagos Islands, where there are no woodpeckers at all, a local species of finch has been recruited into the woodpecker niche by evolution (Fig. 21–1).

Fig. 21–1 The woodpecker finch of the Galapagos Islands has evolved to fill the empty woodpecker niche not by an exaggerated morphological adaptation, but through the behavioral ploy of using cactus spikes to probe for insects.

There are far fewer than a million niches, but then there are many communities or habitats. A habitat is an animal's address. As long as two habitats are sufficiently isolated to prevent interbreeding, there is no reason why any particular niche should not be occupied by a different species in each habitat (Fig. 21–2);

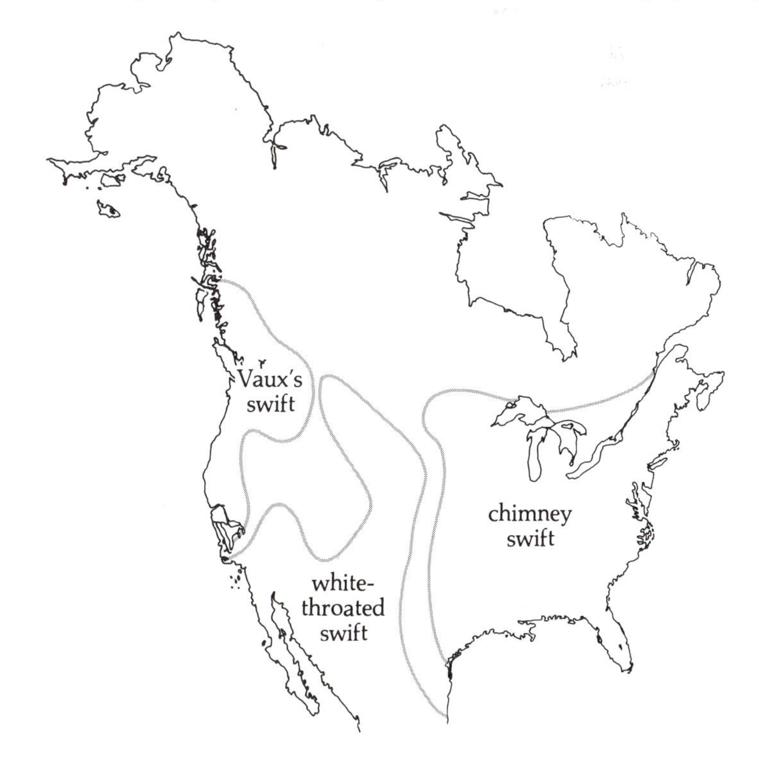

Fig. 21–2 When it is geographically isolated, the same niche can be filled by different species in separate habitats. Here are the distributions of three species of swifts in North America.

and of course competition, natural selection, and habitat changes ensure a regular turnover of the precise species occupying any given niche in any particular habitat.

The question of how many niches or isolated habitats exist is a difficult one. Much of the answer depends on the mobility and environmental tolerance of a species: the impenetrable barriers presented by mountains, oceans, and climate to a leaf-eating primate are as nothing to long-range birds or windborne insects. In a very important way, niches and habitats interact. Just as in human communities, a large habitat will support more "professions" than a small one—a small town barely supplies a niche for a general practitioner, whereas a city permits specialization into narrow niches for surgeons, oncologists, pediatricians, allergists, psychiatrists, radiologists, gynecologists, dermatologists, anesthesiologists, gastroenterologists, urologists, ophthalmologists, neurologists, and so on. The "size" of a habitat, of course, depends not merely on its physical area, but on its stability and productiveness as well: Death Valley, though large, supports few physicians. In ecological jargon the degree of specialization is known as "niche width," and it is clear that for any given resource, a GP or other generalist cannot outcompete animals specialized to exploit it. Specialists are simply more efficient because their anatomy, physiology, and programming are optimized to do one thing well, while a generalist is by necessity a jack of all trades.

The degree of specialization that is possible in any one habitat can be startling. Koala bears, for instance, eat only eucalyptus leaves; sloths restrict themselves to mulberry leaves; the everglade kite harvests a single species of snail; Tinbergen's species of digger wasp captures only honey bees. Each of these specialists leaves plenty of room in the environment for other creatures. In fact, except for the kite's snails, each of these resources is either invulnerable, toxic, or dangerous to the unspecialized world at large, and the specialists' primary modifications have been to overcome the particular defenses of their food resources. Specialists, although they outcompete generalists for a particular niche, pay for their success by being supremely vulnerable to habitat change. Because extreme specialists cannot survive outside their narrow niches, they are dependent upon the precise part of the habitat or the one or two other species with which they interact.

DARWIN'S FINCHES

The Galapagos, those barren, isolated islands which formed the high point of Darwin's pilgrimage on the *Beagle,* offer a spectacular case history of evolution. Darwin was the first to realize their value as a microcosm of the earth's development—a value which has made them a shrine sacred to students of the natural order. Thrust up from the ocean floor by volcanic forces rather than split off from some established continent, these tiny fragments of land offered virgin, competition-free resources to the first plant and animal species that reached them.

But this superabundance did not last for long. The chance invaders quickly multiplied to the carrying capacity of their niche. Rapid population growth encouraged animals to explore new food sources, and any members of a species that were even marginally better equipped to exploit a new food niche found their lot improved. Perhaps new frontiers would have been opened to them as their part of the population became increasingly better adapted to utilize this new resource, and the reproductive isolation which their unique adaptations would have encouraged led inexorably to speciation.

The isolated position of the Galapagos and their geological infancy make it possible to map these hypothetical developments with a precision and clarity generally unobtainable in the more clouded waters of evolutionary history. As Darwin himself phrased it, the Galapagos' relative youth brings us closer to "that great fact—that mystery of mysteries—the first appearance of new beings on this earth." Indeed, it was a small, insignificant finch native to the islands that gave Darwin his first insight into the logical progression of species, and was eventually to inspire his theory of evolution.

While visiting the Galapagos during his famous voyage on the *Beagle,* Darwin collected several specimens of small, gray-brown finch which John Gould, the renowned ornithologist, recognized as an entirely new group of birds. Similarities in eggs, anatomy, nests, and coloration suggested to Darwin that all shared a common ancestry, although their widely varying sizes and beak shapes suited them for several different "professions." One displayed the size and behavioral characteristics of a warbler; one had the thin, probing beak of a woodpecker (though it used a sharp cactus thorn as a tool to probe for the deeper insects); different groups were remarkably well adapted to eat either small, medium, or large seeds or small, medium, or large buds and berries. It seemed to Darwin as if, in the absence of already-specialized competitors, one "colonist" species had spread and adapted to fill all the unoccupied specialist niches available on the unpopulated archipelago.

At first resources would have been plentiful for the founder finches. Whatever they ate, there must have been plenty. But the inevitable pressures of increased population would have favored those individuals best adapted to harvest less popular and therefore less depleted foods. To operate, natural selection requires variation—morphological alternatives to choose from on the basis of competition. But in an environment with a wide range of foods available, many "answers" may be correct at once (Fig. 21-3). Under these circumstances, taking the naturally occurring variation in the finches' bill size as an example, evolution finds opportunities for the "oddballs." Perhaps, in the microcosm of the Galapagos, finches with larger, tougher bills came to specialize on those foods they alone could best eat. Birds with longer, more slender bills found a range of insect life open to them which was relatively inaccessible to their conspecifics. Simultaneously, the different types of food available on various islands would have "selected" for locally specialized variations in food-gathering apparatus. In this way populations became regionalized; specialized body adaptations and behaviors increased as natural selection favored ever more efficient animals; sexual selection pressures, by which animals mate preferentially with those which display certain "desirable" characteristics, must have become increasingly rigid; and eventually the various specialists would have become separate species, no longer able for myriad reasons to interbreed. In fact it does seem that, with all the similarity of form and color, it is precisely as Darwin predicted: the size and shape of the bill is the signal the birds use to determine that they are courting or being courted by the correct species. Whether the mechanism is a hardwired releaser or is based on parental imprinting is not known.

With this sort of division of the spoils, the total food supply in any habitat is efficiently exploited and the system is relatively stable. What is perhaps most impressive in this stability is that evolution has managed to generate it and the widely divergent behaviors and morphologies of these dozen species of finch from only two key ingredients: the modest genetic variation present in perhaps only one small errant flock of finches blown by chance six hundred miles or more into a birdless environment, and time out of mind.

Fig. 21-3 These three species of finches coexist on one island in the Galapagos and all eat seeds. Since the size of seed most efficiently handled depends on beak size, these species do not compete directly with each other.

LIFE STRATEGIES

Unstable habitats or unpredictable environmental changes, therefore, favor generalists, at least until the habitat stabilizes again and new specialists begin moving in from outside or evolving from the inside. This strategy for dealing with change leads to two general evolutionary extremes, which express themselves in conservative or liberal lifestyles among the animals they control. The dichotomy is derived from the usual population growth curve (Fig. 21–4). Any small population introduced into a favorable habitat begins growing at some species-specific maximum rate (r) until the habitat begins to become saturated as the availability of some crucial resource becomes limiting. The rate of growth slows and, in this idealized situation, asymptotically approaches the carrying capacity of the habitat, K. If N is the population size at any instant, then the rate of population increase with time is dN/dt. In the simple world of Figure 21–4, $dN/dt = r[(K-N)/K]N$. When N is low, dN/dt is close to the natural or "intrinsic" rate of increase r, while when N is close to K, the rate of increase goes to zero.

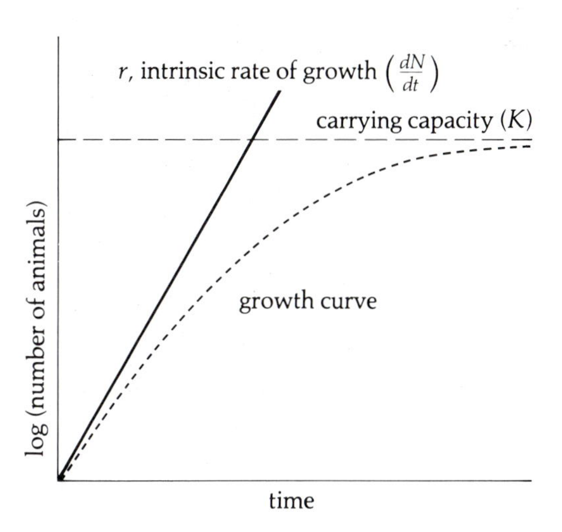

Fig. 21–4 The usual growth curve of a population in a favorable habitat begins at some natural (i.e., unchecked) or intrinsic rate r, but slows as competition with the population for limiting resources increases. Ultimately, provided the species has some social or physiological mechanism for achieving a balance, the carrying capacity K of the habitat is reached.

Conservative or "K-selected" species are adapted to stability, and are almost always near the carrying capacity of the habitat (Table 21–1). They typically live in habitats with relatively constant or at least predictable climates. Competition among individual members of a species, particularly for that resource which sets the value of *K*, is keen. Individuals of *K*-selected species, because of the predictability of their worlds, can afford to develop slowly and wait until they are at their competitive peak to reproduce. Because the environment is so nearly saturated parents usually raise few offspring, but they invest heavily in their young to increase their chances for survival. Most often *K*-selected species are relatively large and live relatively long lives. Elephants are an extreme example of the *K* strategy, bearing a single calf after almost two years' gestation. The young elephant, in turn, develops slowly, requiring much care before it can begin to reproduce.

TABLE 21–1 *General Trends in Life Strategy*

Strategy	K-*selected species*	r-*selected species*
Population	at carrying capacity	below capacity
Environment	constant or predictable	unstable
Intraspecific competition	keen	lax
Development	slow	fast
Body size	large	small
Investment per offspring	large	small
Number of offspring	few	many
Niche	specialists	generalists

Liberal or "*r*-selected" species have evolved to exploit change and instability, and are usually well below the saturation level of the environment and growing quickly (Table 21–1). This lifestyle is recommended by unstable or unpredictable climates, into which these species are adapted to move to exploit the resources which instability denies to specialists. Exploitation is by its nature a boom-to-bust growth strategy, typified by a sudden and perhaps temporary abundance of a resource relative to the ability of consumers to utilize it. Hence, *r*-selected animals need not compete for the resource, but rather are free to concentrate their energies on rapid development and early reproduction in an effort to be first in the race to get their offspring off and running. Since the habitat is well below the carrying capacity and has a superabundance of food for a limited period, time is of the essence: the best approach is for an animal to produce as many offspring as quickly as possible, and to invest as little in them as necessary.

These considerations point toward a small, quickly achieved body size and a short, frantic life. Mosquitoes are a typical example. When warm weather arrives the few larvae which have survived the freezing or drying out of their aquatic habitat emerge into a world rich with prey. They reproduce rapidly, running through many generations in a season, each early female potentially spawning billions of great-grandchildren, hungry for mammalian blood. And yet despite our subjective impressions, the environment remains unsaturated: how many of us have seen animals drained of their blood by ravenous mosquitoes? When the cool weather of fall arrives, the unconstrained boom comes to an abrupt end.

The r/K dichotomy is actually a continuum on which every organism has a place. Our species' intrinsic rate of increase, for example, which allows a doubling every seven years, puts *Homo sapiens* toward the K end, though not far enough to prevent our present headlong rush toward mass starvation; while bacteria, which can double as often as every twenty minutes, are much closer to the r extreme than even mosquitoes. A number of species hedge their bets by adopting both strategies. Many ferns, for example, invest simultaneously in a few metabolically expensive K runners and millions of microscopic airborn r spores (Fig. 21–5). Many aphids and rodents can alternate between generations of K-type stay-at-homes and r-like dispersers which seek out new habitats to exploit. And, of course, lemmings and locusts are the archetypal examples of unpredictable r-style migrations to locate unexploited resources.

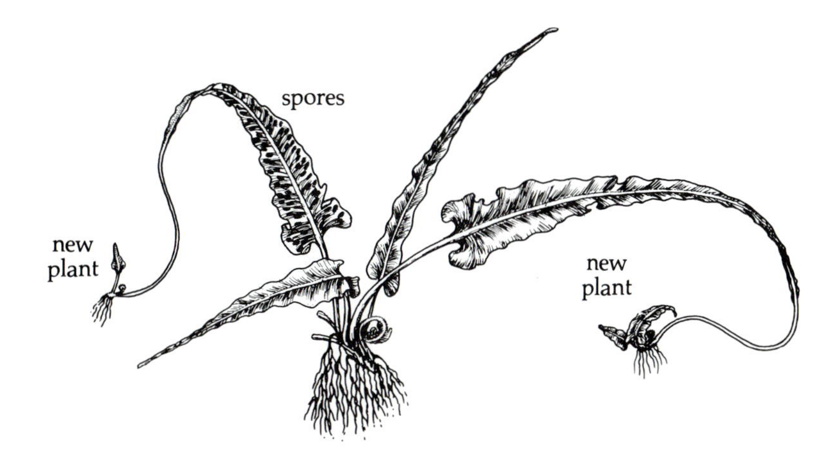

Fig. 21–5 Ferns often pursue both r and K parental investment strategies simultaneously. Walking ferns (shown here), produce and nourish new plants at the ends of their fronds, while also generating millions of airborne spores under their leaves. Maidenhair ferns invest in new offspring through both spores and plants produced from underground runners. In both cases the K offspring are exploiting what experience has shown to be a favorable environment, a strategy which allows very limited opportunities for spreading of the genes, while the spores are sampling a tremendous variety of distant habitats which, however, will only very rarely prove suitable. [From *Field Guide to the Ferns,* by Boughton Cobb. Copyright © 1956, 1963 by Boughton Cobb. Reprinted by permission of the publisher, Houghton Mifflin Company.]

PREDATORS AND PREY

Perhaps the most powerful selective forces giving rise to the r–K continuum are the controlled interactions between predators and prey. A prey species has two general strategies available to it for maximizing its chances of survival. It may invest heavily in avoiding predation, perhaps thereby inviting prospective specialists to make it their own particular niche. Such a species, if successful in thwarting predation, will rapidly become resource-limited. The other strategy prey species may adopt is that of fashioning an evolutionary compromise, investing only enough to make predation difficult but not impossible. In many cases the predator has become almost as essential to the prey's survival as its survival is to the predator. Such prey become predator-limited. Again, the contrasting strategies of fighting and accommodation, hawk and dove, are seen to be two ends of a spectrum, an array of cat-and-mouse stories of ruthlessness and cunning, and an often bewildering complexity.

The most straightforward way for a species to prevent the evolution of efficient specialists is to starve any that try to specialize on it. Most insects, for example, are absent or at least unavailable during the colder part of the year in temperate climates, but their specialists have accommodated. Insectivorous birds avoid starvation by flying to warmer climes for the winter where prey still abounds, while mammals such as bats hibernate. Periodic cicadas have been considerably cleverer. These root-feeding insects live in regional populations which spend nine, thirteen, or seventeen years underground. They emerge synchronously in enormous numbers at their appointed time. The standing bird population is completely unable to cope with the enormous supply of prey, and the supernumerary offspring this surplus allows the birds to raise will starve in the return to normal food levels during the following years (Fig. 21–6). The same strategy starves any underground predators—moles, for example—by bequeathing them a year of starvation after years of plenty. Similar logic underlies the "mast year" cycles of trees: more seeds are produced one year than predators—squirrels, say—can possibly consume, but the next year conditions are too lean to support more than the usual population.

Another "no compromise" plan on the part of prey species is

Fig. 21–6 Periodic cicadas are prey to birds when they are adults and to moles when they are larvae. Ideally, they discourage specialization by starving their predators from time to time. This neat picture is flawed by the existence in most habitats of two or more distinct generations developing underground out of synchrony, so that moles may not have such a hard time of it after all.

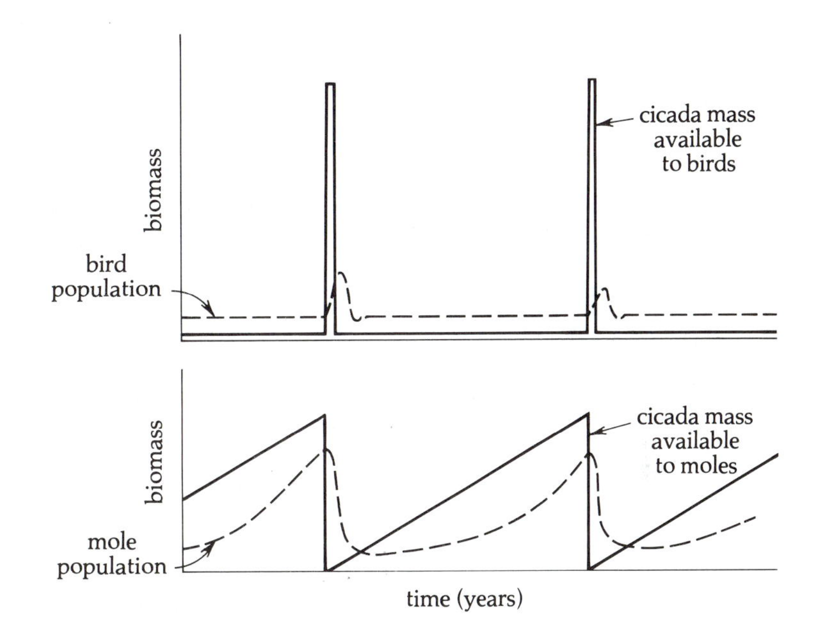

Fig. 21–7 Monarch butterflies *(above)* are strikingly colored black and yellow or orange. They contain toxic substances which make them distasteful, and create a memorable nausea in most birds. Viceroys *(below)* on the other hand, are perfectly palatable, and gain protection from their mimicry of the monarch coloration. They are also smaller, shyer, and less numerous.

to invest in being awful to eat. Monarch butterflies, for example, feed on noxious milkweed, and store compounds which make animals memorably, although not fatally, ill (Chapter 17). To enhance the effects of this lesson they have evolved striking color patterns on their wings for predators to learn, and to recall in future encounters. Metabolically speaking, this strategy must cost the monarchs dearly. Probably their cost lies in having to use less efficient biochemical systems which are immune to milkweed, as well as having to lug around a load of toxins. Such a system encourages cheating, of course, and the viceroy is a con artist *par excellence.* Viceroys look just like monarchs, tarred with the same memorable brush, but underneath it all they would taste fine—if a predator would eat them (Fig. 21–7). They depend on the monarchs, though, to teach the predator population its lesson, so their life cycle must ensure that the average bird has encountered a monarch first. Without this important caveat the viceroy population would be decimated. The genes of each species, then, are carefully attuned to the nuances of this propagandistic warfare. Viceroys are quick, shy, and less numerous than their noxious cousins. They die immediately after laying their eggs, while monarchs not only pad the population by continuing to live but, their driving goal of reproduction past, they live more conspicuously than ever. In this way they continue to impart their lesson to naïve predators and to serve

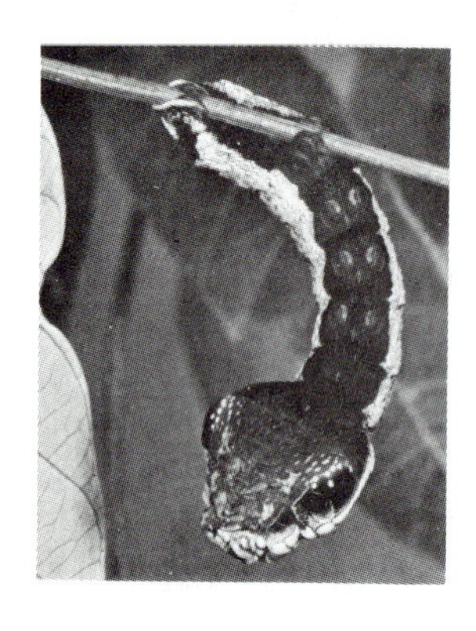

Fig. 21-8 The tail of this caterpillar has evolved to look like a snake and, presumably, frightens off nervous predators.

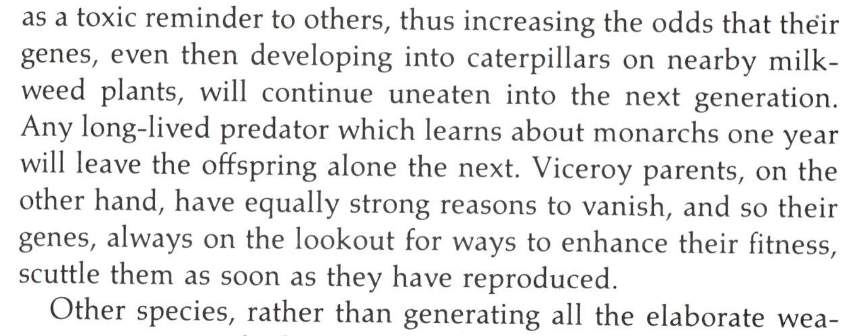

as a toxic reminder to others, thus increasing the odds that their genes, even then developing into caterpillars on nearby milkweed plants, will continue uneaten into the next generation. Any long-lived predator which learns about monarchs one year will leave the offspring alone the next. Viceroy parents, on the other hand, have equally strong reasons to vanish, and so their genes, always on the lookout for ways to enhance their fitness, scuttle them as soon as they have reproduced.

Other species, rather than generating all the elaborate weaponry necessary for being awful, are designed with a morphology and behavior appropriate for merely *seeming* awful (Fig. 21-8) or irrelevant (Fig. 21-9) or like part of the woodwork (Fig. 21-10). Still others make their living from seeming to be what they are not (Fig. 21-11).

A

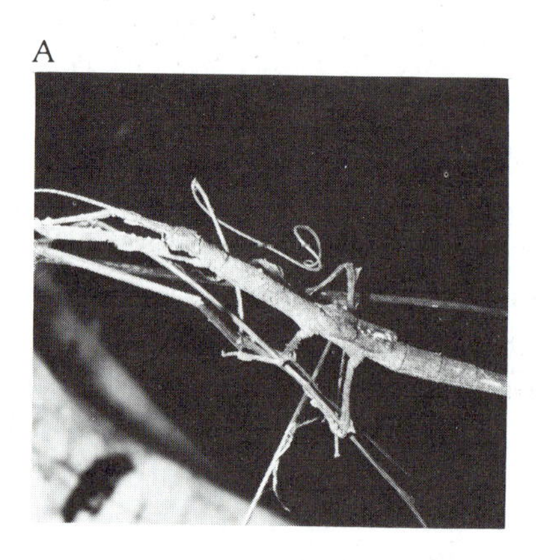

B

Fig. 21-10 Walkingsticks are famous for their resemblance, in color, morphology, and behavior, to twigs (A), and katydids to leaves (B), and so must pass unnoticed by many predators.

Fig. 21-9 This larva closely resembles a bird dropping, a likeness which must reduce its apparent desirability to birds.

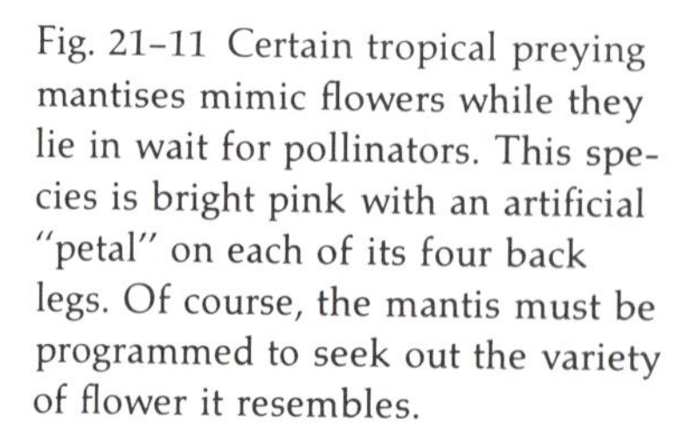

Fig. 21-11 Certain tropical preying mantises mimic flowers while they lie in wait for pollinators. This species is bright pink with an artificial "petal" on each of its four back legs. Of course, the mantis must be programmed to seek out the variety of flower it resembles.

THE BOMBARDIER BEETLE

The number of ingenious ways which nature has developed for prey species to avoid extinction by their ever-watchful predators is practically limitless. The stinkbug's stink, the armadillo's armor, the porcupine's quills—all are variations on the theme of survival, all more or less "costly" to the animal which chooses to defend itself instead of hiding or running.

One of the most elaborate arrays of defensive weaponry is that of the bombardier beetle *Pherosophus agnathus*. Not only is this living battleship encased in heavy armor plating, but it also carries a load of destructive acid ready to spray at attackers. To keep this acid, which is almost boiling hot when it hits the predator, from destroying its own insides, bombardiers must carry the individual components (hydrogen peroxide, bitter quinones, and the enzymes which make them explosively active), in separate, plated tanks, ready to mix and expel at a moment's notice. It can be an extraordinarily effective defense: most predators are hurt, disabled, or even killed by the blast (Fig. 21–12).

Not surprisingly, the bombardier's heavy artillery gives the beetle an unwieldy, lumbering gait, and the terrestrial agility of a landlocked ironclad. As

Fig. 21–12 Bombardier beetles defend themselves by aiming a boiling, caustic spray at would-be predators.

any student of evolution might have guessed, predators have evolved which are programmed to cope with the defenses of this otherwise desirable prey, and several species of beetle have developed which take advantage of the bombardier's defense budget without paying the costs of the weighty armaments and compromises in nervous system finesse necessary to run the system.

While imitating the lumbering gait and odd angular posture of their well-armed cousins, the mimics lack their weaponry and, if attacked, will abandon their slow-paced act to run or fly rather than stand and fight. In addition, they lack the telltale odor that alerts the bombardier's most threatening nocturnal predators to its presence.

Recent experiments show that all skunks, even those which have never before encountered beetles, respond automatically to bombardier and bombardier-like beetles by closing their eyes tightly and rolling the beetles rapidly back and forth on the ground until they have discharged their load of ammunition. In fact, statistical studies indicate that skunks show no avoidance whatever of the bombardier, which is one of the larger, more nourishing beetle species. Certain species of mice, too, actually prey on bombardiers preferentially, holding the beetles' spraying apparatus firmly into the ground and eating from the head down.

Thus at least three separate evolutionary tracks have become interwoven in this one defense strategy. As the bombardiers evolved their artillery the threat from generalist predators waned, but in time some vertebrate predators have come to have specialized evasive tactics which allow them to exploit the bombardiers with little competition. Simultaneously the mimics, the inevitable spongers on any well-ordered system, moved in to take advantage of the bombardier's hard-won protection.

This fabric of interdependent species offers us a valuable insight into the workings of nature, which smiles on predator and prey, straight man and mimic alike. There can be no ultimate, failsafe defense for prey, no effortless mealticket for predator. The cheater mimic may be getting a free ride at the evolutionary moment, but who is to say that the beetle-rolling skunk, or the intrepid mouse, are not on the road to specialization—a specialization which will impose a sort of evolutionary justice on the system as it robs the mimic of its advantage, or perhaps even makes it a preferred target?

MUTUAL DEPENDENCE

But more typical of predator-prey interactions in general are those in which prey are good to eat and regularly available, but not without considerable effort or ingenuity on the part of the predator. Often, however, like some sort of evolutionary joke, prey have come to depend on their hungry predators for their own well-being, although this ironic interdependence may become obvious only when one of the two adversaries is removed.

Around 1900, while Lake Superior was frozen, some moose chanced to walk the fifteen miles across the ice to the previously mooseless Isle Royale, a 2150-square-mile island. In the absence of predators the population grew steadily to about three thousand in 1935. In the process, the moose devastated all the island's vegetation, and suddenly nearly 90 percent of the moose population starved. With fewer mouths to feed the grass and bushes soon recovered and the moose, again feeding without restraint, again increased in number. By 1948 they were up to three thousand again, and once more the population crashed. In 1949 a handful of lost timber wolves wandered across the ice and began to prey on the moose. There are now about two dozen wolves, which cull out the unluckier young, the old, and the sick moose at the rate of about one every three days. They continually test the herds, and only about one chase in every ten is successful (Fig. 21–13). The result is a stable population of

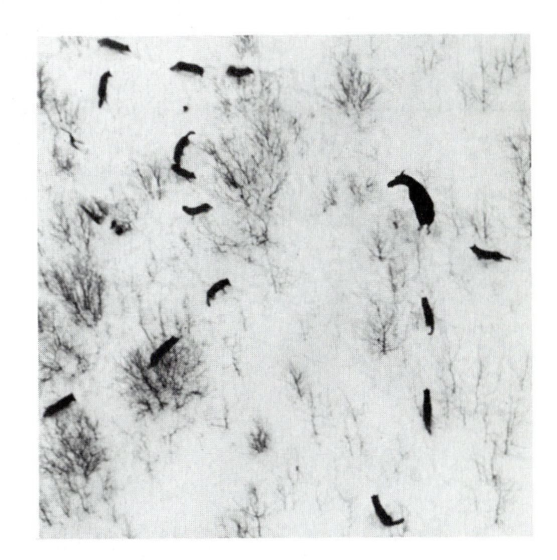

Fig. 21–13 A moose hunt on Isle Royale. The wolf pack has isolated this potential victim from his group, but to no avail. As in about 90 percent of the hunts, the prey is sufficiently large and healthy that the wolves dare not press their advantage.

about eight hundred trim and fit moose, and an equally healthy crop of grass. In some sense, the wolves are doing the moose (and the grass) an unappreciated favor by introducing a necessary balance and stability into the system (Fig. 21-14) since the moose seem to lack the internal mechanism to hold their own numbers at the carrying capacity.

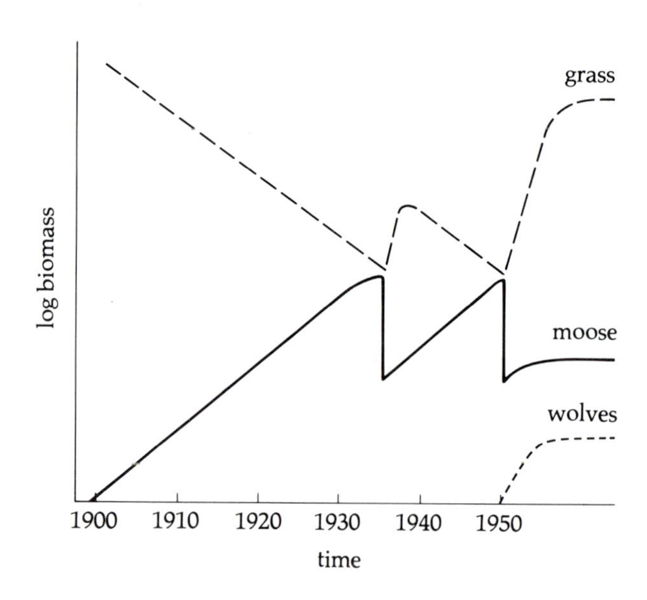

Fig. 21-14 Hypothetical reconstruction of the population census of the moose, the grass, and the wolves of Isle Royale. The moose arrived about 1900 and increased without restraint until they overgrew their food supply in the early 1930s and starved in large numbers. The pattern was repeated, the next mass starvation occurring in the late 1940s. A few wolves arrived shortly thereafter and instituted a natural balance among grass, moose, and wolves which has persisted ever since.

Clear-cut cases of mutual dependence and coevolution—"symbiosis"—are less common than the sort of grudging compromises illustrated by the moose and wolves, but they better illustrate the full potential of natural selection to exploit every opportunity in the most pragmatic fashion imaginable. The mimosa girdler beetle, for instance, lays its eggs in slits which it cuts into the ends of mimosa branches. The slit must heal to protect the developing larvae, and yet, given the tree's chemical defenses against just such predation (the odor of which may be the sign stimulus by which the beetle locates the tree), the larvae cannot grow in live wood. The beetle solves this problem by chiseling a neat trench in the bark all the way around the branch just an inch or so in from the eggs. The branch tip dies in a few days—after the egg-slit wound has healed—and later drops to the ground. This pruning, far from killing the mimosa, is essential to its longevity. Without it, the tree overgrows its foundation and dies after about twenty-five years. The constant pruning that the girdlers provide is the balance which doubles both the tree's lifespan and its fitness. The mimosa might once have had its own independent system of self-control, but with the evolution of a specialist predator and its dependable attacks, this system with its inescapable costs would have lost its value and disappeared. Each species now depends on the other.

In a similar but more extreme case, there is a species of ant that grows and eats one species of fungus for food (see Chapter 25). Without that fungus the ants will starve. On the other hand, the fungus depends entirely on the ants for its food (it must be provided with decaying leaves or starve) and cultivation. Which is the predator and which is the prey? Similarly, it is difficult in the familiar interaction between ants and aphids to assign the black hats and the white. Aphids prey on plants, drawing out and digesting the carbohydrate-rich juices. To a first approximation, animals use protein for growth and carbohydrates for fuel. The limiting resource for the normally sedentary aphids is protein. To get what they need, they must process large amounts of nearly worthless sap and excrete the sugary waste as "honeydew." Ants, which often harvest insect prey as their protein source, have evolved to leave the aphids alone and consume the honeydew instead (Fig. 21–15). In fact, the continual presence of the ants seems to protect the aphids from other predators, and the two groups now depend on each other. In fact, there are even suggestions that the aphids are doing the *plants* a favor by dumping carbohydrates on the ground—carbohydrates which may be consumed by the nitrogen-fixing bacteria in the soil which supply the plant with the nitrogen essential to its growth.

Fig. 21–15 Aphids produce enormous quantities of surplus "honeydew"—plant sap strained of its amino acids—which many species of ant are adapted to harvest: not only do the ants refrain from eating the aphids, they even protect them from predators such as ladybugs. In addition, they have mastered specialized stroking motions which cause the aphids to disgorge the carbohydrate-rich waste product on which the ants depend.

A

B

Fig. 21-16 The hospitable acacia provides ants with roomy hollow thorns to live in (A) and special, protein-rich, nonreproductive "fruits" (B) to eat. In turn, the ants guard the tree from grazing mammals and parasitic insects, and defoliate competing plants. Even the ants' droppings are thought to be absorbed and used by the tree in this balanced symbiotic relationship.

Another intricate example involves the interaction of ants and acacia trees. At first glance ants seem to be exploiting acacias. Not only do they harvest the tree's leaves, but they burrow into its thorns and stems to make their nests. Nevertheless, a closer look suggests that the hospitable acacia has evolved to make the ants comfortable. Its thorns are unusual—hollow and enlarged—and the parts of the leaves that the ants use are oddly specialized and extraordinarily nutritious (Fig. 21-16). Comparisons between "parasitized" and unparasitized acacias reveal that the ants are essential for the trees' survival. The ants keep acacias relatively free from plant-eating insects, and swarm out at the slightest disturbance to repel browsing mammals. Further, they prevent vines from getting a stranglehold on the tree, and defoliate any competition that might shade their host. Again, which is the predator in this interaction?

The lesson to be learned from the ants, aphids, and acacias is that natural selection has fashioned the behavior of each species to make its living in its own peculiar way of life. Evolution must equip each animal with feature detectors that recognize what to eat and what to avoid eating or how to learn this distinction, and must build and calibrate the motor programs the animal needs to obtain its food. Evolution must optimize each animal's behavior to match its prey's lifestyle: its habitat, its dispersion, its behavior, its circadian and annual rhythms. At the same time, each animal must be tuned to recognize and escape its own predators, whether by hiding, fleeing, confusing them, or staying out of synchrony with their predators' behavior. The raw behavioral materials for selection have been the releasers and motor programs evident even in single-celled animals, each tested, focused, and refined generation after generation by evolution to form a mosaic of species interactions, a mosaic which is relatively stable yet intricate enough to tax our human imaginations.

SUMMARY

Heritable variation and natural selection combine to give rise to the great and slowly evolving variety of life which shares this planet. Species exist and coexist by not directly competing for the same food, and by not overeating their food supplies. Direct and complete competition leads to the extinction of the less efficient species (and hence subtends the Niche Rule), while overeating leads to the extinction of the prey, followed shortly by that of the unwise predator. Species fall on the continuum which runs from specialists—animals whose morphology, physiology, and behavior have been optimized for harvesting a narrow range of food—to generalists, with their varied diets but relatively inefficient harvesting. Species also range over the life-strategy scale from *r*-selected exploiters of unstable environments to the *K*-selected individuals which saturate more predictable habitats. Finally, species subject to predation differ in their strategies for avoiding being eaten. The results of these various evolutionary choices leave almost every species affected in some way by the choices of others, a mutual dependence which forms the cornerstone of such stability as there is in our ever-changing world.

STUDY QUESTIONS

1. Periodic cicadas come in nine-, thirteen-, and seventeen-year varieties. Occasionally a substantial portion of the seventeen-year brood will emerge after thirteen years, but may return to the seventeen-year interval thereafter. (Similarly, a thirteen-year group may rarely emerge after nine years.) What do these observations suggest about the timing and synchronization systems of cicadas and the existence of several simultaneous broods?

2. A small proportion of monarchs do not lay their eggs on the poisonous variety of milkweed, and so their progeny are not toxic. How can this be? How do you suppose monarchs recognize their host plants, or each other? How could you determine whether the nontoxic monarchs represent an accidental failure to locate the right sort of milkweed or an inability? What does the future hold for each of these two possibilities?

3. How do you suppose the two sexes of the mantis of Fig. 21–10 find the appropriate flowers or each other?

4. Can you think of cases in which generalists would have the edge in stable habitats or specialists in certain sorts of unpredictable ones?

5. In the theory of island biogeography that was formulated by E. O. Wilson and the late Robert MacArthur, the authors propose—and their proposal has since been largely confirmed—that the number of species increases as the square root of island size. Assuming for the moment that the habitats of islands are identical, and no barriers to movement within an island exist, why ought this relationship to hold?

FURTHER READING

Degabriele, Robert. "Physiology of the Koala." *Scientific American* 243, no. 1 (1980): 110–17.

Diamond, Jared M. "Niche Shifts and the Rediscovery of Interspecific Competition." *American Scientist* 66 (1978): 322–31.

Lack, David. "Darwin's Finches." *Scientific American* 188, no. 4 (1953): 66–72.

Levins, Richard; Pressick, Mary Lou; and Heatwole, Harold. "Coexistence Patterns in Insular Ants." *American Scientist* 61 (1973): 463–72.

Milne, L. J., and Milne, M. "Behavior of Burying Beetles." *Scientific American* 235, no. 2 (1976): 84–89.

Opler, Paul A. "Oaks as Evolutionary Islands for Leaf-Mining Insects." *American Scientist* 62 (1974): 62–73.

Wecker, S. C. "Habitat Selection." *Scientific American* 211, no. 4 (1964): 109–16.

CHAPTER 22

Mating and Territoriality

As we saw in the last chapter, the goal of behavioral ecology is to show how nature has fashioned the particular lifestyle of each species in accordance with its environment and the laws of natural selection. When practiced as a separate discipline, behavioral ecology is far more interested in the workings of evolution than in the physiological mechanisms—releasers, motor programs, drives, and programmed learning—which underlie behavior. Only ethology attempts to compass both levels of analysis. It assumes that the environment and evolution have combined to construct the behavioral clockwork, and that it is this machinery which evolves. In the past chapter we saw how nature manages to pack the world with species. Let us look now at how the same evolutionary forces operate on the behavioral machinery to determine how the members of any one species will divide up their habitat and how they will interact with each other.

Except for the mating season, most animals—most *r* animals at least, and they account for most animals—distribute themselves and behave with little or no regard for other members of their own species. Pure *r* animals are racing against time rather than against each other, in order to take advantage of a temporary or unpredictable abundance of food or other necessity, and to reproduce. The *K* strategy, though less common, has given rise to a variety of increasingly intricate social systems whose usual function is to circumvent the laws of Malthusian economics by taking matters out of the hands of chance—either by preventing the population from exceeding the carrying capacity of the habitat or by distributing important resources such as food, nest sites, or mates in some systematic way.

Simple population control which would keep members below the "nearly everybody starves" level of Malthus can be accomplished in a variety of ways. Some involve external agents: an early frost, a drought, or a drastic change in habitat like a fire are the typical controlling factors of *r*-selected species, and do not depend on the density of the population. Other species, however, are subject to population controls which depend strictly, though often passively, on density. Predation, parasitism, and disease are clear examples. There is in each case a "critical mass" at which specialized predators or diseases—measles, malaria, and (until recently) smallpox, for example—can turn a profit and become self-sustaining. The result is that the population level is held well below the carrying capacity of the environment. Some species, like lemmings and honey bees, come programmed to disperse when population density gets too high. Still others control density by murder or cannibalism. Put one pregnant guppy or fifty adults in a five-gallon tank, and in either case the population will stabilize at about nine. Adults correct the surfeit by fighting, and by eating extra offspring.

THE USES OF TERRITORIES

Still more ingenious ways of controlling population size and density are those by which some groups forcibly exclude certain adults from specific resources. This strategy is quite common, and familiar to us all in the human context in the form of some trade unions like the American Medical Association. If everyone who was qualified and motivated to become a doctor could, then the denizens of this niche would increase until there were barely enough resources—patients—to allow individual M.D.s to survive. This is clearly *not* the observed situation. A Malthusian fate for doctors has been circumvented by an ingenious system which restricts the number of doctors primarily by limiting the number of spaces in medical schools, thereby assuring an abundance of potential resources per individual. The system works, needless to say, to the benefit of the successful, but returning to the animal world, why the genes of the excluded should put up with such a system remains a question. It is tempting to ascribe such sacrifices of fitness, this apparent willingness to be excluded, to *altruism*—a sacrifice of one's own interests for the benefit of another. As far as we can tell, however, genes subscribe to no ethical code, so the existence of such systems in the natural world indicates that there must be something in it for everyone.

The most common form of the exclusion strategy involves dividing the world into defended territories. Each territory, or at least the hotly contested ones, either has a local abundance of the resources necessary for raising offspring, or encompasses one of a limited number of favorable breeding sites. The territorial system of many songbirds exemplifies the feeding territory strategy. Males arrive early in the season and contest for space, singing both to warn off neighbors and landless rivals, and to attract females. That perfectly fit male songbirds really are being excluded from the breeding population was demonstrated by a grisly "shootout" experiment, which showed that when resident territory holders are killed, new "floater" males take over promptly. Floaters arrived nearly as fast as the experimenter could create vacancies with his shotgun, a contingency which suggested that the population of disenfranchised males is either very large, or very watchful, or both.

Tawny owls provide an especially compelling example of the anti-Malthusian nature of territoriality, and epitomize its potential for stability. These owls pair-bond for life, and make their living by catching mice and other small nocturnal rodents. By capturing and banding the highly territorial wood mice that are their prey, H. N. Southern showed that these birds live and hunt on discrete, well-defined territories. Southern sifted out the indigestible bands which accumulated below the feeding roost of each pair of owls and compared the mouse numbers with where they had been banded. In this way Southern was able to define the tidy array of territories that comprises the owls' habitat (Fig. 22–1).

Since any particular pair of owls is likely to live for many years, it would be to their advantage to decide each season when and how much to invest in offspring. Although there is little evidence that most species, including part of our own,

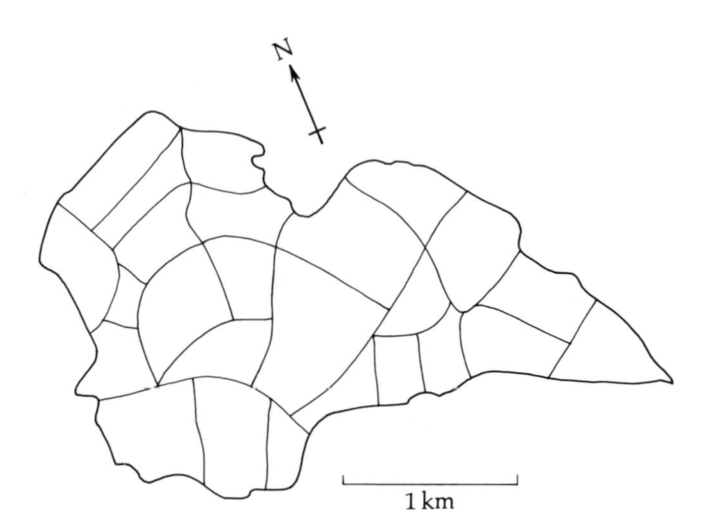

Fig. 22–1 Pairs of tawny owls have divided this patch of habitat into a matrix of well-defined territories whose number and boundaries are relatively stable from year to year.

exercise much judgment and restraint in this endeavor, it is clear that tawny owl couples regulate their reproductive output according to the availability of resources in their territory. They achieve this control in a variety of ways. In an average year 30 percent of the pairs simply refrain from breeding, and in one particularly poor season none of them bred. In addition, unless the year is especially good almost none of the pairs that do breed lays the maximum number of eggs. In each case something in the owls' behavioral clockwork is deciding whether to invest in eggs or to defer egg laying until later in the season, or even to forego reproduction altogether. Even when eggs ("clutches") have been laid, a decrease in the supply of prey on the territory can incite the birds to cut their losses by not incubating their eggs. In an average year this may happen to nearly half the clutches, but if the food supply increases again the owls may start a new clutch. In a poor season the parents may choose once the eggs hatch not to waste their efforts trying to feed the insatiable young. Even in an average year this sort of hardlining accounts for the loss of a quarter of the hatchlings. The result (Fig. 22–2) is a relatively precise control of offspring in any given tawny owl population.

The driving force behind the tawny owls' restraint is its highly *K*-selected lifestyle. In one well-studied population there are 50 adults and 25 breeding and feeding territories. In an average year, only 11 adults die. Hence an average pair cannot hope to produce more than one successful hatchling every second year. Given that they are long-lived birds, their best strategy seems to be to wait for a period of local plenty to invest in the extraordinarily demanding, expensive, and exhausting business of raising successful offspring. In an average year 18 of the potential 100 young owls are fledged (Fig. 22–3), and unless an unusually large number of adults have perished in the winter even this is too many: seven of them must leave to seek their fortunes in less favorable habitats where their chances of survival seem, from this study, to be almost nil.

Seals are also highly territorial, but in a very different way. Their territories have nothing to do with food: seals, after all, prey on fish in an undefendable ocean. In fact, the limiting resource in their case is not food, but breeding sites. The best place for these aquatic mammals to mate, give birth, and raise young is a small, secluded beach, inaccessible both to aquatic and terrestrial predators. That males contest vigorously and often fatally for these special places is understandable, since controlling one is the only way a male's genes can avoid extinction.

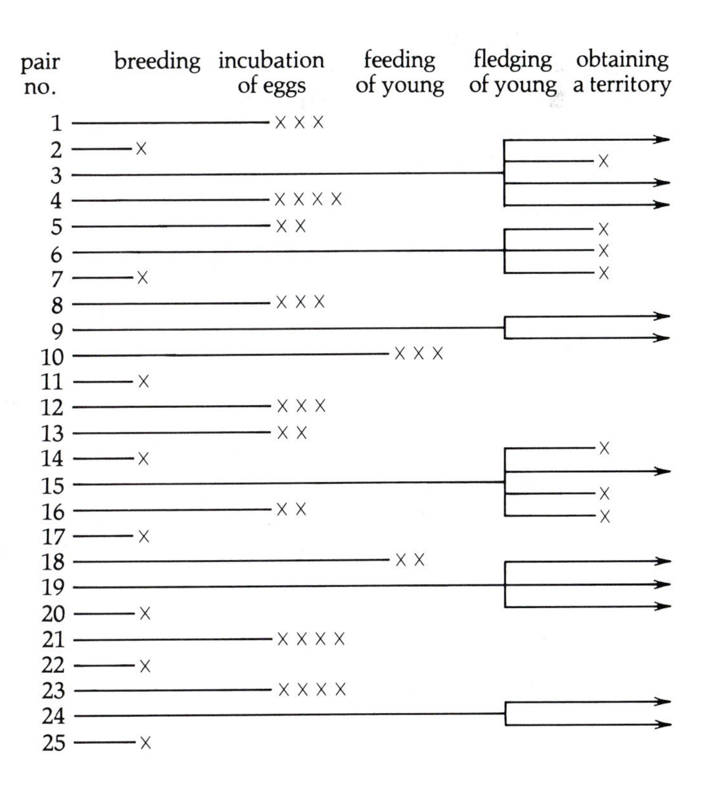

Fig. 22–2 Tawny owls regulate their population in a variety of ways. In an average year, 8 pairs do not breed (as indicated by the × which terminates their line), most do not produce the maximum number of eggs (4), another 9 pairs do not incubate the eggs they have laid, and 2 more refuse to feed their hatchlings. Thus, out of the 100 possible offspring, only 18 are fledged. Even so in an average year only 11 vacancies occur, forcing 7 of the young to leave.

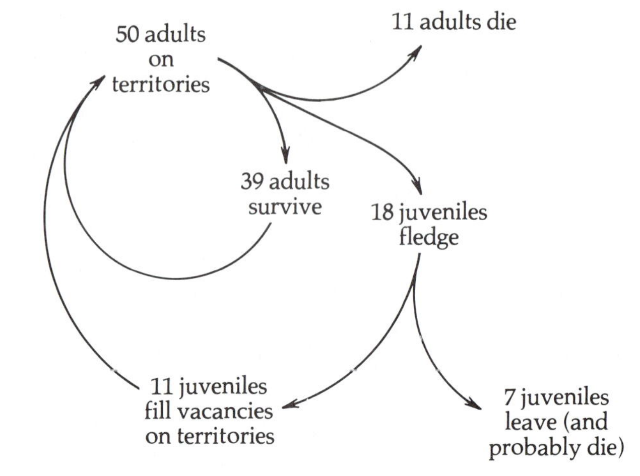

Fig. 22-3 The data of Fig. 22-2 can be summarized into this flow diagram for an average year.

BEHAVIORAL POLYMORPHISM

One of the most vexing observations ethologists have made is that an animal in a particular circumstance may behave one way this time and quite another the next. A bee will spontaneously choose a yellow flower over a blue one today, but pick the blue one tomorrow; a wren may search for caterpillars on leaves one day, on bark the next. Just as often two animals of the same species, tawny owls for example, will choose completely different solutions to the same problem. It may be that factors and contingencies unknown to us are at work, that in reality behavior is as predictable as the motion of Newtonian particles. On the other hand, the overwhelming impression of ethologists is that much of behavior is probabilistic, and that external events serve simply to shift the balance among a preordained set of behavioral responses available to the organism. Initially, this view seems at odds with our notions of instinct and the optimal programming we have come to expect from our genes: surely one unique best solution exists. But if we think about it, behavioral variability and even unpredictability may have its advantages: new flowers to be discovered come in all odors, and caterpillars may be found on either leaves or bark.

John Maynard Smith has turned our intuitive sense of the value of variation into a mathematically defined theory of evolutionarily stable strategies. Smith's analysis shows that the crucial factor for the genes in such behavioral polymorphisms is choosing the right odds for the individual—the probabilities most in concert with the immediate contingencies of nature. In fact, evolution will act to stabilize the behavioral probabilities at some optimum balance (i.e., the "real world" ratio of floral colors or caterpillar locations). One of the most exciting applications of Smith's theoretical analysis is to the range of behaviors we see in aggressive encounters between conspecifics, encounters which form so large a part of social interactions among animals. What could be the optimum program for aggression?

We can easily show that in the face of the unpredictability of others, there is no single best strategy, and that polymorphism is just what we would expect to observe in such a situation. Let's assume two hypothetical postures which animals might assume in aggressive exchanges—hawk (which will always fight, until he gets hurt) and dove (which threatens, but will run away if attacked). Next we will assign a set of "fitness points" to various components of the aggressive interchange: 50 points for winning a fight, 0 for losing, −100 for getting injured, −10 for wasting a lot of time. If everyone were doves, every encounter in this idyllic world would lead to exhaustive threats and counterthreats until one side of the dispute gave in. The winner would score +50 for winning, but −10 for taking so long to do it, for a total score of +40; the loser would end up with a score of −10. Assuming everyone wins half the time, the average dove gene is ahead +20.

Now let's add a hawk gene to our best of all possible worlds. It quickly sweeps the field, winning every encounter for a fitness score of +50, while the doves are left with zero. Hawk genes will skyrocket in frequency in the population, but will never reach 100 percent since hawk vs. hawk encounters, which increase in likelihood, are totally unfit: the winner scores +50 and the injured loser −100, yielding an average score per encounter of a miserable −25. In fact selection, that bookmaker par excellence, will stabilize this hypothetical population at 58 percent hawks and 42 percent doves, with a compromise average fitness of +6.25 per encounter.

It is important to understand, though, that few individuals are purely one behavioral "morph" or other. Anyone who has seen doves at a feeder in winter knows that squabbles are a way of life, and even hawks have their pacific moments. The results of Smith's formula are the same stable set of numbers if, as we see in nature, animals behave one way at one moment and another the next. Behavioral alternatives are, in fact, an essential feature of the well-programmed animal. The natural world operates as if animals were determining which behavioral mode to adopt by the flip of a coin—in our aggression example, a coin weighted for 58 percent hawklike behavior and 42 percent pacifistic. For each set of behavioral alternatives an animal's genes make sure it divides its time between the possibilities in a way which gives it the best statistical chance of succeeding in its immediate social grouping and physical environment. Thus the workings of our amoral genes ensure that an unsavory mixture of liars, cheaters, philanthropists, and bullies (or individuals with all these tendencies competing within them) will form even the most idyllic animal society.

JUDGING "FITNESS"

The general logic of this "exclusion system" whether in seals or tawny owls rests on competition for control of whatever resources make successful breeding possible for that particular species. Among *K*-selected animals the usual pattern requires that males do the competing, but that they be relatively unselective about choosing mates. It is the female's prerogative to choose among all the competing males the one most fit to perpetuate her genetic stock. For, in the end, it is the female which has most to lose. In general it is the female which produces metabolically expensive, protein-rich eggs: male sperm cost little. Often the very act of carrying, bearing, and caring for the young, which usually rests with the female, exposes her to risks which she must amortize as best she can by choosing the male whose genes are most likely to survive the process. The mere ability of the male to win whatever competition or rites of passage his species requires may be an indication of the fitness of his genes, and therefore of the desirability of investing in recombination with them. The males, for their part, provide the females—often free of charge—with favorable and exclusive situations for nesting and feeding, and thus better the odds that her investment will ultimately yield high returns. Although the male will usually do more than his share of defending, his investment is generally too low to justify his being very discriminating when it comes to choosing his genetic partner.

Judging a potential male partner's fitness is a matter of such import to females, moreover, that judgment seems to be the driving force behind the social organization of a variety of animals. For many animals, the moose for example, private nesting sites are impracticable and defensible feeding territories are impossible. Here again female metabolic investment is high, so while males devote themselves to attracting females it is the females which make the choice of mates. But how is the female to recognize the fitness of an array of genes if the males are not slugging it out among themselves for territory? In fact, perhaps these males *are* competing, although in a way less obvious to observers outside their species. From the awesome though weighty antlers of the moose to the awkward claws of male fiddler crabs, and even in the unwieldy plumage of peacocks, males are carrying the burdens of sexual releasers to which females are programmed to respond (Fig. 22–4). The larger or more obvious the stimulus, the better are a male's chances of attracting the attention of a female, but also proportionally greater are his chances of starving or of being captured by a predator. The male's ability to survive in the face of these flagrant disadvantages, these genetically programmed disabilities, may be as sure a sign of fitness as the winning of a territory, and so may account for the females' preference for these otherwise pointless supernormal releasers. On the other hand it could be that the sort of one-way selection discussed in Chapter 3 which leads to ever more stimulating releasers is responsible, and that the fitness of the males with their species-specific handicaps is irrelevant. If this is the case, these species may be caught in a genetic positive-feedback loop, an evolutionary dead end in which a sensible restraint is equivalent to genetic suicide. Since genetic fitness is tallied generation by generation in the form of viable, competitive offspring, there may be no way for the genes of any one individual to become more fit by taking the long view of what may be best for the species.

In another common form of courtship males provide females directly with some fitness-enhancing resource: food that the male has caught (Fig. 22–5), or a nest that he has built. Some

A

B

Fig. 22–4 Sexual dimorphisms are prominent features of male display and ritualized combat. A. Peacock. B. Uganda kob antelope. See also Fig. 3–4.

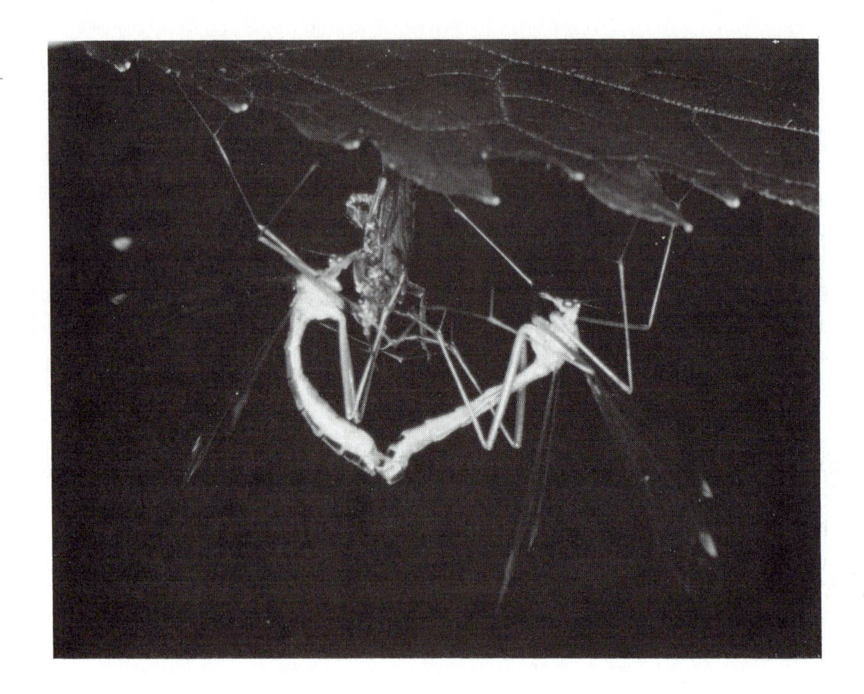

Fig. 22-5 Males of this species of hanging fly court females with offerings of food—a dead fly in this case. This system is susceptible to a variety of tricks and deceptions, ranging from transvestism (males may pose as females to steal the nuptial present) to fraud (males may offer their prospective mates only the dry exoskeletons of prey, having sucked out all the nutritive juices themselves).

male spiders present the females they are courting with silk-wrapped prey, which the female consumes while the males mate with them. This "food display" provides the female with some of the metabolic resources necessary to produce eggs and may, incidentally, distract her from eating the male—a behavior common in the insect world. Male praying mantises, in fact, present *themselves* as the donation, copulating with the female while being eaten (Fig. 22-6). Indeed, it is the loss of the head, and with it the circuits which inhibit the FAP that is mating, that releases mating behavior. If the daughters of such social organizations were selective in no other way, mating unhesitatingly with anyone who offered them a meal (whether of prey or of themselves) and turning that offering directly into eggs, then the males who caught the most and best prey or who were the most nutritious meals themselves would father a larger percentage of the next generation. The females' genes would be recombining preferentially with the presumably more fit genes of the most successful males—either the best hunters, or the largest and therefore the best eaters.

Although when we look intently we rarely find genes to have been foolish, there are instances in which they *appear* to have been deceived. There are certain displays which appear to us to

Fig. 22–6. In some species of mantis the males are themselves the food offering. In the grisly ceremony pictured here, the male's head has been eaten by the female while the remainder of his body mates with her. After mating has been completed, the female will consume the remainder of the male.

be "empty," in which the fitness correlation of the male releaser seems to have been lost without the females' genes having "caught on." For example, the males of several species of fly, like the spiders mentioned earlier, present females with a meal wrapped in a silken "balloon" as the price of copulation. A few species, however, present the females a nonnutritive package of insect exoskeletons, and some omit even this nicety and court with empty silk balloons. How have the female genes come to permit this deception? The two possibilities discussed earlier come to mind again: either there is male fitness information hidden somewhere that we just can't see, or the genes simply have no choice. We can imagine that the female's genes evolved—that is, wired her feature detectors—to recognize the encased prey as the releaser for mating, taking a shortcut, as it were, past a lot of fancy decision-making apparatus. Once the balloon itself became fixed as the releaser, the way was open for genes which encouraged cheating: males which extracted all the juices before wrapping their prey might have come into being. For the female genes, now dependent only on the balloon releaser, there would be no obvious way to select against males which offered just "dry bones." From there it would be a short step to the complete deception of an empty ball.

The next move in this game of evolutionary chess may be for the females to modify their behavioral programming so as to delay copulation until after the present has been inspected. And it's not just the females which are subject to exploitation: on the other side of the coin, males of at least one species which still serves up meals to prospective mates will often pose as females and attempt to rob their suitors of their hard-earned offerings. These imposters then have the choice of eating their ill-gotten spoils or using them in their own courtship.

DOMINANCE HIERARCHIES AND LEKS

Instead of forming a lineup and allowing females to compare and choose from the array of releasers and displays, males of a variety of species settle the matter among themselves and enforce the results on the other sex. The most common form of this chauvinistic system is seen in dominance hierarchies. Here males with no defensible resource to argue over compete with one another for social status. Mountain sheep, for example, sort themselves out by a curious ritual in which they ram each others' heads at full tilt (Fig. 22–7). The animal which proves

Fig. 22–7 Male mountain sheep establish a dominance order through spectacular trials of strength. The contest begins as the combatants run on their rear legs until their back-curved horns crash with tremendous force. Note that both contenders have turned their heads to the left to assure a safe, solid impact. Following the collision, both participants step back and look to the right. The male which wins the long, wearying series of preliminary contests and playoffs is rewarded with nearly exclusive mating rights.

dominant has first rights to any reproductively ready female, and may follow her around unmolested by other males until she is ready to mate. Only when two females are sexually receptive at the same time can even the second-ranking male expect to mate. Females honor the results of the contests in this system either because it provides them with the best male genes or because selection has found that the potential damage which might result from being caught in the crossfire between contending males is a strong argument in favor of passive acquiescence.

These examples illustrate the range of behavioral solutions to controlling resources and judging fitness. As we have seen, however, sorting out the *exact* motives behind the behavior of all the actors in something so seemingly straightforward as mating is actually very complicated. Intuitive guesses have proven to be unreliable in disentangling the competing functions of territory in feeding and breeding, or in sorting out the significance of behavioral and morphological displays, kinship, or coy vs. passive female behavior. Even whether a female chooses a male for himself (his genes, really) or for his territory is rarely clear from observation alone.

Practiced in isolation, then, behavioral ecology runs the risk of being a mere exercise in justifying the ways of evolution to man. A clever mind can imagine an adaptive significance for essentially any behavior or social system. To be more than an intellectual parlor game, behavioral ecology must be capable of making relatively precise, testable behavioral predictions, and, moreover, of predicting exceptions to its own generalizations. One of these exceptions, and perhaps the most extreme of evolution's answers to sexual selection, is the lek. An eccentric, infrequent solution to selection's pressures, the lek is the mating system of two species among the seventy or so of African antelope, two species of grouse, two species of tropical birds, one species of the thousand or so Hawaiian *Drosophila,* and many other unrelated species. In these curious groups males assemble to form an integrated matrix of tiny individual territories painstakingly acquired and militantly defended. The lek itself has several unique characteristics. The matrix is located on one seemingly arbitrary and undistinguished spot in the home range, which is nevertheless usually traditional, remaining the same from year to year. Females, whether Uganda kobs or prairie chickens, come there only to breed: the lek offers no food and no place to nest. Mating success is highly skewed toward the one male in the center which, in kobs and grouse, generally performs 50–75 percent of the copulations. His immediate neighbors may account for the remainder of the total matings, while all the rest—dozens, even a hundred other males—do not mate at all.

But alas, even in the simplest of territorial systems it is difficult to sort out which factors have selected lekking behavior on the part of either sex. Males compete vigorously for a spot on the matrix, even one which offers no current chance of mating, so the ones in the center are constantly contesting for their preferred place. Never daring to relax their guard, and unable to leave for food or water as long as the lek is in force (which may be all day and well into the night for weeks on end), they are consequently under the greatest stress. Their ability to gain and hold the center is presumably a clue to their genetic fitness, and the behavior of the females as they proceed resolutely to the center of the lek to mate suggests that their genes recognize the system's built-in selection pressures.

How can we test this mere surmise? And how could a system like the lek have arisen in the first place? Consistent patterns of habitat or niche among lekking animals are hard to find. The lek may be near one relatively abundant but indefensible limiting resource or randomly placed in a region of indefensibly dispensed resources. In either case, though, other species of the same animals have evolved more conventional solutions to the problems of courtship and mating. One tempting supposition is that the lek is the result of some one-way evolutionary path which created a conventional territorial system based on cues and controls, a path that disappeared in a subsequent habitat shift. Leks, then, could represent a state of collapsed territorial structure, a behavioral cul de sac in which the species are perhaps permanently trapped. Hypotheses like this have a seductive charm, partly because they are so marvelously untestable, depending as they do on the yet-unsolved details of behavioral genetics.

But other questions arise, even out of our untestable theories—questions which are themselves compelling, and whose solutions are perhaps accessible. Why, for instance, should the genes of the excluded, always on the alert for the most efficient ways to propagate themselves, put up with systems like the lek?

The next chapter will explore how the metaphysics of this question may become, after all, only physics.

AGGRESSION

One of ethology's most inescapable questions concerns why animals are aggressive, and whether aggression is part of the biological heritage. Since members of *K*-selected species inhabit the same niche and compete for population-limiting resources, it should not be surprising that these animals regularly fight with each other for control of those resources. Among *r*-selected species, on the other hand, fighting would be a waste of their most precious commodity: time. The occurrence of intraspecific aggression matches these ecological predictions, and we can be certain that the drive to fight and prevail which is aggressiveness is an innate and adaptive behavior. It is in truth motivated not by cultural conditioning but by hormones and releasers, although the extent to which social conditions such as crowding or competition may interact with hormone production and supply is not yet well understood.

The ultimate source of aggressive behavior is, of course, the genes, those masters of behavioral economics. Genes seem to have adjusted the programming of aggression to suit the risks and benefits in each species: sea lions risk a fight to the death for mating privileges, since failure is tantamount to genetic suicide, but mountain sheep engage in ritual duels from which the "loser" retires gracefully, in full expectation of better odds in future seasons. Most animal social systems encourage this latter sort of encounter, whose orchestrated restraint ensures that the combatants will suffer no serious injury. Specialized weaponry developed for protection from or predation on other species is rarely used on conspecifics in any effective way: poisonous snakes wrestle without striking; fish lock jaws but do not bite; antelope push and fence with their horns but will not stab. Considering the risks of unrestrained fighting even to the probable winner, the selective advantage of this "ritualization" (which is nothing more than programming) of aggression is clear: what would be the use of winning ten contests a day if each left an animal exhausted, injured, or an easy prey to watchful predators?

The roots of ritualized aspects of aggression go deeper still. Many species take advantage of a motor program which serves them as a sign stimulus for surrender, and as such instantly halts further aggression. Lizards crouch, cichlid fish retract their fins, sticklebacks adopt a vertical posture, dogs expose their most unprotected spot, their bellies, while gulls offer the backs of their necks to their opponents. Each of these maneuvers has the effect of signaling that the contest is over, and both animals are able to retire from the fray as the winner declines to take advantage of his opponent's vulnerability.

This painstaking orchestration of aggression has enormous advantages. In the working out of territories, for example, it serves to reduce the risk of injury, and in more social species it actually lessens the frequency of fighting and stabilizes social groups. Social animals often work out elaborate dominance hierarchies through species-specific duels or rituals. With practically no injury the animals in a group sort themselves out into a religiously observed chain of command. Once formed, the hierarchy remains stable and fighting virtually disappears. Questions of who has priority when it comes to food or space or mates are settled by consulting the previously established dominance order. Every individual knows its place, and is aware of the probable outcome of challenging anyone higher or lower on the scale.

And so the day-to-day workings of societies, particularly mammalian societies in which the capacities of individual recognition, memory, and subtle signaling among members of the group reach their height, go on smoothly. Overt aggression in such species is generally reserved for rival groups and, to all appearances, mutual defense of the group's resources helps further to stabilize that group's social bonds.

Intraspecific aggression in animals, then, usually lacks that aura of violence and disruption that we tend to associate with fighting. Fighting, in the natural world at least, is for the most part a purposeful ritual. Carefully choreographed and disciplined, it is a behavior which both minimizes injury and diminishes further conflict. Far from being disruptive, programmed aggression provides the cement for the social order, and so catalyzes groups and enhances their fitness by minimizing the time wasted on needless conflict.

What then of our species? Are we programmed to work out group hierarchies, driven to engage in intergroup contests? According to Konrad Lorenz, master of the behavioral analogy, we are; but modern society denies us the small, natural groups to which we evolved, and thus thwarts our adaptive and restrained aggressive drives. Provocative and imaginative as ever, Lorenz speculates that intraspecific war and racial conflict are two of the consequences of our unnatural societies, but that we have also unwittingly developed alternative ritualized "safety valves" for releasing the pent-up motivational waters, outlets such as group sports, boxing matches, hunting, and sex. Although such intriguing ideas add more spice than substance to our consideration of human ethology (Chapters 29 and 30), they do serve to underscore the potential value of an understanding of the role of programmed aggression in social animals.

SUMMARY

In order to remain at a stable, near-saturation level in their habitat and yet still have sufficient resources to raise offspring, most *K*-selected species have evolved mechanisms to restrict access to limiting resources such as food, and to breeding or nesting territories. The most common exclusion mechanism is territoriality: one sex (usually the males) competes for suitable space and the defeated are at least temporarily excluded from the breeding population.

Females, on the other hand, normally devote their attention to selecting the fittest males with which to mate. Their choice (or acquiescense) may be based on the outcome of these male-male contests for place, on the size of male sexual dimorphisms, or on the quality of the resources the males control—food, nests, or space. In observing these crucial interactions, however, it is often hard to sort out the "motives" or the costs and benefits to each of the participants. Speculation in the absence of experimentation or of some accurate knowledge of the mechanisms involved is a popular but generally barren intellectual game.

STUDY QUESTIONS

1. How could you determine experimentally whether a female chooses the winner of a ritualized contest because of his evident fitness or, instead, might be acquiescing merely to avoid bodily harm from the winner should she select a loser?
2. Along the same line, how could you establish whether females choose males on the basis of the fitness required to survive under the handicap of a large sexual dimorphism, or might be "locked in" behaviorally on a sexual releaser whose correlation with genetic fitness is nil? What would be the likely genetic fate of a female which ignored sexual releasers altogether?
3. There are several species in which *females* are territorial and seem to run things. What do you think could account for this unusual social organization?

FURTHER READING

Bernstein, I. S., and Gordon, T. P. "The Function of Aggression in Primate Societies." *American Scientist* 62 (1974): 304–11.

Eibl-Eibesfeldt, Irenäeus. "The Fighting Behavior of Animals." *Scientific American* 205, no. 6 (1961): 112–18.

Gilliard, E. T. "Evolution of Bower Birds." *Scientific American* 202, no. 2 (1963): 38–46.

King, J. A. "Prairie Dog Behavior." *Scientific American* 192, no. 4 (1959): 128–40.

Myers, J. H., and Krebs, C. J. "Population Cycles in Rodents." *Scientific American* 230, no. 6 (1974): 38–46.

Thornhill, Randy. "Sexual Selection in the Black-Tipped Hanging Fly." *Scientific American* 242, no. 6 (1980): 162–72.

CHAPTER 23

Individual Selection

ALTRUISM

The ultimate beneficiary—or victim—of natural selection is not the species, or even the individual, but the genes within the individual. Yet we commonly hear highly respected scientists—even Lorenz—talk about behavior "for the good of the species." Indeed much of behavior, that of highly social animals in particular, does seem at first glance to be directed toward the benefit of others. Parents sacrifice themselves to feed and protect offspring; small birds "mob" passing hawks; honey bees commit suicide by stinging inquisitive mice near their hive; antelope advertise themselves to predators as they warn the rest of the herd of potential danger; subordinate male sage grouse join in communal displays to attract females to a central male although they stand little chance themselves of mating. More perplexingly, some species may be altruistic—that is, display these or other self-sacrificing behaviors—while closely related species are not. In some instances only one sex may be altruistic, and this altruism may come and go with age.

What is the guiding principle in these cases? In each interaction the fitness of one animal appears to be sacrificed or at least risked for the benefit of others. How can such altruism be possible if selection really operates on individuals instead of groups? And if evolution *were* to work on groups, what forces could stop the spread of cheaters—animals whose genes for altruism have been inactivated and who are, therefore, supremely fit to survive in a society of self-sacrificing altruists?

The general answer to the conceptual difficulties raised by altruism is that such charity is at heart inevitably selfish, that an animal's genes are making a net profit by their generosity. Recall that the one goal of a set of genes is to survive, to endure through generation after generation of the bodies they inhabit, and that the measure of their success is "fitness." The simplest and most readily accessible example of the selfish altruism we are trying to define is parenting. Parents entrust the future of their genes to their offspring. Faced with the familiar choice of quantity versus quality, parents must either produce more offspring or put more effort into raising a limited number. This is, again, the basic *r* vs. *K* dichotomy, for which the contingencies of the niche have determined the most adaptive—"fittest"—strategy. *K*-selected parents, "selflessly" working their fingers, claws, or whatever to the bone to help their offspring, are simply congregations of selfish genes investing in their own immortality. Although parents may appear to be sacrificing fitness to help their sons and daughters, their genes are simply enhancing their own ultimate fitness.

Ultimate or "inclusive" fitness—the survival value of a gene summed up over all its current bodily accomomodations—is the real measure of how well a gene is doing in its continuing battle to endure; and each animal, each gene package, must decide how best to enhance the fitness of its descendants in the next generation. Not only does the "correct" answer vary among species, most obviously along the *r*–*K* continuum, but often between ages and sexes. From the moment of conception there is an asymmetry in the sex-specific investment: eggs are large and expensive to make, sperm are small and cheap. As many genes may be active or inactive, depending on which sex they find themselves in, each sex must have its own strategy for investing its time and effort. Females, as mentioned in the previous chapter, with their expensive egg investment to protect, have much to gain by taking care in choosing a mate, while males have little to lose and tend rather, all things being equal, to mate as often as possible with whichever female will have them.

PARENTAL INVESTMENT

But things are not always equal, and the social systems which have evolved to control mating vary from the free undirected release of gametes practiced by many aquatic invertebrates to the elaborate pair-bonding systems of geese, the prides of lions, and the rigid dominance hierarchies of many antelope and primates. What contingencies of physiology, niche, and habitat provide the selective pressures which have fashioned these particular systems, and the differing roles they subtend for each of the sexes? For the most part the problem arises only with the more strongly *K*-selected species. Here, each sex must determine how much to invest in raising offspring. Given the usual case, in which females are stuck with a large initial investment in the form of eggs to generate or helpless fetuses to nurture, most of the variation is to be seen in the behavior of males. In each case, will a male's genes profit more by instructing their bearer to stay with the female he has inseminated and help, or by programming desertion? Birds' eggs must be incubated, for instance, and the male's surveillance when the female must be absent may greatly enhance the offspring's chance of survival. Then, too, once the eggs hatch roles are equalized as both parents struggle to satisfy their ravenous young. In fact, 93 percent of birds are monogamous: male genes in this instance make themselves more fit by helping at home than by attempting to attract other females.

Male mammals, on the other hand, usually face the opposite investment prospect. With their developing offspring tucked safely away in the mother's uterus there is no nest to guard or eggs to incubate, and newborn young must be suckled by the female. With so little to gain, it is hardly surprising that only 10 percent of mammals are monogamous.

Within this bird/mammal, monogamous/polygamous dichotomy, however, there are many exceptions, most directly traceable to ecological factors such as food dispersal in time or space, or predator pressure. One such instructive exception is the redwinged blackbird. Males set up small territories in marshes where there is a dependable supply of emerging insects. Despite the constancy of food supply, which would normally encourage monogamy, redwings are polygynous. As it happens, in this one system the best territories are so much better that it pays for females to nest where the insects are regardless of how many other females already share that territory, rather than to choose a bachelor on a poorer piece of marsh. As Figure 23–1 shows, the females' genes seem to be perfectly aware of the advantage, and saturate the best territories. In fact, it seems clear that the nesting territory in the graph which shows six females has such rich resources (translated in the figure directly into number of offspring) that two more nests could be accommodated before the food per nest would drop to that of the five-female territory, and could accommodate fifteen nests before the fitness per female would drop to the level of the one-female territories. That females do not continue to flock to the best neighborhood rather than settling for the low-rent districts implies either that they are not infallible at judging territory quality, or more likely that the females already in residence are collectively excluding newcomers.

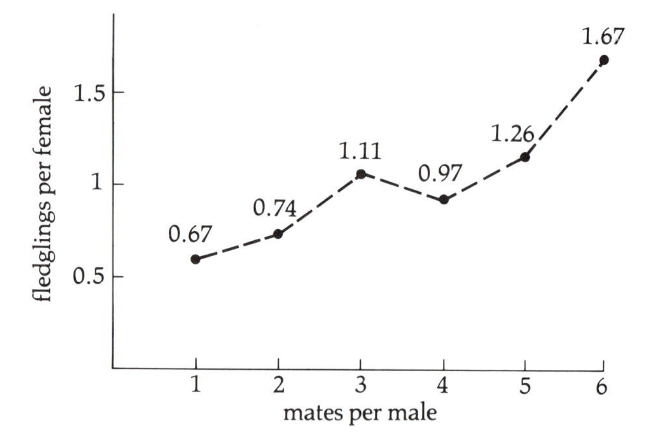

Fig. 23–1 One measure of fitness is the number of offspring reared. For female redwing blackbirds, food is the factor limiting number of offspring. The best territory in a marsh will support about ten baby birds, while the worst cannot maintain even one reliably. Females do better, then, if they nest on the best territories and share the resident male rather than opt for a bachelor male on a marginal piece of marsh.

Parenting and mate selection, as we can see, are coldly calculated episodes of genetically programmed selfishness, and whatever fond emotions may fuel these behaviors in animals, are doubtless part of the fitness-enhancing clockwork. But what then of the other examples of individuals which apparently sacrifice their fitness—perhaps foregoing reproduction altogether—with no benefit to any offspring of their own? In any given season most male sage grouse, for example, will never mate, choosing instead to help make the lek more attractive to

Fig. 23–2 Male sage grouse perform a "booming" display in leks. The drab females visit the leks for the sole purpose of mating. They proceed to the center and mate with the central male or one of his immediate neighbors. So important is place on this featureless but traditional matrix that males begin competing for position months before the females become reproductively ready, and during the mating season man their stations daily in the 4 A.M. darkness, hours before any females appear.

passing females by participating enthusiastically in the communal display on the "booming grounds" (Fig. 23–2), thus increasing the number of times the favored central male can mate. Turkeys have a similar system. Why should the genes of the excluded males put up with this system? Why are they so passive? Why not risk all in an attempt to mate themselves?

There are two feasible answers to this question. In some species that share this system, patience may be rewarded: an excluded male's chances of mating usually increase with age and experience. In these cases it makes sense for them to wait until the odds are in their favor. This explanation will not work for short-lived animals, however, or for species in which the mating frequency is too highly skewed toward a small proportion of the males. Excluded sea lions, for example, stand little chance of ever mating, and so often battle desperately, even to death, for a breeding territory. But the grouse and turkey displays are all show.

KIN SELECTION

For turkeys, and a surprising number of other cases of "altruism," the apparent method in their madness lies in the concept of kin selection. The ultimate criterion for evolutionary fitness is how many of an animal's genes make it into the next generation, but selection makes no distinction between animals which flood the environment with their own offspring and those which invest their energies in nurturing the offspring of their near relations, as long as the frequency of the gene in question remains the same. All turkeys on the lek are thought to be relatives—perhaps brothers—so statistically the genes of the excluded are doing better at getting into the next generation by way of their sexually successful brother in the center, by devoting themselves to a group display to attract females to him, than if each went out and tried to find willing mates on his own. Stiff competition for females may have led to these supernormal displays, particularly where the animals' density is so low that a lone displaying male might be easy to miss but a group would be obvious. If the conspicuousness or attractiveness of two males displaying is more than twice that of one, the group system will be selected for, but the attractiveness must increase as the square of the number of displaying males before chastity on the part of all but one can work.

The basic equation for kin selection specifies that the benefit in fitness to the recipient of any altruistic act must exceed the cost to the donor by more than the ratio of their relatedness. Relatedness is nothing more than the proportion of genes they have in common or, put another way, the probability that any particular gene will be shared. The formula for any altruistic act, then, is (cost to donor)/(benefit to receiver) $\leq r$. Hence, the threshold for altruism must depend on how closely related two animals are—half of the genes are shared in the case of siblings or between parents and their offspring, a fourth among aunts or uncles and their nieces and nephews or between grandparents and their grandchildren. Hence, for kin selection to work, a celibate chorus turkey must attract more than twice as many extra females to the lek to mate with his brother than he could attract on his own.

The most dramatic example of this selection is to be found in the Hymenoptera—wasps, ants, and bees, the so-called social insects. True sociality has arisen independently on eleven occasions in the insects, and ten of those times were among the Hymenoptera. A quirk of Hymenopteran genetics has been the motive force behind this tendency toward a social way of life: male Hymenoptera are haploid. With only one set of chromosomes, all sperm cells from a given male are genetically identical. Hence *all* his daughters have *all* his genes instead of the half that the usual diploid male's offspring possess (Fig. 23-3). As a consequence, his daughters have more genes in common with each other than normal: the 50 percent of the genes they have inherited from their father are identical, while on the average only half of their maternal genes are the same. Hymenopteran sisters, then, share three-quarters of their genes, while normal sisters have only half their genes in common. The most important consequence of this is that worker bees are more closely related to each other than they would be to their own offspring—three-quarters vs. one-half. Hence, their genes do better by helping their mother to raise their siblings than they would by becoming queens themselves.

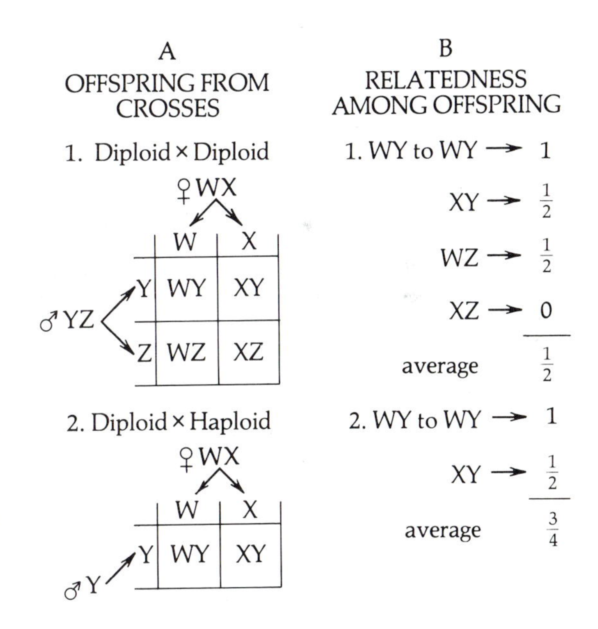

Fig. 23-3 In most species, both parents are diploid and so produce gametes which contain only half their genes. This is indicated in the matrix (A1) as an imaginary one-gene organism carrying a different version (allele) of its gene on each chromosome (W and X) mates with another which is carrying two other alleles (Y and Z). The four classes of possible offspring are indicated in the matrix (A1). Any one recombination, WY for instance, is related to the various genetic possibilities by ½ on the average (B1). (The same mathematics works for multigenic animals.) In the Hymenoptera, however, the male is haploid and so can contribute only one genetic possibility to the matrix (A2). As a result, the average relatedness of any one offspring to the others is ¾ (B2). Since the diploid female—the queen—is only related to her daughters by ½, this means that daughters do better—multiply their genes at a higher rate—raising sisters rather than daughters of their own.

Although most dramatic in the Hymenoptera, kin selection is by no means limited to the social insects. In ground squirrels, for example, unscrambling the maze of kinship causes much of the apparent behavioral arbitrariness and unpredictability of their social system to vanish, and provides evolutionarily sound, perfectly selfish reasons for behaviors which could formerly be interpreted only as altruistic, pointless, or aberrant.

Fig. 23–4 A ground squirrel gives an alarm call—one of two classes depending on whether the predator is airborne or territorial. The alarm call increases the chance that nearby squirrels will avoid predation but also increases the calling individual's own conspicuousness and risk. Alarm calls, however, are rendered primarily by females with close relatives nearby and so probably represent a variant of kin selection.

Ground squirrels live in burrows and hibernate nearly two-thirds of the year. During their short summer they eat grass seeds and reproduce. Females defend territories, but only against certain squirrels, ignoring others of the same age and sex. Despite their territoriality females will hibernate in small groups, while the nonterritorial males pass the winter alone. When predators approach, some individuals consistently give "altruistic" alarm calls (Fig. 23–4), thus increasing their chance of being attacked, while others will not. Males fight over estrus females, and the frequency of mating is highly skewed toward a few older males. Males which enjoy a high rate of reproduction, however, will abandon the scene of their triumphs and try their luck during the following years against unknown males in areas with untried food supplies, while unsuccessful males will perversely remain where they have encountered only failure in the past. Finally, when the time for hibernation approaches, juvenile females will settle in burrows close by where they were born, while males disperse ten to twenty times farther away.

None of this demographic jumble made much sense until behavioral ecologist Paul Sherman marked hundreds of squirrels and their young individually year after year. By keeping track of who was related to whom he was able to sort out their seemingly arbitrary behavior, and found it to be almost entirely explicable on the basis of kin selection.

The first step in any such analysis is to construct a life history of the animals in question to see whether they are *r* or *K*, what limits their numbers, and so on. In the case of the squirrels the major source of mortality is the winter. The squirrels must store enough reserves to survive the unpredictably long and cold winters of their mountain habitat. In any year a quarter to two-thirds of the adults fail to accumulate adequate supplies and die in hibernation. The situation for the juveniles is even bleaker: from 50 percent to 95 percent die each winter. A second source of mortality is predation: 5–10 percent of the squirrels are taken yearly by coyotes, weasels, badgers, bears, and hawks. Fighting between males reduces their life expectancy even further, and each season adult squirrels kill as many juveniles as the predators.

From these data we can guess that surviving the first winter, controlling food resources, guarding against predators and bloodthirsty conspecifics, and reproducing early are the elements crucial to fitness in ground squirrels. Since the males pursue the typical mammalian pattern of low parental investment, fitness is largely in the hands of the females. An estrus female accepts whichever male is able to fight off the other males long enough to mate with her. In general these males are older and larger—the 50 or so percent that have survived to the ripe old age of three or four—so we may suppose that the winning male's genes have thereby proved their fitness and thus justify the female's passive acquiescence. The females, as mentioned above, show a curious variability in their behavior, sharing territories and food with some squirrels and vigorously chasing out others. Sherman's records show that, in fact, the females are sharing with their relatives—sisters, mothers, daughters, and their offspring—and attacking all others. The squirrels learn to recognize relatives as they grow up, probably in their first few days above ground. We know this because young which are introduced into the burrow of an unrelated female before the parent has learned to recognize her own offspring will later share and cooperate with their adopted mother and her kin, but attack their own true relations. Presumably this process is a special variation of imprinting, made possible by the strong tendency of daughters to settle near their birthplace. This assures a high degree of relatedness in any one area, so that any squirrel a young squirrel encounters regularly on its mother's territory under peaceable conditions is almost certainly a relation.

Females enhance their fitness by aiding their kin selectively in a number of ways. Sherman found the limiting factors on fitness to be winter survival, food, predation, and infanticide. Group hibernation is clearly an efficient way to survive the winter, and the programmed toleration of kin means that only females which are related cooperate in this way. This may be particularly important in helping juveniles through their first difficult winter. Only relations are allowed to share food during the warm season. Warning calls are produced much more frequently by older females with living kin than by males or childless females, and the advantage of such warning to a female's nearby relations outweighs the risk she herself incurs. Finally, Sherman found that infants are never killed by their own relatives, but rather by homeless squirrels which would benefit by a reduction in competition. Hence, related females cooperate in driving off unrelated animals.

The very different dispersal behavior of males appears to be directed toward minimizing deleterious inbreeding. By migrating far away as juveniles, the males avoid mating with their relatives. As Sherman reported, highly successful males keep moving, reducing the possibility that they might court their daughters in the following season. In one way or another evolution seems to have programmed this social species with a strict eye on kinship.

RECIPROCITY

Although kin selection is being found to be very widespread, not all altruism depends on relatedness. Another important reason for altruism is reciprocity—the expectation of ultimate repayment. Grooming in social animals is a typical example of reciprocity but one which, like all examples in this class, requires some social selection against cheaters: that is, an ability to recognize and discriminate against those which chronically refuse to repay favors. This requirement suggests that reciprocal altruism will be found only in small groups of animals capable of recognizing each other individually, and as far as the phenomenon has been studied this seems to be the case.

Recognizing that at least this much of altruism is selfish, and that its impact on social systems has been and continues to be enormous, ethologists have intensified their study of the behavioral interactions of identified individuals, keeping track of who is related to whom and who does what to whom. From this perspective has come both a new understanding of the whys of behavior, and a new set of *how* questions. These problems will be explored in the following chapters, first in some social insects where everything is, as usual, fairly clear, and then in several groups of mammals.

SUMMARY

Altruism—the apparent self-sacrifice of fitness by one animal for the benefit of another—takes three general forms. In the most obvious form parents help their offspring, or more precisely, genes invest in their own immortality. The cold, calculating, unromantic nature of this exercise in genetic economics is well illustrated by the variety of thoroughly pragmatic investment decisions made by various animals, decisions which include rape, adultery, nest parasitism, desertion, and infanticide. The second class of altruism is kin selection: helping out genetic kin in proportion to their genetic relatedness. Because of a genetic quirk, this form of sham altruism reaches its height in the Hymenoptera. The other version of altruism is reciprocal and depends on the probability of repayment in the same or in some other form. This latter variation provides much of the social "cement" which binds small groups of higher animals. The basic lesson from careful study of apparent self-sacrifice is that altruism in nature is basically selfish, a profitable programming ploy by which the genes of one animal contribute to the survival of their twins in other animals.

STUDY QUESTIONS

1. If fitness is measured in terms of how many of its genes an animal gets into the world, why not reproduce through parthenogenesis, and generate identical copies of oneself?
2. On the basis of relatedness, should male Hymenoptera invest in helping to raise sisters or in seeking a mate and fathering offspring? Are the males as closely related to the workers as the workers are to them?
3. Consider how the interests of offspring and parents might differ when it comes to the production of additional siblings. Should the firstborn help or hinder? What about father vs. mother?
4. Consider the slime molds of Chapter 15. The individual amoebae cooperate to create the slug and then the fruiting plant. And yet only a fraction of the cells form the fruiting body on top, and it is these cells which create the reproductive spores. Might this be altruism at the microscopic level?

FURTHER READING

Bertram, B. C. R. "The Social System of Lions." *Scientific American* 232, no. 5 (1975): 54–65.

Heinrich, Bernd. "Bumble Bee Foraging and the Economics of Sociality." *American Scientist* 64 (1976): 384–95.

Sherman, Paul W., and Norton, M. L. "Four Months of the Ground Squirrel." *Natural History* 88, no. 6 (1979): 50–57.

Smith, J. M. "The Evolution of Behavior." *Scientific American* 239, no. 3 (1978): 176–93.

Part VII

SOCIAL BEHAVIOR

CHAPTER 24

Honey Bees

Bees and ants, the subjects of this chapter and the next, both evolved from their common ancestor, the wasp. The basic difference between bees and wasps is that bees are vegetarians: unlike most wasps, which obtain protein by consuming other insects, the bees' protein source is pollen (Fig. 24–1). There are twenty thousand known species of bee, with lifestyles ranging from the completely solitary to the unabashedly social. Like Tinbergen's species of digger wasp, the life cycle of a solitary species of bee begins with a mated female which endures the winter underground, then finds, digs, or builds a nest, lays and provisions an egg, seals the nest, and begins again. Offspring hatch out, females mate with the short-lived males, and so the generations continue. A variant on this theme is "mass provisioning" in which, like Baerends's digger wasp, the bee feeds her larvae day by day, rather than sealing them in with their larders full.

SOCIALITY IN BEES

The bumble bee is a familiar example of a modest level of sociality (Fig. 24–2). The mated bumble bee female overwinters,

Fig. 24–1 A forager bee landing on a crocus flower. A lump of yellow pollen is packed into the "pollen basket" on her right rear leg. The pollen-gathering strategy is simple: as a forager rubs against a flower's stamen she covers herself with the sticky pollen. Then, often while hovering, she cleans her head and furry thorax with the specialized comb-like array of hairs on her front legs, uses her middle legs to brush out her abdomen and collect the combings from the front legs, and then passes the sticky wad to a special pollen press on the rear legs which packs the pollen into a large, basket-like array of hairs.

Fig. 24–2 Bumble bees usually nest in the ground, often in holes dug and abandoned by mice or other rodents. The queen *(upper right)* builds the first cells, provisions them with pollen, lays her eggs, then seals the cells and keeps them warm while the larvae grow and pupate. When they emerge the daughters, which are smaller than the queen, take over most of the building, foraging, and provisioning. The sealed cells in this nest contain larvae or pupae, while most of the open ones are being used to store either pollen or honey. At the end of the season all of the stores will be used to raise a new generation of reproductives.

builds a nest, lays some eggs, and begins feeding the larvae. However, when the larvae pupate the female remains to guard the developing pupae and to keep them warm. The haploid/diploid sexual system of the Hymenoptera gives the queen total control over the sex of her offspring. Early in the season she fertilizes the eggs as she lays them with some of the sperm stored in her spermatheca, thus producing the diploid workers the colony needs for growth. These workers are perfectly capable of laying eggs of their own, but their mother, now properly called the queen, prevents them from doing so. Her dominion is established by a mixture of physical force, psychological warfare (threats), and the selfish, kin-selected programming of these Hymenoptera. Any eggs laid by the queen's daughters are promptly eaten, a behavior widespread in semisocial bees and doubtless the source of the curious "trophic egg" ritual in some highly social bees and ants, whereby workers lay infertile eggs for the queen to eat as her source of nutrition (Fig. 24–3). The bumble bee sisters, thwarted in their egg laying, begin doing all the work of building cells, tending young, and gathering food, while the queen contents herself with laying eggs and maintaining the hierarchy. As the end of the season approaches the queen begins to lay unfertilized eggs which will result in haploid drones, and the colony raises a generation of reproductives—males and future queens—to ensure the continuance of the colony's genes. What accounts for the self-sacrifice of the

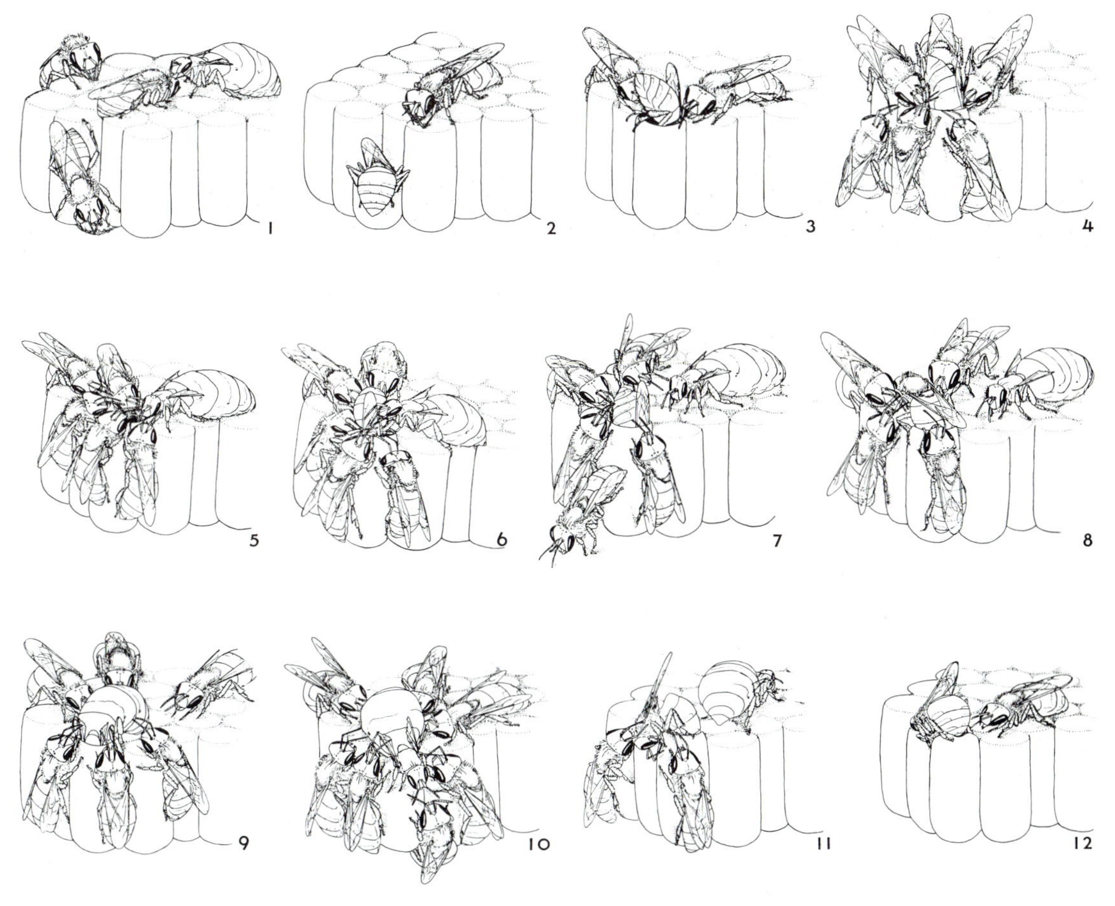

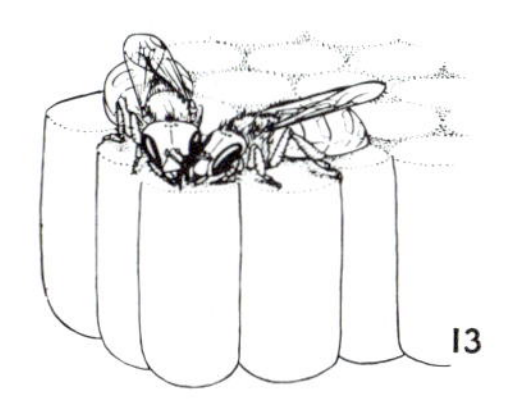

Fig. 24-3 "Trophic" (nutritive) egg ceremony of a stingless bee. Workers construct a new brood cell (1–3), but the queen (the large bee in 1) ignores their activity. When the cell is completed the workers fill it with larval food (4–7) which attracts the queen. She stands by as the cell is filled and one worker (8) lays a small trophic egg. The queen eats this offering (9) as well as some of the larval food, and then lays her own fertile egg in the cell (10). The cell is then sealed by the workers (11–13) and the egg left to develop.

bumble bee workers, as they grudgingly help to rear their mother's offspring rather than their own? The answer, as we saw in the last chapter, is probably a barely tolerated, forcibly imposed version of kin selection. Bumble bees seem to be a species only halfway along the road to full sociality espoused by the honey bee, whose smoothly programmed social system turns the self-sacrifice of the many to the benefit of all.

HONEY BEE SOCIETY

Honey bees evolved in the tropics, where two primitive species are still to be found. The colonies, consisting of ten to a hundred thousand bees, are perennial, and reproduce by raising a new queen and fissioning into two colonies. Tropical honey bees build sheets of comb which hang exposed from tree limbs (Fig. 24-4), and the individual cells of which the comb is made are used to raise new bees and to store honey and pollen. The adaptation which has permitted our familiar honey bee, *Apis mellifera,* to penetrate into the temperate zones as far as southern

A

B

Fig. 24-4 A. The dwarf honey bee, *Apis florea,* lives on single sheets of comb hanging from tree limbs. The warm climate of the tropics permits the colonies to subsist in the open year round. B. In this picture the bees have been removed to expose the comb and the horizontal dance floor on top.

Canada in America and northern Germany in Europe is its habit of building its comb inside hollow trees. Hence, our bees can endure periods of very cold weather by allowing their mass metabolism to warm the cavity.

Fig. 24–5 The queen is much larger than the workers, and devotes her life to laying eggs. Young "nurse" bees constantly attend the queen, grooming and feeding her. They are attracted by a pheromone, and in the process of licking her to keep her clean, ingest chemicals which repress their ovaries. This substance is passed throughout the hive through food exchange or "trophо-laxis."

The life cycle of temperate honey bees begins as the colony, perhaps ten thousand strong, starts to forage in the early spring. The resources are almost exclusively theirs in this season since their only competition comes from the relatively small numbers of surviving "seed" queens of annual species which must start their colonies anew each year. During this period of surplus honey bees invest heavily in raising new bees. The queen is little more than an egg-laying machine, capable of turning out up to three thousand eggs each day (Fig. 24–5). She is able to manage this feat because none of her time is required for the routine tasks of foraging, nest building, brood tending, or guarding—chores which are taken care of by her daughters. She does not even have to spend time actively dominating the workers, like her less socialized counterparts must. In what must be the ultimate example of ritualized dominance, she produces a pheromone which automatically inhibits her worker daughters from raising any of her eggs as rival queens, and keeps their ovaries from functioning.

Fig. 24-6 A. Bees use the honeycomb both to store food and to raise brood. Larvae, tightly curled and almost ready to begin pupating, occupy several cells, while pupae are developing in the capped compartments. B. The entire developmental cycle runs from an egg *(far left)* to a pupa about one day before emergence *(far right)*.

An egg hatches in two days and the resulting larva (Fig. 24-6) is fed more than once a minute, day and night, for the next week. Just what a larva is fed depends on its caste—queens and workers are both raised from the same diploid eggs, but a larva destined to become a queen is fed "royal jelly," a secretion rich in protein and chemical signals produced by the mandibular glands of workers. (The adult queen is fed a steady diet of royal jelly as her sole food source.) In the spring, male bees or "drones" are also reared in large numbers (Fig. 24-7). The colony grows until there is no room for new bees, or there is simply no longer enough queen pheromone to keep the workers from raising new queens.

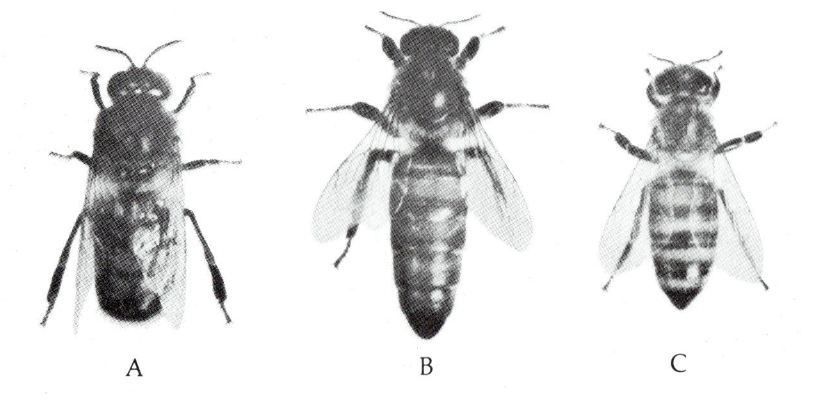

Fig. 24-7 (A) Drone. (B) Queen. (C) Worker.

Once queen cells have been started, the worker bees tending the queen stop feeding their mother, and as she slims down enough to fly, her egg-laying ability gradually declines. As the new queens begin to mature in their cells the old queen will occasionally "pipe." Pressing her body against the comb she vibrates her thorax with her flight muscles, producing a pulsed tone (Fig. 24-8). Any mature queen in a cell responds in like form, making a "quack." This communication establishes that a new queen is ready, and the colony can divide. It also indicates to the new queen that she must remain a while longer in her cell or risk a fight with the old queen.

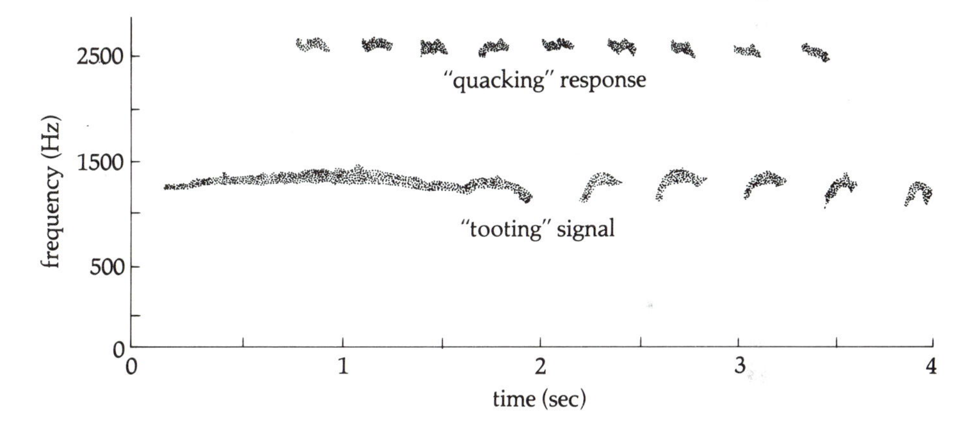

Fig. 24-8 The ruling queen and her potential successors communicate through a pulse-coded auditory exchange. The old queen's "tooting" signal elicits a "quacking" response from new queens mature enough to emerge from their cells, and serves to inform the hive that swarming is now possible. When the new queens' quacks fail to elicit a toot in reply, they know it is time to emerge.

As soon as the weather is favorable an eerie calm prevails in the hive as foragers stop foraging, and the fanning and comb-building activity of workers in the hive ceases. Suddenly a worker begins running through the crowded comb producing a peculiar buzzing signal, and the colony explodes in a frenzy of madly rushing bees. Within seconds the old queen and about half the colony leave in a swarm and form a cluster on a nearby branch. The new queen is free to emerge and, often using the pipe-quack system, locates and kills any other developing queens. Should two queens emerge simultaneously they fight

Fig. 24-9 Attracted by her pheromones, drones approach the queen from downwind. This caged and suspended queen's failure to move interrupts the drones' mating program.

Fig. 24-10 A large swarm of bees clustered on a limb of a tree.

until one or the other is killed. A day or so later the new queen will fly out, mate with several drones, and return to the hive to begin her lifelong career of egg laying.

Drones have an equally routine life. They spend most of their time in the hive doing nothing. Only two hours in the early afternoons of warm, sunny days serve to justify their existence in the hive. Then they fly out to the mysterious special mating areas described in Chapter 13 to wait for queens. When one arrives the drones behave much like Tinbergen's digger wasps, tracking their quarry's pheromone upwind until they see her, then following her visually until one of them catches her. The successful male checks the queen he has caught for some olfactory or tactile cue that releases mating (Fig. 24-9). Copulation for drones is fatal, and only about one in a thousand ever mates. When fall comes the worthless drone population is unceremoniously evicted from the hive and allowed to starve.

The swarm, meanwhile, has been searching for a new place to

live. While the swarm hangs packed around its queen in some nearby tree (Fig. 24–10), scouts reach a consensus of "opinion," and the swarm flies off to a new cavity to set up housekeeping. Within a few days a comb is started, eggs are laid, food gathered, and a new colony is begun.

Now, though it may be only June, the hive's emphasis shifts to preparing the colony for winter. New bees are raised, but at a lower rate, and the colony size increases steadily, but slowly. The bees devote the better part of their energies to collecting and storing the food which will sustain the colony throughout the winter (Fig. 24–11).

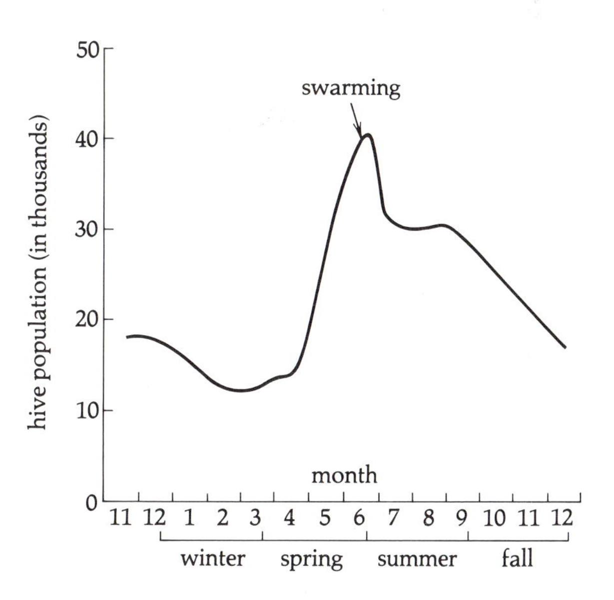

Fig. 24–11 The usual population curve in bees shows a gradual decline through the winter as "insulation" bees on the cluster freeze and die. Rapid growth in the spring follows, then there is a dramatic drop as the colony swarms, followed by stability or a gradual decline until the cold weather sets in again.

The burden of virtually all that a colony must accomplish falls on the worker bees. Each worker passes through a series of age-dependent tasks—cleaning cells, nursing larvae, building comb, and guarding the hive from predators and foragers from other hives, respectively—until at three weeks of age they begin to forage (Fig. 24–12). Those bees that escape mishap end their six-week lives simply by wearing out, falling unnoticed to the ground on some foraging trip with perhaps a thousand miles of flying behind them—all to collect less than an ounce of nectar.

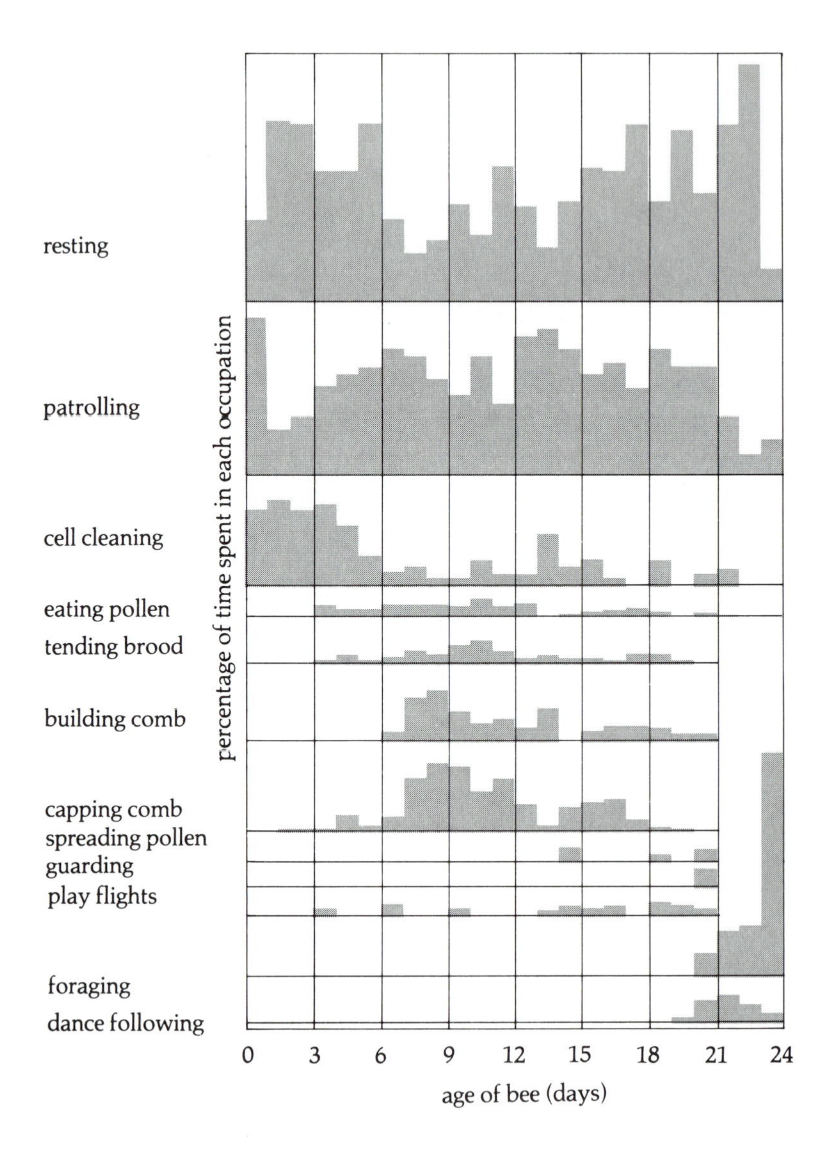

Fig. 24-12 Worker bees go through a series of occupational specializations, as indicated by this record of the first 24 days in the life of one bee. Cell cleaning is the young bee's first chore, followed, as the worker's mandibular glands mature, by feeding the brood and queen. When the wax glands in her abdomen begin to function the worker begins building comb and capping cells. Finally the bee begins outdoor foraging, which she continues until she wears out, at about six weeks of age. Note the extensive amount of time spent resting and wandering around in the hive.

KIN SELECTION AND SOCIALITY

For an ethologist, the "how" questions of behavior are only half the story. *Why* are bees programmed as they are? What is the evolutionary logic behind their unique behavior and social organization? The first thing we must realize about honey bees is that the *colony*, not the individual insect, is the animal and subject to selection. Drones and queen are the sex cells, the workers the somatic cells. Why do all these sixty thousand or so individuals cooperate to create that superorganism which is the colony? The answer may almost certainly be found in kin selection. As discussed in Chapter 23, the genetics of the Hymenoptera are such that sisters are more closely related to the new sisters they help the queen raise than they would be to their own offspring. As the ancestors of honey bees evolved toward that intermediate level of sociality exemplified by bumble bees, any female whose genes programmed her to stay with her mother and raise the same number of offspring that she would have been able to on her own would be 50 percent more fit—that is, 50 percent more efficient at getting her genes into the next generation. The frequency of such a gene would initially double every second generation, even assuming no increase in efficiency from sociality itself.

This is all very neat, but enormous problems with strict kin selection theory, that basic foundation of sociobiology, become apparent when we look closely at honey bees. Most obviously, the theory assumes that the queen mates with only one male. As it happens, though, honey bee queens mate on the average about seven times. The reasons that a queen which has been programmed for multiple matings ought to be more fit are clear. First, if by chance she should happen to mate with one of her brothers, there is a 50 percent chance that half of her offspring would be inviable, since the haploid/diploid sex-determination of the Hymenoptera is actually based on a single sex gene which comes in at least a dozen versions (alleles). A bee homozygous for this gene is a male. Haploid eggs are, obviously, homozygous, and develop into drones. A new queen carries one of her mother's two sex alleles, and so do each of her brothers. If by chance she were to mate with one of her brothers (half of which have the same sex allele as their sister), half of her diploid eggs would be homozygous and would develop into sterile, useless drones (Fig. 24–13). Multiple matings provide an additional way for the queen to hedge her bets. The whole phenomenon of drone areas where males from many hives may congregate is a strategy for avoiding inbreeding. A second justification for her promiscuity would be to obtain more sperm. The queen is equally related to her offspring regardless of which drone may be their father, and the longer she can continue to lay fertile eggs the fitter her genes will be.

All this makes sense from the queen's point of view, and it is certainly nothing new to discover that parents may have self-

Fig. 24-13 The haploid/diploid sex system of the Hymenoptera is actually based on whether or not an individual is homozygous for a sex allele. Haploid insects are by definition homozygous and so develop as males. But diploid males are possible too, if the mother shares one sex allele with the father. In this example, a queen and drone mate and produce eventually new queens and drones. These sexual offspring are allowed to interbreed and the diploid offspring of these new queens are shaded in the matrix. In half of the cases, 50 percent of the progeny are sterile diploid drones. Such colonies cannot survive unless they assassinate their ruler and rear a new queen—a last-ditch attempt they generally will make.

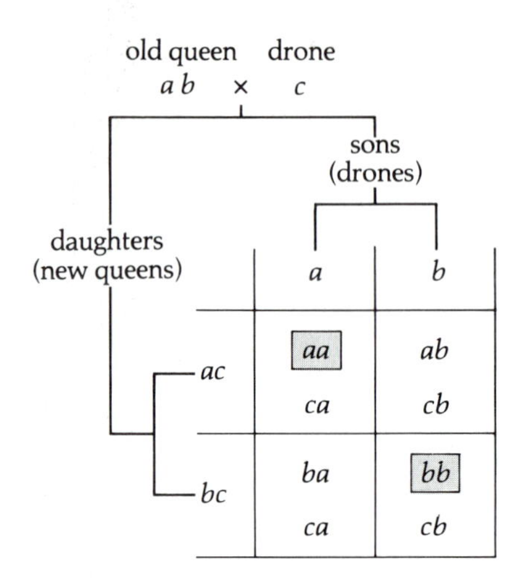

interests different from those of their offspring, but it raises questions in the whole kin-selection argument which requires that daughters forego reproduction for their own genetic good. Ironically, kin selection was originally formulated to explain honey bee sociality, but we are now in the position of having to look more deeply into the evolution and natural history of bees for a way of saving the hypothesis. For example, we could imagine that the sperm from different drones might remain separated in the queen's spermatheca. Thus the interrelatedness of workers at any one time would remain high, even though the father would change every year or so. Alternatively we could suppose that queens remained monogamous until the development of a sterile worker caste carried the honey bee beyond the point of no evolutionary return. This development would have given the queen the freedom to switch her reproductive strategy to increase the fitness of the hive.

The one-way nature of this hypothetical evolution is hard to put to the test, however, and an experiment has been provided by nature which ought to make us at least uncomfortable with this line of reasoning. When the first European settlers reached the Cape of Good Hope they brought with them a few colonies of honey bees. Now, only a few hundred years later, those isolated bees form a separate race, *A. m. capensis*, with the race-specific peculiarity that its workers are so much less sensitive to the queen's sterilization pheromone that their ovaries are large and potentially functional. In the absence of the true queen they can begin producing the queen pheromone themselves, and lay fertile diploid eggs. Since these eggs are produced without the worker queen's ever having mated, her offspring's genes come *entirely* from her. It is clearly within the scope of evolution, then, to dispense with much of the sterile caste system even after it has been so elaborately worked out, and genes for doing so might be expected to spread, at least initially. We might ask why this sort of genetic cheating isn't rampant in social insects. But, alas, when it comes to studying these fascinating societies, the more we learn, the less we seem to know.

THE DANCE LANGUAGE

At every stage of a bee's life its behavior, which is so engrossing yet so melancholy, is a result of its unique sensory perspective and intricate programming. The whole social fabric of the hive is determined by genetically specified behavior patterns. Each caste has its own releasers and motor programs. As we have seen, the bees recognize flowers innately, and their programming allows them to acquire individual, species-specific bits of knowledge with almost split-second precision. Drones come equipped with a modified "digger-wasp hunting program" for finding queens of their own species, and queens come programmed to communicate with each other acoustically without the mediation of complex thought processes. Workers, programmed to recognize the queen, drones, each other, and the larvae of each caste by odor, treat each accordingly. When the hive gets too hot, bees are programmed first to fan it and then to gather water to cool it by evaporation. Stinging, too, is carefully programmed, and even dead bees are designed to exude the olfactory releaser that ensures their removal from the hive. At present the major unknown in bee behavior is the elaborate programming tour de force which allows for their navigation (Chapter 13) and that most essential ingredient for sociality, the high degree of communication which becomes necessary to coordinate the behavior of the individual members of the society.

Perhaps the most impressive example of the astonishing potential of evolution to solve the problem of communication outside our species is the dance communication system of honey bees. It has been known for some time that bees communicate about food. Indeed, more than two thousand years ago Aristotle noticed that the first bee to find food (the "forager") enabled others (the "recruits") to locate the same spot.

Aristotle guessed that arriving foragers actually lead the recruits from the hive to the food. Maurice Maeterlinck tested this hypothesis around 1901 by marking the forager at the food source and then capturing her as she left the hive on her second

Fig. 24–14 Returning foragers often perform dances that take either of two general forms: (A) the round dance or (B and C) the waggle dance. In the latter version, the dancer vibrates her body during the straight-run segment. Potential recruits crowd around.

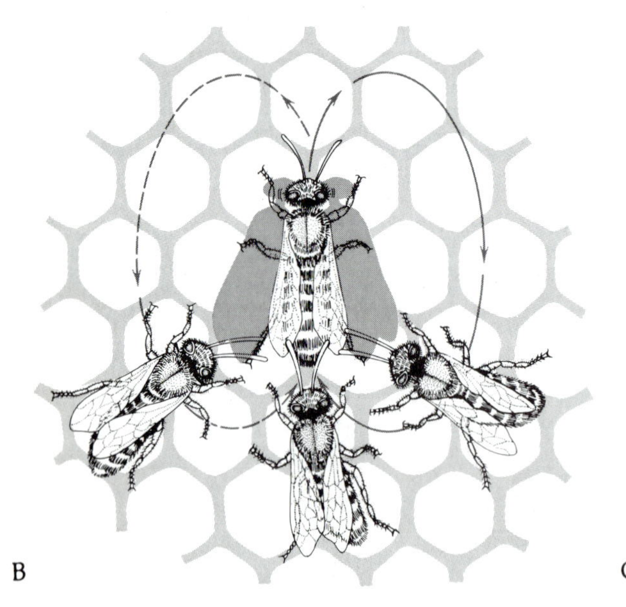

trip. Recruits were nevertheless able to locate the food. Around 1915 von Frisch took up the problem, and observing the behavior of the returning foragers through the glass wall of an observation hive, saw that they performed *dances* which attracted the attention of other bees. These attenders would then rush out of the hive. By a series of clever choice experiments, von Frisch was able to show that the recruits had learned the odor of the food from the dancer and then left the hive in search of it.

Von Frisch actually noticed *two* dance forms: the round dance (Fig. 24–14A) and the waggle dance (Fig. 24–14B). Since his foragers (the round dancers) were only gathering sugar solution, while many of the waggle dancers were clearly collecting pollen, von Frisch supposed that bees have a two-word language: round, he thought, meant nectar, and waggle would mean pollen. In 1944 he discovered that he was wrong. Following up a chance observation, he trained some foragers to a distant feeding station (see Fig. 24–15) and set out dishes offering the food

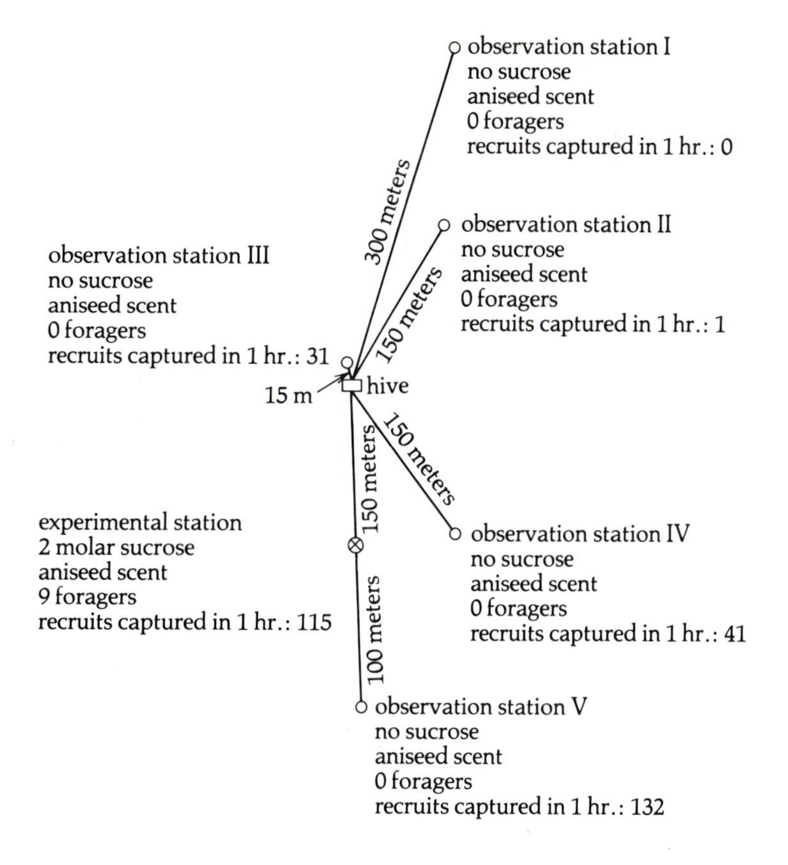

Fig. 24–15 When Karl von Frisch trained foragers to an experimental station 150 m south of the hive and set out scent plates with the same food odor at various spots around the hive, recruits showed a strong preference for the vicinity of the experimental station. Von Frisch interpreted this to mean that the dance the trained foragers performed had communicated distance and direction information. The design of his experiments, however, left it possible that the recruits might be looking for odors—the unique hive odor bees constantly exude, the recruitment pheromone they broadcast at rich food sources, or the odor of the locality which is infallibly absorbed onto the waxy hairs of the foragers' bodies.

odor in a variety of directions and at different distances. Recruits showed a clear preference for the general area being visited by the foragers, and when von Frisch looked at the dances he discovered them to be waggle dances. Very soon thereafter von Frisch discovered the now-famous (and still astonishing) true dance correlations: the length of each dance cycle increases as the food source is moved farther away (Fig. 24–16), and the angle between the waggle phase of the dance and vertical matches the angle between the food and the sun's azimuth (Fig. 24–17). The round dance turns out to be nothing more than a limiting case for distances so short that they encompass no waggles at all.

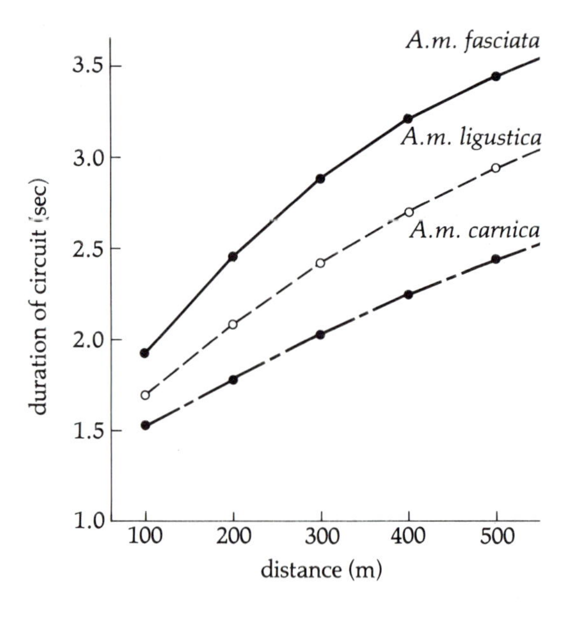

Fig. 24–16 The number of waggles (and the duration of waggling) in a forager's dance correlate with the distance to the food. The exact correlation, however, depends on the race of bee. Shown here are three races, the Egyptian *(fasciata)*, Italian *(ligustica)*, and German *(carnica)*. (The durations are for whole circuits, and so include the semicircular return.)

Von Frisch called the dance communication system a "language" because it depends on abstract, symbolic representations, by far the most elaborate such system known outside *Homo sapiens.* A language, so defined, consists of a collection of symbols that are arbitrarily assigned meanings within a culture. For bees, the conventions which define vertical as the direction of the sun and which assign a value of so many meters to each waggle are shared within each hive, enabling foragers to encode their messages and permitting recruits to decode them. The arbitrary nature of the language is revealed by the subsequent discovery that different races have different conventions—dialects, in fact—for converting distance to waggles (Fig. 24–16). In addition, there are at least three other complex but arbitrary rules to prevent misunderstanding.

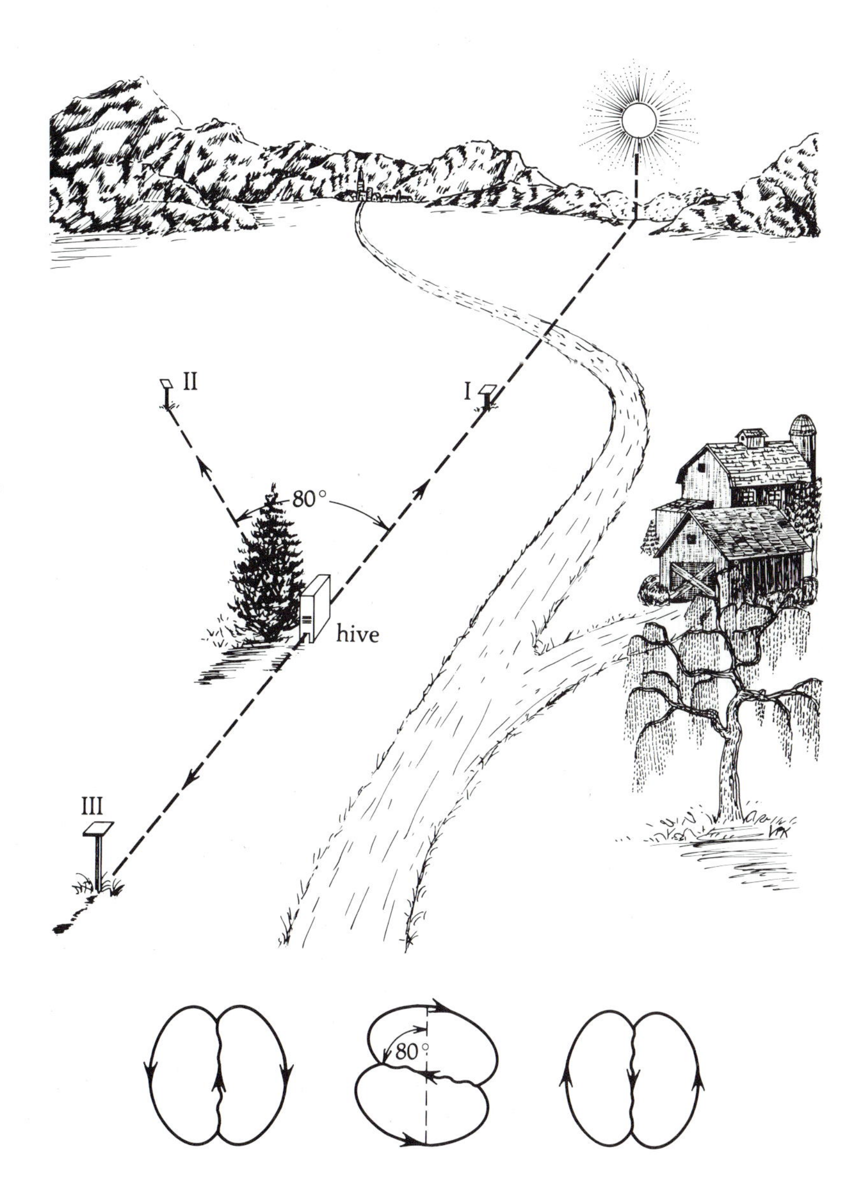

Fig. 24–17 The direction of the waggle run of the dance on the dark, vertical sheets of comb correlates with the direction of the food. If the food is in the direction of the sun (Case I) the dance is directed upward, while if it is opposite the sun the dance is aimed down (Case III). When, as in Case II, the food is 80° left of the sun, the corresponding dance is 80° left of vertical.

BEE DANCE CONVENTIONS

Any language depends on conventions shared within a culture. For bees, arbitrarily defining "up" as the direction of the sun is one such convention: they could equally well have chosen down or, for that matter, 13° to the right. Using the sun as their main reference, the "root word" which determines all other units of the language, rather than north or whatever direction the hive might face, is another. The distance dialect, too, is a linguistic-like convention. Recently, three more dance conventions have been uncovered which must further enhance our respect for the dance language.

The primordial honey bees danced on the horizontal platform created by the top of their comb, which hung suspended from the branches of trees. Foragers oriented to the partial view of the sky which filtered through the leaves of their home tree. The horizontal-hive experiments we discussed in Chapter 13 take advantage of this ancient behavior. But consider the difficulty which the bees face: they all agree to use a solar coordinate system, but where in the crazy quilt of sky patches they can see is the sun? The resolution of bee vision (Chapter 6) is a few degrees, while the sun itself is a 0.5° disc. Since they are clearly unable to see the sun as a disc, how could they definitively rule out a bright cloud? A variety of possibilities for making this discrimination, however, remain: the sun's elevation, its complete lack of polarization, its small size, its color, its brightness. Bees, in the best ethological tradition, have chosen from all these possible diagnostic characteristics a single sign stimulus: color. Bees look for a spot in the sky which has less than about 15 percent UV light, and ignore everything else. The probable IRM which completes this circuit, a color-opponent cell which weighs the input from UV rhabdomeres against those from green and blue cells, has already been found.

But what if a dancer with an urgent message to communicate finds that the sun is not visible, hidden perhaps by a tree or cloud? Although, as we saw in Chapter 13, the bees can fall back on polarized light, the sky is full of polarization ambiguities. Although the polarization in a patch of sky at any elevation has an angle which is determined by the sun's location, the bees cannot run the calculation in reverse. The sun usually generates *two* patches with the same angle of polarization at any elevation: how are the bees to decide which it is they are seeing (circled in Fig. 24-18)? There are several possible cues available in the polarized sky—degree of polarization, color, gradients of polarization—but the bees ignore all these and resolve their dilemma by selecting arbitrarily one alternative and ignoring the other. They agree to pretend that the ambiguous patch is always the alternative *farther* from the sun. Since they are consistent both in dancing and in interpreting the dances with respect to one alternative, the communication system works.

There exist other pairs of possible sky patches, however, for which this rule

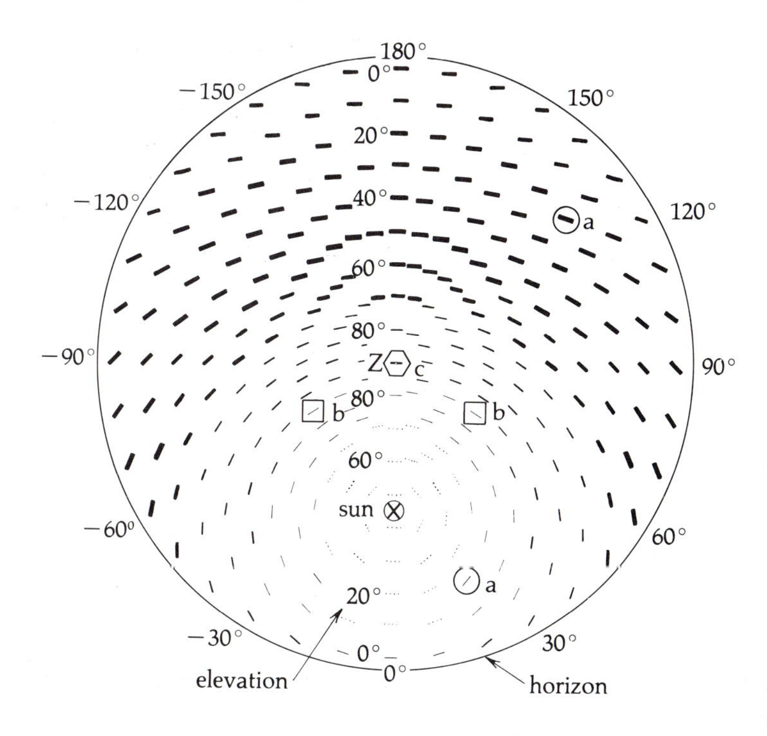

Fig. 24–18 The pattern of polarized light in the sky depends on the location—the elevation and azimuth—of the sun. In this example, the sun is at an elevation of 45°. Bees are able to judge the location of the sun from a small patch of sky, but often make consistent errors when, as is often the case, the angle of polarization in the patch exists in another patch at the same elevation. Two examples are illustrated here: points A, the usual case, and points B, the only pair which have the same relative azimuth to the sun. Point C is the zenith.

will not work (enclosed by squares in Fig. 24–18). Here the two polarization patterns are both vertically polarized, are at the same elevation, and are at the same distance on either side of the sun. The bees deal with this contingency by assuming that the patch they see is the one to the *right* of the sun. The "right-hand rule" again results in the bees' misidentifying the patch half the time, but since they are consistent in their misconceptions the language works. If both "speaker" and "listener" have the same wrong definition of a word, communication is still achieved.

Finally, there is the zenith, a point which is neither left nor right of the sun, and is equidistant from the two possible locations of the sun. For this singular patch bees have no rule at all, and their dances, when the zenith is the only referent available for orientation, indicate both possibilities.

The lesson of these conventions is by now a familiar one: from a surfeit of information, the animals choose only one or two key elements to attend to; and in the case of such sign stimuli in social interaction, uncertainty is not permissible—everyone must be wired in exactly the same way and live by the same set of rules or social life would turn to anarchy.

A

B

Fig. 24–19 In order to see if distance might be judged on the basis of effort, von Frisch equipped bees with 55-mg lead weights (A) or drag-producing tinfoil flaps (B). Both treatments caused the bees to overestimate distance.

INFORMATION PROCESSING

Von Frisch's subsequent investigation of the dance was characterized by a self-perpetuating series of classically simple and straightforward experiments. Puzzling over how animals with such poor vision could possibly measure distance, he noticed that bees flying either uphill steeply (during an experiment to discover whether bees have words for "up" or "down") or upwind overestimate the distance to the food while those going downhill or downwind consistently underestimate. Could they be measuring *effort* on the outward flight? To confirm his guess, von Frisch outfitted his bees with small lead weights or with tiny drag-producing flaps, both of which served to increase the distance message (Fig. 24–19). In a classic though equally simple series of choice experiments in which foragers danced for the location of their training stations and searching recruits were monitored at a variety of locations nearby (Fig. 24–20), von Frisch demonstrated that recruits probably make use of the information gained from the dance.

The honey bee's dance has providentially provided the key to much of its programming. Von Frisch, for example, began his studies of honey bee orientation in earnest when he discovered that foragers would dance on a horizontal surface and, given a view of the sky, would point directly at their goal. By manipulating their view of the sky at will, von Frisch could learn which cues are important to the bees and to some extent how they are used. This ability to orient on horizontal surfaces to visual cues is in fact exactly how the two tropical species communicate, and thus the gravity-referenced dances of our temperate-zone honey

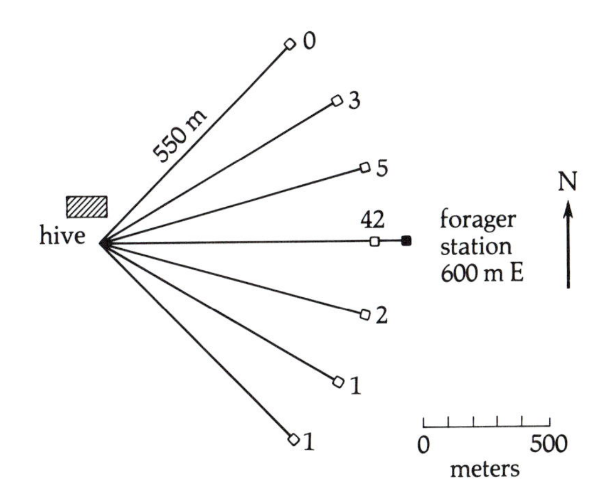

Fig. 24–20 In a series of "fan" experiments, von Frisch trained foragers from a hive (St) to a forager station (F) and then set out scent plates with the food odor in an arc. The numbers represent recruits approaching these plates. Since these visitors were not captured, bees may have been counted more than once.

bees are a secondary adaptation to living in cavities and having to dance in the dark.

Virtually everything we know, too, about the economics of foraging has come from studying the dance. When foragers find a food source, can they determine whether they will turn a profit on it—that is, collect sufficient nectar of a high enough quality to pay for their fuel going out and back, the energy necessary to convert it to honey, and the depreciation on the forager? Using whether or not the forager dances on its return to the hive as a criterion, it becomes clear that the quality of the food (sweetness) must become greater as the forager is forced to fly farther and, hence, to "pay" more (Figs. 24–21 and 24–22).

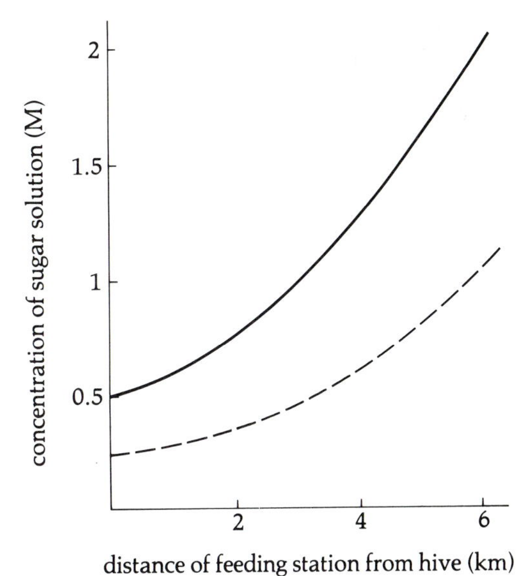

Fig. 24–21 The farther a food source is from the hive, the sweeter it must be to inspire dancing and elicit recruitment. Such judgments, however, depend on the general quality of food available in the environment. Hence on the day described by the lower curve food was scarce and foragers advertised even a weak ¼-M sugar solution near the hive, but were willing to fly 3½ km for a ½-M source. On a day with more concentrated nectar available from natural sources (upper curve) the solutions had to be approximately twice as sweet at each distance to generate dancing.

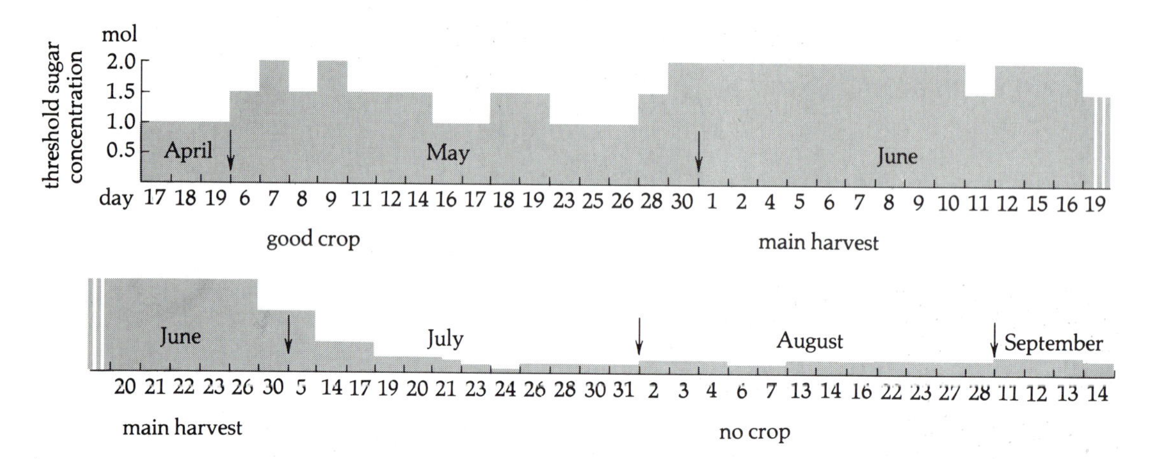

Fig. 24–22 The quality of natural food sources can be monitored by determining the threshold for dancing in the hives of bees foraging at experimental feeding dishes. In late spring and early summer floral competition is so stiff that a 2-M sucrose solution (almost as concentrated a syrup as it is possible to make) is barely sufficient to stimulate dancing. In late summer and early fall, however, bees fall on such hard times that even ⅛-M food, whose sweetness is barely perceptible to honey bees, may elicit frantic dancing.

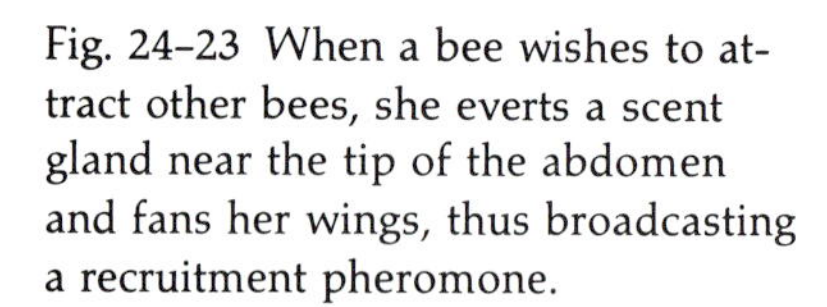

Fig. 24–23 When a bee wishes to attract other bees, she everts a scent gland near the tip of the abdomen and fans her wings, thus broadcasting a recruitment pheromone.

The dances also tell us how bees are programmed to choose a cavity when they swarm. As we saw in Chapter 18, scouts search for and inspect various possibilities, and then return to the swarm and advertise their finds (Fig. 24–23). Other scouts are dispatched by these dancers to examine the possible homesites in question, and dancing scouts will even attend other dances and fly off to look over the competition, changing their votes if the new one seems better. A consensus forms about which cavity is best, and once virtually all the dancing indicates the same site (perhaps days later) the swarm flies off to it and sets up housekeeping (Fig. 24–24).

Finally, the dances provide intriguing clues—as well as questions—about the logic of bee evolution. Why, for example, are

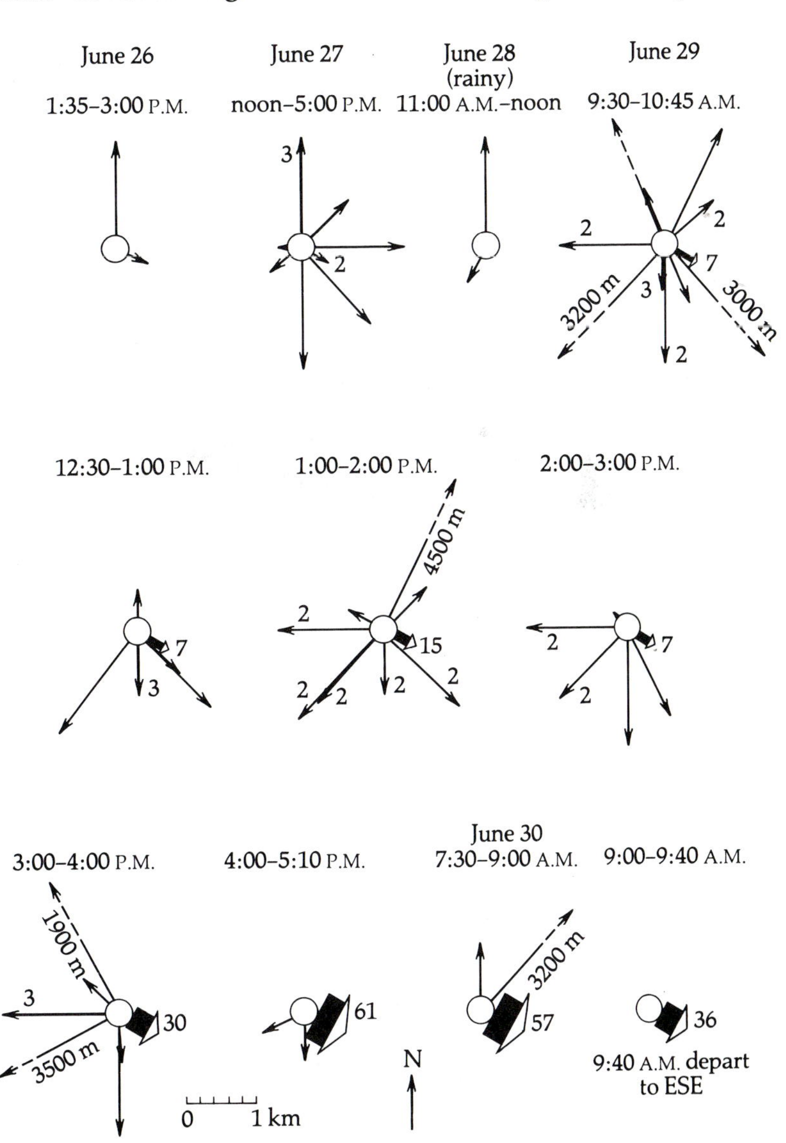

Fig. 24–24 Scout bees' dances on a swarm cluster. The hive swarmed at 1:35 P.M. on June 26. Dances the first three days announced the discovery of eight potential nest sites. (The direction of the site is indicated by the direction of the arrow where north is up, while the distance to the site is represented by the length of the arrow. The width of the arrow and the number near the tip indicate the number of dances to each site.) On the third day some 14 new sites were found, but a nearly unanimous concensus began to form in favor of the cavity 350 m to the southeast. After almost four days of evaluating the market the swarm departed for their new home.

there dialects? Why is it that German bees signal 100 m by two waggles while Italian bees use four and Egyptian bees ten? Why is dancing less accurate for closer targets than for more distant ones? Why do swarms, offered a choice between a nearby cavity and a much more distant one, inevitably choose the farther? Von Frisch, in his autobiography, likened bees to a magic well—the more that is drawn out, the more there is in the well. But, then, this is the perpetual theme of scientific inquiry, and a great comfort to practitioners of science who, if their problems admitted of ultimate solution, would lose the tantalizing opportunity of imposing order on intellectual chaos.

THE DANCE CONTROVERSY: TESTING HYPOTHESES

Every scientist approaching a phenomenon brings with him a collection of personal biases and expectations born of experience and analogy. In the best of cases these provide the basis for intuitive guesses, hypotheses, which provide essential direction for what would otherwise be aimless research. On the other side of the coin these same expectations can be blinding or misleading.

The essential role such expectations play in the vital pursuit that is science is most dramatically illustrated in scientific controversies. One we have already explored is the nature/nurture debate. Another involves the honey bee. In the animal world, the dance-language system of bees is a clear anomaly: there is simply nothing like it. The unprecedented, unaccountable complexity of the system accounts for von Frisch's having studied bees and their dances for thirty-five years before he spotted the distance and direction correlations. On the other hand, as Adrian Wenner and his colleagues began to suspect in the mid-1960s, perhaps the correlations were merely artifacts, phenomena which, though very real in themselves, are secondary, functionless consequences of the evolutionary process, rather than primary agents of cause and effect. Most insects, for instance, will take whatever direction they might be traveling with regard to the sun and turn it into an angle to walk with respect to gravity should the design of an experiment now require them to walk on a vertical surface, but no communication is involved. Doubters saw the bees' anomalous "language" as a product of romantic and anthropomorphic fancy. By analogy with other social insects, honey bees ought to be able to forage using odors alone; and upon close examination no single experiment by von Frisch really attacks this alternative directly. Subjective experience had convinced von Frisch that odors had only unimportant local influences; and besides, the dance correlations were so remarkable, so obviously adaptive a piece of programming, that they must, he thought, have purpose: that purpose he theorized ought to be communication.

As we have seen, however, evolutionary guesses are notoriously dangerous, and skeptics of the dance hypothesis were able to show that under certain experimental conditions honey bees clearly used odors to locate food. Many people see insects as simple creatures, programmed to do things in just one way: they must either use odor or language to recruit each other to food sources, and if they use odor they must lack language. But the lesson of backup programming in navigation should alert us to the difficulties with this line of reasoning. Clearly the only way to settle a dispute over the bees' recruitment system would be to try to pit odor directly against language in some unambiguous experiment.

The trick by which this was managed takes advantage of two quirks in bee behavior. The first is that even on a vertical comb dancers will orient their dances to the sun (or a suitably bright light) if it is visible, or to gravity if it is not. Of course, in such cases attenders interpret the dance with regard to the same cue and so no confusion results. The second is that a bee's sensitivity to light is controlled by the brightness of its surroundings as perceived by its three ocelli, which are small, simple eyes on the top of the bee's head between the compound eyes. If we paint them over, a bee becomes about eight times less sensitive to light. The trick, then, is to paint over the ocelli of foragers and then expose the vertical hive to an artificial sun, carefully adjusted to be just invisible to the ocelli-blackened bees but perfectly obvious to unpainted ones. The foragers, then, dance to gravity. The recruits (whose ocelli are normal) will, if they use the dance correlations at all, interpret the dance with respect to the light. The two different frames of reference we have engendered ought to create a miniature Tower of Babel inside the orderly hive, and result in sending the recruits, who now speak a language quite different from that of their informants, to some spot well away from the feeder actually being visited by the foragers. In fact, we should be able to predict just where the recruits should turn up as a result of their misinterpretation. Indeed this is precisely what happens, and by adjusting the angle of the light we can make the unwitting foragers send their recruits virtually anywhere we wish. Hence, since the foragers can be made to "lie," the bees must have a language.

How then did the dance-language critics come by their results? Apparently odor acts as a *backup system* for the dance, and so when the experiment to test the role of the dance by eliminating it was run, the bees still did perfectly well, just as bees which have been denied the sun will substitute polarized light and continue to orient almost as accurately. The bees tricked these researchers, just as they had von Frisch countless times before. All of which underscores a point made earlier: science must test its hypotheses against all alternative explanations. Only in this way can we ensure that our results are likely to reflect the closest approach to truth we can make, and that neither our intuition, nor the cleverness of our subjects, is misleading us.

SUMMARY

Honey bees live a well-ordered, highly social life which depends on elaborate systems for communication and division of labor. Except during mating and occasional episodes of swarming, the queen is an egg-laying machine. Drones exist only to spread the colony's genes by mating with new queens. The workers do everything else, passing sequentially through a series of tasks, first cleaning the hive, then nursing the larvae and queen, building cells, guarding the hive, and finally foraging for food. Although by no means the most complex, the most remarkable piece of honey bee behavior is the dance language by which returning foragers specify the location of food. The dance has proved to hold the key to understanding much of the bees' programming for navigation, economics, and nest choice. Because of the dance, their seeming domestication, and their willingness to live in glass houses, we know more about bees than about any other multicellular species on the planet.

STUDY QUESTIONS

1. The "father of sociobiology," E. O. Wilson, says in his book *The Insect Societies* that the rapid development of the honey bee queen (two weeks from egg to adult vs. nearly three weeks for workers) is very odd. Do you find it so?
2. Persuasive evolutionary arguments demand that a diploid organism must invest equally in male and female offspring. In honey bees, the investment ought to be 60 percent in females and 40 percent in males. Is it?
3. Are honey bees *r*- or *K*-selected?
4. How could you test the clumped-sperm hypothesis under natural conditions?
5. Why should a swarm prefer the more distant of two cavities? Recent tests suggest that different races have different cavity size and distance preferences. Italian bees, for example, prefer sites nearer and smaller than do German bees. Can you formulate a hypothesis which explains this, as well as racial differences in dialect?

FURTHER READING

Esch, H. "Evolution of Bee Language." *Scientific American* 216, no. 4 (1967): 96–105.

Gould, J. L. "The Honey Bee Dance-Language Controversy." *Quarterly Review of Biology* 51 (1976): 211–44.

Heinrich, B. "Energetics of Bumble Bees." *Scientific American* 228, no. 4 (1973): 96–102.

Morse, R. A. "Environmental Control in a Beehive." *Scientific American* 226, no. 4 (1972): 92–98.

von Frisch, K. "Dialects in the Language of the Bees." *Scientific American* 207, no. 2 (1962): 78–87.

———. "Decoding the Language of the Bees." *Science* 185 (1974): 663–68.

Wenner, A. M. "Sound Communication in Honey Bees." *Scientific American* 210, no. 4 (1964): 116–24.

CHAPTER 25

Ants

There are more than twelve thousand varieties of ants in the world. According to E. O. Wilson, there are more species of ants in a square kilometer of Brazilian rain forest than there are species of primates in the entire world. The proliferation of ant species has ensured that for virtually every available habitat, there is a type of ant peculiarly adapted, both in behavior and morphology, to fit it. Yet omnipresent as they are, stunning in their variety and idiosyncrasy, no single species of these "wingless wasps" is understood in any significant depth: there is no honey bee of ants. But this incredible variety provides limitless opportunities for discovering not only *what* evolution can do, but how.

THE BASIC ANT

Ants, even the most bizarre, are all elaborations on the basic life strategy typified by the plain-vanilla pavement ants we see every day on sidewalks or at picnics. For the most part, ants are denizens of dark, underground nests, and are constrained to move about on foot through massive, dimly lit forests of grass and leaves, with rarely a view of anything more than a few millimeters away. Unlike bees, they are more creatures of smell and touch than of vision. Indeed many ants are blind, including some species which spend their whole lives above ground. Concentration on olfactory communication has turned ants into walking pheromone factories (Fig. 25–1), and allows them to dispense any of a wide variety of "messages." For example, they typically recruit each other by means of odor trails deposited on the ground as they return to the nest from a food source (Fig. 25–2). Alarm, too, is usually communicated by pheromones, and because a concentration gradient is created centered on the alerted individuals, directional information is also present. Pheromones may also play a small part in concocting the odor unique to each ant colony which serves to identify the nest and its members. Other odors from the specialized pheromone glands of ants leave more subtle messages which we have yet to decode.

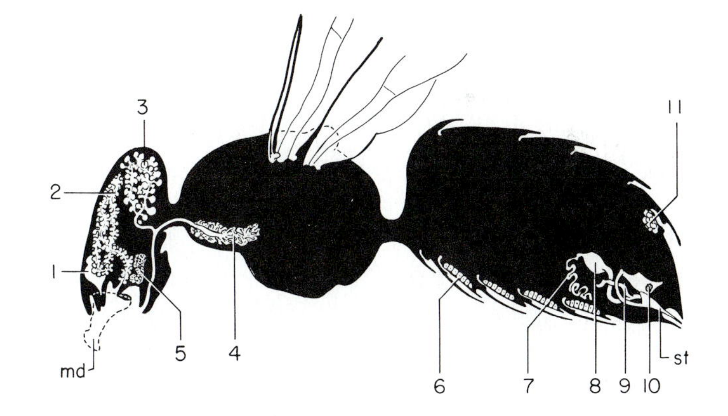

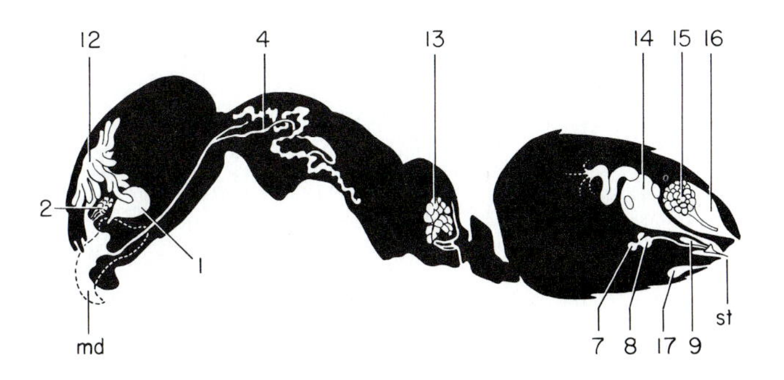

Fig. 25–1 The chemical language of social insects is based on an array of specialized glands. A. Eleven exocrine glands have been identified in the bodies of worker honey bees. (A twelfth, found in the feet, is used to leave trails.) The mandibular glands (1), for example, produce royal jelly, the substance which determines the caste of female larvae. The other glands are the (2) hypopharyngeal, (3) head labial, (4) thoracic labial, (5) hypostomal, (6) wax, (7) poison, (8) poison vesicle, (9) Dufour's, (10) Koschnikov's, and (11) Nasanov's. "St" is the sting, and "md" are the mandibles. B. Worker ants share many glands with honey bees (indicated by the same numbers), but have in addition the (12) postpharyngeal, (13) metapleural, (14) hind gut, (15) anal, (16) anal reservoir, and (17) Pavan's gland. Other glands doubtless exist whose anatomy and function are yet to be discovered.

A

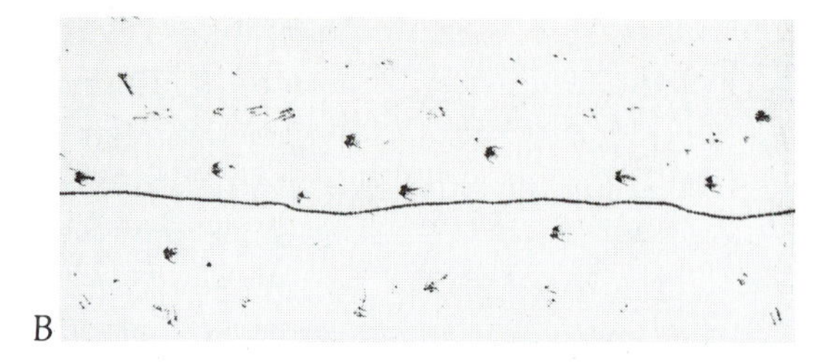

B

Fig. 25-2 Most species of ants leave a pheromone trail from food sources to their nests. A. A fire ant lays such trails by dragging its sting along the ground. B. The trail (here seen scratched on a smoked plate) guides recruits to and from the food.

One of the best tools, however, for "getting at" the olfactory and tactile language of ants is to examine how the many parasites and harmless symbionts of ants manage to survive in their midst. Ants not only guard their nests with great care, but they also have a tendency to carry off and consume anything that moves. Yet they tolerate—even to the point of carrying them gently into the nest—a variety of arthropods ranging from mites and beetles to wasps, flightless flies, and millipedes (Fig. 25-3). The key to the success of this motley collection of parasites is either their successful mimicking of the odors and behavior

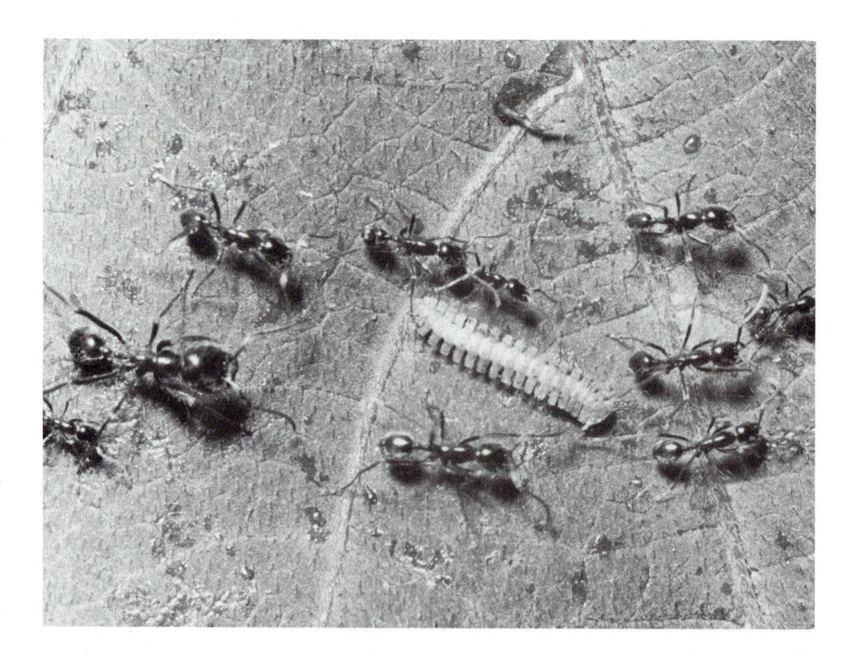

Fig. 25-3 A variety of symbionts live peaceably with ants. Here a millipede follows the pheromone trail of its host, a species of army ant. These millipedes may even be picked up and carried by the ants. To all appearances the millipedes do the ants no harm, and may even aid in keeping nest sites clear of debris.

appropriate to larvae or adults (Fig. 25–4), or the production of "appeasement" substances whose exact workings on the behavior patterns of adults are not understood. As in other such cases, comparison of model and mimic reveals which features are salient to the ants without, however, revealing exactly how or why.

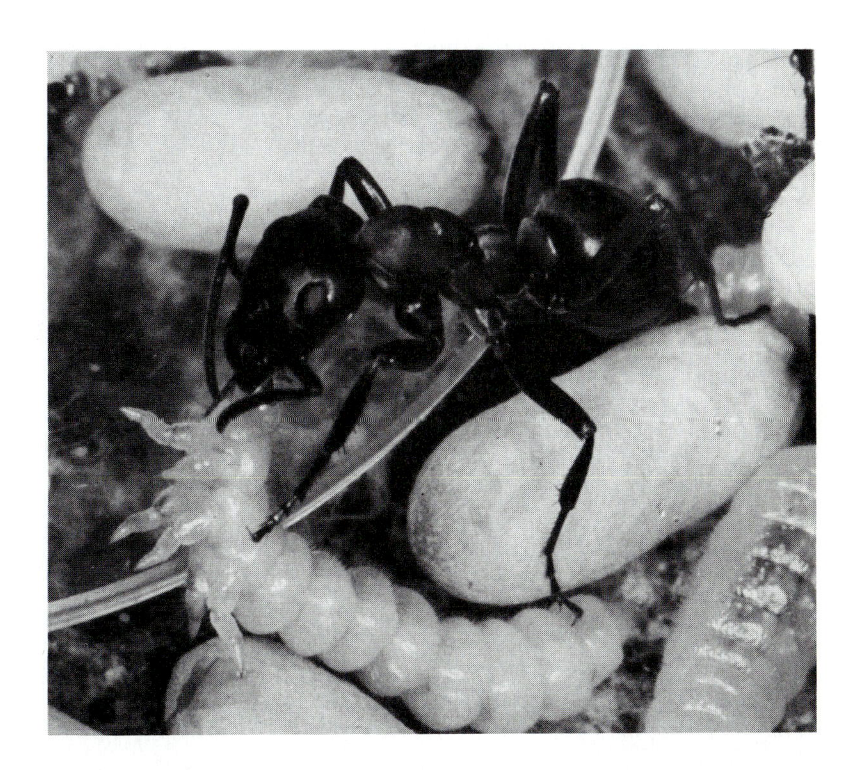

Fig. 25–4 Some of the intruders in ant colonies are less innocuous than the millipedes. Here a beetle larva, equipped with glands which mimic the brood odor of ants, demands and gets a meal from a worker. When the beetle larva is not being fed by adult ants it dines on nearby ant larvae.

The life history of most ants is something like an extended bumble bee cycle. In the late summer colonies produce winged reproductives—males and queens—which congregate to mate over prominent landmarks. A queen sheds her wings after mating and digs a hole in the ground in which she spends the winter. When spring comes, the queen lays a round of eggs which she may feed from her own fat stores and degenerating flight muscles without ever leaving her sealed cell. In many cases these developing daughters do not reach the adult stage until the following spring, thus extending the founding queen's dark, solitary vigil to eighteen months or more. Once the workers emerge, however, the colony begins to grow as the young workers gather food and extend the nest. Workers of many species of ants also produce infertile trophic eggs, which they feed both to the queen and to developing larvae. In other species, the larvae feed directly on captured prey (Fig. 25-5), on other food left near them, or indeed on each other. Unlike bees, which reproduce annually, an ant colony may take five or ten years before reaching the size necessary to begin indulging in yearly reproduction. Like the annual species of bees and wasps, the ants' usual pattern is to produce a batch of reproductives which go off to mate, after which each queen begins her own nest. Some species of ants, on the other hand, undergo a terrestrial version of the honey bees' colony fission or swarming.

Fig. 25-5 Primitive ants do not feed their larvae directly, but rather leave prey nearby for the larvae to snack on *(upper left and lower right)*.

Fig. 25-6 Caste dimorphism is often extreme in ants. Here the enormous queen of a fungus ant colony stands in her garden, surrounded by minors and medias.

Many ant species produce workers of varying sizes (Fig. 25-6). Sometimes the workers seem randomly distributed over the size range, but most often there are two or three obvious size classes typically referred to as majors, medias, and minors. Whereas many social bees and monomorphic ants (ants with workers of only one size) go through a series of age-dependent

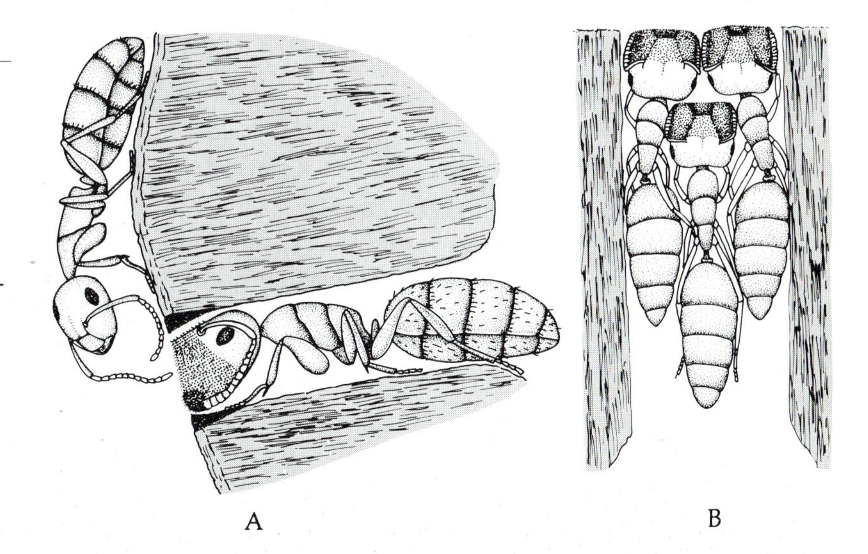

Fig. 25-7 The largest worker caste is usually specialized for colony defense. In this species of European ant the soldier caste is morphologically adapted to act as living barricades.

tasks, the function of the striking polymorphism of some ants represents more efficient behavioral specialization. For example, the smallest may tend eggs and larvae exclusively, while the middle-sized or "medias" gather food and the largest guard the nest. Beyond just size, other morphological and behavioral adaptations specific to a worker caste may further enhance its efficiency (Fig. 25-7).

GARDENERS AND LIVESTOCK MANAGERS

The enormous variety of ant lifestyles illustrates the range of possibilities open to exploitation by ants and the levels of interdependence which can ultimately evolve. One of the most commonly used examples of this mutual reliance is the relationship between ants and aphids. As described in Chapter 21, aphids suck the sap of plants and excrete a carbohydrate-rich substance known as honeydew. Many species of ants collect this aphid waste, and the aphids benefit from a measure of protection that the hungry ants incidentally supply. This simple description fails to make clear, however, the extent of the behavioral and morphological accommodations which each species makes to the other. Aphids produce such enormous quantities of honeydew that they must dispose of the sticky substance or be trapped by it. Some aphids flick away the drops and others shoot them from their abdomens. In the presence of ants, however, some aphids retain the excess until an ant "milks" them, stroking the aphids until a drop oozes out. Some aphids have even developed a bowl-like set of hairs as long as their bodies to hold large drops of the honeydew until the ants can empty them.

The protection supplied by ants is apparently real. "Tended" aphids have larger, more stable populations, they process plant sap faster, and in some species they even lack their normal if rather ineffectual defenses. In the face of danger, some species will scramble onto the backs of the ants for protection (Fig. 25–8), but one of the most extreme cases of dependence is found in corn-root aphids. Cornfield ants bring these aphids into their nests for the winter where the aphids lay their eggs among those of the ants. The ants care for the newly hatched aphids, placing them on the roots of weeds and tending them. Later, they transfer the aphids to corn roots. In some species of ant each queen carries a dowry of aphids with her on her mating flight to start her own domesticated aphid stock wherever she establishes her new nest.

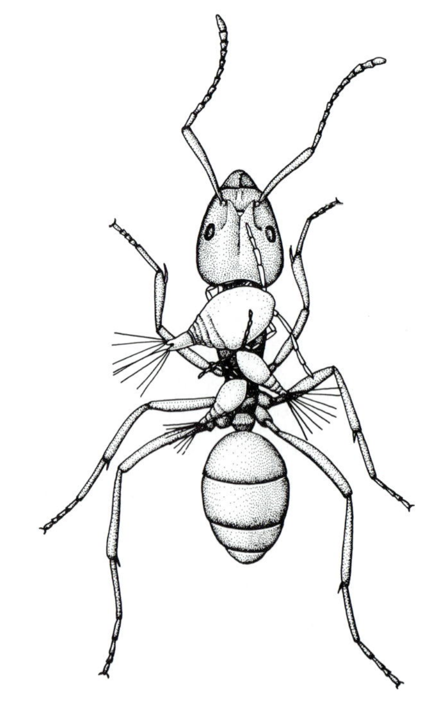

Fig. 25–8. This species of aphid-like Coccida escape from danger by climbing onto the backs of the ants which tend them. Long hairs on the tail hold the aphids' sugary secretions in a discrete drop until an ant arrives to collect them.

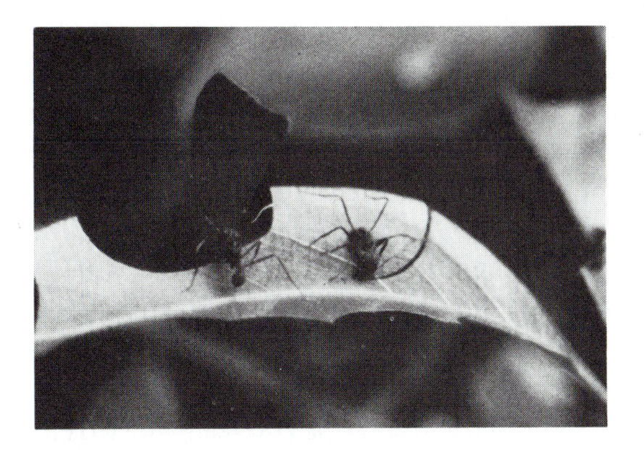

Fig. 25-9 Leaf-cutter ants slice semicircular pieces of leaves and, marching in long columns, carry them back to the nest. There the leaves are used to culture fungus.

Another class of astonishingly well-adapted, specialized ants are the leaf cutters, which typically cut out leaf fragments to carry into their nests (Fig. 25-9). In this species, the medias gather the leaves but the minors "ride shotgun" on the fragments, defending the hardworking medias from a common species of parasitic fly. Inside the nest the ants chew around the edges of the leaf bits, fertilize them with an anal secretion, and "plant" them. Other, older leaves in the vicinity which are already covered with growing fungi are used as cultures. Small patches of these fungi are uprooted and grafted onto the new leaves where they thrive and multiply. Needless to say, this fungus is the major source of food for leaf-cutting ants, and the fungus in turn probably depends on the ants for its survival. In addition to sporophores, this fungus produces nutritious fruits ("gongylidia") which the ants consume. They tend their fungus gardens by weeding out competing species of fungi and selectively fertilizing the one they want (Fig. 25-6). Queens of this species too take a fungal hope chest with them as they fly out to mate, and they set up new gardens when they excavate their initial chambers.

By all standards the fungus-growing strategy is enormously successful. When analyzed, a 6½-year-old nest was found to have required an excavation of some 40,000 kg of soil, and calculations suggest that the colony had consumed nearly 6000 kg of leaves. Typical ultimate population sizes are on the order of two million, and nests often extend 5 m into the earth. A colony this size can send thousands or even tens of thousands of reproductives out each year. The key to this success is the ability to get at and make use of the food value of vegetation. Trees, bushes, and grass have gone to great lengths to make cellulose a nearly indigestible substance. A variety of animals eat leaves and grass, but most attack cellulose by brute force, relying on massive forces of enzymes and repeated bouts of mastication to achieve the final breakdown. Leaf-cutter ants have hit on the canny trick of letting the fungus consume the cellulose, and then eating the fungus. As we will see in the next chapter the other major trick, discovered by termites and ungulates, is very similar: let bacteria or protozoa do your digestion, and then eat them.

Not all ants are constrained to live in the ground. Weaver ants, for example, have evolved to construct their homes in trees. Although many ants will painstakingly excavate rotting tree trunks, weaver ants build their homes by remodeling the leaves. This they do cooperatively, folding over each leaf and

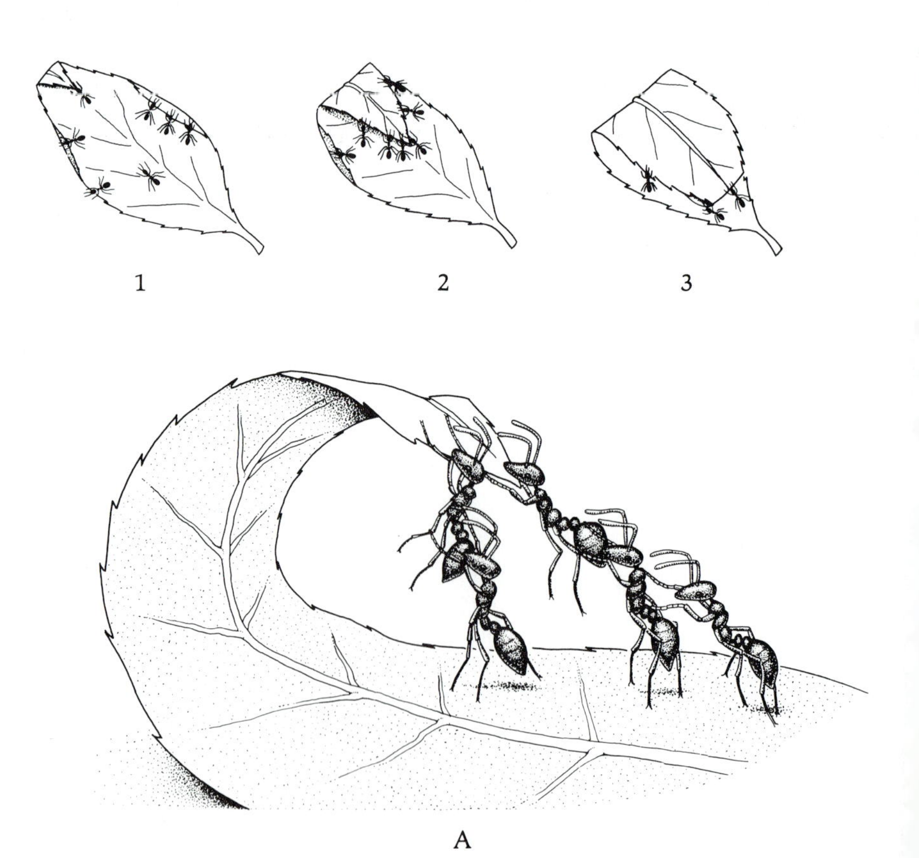

then stretching its tip to its base. The resulting one-leaf compartments are then woven together to form long enclosed runways and tentlike galleries. The way the ants meld the leaves into a nest is remarkable (Fig. 25-10). While one group of workers holds two pieces together, other workers carry larvae back and forth across the gap to form a joint. The larvae, on signal, produce the silk normally used in making pupal cocoons, and in this way weave the elaborate chambers and passageways which form the nests of the weaver ants.

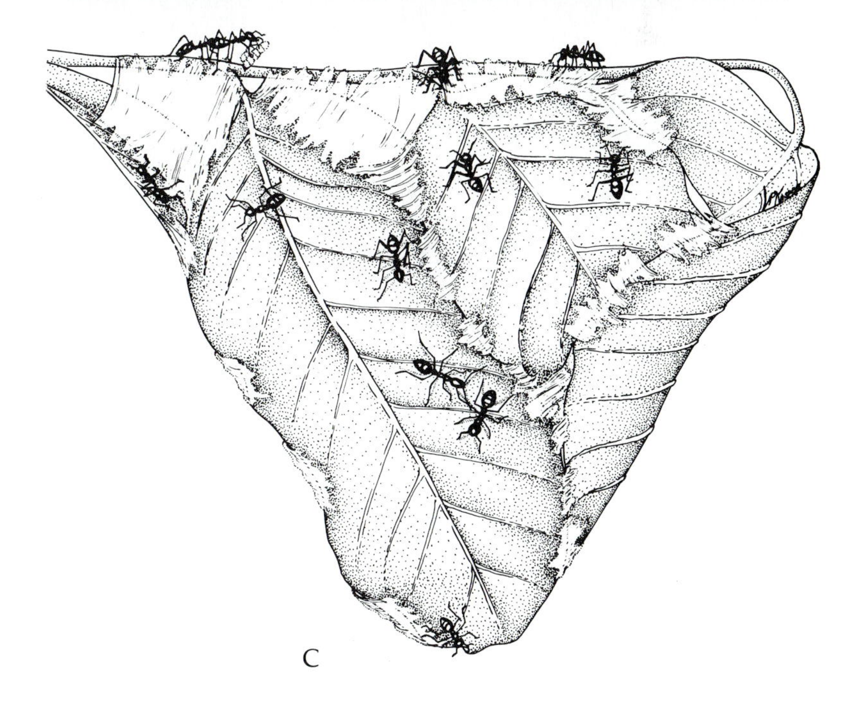

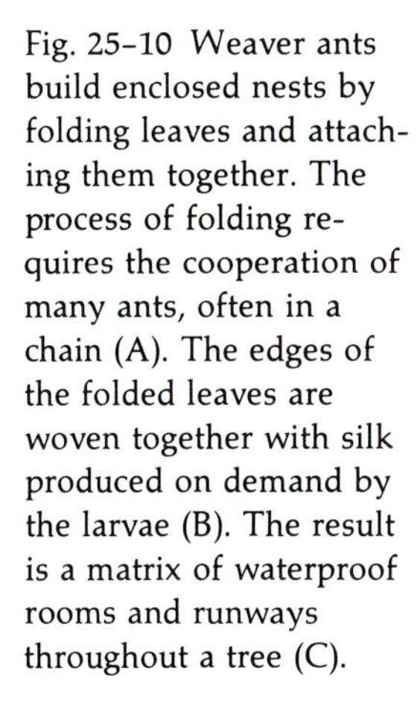

Fig. 25-10 Weaver ants build enclosed nests by folding leaves and attaching them together. The process of folding requires the cooperation of many ants, often in a chain (A). The edges of the folded leaves are woven together with silk produced on demand by the larvae (B). The result is a matrix of waterproof rooms and runways throughout a tree (C).

ARMY ANTS

Another group which often nests above ground is the legendary army ant, whose fame comes from the massive and devastating raids carried out during its nomadic phases. Despite their awesome efficiency, the raids are quite disorganized. As day breaks the swarm of perhaps a million or so ants which have linked themselves together into a cluster around the queen and her young for the night begins to dissolve (Fig. 25-11). Depending on the species, and there are about 150, raiding columns or fronts form in what appears to be a haphazard fashion. Their initial direction seems to be chosen along the path of least resistance. Although individual ants can move about 2-4 cm/sec (70-150 m/hr), the column or raiding front progresses at only 20 m/hr at best. No leaders or scouts are apparent. Slow

Fig. 25-11 A. In the nomadic phase, army ants form a nocturnal bivouac cluster which contains the queen, brood, and spoils of the day's raid. This colony of *Eciton hamatum* is about 40 cm across. B. The cluster consists of ants hanging onto each other. The cluster is formed as ants run down from above, hanging on head-downward as they reach the bottom.

and undirected as they seem, these essentially sightless ants capture and kill virtually everything they can, even "treeing" and then capturing flightless victims or invading insect nests and taking those that refuse or are unable to flee (Fig. 25-12). A few species of army ants are relatively indiscriminate in their

Fig. 25–12 Raiders from an army ant colony have discovered a wasps' nest. The adult wasps fled as the ants began to ransack the nest for its defenseless larvae and pupae.

raids, but most species attack primarily other ant colonies. Eggs, larvae, pupae, and slain adults are carried back and consumed. Army ant colonies are said to avoid each other, however, and prey only on nest-building species. The booty is systematically and cooperatively carried back to caches for storage (Fig. 25–13). As night falls, the raiders begin ferrying their own larvae and the day's spoils along the raiding trail to a new site, where the colony dines and camps for the night (Fig. 25–14).

A

B

Fig. 25–13 Automatic cooperation is essential to the raiding strategy of army ants. A. Five workers march in lockstep carrying the tail of a scorpion back to their booty cache. B. A living bridge of workers allows other raiders to cross a gap in one of their trails.

Fig. 25–14 Each night during its nomadic phase the colony moves to a new bivouac, usually at the site of one of the day's food caches. A. During these emigrations the soldiers stand flanking the column, apparently acting as guards. Note their specialized sabre-like manidibles. Workers carry their larvae to the new bivouac. B. In this side view, workers can be seen carrying larvae slung underneath their bodies. The ants at the top left are transporting a wasp larva captured earlier that day.

Army ants spend their lives alternating between two completely different behavioral states: a nomadic phase characterized by these notorious raids and nightly migrations, and a relatively quiet "statary phase" (Fig. 25–15). The controlling element in the army ant colony appears to be the queen and her endogenous rhythm of egg production. In a manner analogous to the menstrual cycling of mammals, she undergoes a 35-day (±2) ovarian cycle which seems unrelated to external cues. At what we will arbitrarily define as the beginning of a cycle, the queen begins to swell with eggs (Fig. 25–16). Simultaneously, the nightly ritual ends and the three-week statary phase begins. After about a week, the queen begins to lay an enormous batch of eggs at a rate of about 10,000 a day for seven to fourteen days. The eggs hatch and minute growing larvae are fed with the spoils of the small but ever-larger daily raids that have again resumed. In the meantime the pupae from the previous cycle are about to become young adults (called "callows"). When this

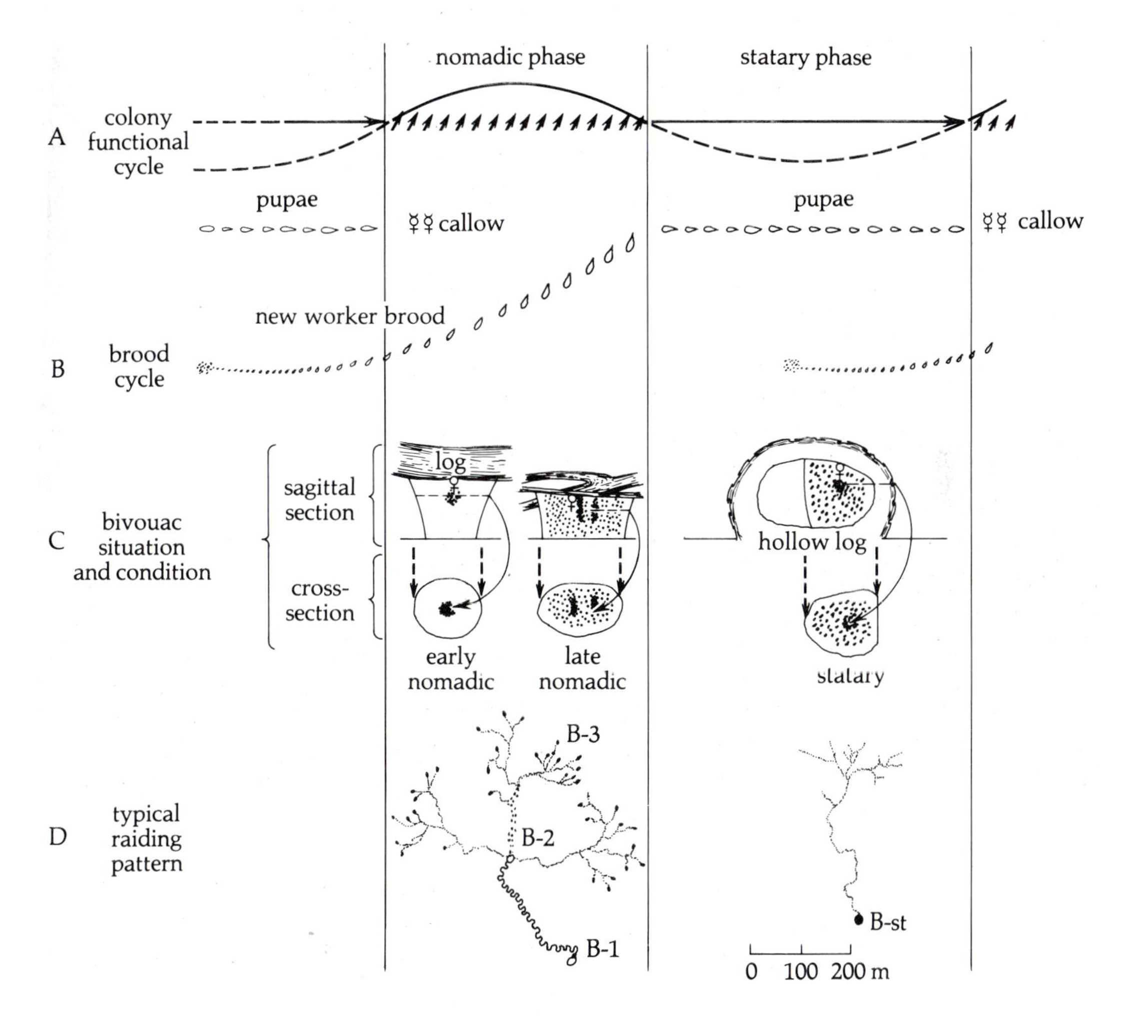

Fig. 25-15 The endogenous cycle of the army ant *Eciton hamatum* is illustrated here as it alternates between nomadic and statary phases. The statary half of the cycle is characterized by a stable bivouac (B-st) in a protected spot while the colony's 100,000–200,000 pupae develop. During this period the queen lays a similar number of eggs and the resulting larvae are fed by small raids. When the pupae hatch into callow workers the nomadic phase begins with its massive raids and nightly emigrations with the growing larvae to new bivouac sites (B-1, B-2, and B-3).

Fig. 25-16 At the beginning of the statary phase the army ant queen swells with eggs, and then lays 100,000–200,000 within two weeks.

event occurs, the colony that is perhaps a sudden 25 percent larger again initiates the nomadic phase with its large daily raids and nocturnal relocations.

The nomadic phase continues until the fully grown larvae spin cocoons (i.e., pupate), at which time the colony's food needs drop drastically. The queen begins once more to swell with eggs and become less mobile, and the young pupae undergoing metamorphosis probably become somewhat more sensitive to damage from mechanical disturbance and extremes of temperature. Although any of these reasons would be adequate to explain why the nomadic phase ends, the mechanism or trigger which shuts off massive raiding and nightly emigrations in favor of hearth and home is not known. The disappearance of some larval pheromone might be the safest guess since the odors of larvae appear to be crucial releasers in many social insects. Honey bee foragers, for example, rarely gather pollen unless there are larvae, the main consumers of this protein source, in the hive. Presumably some olfactory signal from larvae triggers pollen gathering. But honey bees have larvae to feed throughout most of the warm months of the year, and consequently they will gather pollen almost continuously. The many species of ants which produce relatively synchronized broods, however, experience dramatic shifts in food needs as eggs hatch in droves into hungry larvae or as larvae pupate. In this alternation between nomadic and statary phases, army ants epitomize the potential of self-stabilizing behavioral cycling.

The other major behavioral cycle in army ants is the annual ritual of colony division. It is begun during one statary egg-laying cycle when the queen lays an unusual round of eggs. Many either fail to develop, are unfertilized, or are prevented from developing and are therefore haploid males. From this set of eggs two to four thousand males, half a dozen queens, and no workers are produced. This brood almost always appears in the first cycle after the beginning of the dry season, so it is likely that an environmental cue is the switch. By the time the sexual brood have pupated and the next statary phase has begun, the ant swarm has divided roughly into halves, one with the brood—the developing reproductive pupae—and one with the queen. When the sexual brood, two to four thousand strong, emerge, the colony enters the nomadic phase typically induced by the emergence of three hundred thousand or so workers. There is, however, one dramatic behavioral difference: the two halves of the colony form separate raiding groups, and when the time for the first emigration comes the two groups go their own ways. The half with the new queens disposes of all but one or two of the new queens at the bivouac, and the males, which may go with either group, leave over the course of the next few days. Exactly who then mates with the virgin queen—a brother or a male sent out by another colony—is not known, but the general synchronization of sexual broods with the onset of the dry season would certainly facilitate intercolony gene exchange.

SLAVE-MAKING

Odors play an important role in the life of the ant, dividing not only species from species but colony from colony. The use of nest or colony odor as the means of identifying group members, joined to the plunder-and-pillage tactics of many species of ants, probably explains the origin of slave-making, that unique line of ant evolution. Slave-making ants steal pupae from the nests of other colonies (of their own or other species) and, rather than eat them as the army ants would do, allow them to hatch. Accepting the foster colony's odor as their own, the new ants go about their well-programmed tasks oblivious to the fact that they are unrelated to the queen. And having absorbed the colony odor themselves as pupae or callow adults, they are unquestioningly accepted into the society of their kidnappers. All that seems to be required for selection to take over and turn chance into necessity is that the programming of the new ants be appropriate to the needs and the lifestyle of their captors. Two lines of development could take off from this point: the ants might plunder their own species just as honey bees rob other hives of their honey, or they might rob others. The first case is a simple way of increasing both antpower and food supply, but has a limited usefulness. If all the colonies of a species came to depend heavily on other colonies of the same species for their continued existence, the system would collapse in internecine warfare. Hence, a species which captures and enslaves others of its own species is a casual, opportunistic, nondependent slave-maker. The slaves have nothing to offer that the "masters" cannot provide for themselves.

The other strategy is far more interesting, and has intriguing evolutionary possibilities. Here, whole specialized morphologies and behavioral repertoires crafted painstakingly by evolution for the use of one species become subservient to that of another. If the gain outweighs the cost of gathering the pupae, the system can begin rapid selection which should lead ultimately to three alterations in the behavior of the robbers: (1) species-specific selection of the optimal slaves; (2) nondestructive raids which merely strip the besieged colony of pupae, leaving it to produce more for future "harvesting"; and (3) extreme division of labor in the slave-making colony: eventually the "masters" should come to depend on the slaves for food gathering, nest construction, brood rearing—every necessity of life except, perhaps, the raiding itself.

Our hypothetical reconstruction of the development of slave-making is mirrored in nature. The casual, opportunistic level may be seen in many colonies of relatively indiscriminant raiders, since some but not all of these nests contain adults from other colonies which have been ransacked for eggs, larvae, pupae, and food stores. These slaves may be of the same species or of different ones, but only obvious morphological distinctions between the species allow us to detect the phenomenon unless we can actually observe the raids.

At a more advanced (i.e., more thoroughly evolved) level, some species, although perfectly competent to manage without slaves, carry out organized raids on any nearby colonies of chosen slave species. Slaves for these species are a luxury rather than a necessity. The specilizations which they have developed to facilitate their entry into the nests of their specific "hosts" are highly developed and formidable. In one case, for instance, these invader ants have greatly enlarged glands from which they spray what E. O. Wilson calls a "propaganda substance," which apparently mimics the specific flight-alarm pheromone of the host, and serves to clear out the nest defenders before the attack. Later the adults of the raided colony return and begin dutifully to rear another round of larvae and pupae which the slave-makers will doubtless harvest.

Perhaps the high point, if it can be so called, of slave-making comes with species such as the Amazon ant. Here the adult slave-makers are almost helplessly indigent, unable to dig or to rear brood, but as their name implies, exquisitely adapted to reproduce and to raid (Fig. 25–17). Slaves greatly outnumber masters at home, and the species raided may be very specific. In

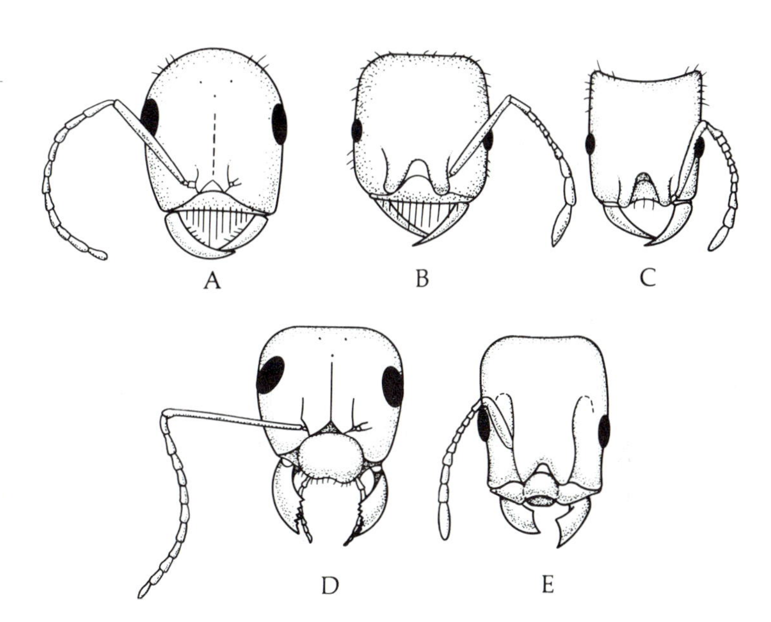

Fig. 25–17 Although all five species pictured here are slave-makers, only one (D) is capable of supporting itself: it is equipped with standard ant mandibles complete with teeth. The other slave-makers, however, are specialized either for clipping off their opponents' appendages (E) or piercing the exoskeletons of defenders (A–C), an operation normally directed at the head. The ant pictured in (A) is an Amazon ant, named for the warlike nature of the female soldiers.

fact some slave-makers take advantage of interspecific polymorphism by gathering an efficient brood-rearing species from one nest and good foragers from another.

The nearly complete dependence of many ant species on the slaves they draft into their service raises the immediate question of how slave-making queens can found new colonies in the first place. Two possibilities come to mind: the queen could be endowed with behavior patterns and morphology for raising a round of brood which, although necessarily few in number, could then set about raiding nearby slave species; or the queens could themselves attempt to take over established slave colonies. As is the usual case with ants, one species or another employs practically any system we can imagine. There are even otherwise non-slave-making species in which an enterprising queen takes over a full-sized colony and lets the slaves raise her own brood. Once the slaves have died off the queen's offspring take over, but do not engage in further slave-making. Needless to say, wherever selection favors rapid colony growth, this species can step in.

The basic task for most dependent or "obligatory" slave-making queens, though, is to penetrate the selected host nest, kill its queen, and gain control. Queens of some species edge in tentatively, hide out inside for a few days (probably picking up the colony's odor), and then attack and kill the resident queen.

Workers recognize their queen by the pheromones she employs to keep them under her control, and the usurping queen must now put off her olfactory disguise as a worker and begin producing the queen odors herself.

Other dependent slave-makers use less subtle tactics. Some queens, for instance, rush straight in and begin killing workers by the dozens until only the brood are left alive. They must then guard the pupae themselves until they emerge and unwittingly begin serving their slavedriving stepmother. In a third strategy, new slave-making queens go with raiding parties from the old nest and remain behind in the besieged colony with a few worker sisters. If the queen's species is an out-and-out slave-making one some of the queen's offspring brought up by the host ants promptly set about replenishing the worker supply from other colonies. For modest slave users, however, the new workers merely take over from the slaves.

Slave-making is a strange form of parasitism, highly evolved and complex in its workings and in its implications. It would be surprising indeed if evolution had not managed to find a short-cut to total parasitism by way of slavery. The need for costly, time-consuming raids would be eliminated if the host queen were suffered to remain more or less as a hostage, and so to produce new slaves at a steady pace. The overmastering colony would live on as guests, off the fat of the land. Indeed, such feudalistic systems do exist, in which the parasitic queen somehow blocks the host queen's production of reproductives and substitutes her own. Workers of the parasitic "nobility" may be absent, or if present, totally incompetent, leaving all the work to the resident serfs.

The ultimate example of this is *Teleutomyrmex*—as its name implies, the "ultimate ant." The parasitic queen actually *rides* the host queen, dispensing her eggs along with her host's, and being fed with her by the host workers (Fig. 25–18). *Teleutomyrmex* eggs hatch into queens and males which mate in the nest. The new queens either fight for a place on the host queen—there is room for several—or depart in search of another host colony. The rate of new queen production must be fantastic since Wilson reports that each parasitic queen produces eggs at the rate of two a minute. The extreme inbreeding necessitated by this parasite's lifestyle has, however, taken its toll, providing us with a glimpse into the science-fiction world of genetic debility that is the result of parasitism and the soft life: glands are gone, the

Fig. 25–18 Queens of the "ultimate ant," *Teleutomyrmex,* ride on the host queen and are fed by host workers *(lower right)* as they dispense their eggs along with those of the host queen. Their offspring, all reproductives, mate in the colony. The new queens must either fight for a place on the host queen or leave in search of a new colony.

sting is absent, the feeding apparatus is suitable only for the liquid diet of an invalid, pigmentation has been lost, males lack the wings that would be necessary for effective outbreeding, and the brain is degenerate. This total specialist species has reached the ultimate in evolutionary dead ends. So thoroughly has it burned its genetic bridges behind it that it cannot even penetrate nests of the same host species in distant locales. So perfect is its match with one geographic race that the minute difference that a slight spatial separation must have engendered in a few host genes is fatal to the parasites, who have become too specialized to adapt.

One of the most exciting questions for ethologists interested in the behavior of ants concerns the basic genetic elements in the behavioral repertoire of the myriad species. The bizarre examples of behavioral and morphological specialization which have formed the basis of this chapter are all evolutionary adumbrations on the Basic Ant, that subterranean, olfactorally dominated creature whose programming has provided the raw material for behavioral selection. With such a proliferation of species, virtually every stage on the route from the typical to the incredible may be found and studied. Beyond the lessons implied for behavioral ecologists by this heroic evolutionary saga, there is a real opportunity to understand how changes in neural mechanisms themselves may account for behavioral transformation. We ought to be able to discover the critical changes in releasers and programming and, ultimately, dissect their component genetic parts. No other system presents us with so bright and accessible a prospect for understanding the real ethological "hows" of behavioral evolution.

SUMMARY

Although ants share a common basic body plan and life history, they have managed, in a sensory world almost exclusively chemical, to invade virtually every habitat and to occupy a multitude of niches. Some, for example, are basically grounded wasps, hunting on an individual basis for insect prey. Others tend aphids or gather seeds or raise fungus. Most live in the ground, but many have evolved to live in trees. One group, the army ants, has species that live exposed on the surface and hunt in massive groups. Finally, one diverse class of ants makes its living by parasitizing or enslaving other ants. All the variety of behavior that is shown by the myriad species of ant illustrates virtually every step in the evolutionary progression from generalized to specialized morphology and behavior down hundreds of separate adaptive paths.

STUDY QUESTIONS

1. Why do you suppose that bees and wasps have maintained their monomorphic jack-of-all-trades strategy? The distinctive worker subcastes of ants seem to be so much more efficient—why didn't evolution get around to inventing them in bees?
2. What could be the trigger for the nomadic phase of army ants? There are several possibilities. What experiments could sort them out?
3. Most species of slave-making and parasitic ants seem relatively closely related to their hosts. Why do you suppose this is?
4. In many species of ants and bees with sterile worker castes, removal or death of the queen will lead to regeneration of the workers' ovaries and production of unfertilized eggs. These eggs give rise to males. In the case of ants in particular, the workers usually cannot raise a new queen and so, although the colony may linger on for a year or so, it is doomed. It may be that the production of worthless males is an artifact of the evolution of haplodiploidy in social Hymenoptera, or it may be a useful feature. Can you think of arguments to support the latter view?

FURTHER READING

Batra, S. W. T., and Batra, L. R. "Fungus Gardens of Insects." *Scientific American* 217, no. 5 (1967): 112–20.

Hölldobler, B. H. "Communication Between Ants and Their Guests." *Scientific American* 224, no. 3 (1971): 86–93.

Hölldobler, B. K. "Communication Between Ants and Their Guests." *Scientific* 6 (1977): 146–54.

Topoff, H. R. "Social Behavior of Army Ants." *Scientific American* 227, no. 5 (1972): 70–79.

Weber, Neal A. "The Attines: The Fungus-Culturing Ants." *American Scientist* 60 (1972): 448–56.

Wilson, E. O. "Slavery in Ants." *Scientific American* 232, no. 6 (1975): 32–36.

CHAPTER 26

African Ungulates

The ultimate goal of behavioral ecology is to forge a link between the "what" and the "why" questions of ethology. In the last chapter we saw in detail how the ants have fitted their behavior to their various niches, and how in the process evolution has generated ever more bizarre specialists. But since all ants share the same basic social system our emphasis was on the various ways behavioral programming manages to fit that system into widely differing niches. In this chapter we will look at behavioral ecology from the other direction: African ungulates all share the same basic niche, and minute variations in that niche create vast differences in the social systems of the animals which live under their aegis. Ideally, we should be able to predict an animal's social system from its ecology, that is, from the exact niche into which it has evolved. Ungulates challenge this approach, since many ecologically similar species of ungulates manage to coexist and yet at the same time to display an enormous range of social behavior.

THE RUMEN

The ungulate group is formally considered to consist of the 125 species of bovids—goats, sheep, cattle, and antelope—although another seventy-five species of hooved mammals such as pigs, elephants, hippos, giraffes, rhinos, horses, camels, and so on are also often included (Table 26–1). Ungulates are basically grass-eating machines which obtain most of their sustenance from

TABLE 26–1 *Mammals: A Listing of Orders*[a]

	Order		
1	Monotremata (egg layers)		[6 species]
2	Marsupialia (marsupials)		[235 species]
3	Insectivora (insect eaters)		[300 species]
4	Dermoptera ("flying" lemurs)		[2 species]
5	Chiroptera (bats)		[790 species]
6	Primates		[190 species]
7	Edentata (sloths, anteaters, armadillos)		[30 species]
8	Pholidota (pangolins)		[7 species]
9	Lagomorpha (pikas, rabbits, hares)		[65 species]
10	Rodentia (rodents)		[1715 species]
11	Cetacea (whales)		[90 species]
12	Carnivora (carnivores)		[250 species]
13	Pinnipedia (seals, sea lions, walruses)		[31 species]
14	Tubulidentata (aadvarks)		[1 species]
15	Proboscidea (elephants)		[2 species]
16	Hyracoidea (hyraxes)		[9 species]
17	Sirenia (manatees)		[4 species]
18	Perissodactyla (odd-toed ungulates)		[17 species]
	Equidae (horses)	[8 species]	
	Tapiridae (tapirs)	[4 species]	
	Rhinocerotidae (rhinoceros)	[5 species]	
19	Artiodactyla (even-toed ungulates)		[205 species]
	Suidae (pigs)	[9 species]	
	Tayassuidae (peccaries)	[2 species]	
	Hippopotamidae (hippos)	[2 species]	
	Camelidae (camels)	[6 species]	
	Tragulidae (mouse deer)	[6 species]	
	Cervidae (deer)	[53 species]	
	Giraffidae (giraffe and okapi)	[2 species]	
	Antilocapridae (pronghorn antelope)	[1 species]	
	Bovidae	[125 species]	
	Goats [10 species]		
	Sheep [7 species]		
	Cattle [16 species]		
	Antelope [92 species]		

[a] Based on E. P. Walker's *Mammals of the World* (Baltimore: The Johns Hopkins University Press, 1964), p. 415.

cellulose, a singularly difficult substance to digest. The group we will focus on, the bovids, all share a special structure known as the rumen, which slowly extracts nourishment from vegetation (Fig. 26–1). It is designed to break down cellulose by releasing the rich cytoplasm from inside the plant cells, thus making available the food value of the cellulose itself.

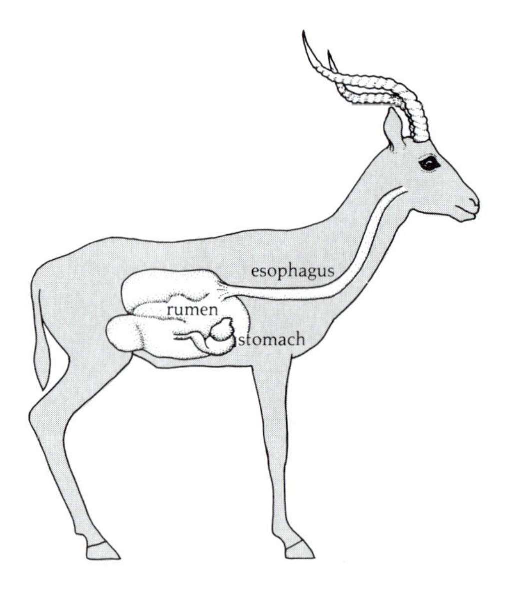

Fig. 26–1 The general body plan of ungulates includes an enormous rumen to ferment the vegetation, followed by a relatively small stomach for digestion.

The rumen, then, is little more than a large antechamber on the way from the mouth to the stomach, which serves three functions. First, it provides a time delay in the system, a holding tank for the time-consuming process of predigestion. Second, it keeps the partially digested vegetation close to the mouth, so that it may be easily regurgitated and chewed more thoroughly at the animal's leisure. Finally, the rumen is used as a fermentation vat in which cellulose is further broken down for the animal by microorganisms. Periodically this vegetable soup flows into the stomach where the animal's own digestive enzymes can readily digest the combination of nutrient-rich cellulose broken down by its own staff of domesticated bugs, and the surplus of microorganisms. The rumen system permits ungulates to exploit the wealth of nourishment offered by vegetation far more efficiently than any other group of vertebrates.

Although all ungulates have similar digestive strategies regardless of their size, they differ in exactly what they eat and where they dine. Generally the largest eat grass on open plains, the smallest live on fruits and leaves in thickets, and those in between browse on a mixed diet of leaves and grass in the transition zone from woods to plains. Though they confront the same general array of predators, the ungulates' behavior in the face of danger is species-specific and ranges from the typical defense of the smallest, which hide from predators, to the strategy of the middle-sized ungulates, which usually run, to the stand-and-fight tactics of the largest. These correlations of body size to defensive maneuvers suggested to Peter Jarman that if social behavior has evolved as a consequence of a species' prey and its predators, a concomitant correlation between social system and body size ought to exist. Surprisingly enough, this seems to be the case.

SOCIAL SYSTEMS

The smallest ungulates are the leaf- and berry-eating duikers and dikdiks (Fig. 26–2). These tiny antelope may weigh as little

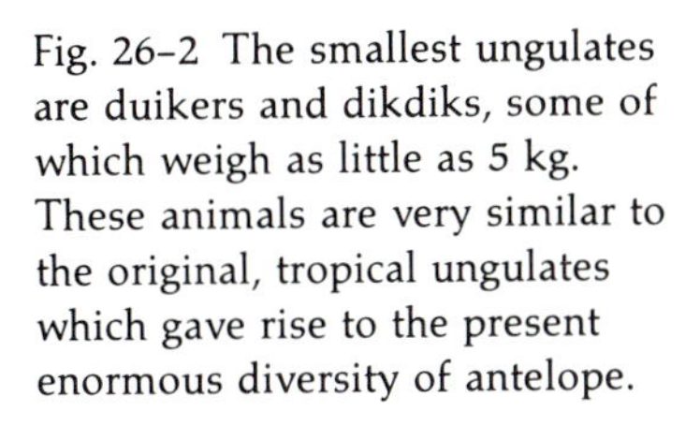

Fig. 26–2 The smallest ungulates are duikers and dikdiks, some of which weigh as little as 5 kg. These animals are very similar to the original, tropical ungulates which gave rise to the present enormous diversity of antelope.

as 5 kg—about the weight of an adult house cat. Like tawny owls, they live on exclusive territories in heavy vegetation and form long-term pair-bonds. Young are raised, then chased out to establish new territories of their own. Although little is known about these miniature creatures, even less can be said with certainty about the somewhat larger antelope (20–80 kg) like the bushbuck. The difficulty in the case of the bushbuck, for example, is that of access: naturally timid and nervous, they live in thickets, leaving them only under cover of night. It is understandably difficult to know what noctural animals do in thickets at night, but the scanty evidence suggests that the males live on vague territories with small groups of females which may or may not be faithful to one male. When bushbucks emerge at night to browse on the bushes and search close to their cover, however, there appears to be a discreet silence in which no claims of territoriality or even of dominance can be observed.

Much more is known about the still-larger ungulates because they are diurnal, and live in relatively open country where they can be more easily observed. Among the smaller of these brush and plains ungulates are impala and wildebeest. Males divide the available pasture into territories which they mark and defend (Fig. 26–3). They display on their territories in an apparent bid to attract the meandering herds of females and young. What really draws these groups into a particular territory, though, is the grass, and any female coming into estrus while browsing on

Fig. 26–3 Territoriality plays a major role in the lives of all but the largest ungulates. Here, two wildebeest males engage in a harmless ritualized duel along their invisible but clearly defined mutual border.

the vegetation on the territory of a resident male mates with him. Hence, a male tries desperately to herd the females, to keep them on his private lawn for as long as possible. However, once they have eaten all the available grass they leave, and the male defends his turf as it regenerates itself.

This territorial matrix system provides a classic example of sexual exclusion. Territories are of varying quality, and the contest to establish control over the richest ground assures that the strongest males have the best chances for mating. Younger or weaker males may either take up undesirable (and, hence, relatively uncontested) territories in the lion-infested bush and high grass (Fig. 26–4), or join bachelor herds in which elaborate dominance hierarchies are established and maintained. In bachelor herds defeated males recuperate and juveniles learn the ropes. Males reaching the top of the herd's hierarchy begin to watch for likely opportunities on the territorial matrix, and if they are successful their once-grand but now-defeated opponents join the bachelor herd near the bottom of the scale.

The wildebeest matrix is by no means fixed in any one place. Once the grass has been eaten the entire group may move on, a new matrix being established at the next place where the grass is acceptable. In fact when the grass is really sparse the herd doesn't stop at all, but eats its way across country at the rate of several miles a day. During such interludes the males set up a traveling matrix, herding nearby females proprietarily and chasing away adolescent male offspring (Fig. 26–5). The bachelor males again form their own herd.

Fig. 26–4 For males unable to win a place in the territorial matrix, two options remain: they may join a bachelor herd and wait for a chance to challenge a weakening territory holder, or take a relatively undesirable (and hence uncontested) spot near the lion-infested tall grass at the periphery.

Fig. 26–5 In their migratory phase wildebeest move by the tens of thousands, eating their way across the plains. Here, too, males may set up territories, albeit mobile ones, and force less fortunate bulls to the periphery where risks are greater.

The very largest bovids like the buffalo are almost completely migratory, and depend on an absolute rather than a spatial dominance hierarchy. The bulls, having sorted the hierarchy out among themselves in some species-specific contest, always give way to the most dominant or "alpha" male. The alpha male exercises his perquisites by following estrus females until they are ready to mate—a behavior known as the "tending bond." Other males stay away, and only when two females are in estrus simultaneously does even the second-ranked "beta" male have a chance. Simultaneous estrus is relatively common in open-country ungulates, however, since the young are slow and awkward, and easy prey for predators. Rather than serving the younger generation up in installments to watchful carnivores, these ungulates prefer, like oak trees and cicadas, to synchronize their production of offspring, literally overwhelming the appetites of their enemies until the surviving young can hold their own. The consequences of this strategy will become clear in the next chapter.

UNGULATE EVOLUTION

Although the social systems of ungulates seem to correlate with ecological, behavioral, and physiological variables (Table 26–2), this type of analysis often fails to sort out cause and effect. For example, does size determine diet or vice versa? What a behavioral ecologist would really like to know is how the social systems evolved, and what were the selective pressures, preadaptations, genetic constraints, and accidents which gave rise to what we see today. In the case of the ungulates this ambitious plan can be achieved in part and the evolutionary drama partially reconstructed. It seems clear that the original ungulates were duiker-like antelope which lived in rain forests and ate low vegetation. Our ancestors were overhead in the trees eating (less efficiently) the higher vegetation.

A habitat change crucial to both groups of herbivores took place about twenty-five million years ago which created three new habitats: dry forest, scrub, and savannah. These new habitats, created by gigantic movements of the continental plates and maintained largely by lightning-generated fires, became available to those tropical forest species variable enough to exploit them. Three major hardships, however, were imposed

TABLE 26–2 *Correlation of Biological, Behavioral, and Physiological Factors in Various Ungulate Social Systems*

Typical weight (kg)	*Usual diet*	*Habitat*	*Social system*	*Examples*
15	leaves and berries	thick vegetation	monogamous pair-bonds on exclusive territories.	duikers
40	leaves, shrub, grass	thickets and forest edge	exclusive male territories; small groups of females.	bushbuck
75	shrub, grass	forest edge and grassland	exclusive male territories; bachelor herds; female herd.	impala, Thompson's gazelle, wildebeest (in stable habitat)
150	grass	grasslands and open savannah	temporary or moving male territories; bachelor herd; female herd.	wildebeest (in savannah habitat)
400	grass	open savannah	male dominance hierarchy; tending bond; mixed herd.	buffalo

Source: P. J. Jarman, "The Social Organisation of Antelope in Relation to Their Ecology," *Behaviour* 48 (1974): 215–67.

on the would-be inhabitants of these new niches: although competition was negligible, the land was dry, the quality of the food was low, and the risk of predation was higher than in the sheltering woods. The rumen system, one of several successful physiological strategies for forest herbivores, represents a chance preadaptation to some of these new constraints. Because it extracts water both from the plant cells themselves and from their breakdown products, it is relatively water efficient. This would have been no particular advantage in its original environment, but was crucial to the exploitation of this new, comparatively arid habitat. Given time, the rumen can extract food value from vegetation more completely than the multiple stomach or "afterburner" strategies of other herbivores, and so with a minimum of physiological change it can cope with the lower quality food that resulted from the habitat change.

As we have seen, then, the basic physiological variable in ungulates is probably body size, and it seems plausible that the function of increasing body size was simply a prerequisite for carrying around a necessarily larger rumen. We can theorize that as the new environments made their appearance, ungulates just large enough to take advantage of the best of a bad lot filled the most desirable niches. Still larger rumens were required to turn a profit on those next, drier habitats with even lower quality vegetation. In fact, a simple "scaling up" of the rumen to take advantage of tougher, drier, less edible food seems to be a plausible explanation for the diet/body-size correlation. Jarman makes the cause-and-effect argument from the other direction: he theorizes that as ungulates got larger, so too did their mouths, and as a consequence they could no longer pick high-quality morsels selectively from the coarser herbage. The increase in roughage would have required a larger rumen.

PROGRAMMING AND BEHAVIORAL ECOLOGY

Although ungulates survived their habitat change, physiological preadaptations and changes are only part of their story. We have seen case after case of animals which, like the "ultimate ant" of Chapter 25, seem to lack behavioral flexibility and are thus caught in evolutionary "traps" or dead ends, and we have invoked the notion of one-way evolution to explain this phenomenon. What behavioral preadaptations might have allowed the primitive, presumably monogamous, territorial duikers (which to all appearance remained unchanged for millions of years in their original habitat) to give rise to the polygamous, often nonterritorial denizens of the open plains? We know that small ungulates forced to live in the open fields of zoos often do *not* set up exclusive territories. The females instead form loose, unfriendly herds, with more or less strictly enforced dominance hierarchies, and the males fight until only one is left.

What is programmed into duikers by their genes is apparently not a detailed plan of how to set up a monogamous territorial system, but either a set of alternative social plans with some habitat-dependent rule for choosing between them, or more likely an edict to seek some particular state or condition that somehow "feels right." In this case we would expect the animal to experiment with its behavior until the "feedback" matched the innate model, just as a child learns to crawl (Chapter 11). The social system arrived at would fall out as a secondary consequence, and its maintenance would be facilitated by imprinting or other cultural factors. Perhaps with regard to this "feedback" hypothesis, female ungulates might come programmed first to *hide* and only secondarily to establish territories, while males would be supplied with the reverse order of priorities. This set of instructions would give rise to a strongly territorial system in heavy vegetation, and in the open might force the animal to hide in a herd. Indeed, as we will see in the next chapter, a herd of antelope is an excellent place to hide as long as you look just like everyone else. Dominance hierarchies within herds must be a consequence of the programming for territoriality which is, after all, simply a form of local dominance in which males work out a hierarchy according to territory quality.

Considering social systems to be consequences of the interaction of habitat and programming rather than absolute givens allows us to explain much of the observed range of ungulate behavioral ecology. Bushbucks, for example, live on territories in heavy vegetation during the day but herd loosely when in the open during the concealing darkness of night. During the mating season at least, impala and wildebeest males must experience a surge of territoriality "drive" at the expense of their inclination to hide since they set up a territorial matrix. Songbirds do precisely the same thing when, as we saw in Chapter 12, an upsurge of androgens causes them to leave the herd-like protection of their winter flocks to compete for place. Females herd, as do juvenile males which have been driven out of the female herds by jealous territorial males. This expulsion is similar to the process by which duikers drive off their offspring. In duiker societies the adult male drives off his sons, and then somewhat later the adult female expels her daughters—a part of the ritual that seems to have been lost in larger ungulates. In the open, excluded males herd together naturally and form a dominance hierarchy. It would be a fascinating experiment to raise wildebeest in a lush forest and to see how they sort things out in that environment.

As the food becomes poorer and a larger area consequently becomes necessary to support each individual—and larger rumens to process the input—there must come a point at which no male can defend a territory sufficiently large, and the social system should collapse. In fact, wildebeest living in relatively good habitats have a territorial system, while populations in poor habitats do not. In the latter case, as described above, the males walk with the ever-moving herd, defending only a small area around themselves. Higher ranking males tend to be found in the center of concentrations of females, while the lowest ranking males and juveniles form the usual bachelor herds. When a patch of good grass is encountered, however, the matrix is quickly, though temporarily, reestablished. It seems clear that the step to the tending bond practiced by some ungulates comes with still poorer food, nearly continuous movement, and smaller herd size, so that the alpha male can practicably monitor the whole group. The smaller herd size is probably the result of the still larger body size necessary to process less nutritious food, a size which now becomes in itself a weapon against many predators. Indeed, the largest ungulates are nearly invulnerable, and the defensive advantages brought by large herds are diminished (Figs. 26-6 and 26-7).

This comparison and contrast of the behavior of African ungulates generates a nice story, but since nearly all the data were available from the start, there is a very real danger that our reasoning may all be circular. The true test of an explantion is not so much how well it explains the existing facts—no explanation is ever that perfect—but rather its ability to make predictions, particularly unexpected predictions, outside of the original body of data. For example, the prediction that there should be no large monogamous ungulates is not very useful since monogamy is so rare in mammals. But a capacity for

Fig. 26-6 By virtue of its size, an African buffalo is at risk only from a group of lionesses, and even a small group of these ungulates is invulnerable. Larger herd size, therefore, is pointless.

Fig. 26-7 Another species of large ungulates, the musk oxen of Canada, take advantage of their awesome size when threatened by forming a defensive circle.

predicting the social systems of, say, American ungulates, would be encouraging. Can our argument account for the peculiarities of mountain sheep of the Canadian Rockies? Although they are classified as ungulates, they do not live in a classic African forest, brush, scrub, or plains habitat; they face a different climate and must deal with a different set of predators. Mountain sheep are able to survive on vegetation of very low quality and they are, therefore, relatively large (though at 75–200 kg they are a bit light by African standards) and range widely. And as would be expected from the African synthesis, male mating rights are organized by a system of dominance hierarchy rather than by territoriality. Their hierarchy is constructed on the basis of a spectacular ritual which serves to measure both strength and skill. Males rise into the air on their back legs and charge each other, slamming their backcurved horns together with a resounding crash. Then, as if to display how little this enormous impact has shaken them, the rams step back majestically and look to the side (Fig. 26–8).

A

B

C

Fig. 26–8 Mountain sheep, like other large ungulates living in small herds, establish a dominance hierarchy. Their ritual begins with a stereotyped two-legged run (A), followed by a shattering crash of horns (B) and an exaggerated head display (C). The ultimate winner of these contests, the alpha male, will have exclusive mating rights.

According to ethologist Valerius Geist, the male mountain sheep with the largest horns does virtually all the mating. His horns are a good indication of his status and also of his previous health; but ironically, the faster the horns grow, the less time the animal is likely to live. It seems that the rams nearly all die with a horn length of about a meter, although the lifespan required to generate such horns may range from seven to fifteen years.

Mountain sheep live in somewhat smaller groups and range less widely than do their ecological counterparts in Africa. This might be due in part to the absence of large, powerful predators and terrain which is unfavorable for the group-hunting strategy of wolves. On the other hand, Geist noticed a behavioral curiosity which may have something to do with this: herds refuse to walk through the forests. This is presumably an antipredator precaution, and one which may explain why mountain sheep have not recolonized areas from which they have once disappeared.

LEKS

Perhaps the most difficult test case for any theory of ungulate behavioral ecology is found in the behavior of the Uganda kob. These antelope fall between bushbucks and wildebeest in weight, but the males organize themselves into a lek—that peculiar system by which males congregate on a small, traditional area, each defending a tiny territory which is so small as to be useless for feeding or tending young (Fig. 26–9). As with other

Fig. 26–9 Uganda kob males establish a tight cluster of defended territories on a traditional but otherwise worthless piece of ground. Females come to this lek and mate with the center males.

leks, mating distribution is highly skewed, with the one or two males on the central, most hotly contested territories having considerably more sexual success than all the other males combined.

The life of a central male, with its constant displaying, fighting, chasing, and mating, must be a strenuous one (Fig. 26–10).

A

B

C

Fig. 26–10 Life for the center males on a kob lek is frantic: they engage in roughly two ritualized but tiring territorial displays per minute. Note the turned-head display (C) which resembles that of many ungulates—mountain sheep of Fig. 26–8C, for instance—and the equally widespread head-down maneuver (B; see also in Fig. 26–3 for wildebeest). As center males tire, they retire to the bachelor herd to recuperate, permitting a gradual centripital movement of males on the matrix.

Not surprisingly, the movement of males is centripetal as central males burn out under the stress and are replaced with their neighbors. A replacement ultimately frees a peripheral territory which is filled by a member of the bachelor herd. When ready to mate, females arrive at the lek, attracted perhaps by the overwhelming concentration of displaying males. They ignore the overtures of isolated males and males in the bachelor herd, and walk through the periphery to the center of the lek. Since this mating system does not seem to expose the females to physical risks if they choose to mate with off-center males, we must suppose that females come programmed to go to the center because it is good for their genes.

But the question for behavioral ecology is how the lek system evolved out of run-of-the-mill ungulate territoriality. What is the special selection pressure on the kobs or the unusual environmental parameter that caused them to depart so radically, and what behavioral flexibility permitted it? The only environmental peculiarity seems to be that a relatively stable water supply is usually nearby, and food seems to be undefendably dispersed but abundant. Is this the critical parameter?

A similar pattern holds for the lesser known Lechwe kob, and it too has a lek system. For this species, the mating season begins when the annual flood waters peculiar to their habitat recede, exposing a temporary but highly productive food source—the floodplain grasses. Males rapidly build up the food reserves that will be necessary for the rigors of fighting and mating. A lek is formed, but it must move to higher ground as the waters begin to rise again. Females, as usual, go to the center to mate, and the central males fight, charge, or otherwise defend their favored positions about a hundred times an hour. If we look at other lek species, we will see the pattern of undefendable seasonal abundance appear again and again, from the berries of evergreen sagebrush which the prairie chicken eats to the enormous (relative to body size) rotting fruits next to which picture-wing *Drosophila* establish their leks.

The question in the case of the ungulates, however, is not how to justify the kob system, but to understand how it could have evolved out of the mainstream of ungulate territoriality. Leks, by their very rarity, challenge the whole enterprise of behavioral ecology. There are no answers as yet, but it may be that the generally ignored females are driving the system by ignoring isolated males and bachelor herds as they head purposefully toward the center of the lek. The behavior of the male kobs, like that of the male lovebirds of Chapter 19, is the product of standard sorts of innate behavior manipulated and molded both by experience in the bachelor herd and by the behavior of the females.

What seems to be missing, then, is any understanding of the behavior of female ungulates. One lesson of field ethology is that male ethologists generally find the females of the species being studied to be relatively passive elements in the social system, while female ethologists find case after case of female-dominated societies. In this regard some ethologists are doubtless guilty of the charge that they anthropomorphize behavior: like human beings everywhere, we interpret what we see out of our own experience. This argues, however, not so much for putting unbiased machinery between ethologists and the animals they study, as for stimulating ever-wider interpretations of behavior and encouraging a second, third, fourth, or fortieth look at even the most familiar of species.

SUMMARY

The basic goal of behavioral ecology is to understand how small but well-defined differences in niche and habitat affect social organization. The African ungulates are an ideal group for this analysis since many diverse species coexist, though all share a rumen-based digestive strategy. Behavioral ecologists have made several observations concerning this coexistence. The smallest ungulates in the ecosystem form monogamous pairs on exclusive territories. Males of middle-sized ungulate species continue to defend territories, but females move in large herds while males which are unable to gain territories form a bachelor herd with internal dominance hierachies. The largest of these animals live in small herds and depend on a male dominance system. These trends correlate with type of food, body size, and predators, so that evolutionary cause and effect are not obvious.

More perplexing, perhaps is the question of how what must once have been common behavioral programming could have been so greatly modified. The vast flexibility in ungulate social behavior that allows them to change social systems as they change habitats suggests a programming strategy which directs the animals to behaviors which will achieve specific goals, rather than specifying exactly what social behavior they will dogmatically adopt.

STUDY QUESTIONS

1. What do you suppose would happen if a small herd of wildebeest were released in a forest? What might be expected in a desert with a small but rich patch of food and water? What might happen to kob under the same circumstances?

2. What are the long-term survival prospects for the two sexes in wildebeest? What does any difference depend on? If the females could control the sex of their offspring would they be wise to skew the sex ratio in general, or in response to environmental change? Would this be helpful in the short run or the long run?

FURTHER READING

Bell, R. H. V. "A Grazing Ecosystem in the Serengeti." *Scientific American* 225, no. 1 (1971): 86–93.

Owen-Smith, N. "Territoriality in the White Rhinoceros." *Nature* 231 (1971): 294–96.

CHAPTER 27

African Carnivores

We saw in the last chapter how the ungulates manage to divide up vegetation in a way that allows many species to coexist, and how at the same time the rules by which this resource division is made determine or profoundly influence each species' social system. This chapter is concerned with how the five major species of African carnivores manage to divide up the ungulates, and how this peculiar division is reflected in turn in the social structures of the predators.

In the Serengeti only three of the dozens of species of African ungulates are sufficiently numerous and accessible to form a major part of the carnivore diet. The three species bearing this dubious distinction—wildebeest, zebra, and Thompson's gazelle—constitute about 90 percent of the ungulate biomass, and are compared in Table 27-1. All three species share the same antipredator strategy: running away. In the face of a continuous selection pressure, zebra, gazelle, and the ungainly looking wildebeest can each sustain a pace of about 45 M.P.H. for about a mile (three times faster for that distance than the fastest human runners). It comes as no surprise, then, that carnivores make most of their living off very young, very old, moribund, or dead antelope and zebra.

The stable coexistence of five predators on three species of

TABLE 27–1 *Comparison of Major African Ungulate Prey*

Criterion	*Zebra*	*Wildebeest*	*Gazelle*	*Other*	*Totals*
Average adult weight (kg)	165	110	12	—	—
Number ($\times 10^3$) of individuals in Serengeti	150	400	200	200	950
Biomass (10^6 kg) in Serengeti	25	45	2	8	80[a]

Source: George B. Schaller, *The Serengeti Lion* (Chicago: University of Chicago Press, 1972).
[a] Represents an annual consumption of about 550×10^6 kg of grass.

prey is a remarkable feat of niche division. It is not uncommon for several different species of predator to coexist in the same habitat, but the common explanation for this phenomenon is that the predators divide up the prey according to body size. Darwin's large-billed finches, we have seen, take big seeds, while the small-beaked ones concentrate on the little ones. This glib explanation suffers a conspicuous failure, however, when confronted with the Serengeti system: except for the cheetah, all the predators, large and small alike, take everything from the diminutive gazelle through the massive zebra. The answer to this ecological conundrum is far more interesting and revealing than any simplistic size correlation. Only a careful comparison of the ecology and behavior of the predators allows the basis for their more-or-less amicable "division of the spoils" to emerge.

LIONS

The most thoroughly studied of the Serengeti predators is the lion. Lions live in prides which occupy large, exclusive home ranges and which typically consist of two adult males, many adult females, and a variable number of juveniles and young (Fig. 27–1). The females do the hunting, but the males drive the females away from their kills and feed themselves first, often allowing some cubs to feed with them. Lions rely almost entirely on group hunting: the females stalk their prey on a wide front, often encircling it by moving from one bit of cover to

Fig. 27-1 This pride of lions is waiting out the Serengeti heat in a rare patch of shade. According to George Schaller, lions spend about 90 percent of their lives resting.

another when the herd they are stalking is busy eating, and freezing in mid-stride whenever a head is raised (Fig. 27-2). Finally, when they are close enough or when their prey becomes wary, one female will often make a rush which, if the lions are lucky, will scatter the herd and send individuals stampeding

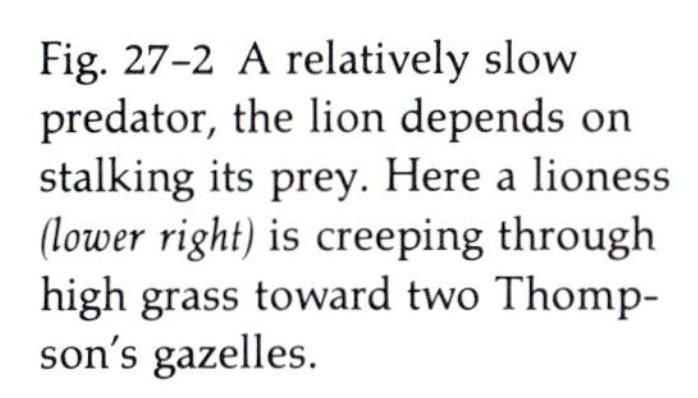

Fig. 27-2 A relatively slow predator, the lion depends on stalking its prey. Here a lioness *(lower right)* is creeping through high grass toward two Thompson's gazelles.

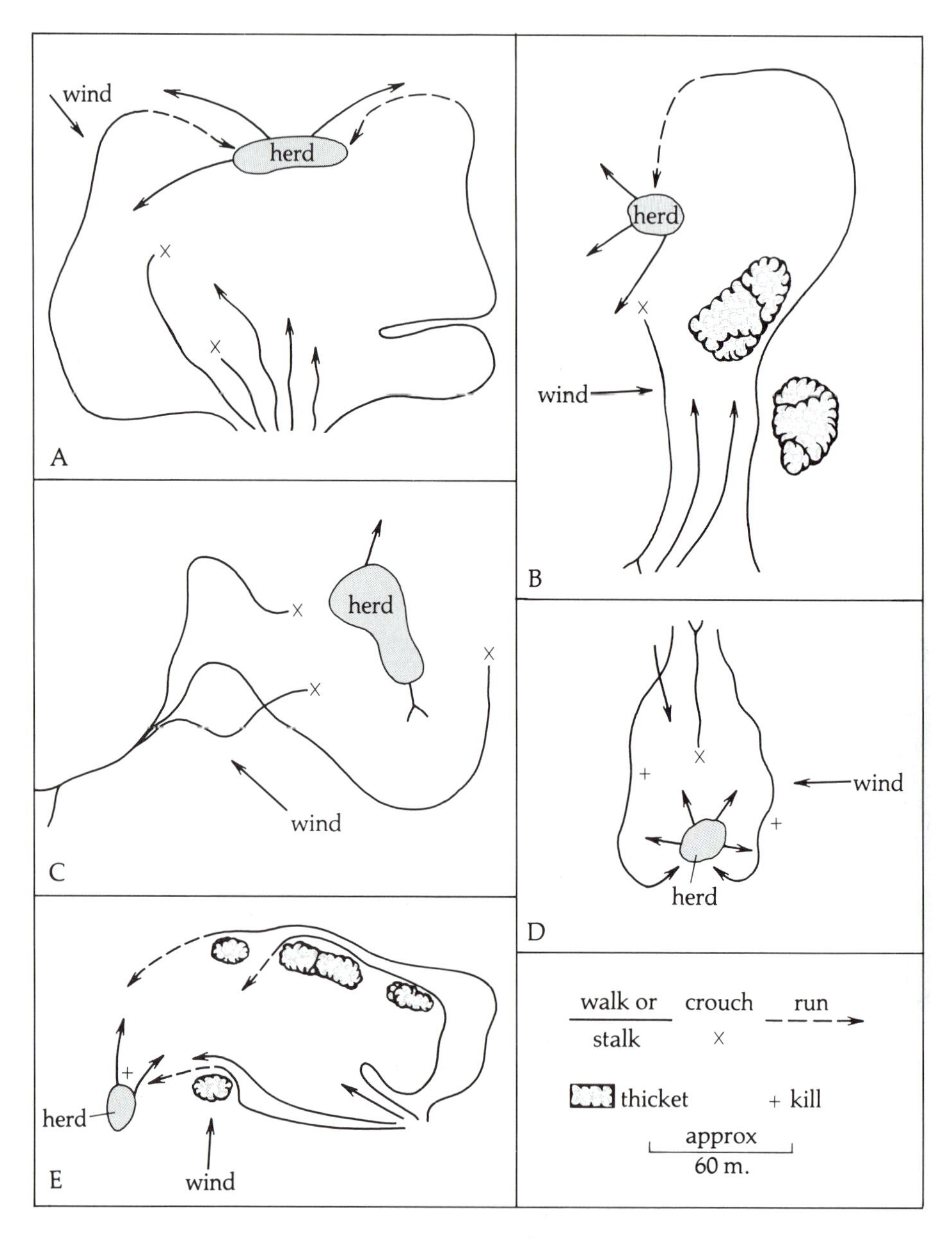

Fig. 27-3 Lions hunt in groups, encircling their prey and stalking warily to get as close as possible. One or two lions rush the prey, and if the herd scatters some of its members may flee within range of one of the hunters lying in wait.

within range of the other hunters (Fig. 27-3). Proximity is crucial since lions, although they can accelerate very quickly, are not as fast as their prey and have little stamina (Fig. 27-4).

The success of the hunt has been shown to depend on the number of females participating (Fig. 27–5A), even though at first glance it looks as if females ought to do best hunting alone. In fact the probability of success does not quite double with two lions, and levels off with four. Certainly females do not hunt in groups out of physical necessity. They are enormously strong, and one lion can easily bring down anything that it can catch smaller than a buffalo. Rather, it may be the dependent males and cubs which make group hunting a necessity, for when even one extra nonhunting mouth is added, the peak of the food-per-hunter graph shifts toward cooperation (Fig. 27–5B). Another advantage to a group-hunting predator is that although lions are large enough to rob other predators with impunity, a group of lions is usually more successful at holding a kill

Fig. 27-4 Lions depend on surprise, fast acceleration, and their great size and strength to capture prey. Here a female lion chases a Thomson's gazelle.

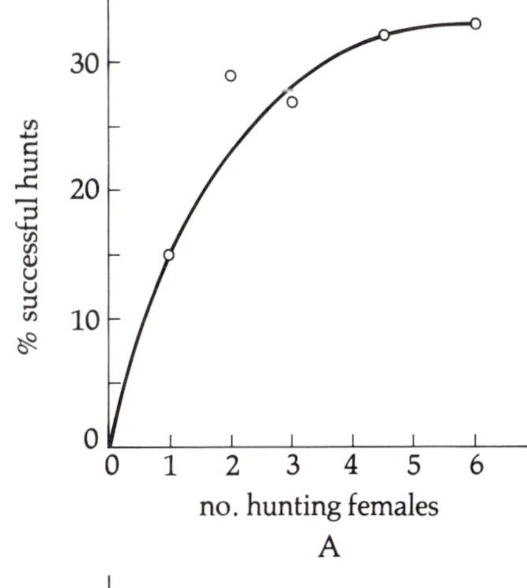

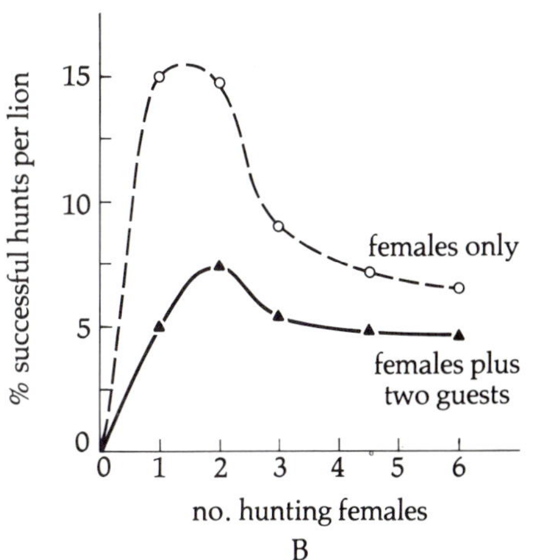

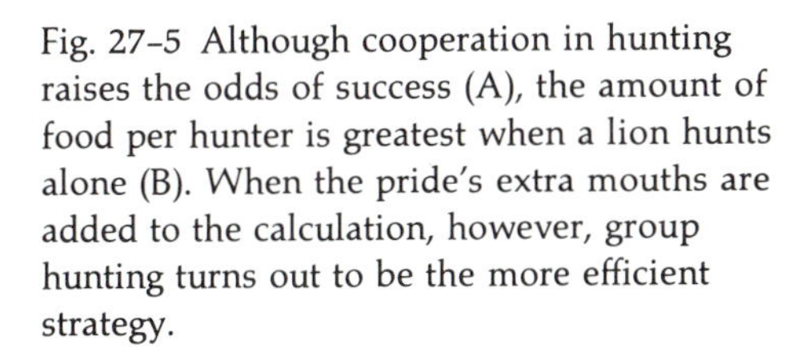

Fig. 27–5 Although cooperation in hunting raises the odds of success (A), the amount of food per hunter is greatest when a lion hunts alone (B). When the pride's extra mouths are added to the calculation, however, group hunting turns out to be the more efficient strategy.

against the persistent and unnerving psychological warfare waged by scavenging hyenas.

What holds lion prides together? The two resident males seem not to compete with each other for food or for females, nor do the females compete among themselves for food. Indeed, lionesses suckle each other's young indiscriminately. Such a picture of gentleness and generosity in this legendarily ferocious beast is appealingly romantic, but other aspects of lion social life are less endearing. When new males take over a pride, killing or driving out the former masters of the harem, they kill all existing cubs as a matter of course, and continue to kill any new cubs engendered by the unseated males as they are born (Fig. 27-6). Most pregnant lionesses, however, undergo spontaneous abortions when a changeover in males occurs. Such turnovers in the ruling clique of males occur every two to four years.

Before the notions of kin selection and parental investment gained ascendancy, it was difficult to make any sense of the lion social system. Now, however, nearly everything can be put into place. By keeping track of individuals, ethologists now know that the males which cooperate to take over a pride are brothers which were forced out of their far-away home pride by their father as they approached sexual maturity, and which have

Fig. 27-6 When new males take over a pride they begin their reign by killing the existing cubs. At the same time, females spontaneously abort any current pregnancies.

been hunting together for years, biding their time and waiting for their chance to take over a pride of their own. Their high relatedness reduces the selective advantage of competition. The females in a pride are also related to each other in a complex hierarchy of sisters, mothers, and daughters. Hence their cooperation makes sense. The asymmetry in the way the two sexes treat cubs—males killing them, females tolerating them; males allowing their cubs to feed first while females make them wait—represents different strategies of parental investment, and even of reproduction. Males have only a limited ascendancy in the pride, and consequently only a brief opportunity to pass on their own genes. By killing all cubs of the previous males they bring lactating females back into reproductive readiness and hasten the day that their own genes will appear in cubs. Ensuring that their cubs feed before the females is again a way of enhancing the short-term fitness of their own genes. Spontaneous abortion, hormonally induced, cuts the female's losses, too, since any cubs of previous males she carried to term would be killed in the end. In this way her genes make the best of a bad lot.

While males are in a sense *r*-selected, the long-lived females are more nearly *K*-selected. They live in a stable, hereditary pride where they will have countless opportunities to reproduce. Their best strategy is to sacrifice as little as they can for their offspring since there will be other chances. Hence even when their pregnancies are uninterrupted they produce litters of very tiny offspring which represent a minimal investment of bodily resources, and feed them only if there is extra food.

LEOPARDS AND CHEETAHS

The two other species of large cat have very different hunting and living strategies. Leopards ambush their prey, often at dusk or during the night as the herds come warily to the communal water holes. Leopards are not nearly as big as lions, and are easily robbed of their kills by lions, or packs of hyenas or dogs. Because of their size leopards must concentrate on smaller prey than lions, and to avoid being robbed of their kills they often drag them up into trees to eat (Fig. 27-7). Small prey size and the lack of any selective advantage to group ambush are consistent with the leopard's solitary social life. Its need to be inti-

Fig. 27-7 Leopards hunt alone and depend on ambushing their prey. They carry their victims into trees to reduce the chances of being robbed of their spoils.

mately familiar with the terrain and with favorable sites for ambushing likewise can account for its territoriality.

Cheetahs are also relatively small, but are unlike leopards in virtually every other way. Their hunting strategy is *speed.* An adult cheetah can accelerate rapidly to about 70 M.P.H.—roughly twice the speed of lions and leopards, and more than 20 M.P.H. faster than the antelope they hunt. Their strategy requires them to hunt in the open and during the day. The physiological adaptations which permit their remarkable speed impose severe limitations on their choice of prey. Cheetahs have very flexible, highly muscular spines, which they flex powerfully as they run, using the spine's spring to increase their speed (Fig. 27-8). Spi-

Fig. 27-8 Cheetahs depend on superior speed to capture prey. To achieve their 70-M.P.H. velocity, however, they have had to sacrifice the strength and the claws necessary to catch anything larger than gazelle.

nal muscles, however, are inefficient and easily exhausted, leaving a cheetah after a 300-m chase unable to manage another, no matter how tempting, for nearly half an hour.

As another sacrifice for speed cheetahs have lost the retractability of their claws. Most cats use their notoriously well-maintained claws as weapons, which allow them to grasp and manipulate prey with their powerful legs. Cheetahs, with their dog-like fixed (and therefore worn) claws that adapt them instead for running, must overcome their prey solely with their jaws and teeth (Fig. 27-9). As a result they are almost exclusively restricted to the smaller gazelles, which are the fastest of the antelope. As gazelle specialists, cheetahs lack territories and

Fig. 27-9 After successfully running down a Thomson's gazelle, this cheetah simultaneously strangles it and drags it toward cover.

follow the herds that in turn follow the seasonal progress of the grass. Cheetahs are only vaguely social, hunting occasionally in small groups, but the females are strictly solitary when they bear and raise their cubs (Fig. 27-10).

The period of cub rearing is crucial to the fitness of offspring in all three cat species. Like the familiar domestic house cat, lions, leopards, and cheetahs come wired with black-and-white, motion-sensitive vision. Innate releasers draw the cubs' attention to moving objects, provoking them to attack and "playfully" disembowel everything from falling leaves to their mother's tails (Fig. 27-11). Later in this period of self-calibration

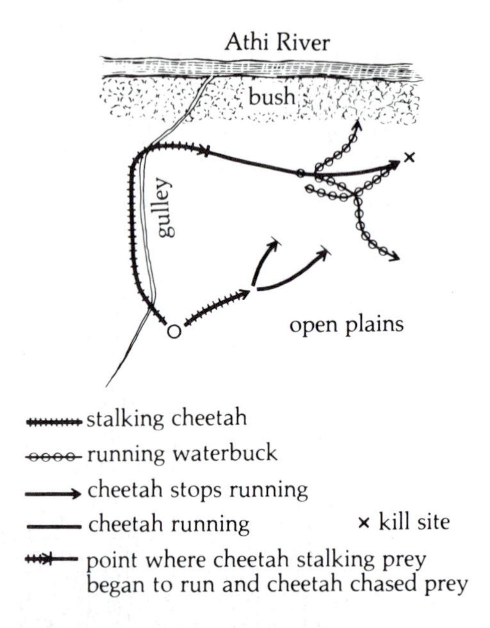

Fig. 27-10. Although cheetahs are not particularly social, mothers do hunt with their cubs. Whether there is any advantage in this potential cooperation beyond teaching the cubs about hunting is questionable. In the incident recorded here, however, the young probably did help. When the mother and her four nearly full-grown cubs spotted a herd of three waterbuck, the female cheetah left her offspring and moved quickly behind the cover of a gully. The cubs and mother then stalked more quickly across the plains until they were seen. As the mother began her run, one waterbuck ran toward the bush along the river while the others ran toward the plains. One waterbuck, now seeing the cubs, cut back toward the bush and was felled by the adult cheetah.

Fig. 27-11 Lion cubs are irrepressibly playful, regularly stalking and pouncing on the tufts of the adults' tails (A) or wrestling with each other (B). The utility of this practice as adults seems obvious.

the cubs accompany their mothers on hunts, learning presumably by observation or trial and error the fine points of their innate stalking behavior, the appearance of their particular prey, the appropriate distance at which to attack each species, group coordination, and so forth. Most important, they have to learn to concentrate on one individual in a herd—to pick it out on the basis of some distinction, preferably some sign of weakness, and to stick with it in the face of the neurally exaggerated movement of other members of the herd as they run this way and that across a lion's or cheetah's line of attack. A cat which switches targets is drawn into a time-consuming turn, and almost inevitably fails. This difficulty is reflected in the data describing the hunting success of cheetahs. When herds run as a unit a cheetah has barely one chance in ten of making a kill, whereas if one individual hesitates a fraction of a second when the rest run, thereby making itself an obvious target, its probability of being killed goes up threefold, and if it fails to rejoin the herd in time it has essentially no hope of escape.

HYENAS

The other two major predators of ungulates are hyenas and hunting dogs. Spotted hyenas are the most numerous of the Serengeti hunters and outweigh both leopards and cheetahs. The impact of these hunters is enormous: they kill about 75 percent of all wildebeest calves each year. Only the synchronization of wildebeest births, which overwhelm even the rapaciously efficient hyena population, allows this species of antelope to persist in its open habitat. The hunting strategy of hyenas is remarkably simple: they chase their prey until they wear it out (Fig. 27–12). They are neither faster than their prey, like cheetahs, nor stronger, like lions—these nocturnal hunters simply have more endurance than anything which is too young, too old, ill, or fails to run in a straight line.

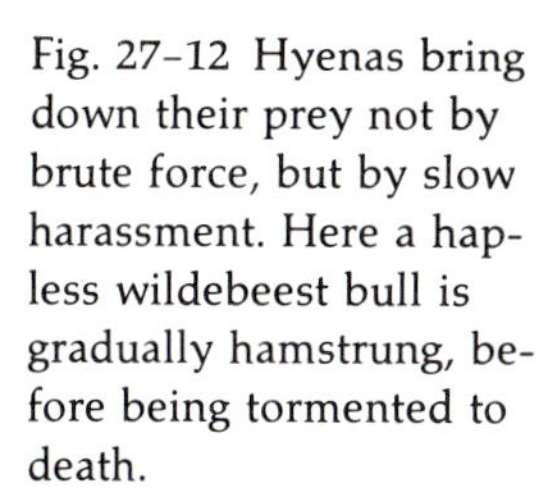

Fig. 27–12 Hyenas bring down their prey not by brute force, but by slow harassment. Here a hapless wildebeest bull is gradually hamstrung, before being tormented to death.

The major problem which hyenas face is agreeing on which animal to chase. They must agree, since the success of the hunt depends on wearing out one particular animal, regardless of how many times they have to flush it from the herd or how many other, less tired animals cross their paths—and this agreement must come without the luxury of any sort of sophisticated communication among the animals. The decision must arise from some peculiarity, morphological or physical, to which all the group is attracted—some anomaly which will permit them to agree on and single out a victim from a herd of perhaps several hundred animals. In search of their target hyenas calmly scrutinize the herd, then test it by beginning a run. It is uncanny how virtually all the hunters will notice the same animal and begin to focus their attention on it. It is no wonder, then, that all wildebeest except the young, the old, or the sick look alike.

Once their prey has been selected the hyenas in the lead exhaust themselves trying to turn the victim. This maneuver allows the rest of the pack to cut across the broad arc the prey is encouraged to take and thus keep up with the target without having to exceed their speed of greatest endurance. As the lead animals begin to fade new leaders take over, and many hunts, after covering one or more miles of open plains, end on virtually the same spot on which they began.

Hyenas lack the claws of the cats and so must bring down their prey by harassment—a nip here, a bite there—until their victim trips or falls. The group-hunting technique is essential to hyenas in bringing down their major prey, wildebeest. One hyena stands no chance of evading the horns of even an exhausted wildebeest. But two hyenas, one on each side, can take turns alternately biting and dodging the horns. And again, only a second hyena permits the separation of a calf from its protective mother: one hyena has about a 15 percent chance of taking a calf while two have a 77 percent chance.

What happens after the hunt is as remarkable as the events of the hunt itself: the hyenas eat everything, skin and bones (40 percent nutrition) included, in a matter of minutes. An entire 100-kg wildebeest can be reduced to nothing more than a pair of horns and a dark spot on the ground in thirteen minutes, and each hyena can, in this short time, consume 14–15 kg. This extraordinary efficiency in hunting, killing, and eating is achieved through the complete coordination and cooperation of the pack, a harmony which is the consequence of kinship (Fig.

27–13). The clans share the protection of underground burrows during the hot days—dens dug and maintained in part by the otherwise unemployed cubs—and live on exclusive territories (Fig. 27–14). The terrible efficiency of hyenas is tempered somewhat by the impractical attachment they show to their homeland, even as their prey moves out in search of fresh grass, making starvation a real possibility.

The packs are led by the females, which are larger than and dominate the males. Each female in this unusual social structure is provided with a sham penis and scrotum (Aristotle thought the species was hermaphroditic), from which we can easily surmise what the releasers involved in social dominance must be. The false genitalia come into play in a "greeting" ceremony. When a pair of hyenas meet, one will face away while the second circles downwind to catch the other's individual scent. Then the second hyena will move alongside the first, head to tail. One will then raise its inside rear leg and expose its genitalia and the two associated scent glands. Far from being a greeting between old friends, this interaction is a dominance ritual seen also after one hyena snaps at another: the less domi-

Fig. 27–13 Hyena den with four adults and four cubs.

Fig. 27–14 Border clash between two hyena clans.

nant individual is exposing its most vulnerable part to the teeth of its superior, a common movement in mammalian appeasement displays.

HUNTING DOGS

Remarkable as the efficiency of a clan of hyenas seems, a pack of African hunting dogs is even more devastating. Wild dogs parallel hyenas in many respects: they live in groups, hunt in packs, and bring their prey down by harassment (Figs. 27-15 and 27-16). Where they differ from hyenas, the dogs seem to have the edge: they hunt by day, migrate to follow the herds, and can reach speeds about 5 M.P.H. faster than hyenas. As a

Fig. 27-15 Wild dogs begin a hunt by testing a herd, looking for any individuals which might be likely targets, especially the young, the old, or any animal that is conspicious in its morphology or its failure to run with the herd. Any individual which stands out in some way is far more likely to become the focus of the wild dogs' attack.

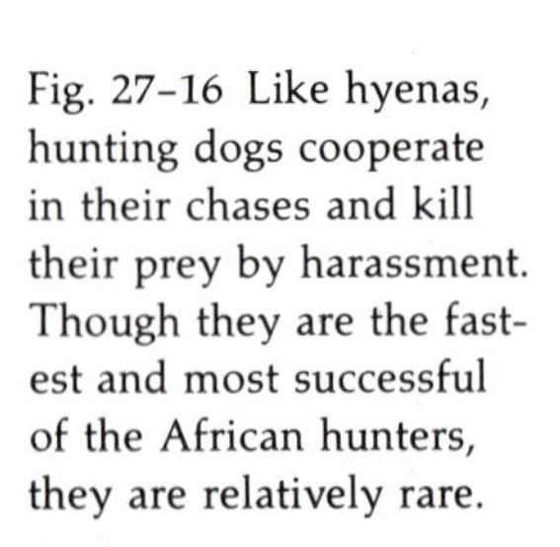

Fig. 27-16 Like hyenas, hunting dogs cooperate in their chases and kill their prey by harassment. Though they are the fastest and most successful of the African hunters, they are relatively rare.

result their success rate in terms of kills per chase is about 60 percent, roughly twice that of hyenas. Yet despite such advantages wild dogs remain only a minor predator, and no one quite understands what limits their numbers. Their food-sharing behavior is that of an animal with high adult mortality: pups are fed first, and because they live in kin groups, they are fed by everyone. This is just the opposite of the behavior of the long-lived lionesses which, although feeding is largely communal, will allow their cubs to starve before they themselves go hungry. Even the rather briefly ascendant pride masters only let their cubs feed simultaneously at best. Wild dogs show all the signs of animals which cannot rely on a second chance to breed. There must be some powerful limiting mechanism, perhaps of parasitism or disease, which makes their lives tenuous despite all their ruthless efficiency.

CARNIVORE NICHES

The remarkable thing, of course, is not that there are so few wild dogs relative to, say, hyenas, but that all five predators are able to coexist at all—that there are five ways which do not directly compete to make a living off three prey species. How does evolution manage such stability? The answer lies in a comparison of the occupational strategies of the predators, as summarized in Table 27–2. The carnivores avoid competition by hunting primarily in different places at different times, and by using different techniques to capture different segments of the prey population. Cheetahs are unique in their high-speed chase strategy, but as a consequence must specialize on small gazelle. Only the leopards use an ambush strategy, which seems to play no favorites in the prey it chooses. Hyenas and wild dogs are similar, but hunt at different times. And the lion exploits the brute-force niche, depending alternately on short, powerful rushes and strong-arm robbery (Fig. 27–17).

These divergent hunting strategies are reflected in the predator-specific behavior of the prey, the most obvious aspect of which is the very different distances to which a herd will allow a specific predator to approach before fleeing. Herds of zebra or wildebeest, for example, will let lions get much closer than hyenas before running, and will react indifferently to cheetahs (Fig. 27–18). Gazelle, on the other hand, give cheetahs a wide

TABLE 27–2 *Comparison of Ungulate Predator Strategies in the Serengeti*

	Lion	*Leopard*	*Cheetah*	*Hyena*	*Wild Dog*
Number	2000	600	250	3000	300
Weight (kg)	110–180	35–55	35–55	45–60	17–20
Hunting Preferences					
Time	night	night	day	night	day
Place	woodland and brush	thickets	plains	plains	anywhere
Technique	stalk, rush at 35 M.P.H.; or rob	ambush	stalk, chase at 70 M.P.H.	high-endurance chase at 35 M.P.H.	high-endurance chase at 40 M.P.H.
Success rate of chases	30%	—	70%	30%	60%
Social?	yes	no	slightly	yes	yes
Home range?	yes	yes	no	yes	no
Prey taken					
Gazelle	30%	64%	91%	35%	40%
Wildebeest	35%	—	—	50%	—
Zebra	35%	—	—	15%	—
Quantity of prey taken (10^6 kg/year)	5.5	0.9	0.4	3	0.25
Efficiency (kg of prey required to fuel each kg of predator per year)	22	36	40	20	45

Source: George B. Schaller, *The Serengeti Lion* (Chicago: University of Chicago Press, 1972).

Fig. 27–17 Because of their size and strength, lions are regularly able to rob other hunters of their kills. Here a cheetah is being chased away from the gazelle it has run down.

Fig. 27–18 Lions are slow predators and without the element of surprise are unlikely to capture healthy adults. These wildebeest seem perfectly aware of the lions' shortcomings, and so are careful to keep only about 60 m between themselves and their predator. Cheetahs, too small to catch wildebeest, are ignored altogether, while packs of wild dogs and hyenas are kept at a good distance.

berth but will, because of their own greater speed, allow hyenas a bit closer. Packs of hyenas or hunting dogs are considered dangerous, but single ones, unlike lone lions, are ignored (Fig. 27–19). This calibrated wariness must be a product of cultural learning and frequent experience, and takes in habitat as well. Ungulates are much more watchful when potential cover for stalking predators is nearby, and are almost hysterically nervous when forced by thirst or other necessity to enter the thickets patronized by leopards. Since ungulates depend on processing a large volume of relatively low-quality food, time and energy spent running instead of feeding must be kept to a minimum, yet each animal must at the same time minimize the risk to itself. The basis for differences in judgment about danger distances results from each prey species' evolution of a "sense" of its predators' species-specific acceleration, maneuverability, and endurance compared to its own reaction time, acceleration, maximum speed, maneuverability, and endurance. These relationships are shown schematically in Figure 27–20.

The hunting strategies of the predators are also reflected in their social systems. Cheetahs and leopards, which have little or

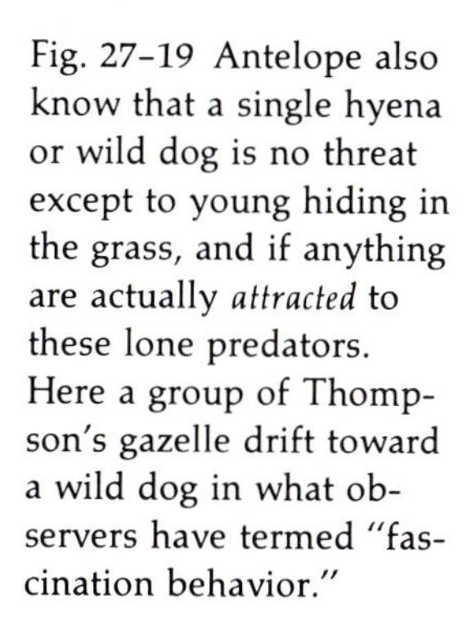

Fig. 27–19 Antelope also know that a single hyena or wild dog is no threat except to young hiding in the grass, and if anything are actually *attracted* to these lone predators. Here a group of Thompson's gazelle drift toward a wild dog in what observers have termed "fascination behavior."

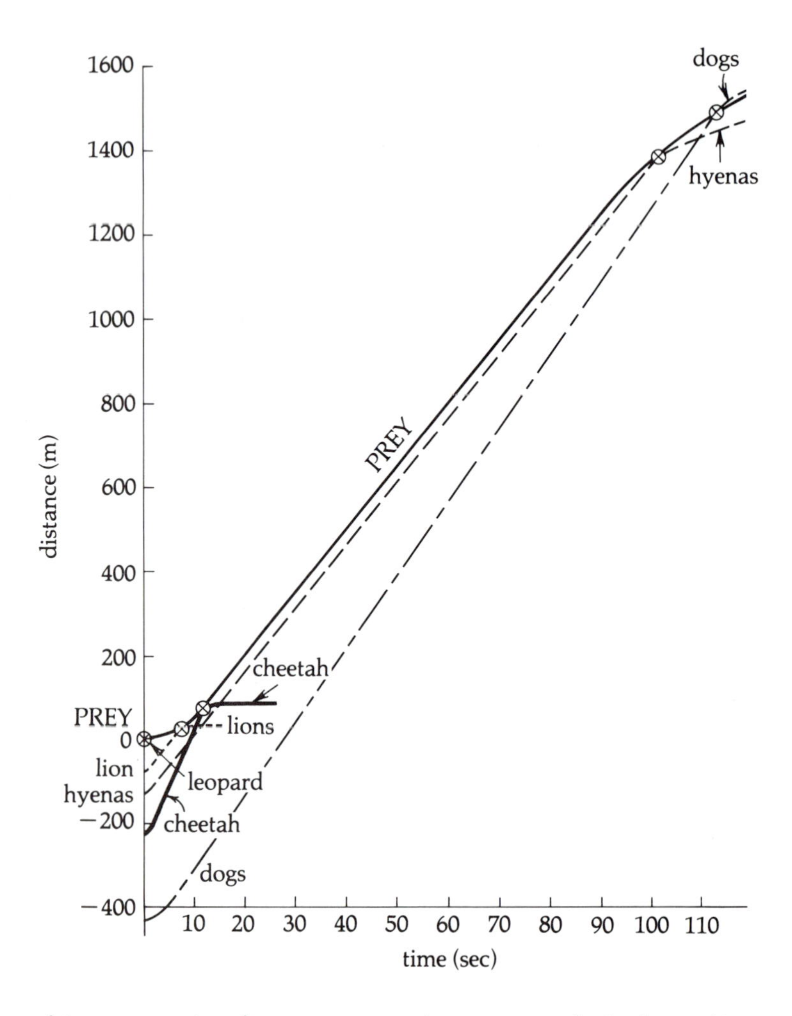

Fig. 27-20 The logic behind the predator-specific flight distances. The prey are shown beginning to run at distance and time zero. They accelerate over about 10 sec to their top speed of 15 M.P.H. Lions are allowed closest, for although they accelerate faster their top speed is lower. Cheetahs on the other hand have a higher top speed, but their limited endurance means that gazelle need to keep more than 200 m between themselves and their predator to be safe. Hyenas, which are not only about as fast as their prey but also have slightly greater endurance, must be kept at an intermediate distance. Wild dogs, with their superior speed, must be kept the farthest away of all. Since hyenas and wild dogs focus on the slowest members of a herd, healthy adults can safely allow these predators to range somewhat closer.

nothing to gain from cooperation, are relatively solitary, whereas the stalk-and-rush strategy of lions and the chase-'em-down strategy of the canids work best with groups of hunters coordinating their actions. In these latter cases kin selection seems to be the crucial factor in restraining the selfish, anarchistic competition and interference which would otherwise prevail. For the carnivores, then, the ungulate-predator niche is subdivided not so much by what they eat, but by how and when they catch it. This resembles the arrangement by which insectivorous birds, wasps, praying mantises, bats, spiders, and so on manage to share the flying insects. And just as with insects, only a modest fraction of the potential prey are consumed each year—about 12½ percent in the case of ungulates. The net result of this predation is fast, healthy, svelte antelope with a slightly lower life expectancy, more cases of ungulate ulcers perhaps, high infant mortality, healthy grass, and fewer vultures and jackals to lower the tone of the neighborhood.

SUMMARY

The five major species of African carnivores manage to coexist in the Serengeti on only three important prey species: wildebeest, zebra, and gazelle. The apparent stability of this division of the spoils results from differences in the segments of the prey population that are taken and the hunting strategy each predator uses. Hunting strategy is, in turn, reflected in the social organization of the predators. Leopards are highly specialized hunters, ambushing their prey from trees in thickets. Cheetahs are also selective, and take only gazelle in the open by means of a high-speed chase. Both species would gain little by hunting in groups and so are solitary. Lions concentrate on stalking larger prey, overcoming them by brute force, but are not above robbing other predators of their kills. Hyenas and wild dogs both hunt in large groups, searching out young or weak animals to exhaust and kill by harassment. Dogs hunt mainly in the daytime, while hyenas favor darkness. These latter three predators are, of necessity, social. The differences, then, between how, when, and what they hunt accounts for the relatively stable coexistence of these five predators.

STUDY QUESTIONS

1. Lionesses take no part in defending the ruling males from challengers. What are the likely costs and benefits of this choice, and how does it relate to their short- and long-term fitness? Does kin selection play a role?
2. How do you suppose the unusual social structure of hyenas might have evolved?
3. Until recently, the sixth major predator in the Serengeti was man: a slow but intelligent hunter using poisoned arrows which kill the confused victim several hours after it has been wounded. Where in the niche division does this sixth strategy fit in? What effect do you think primitive human hunters have had on the prey and on other predators?

FURTHER READING

Bertram, B. R. "The Social System of Lions." *Scientific American* 232, no. 5 (1975): 54–65.

CHAPTER 28

Primates

Except that they lack rumens, most primates might be thought of as arboreal ungulates, making their living off a generalist's diet of leaves, buds, twigs, fruit, bark, and the occasional insect. The remaining few of our order—ourselves excluded—are insectivores. Since they are our closest phylogenetic relations, the evolution and behavior of the two hundred species of primates is of particular interest. Behavioral ecologists, convinced as they must be of the strong and necessary link between niche and behavior, would desperately like to organize the primates along the lines of the neat and tidy synthesis depicted in the chapter on ungulates, thus forging tidy theoretical links between their ecology and their social systems. Out of such an analysis, of course, we hope to find evidence of the evolutionary road along which our species traveled. Two major obstacles stand in our way, however. The first is habitat: whereas the many plains-living ungulates are relatively easy to study, the typically shy, tree-dwelling primate presents enormous difficulties to students of behavior. As a result, little is known about most primates. The other difficulty is that even given what *is* known, primates show only vague signs of having suited their social systems to their ecology, or at least to our perceptions of it. Despite these obstacles, though, there ought to be hope since primates do demonstrate the same wide ranges of social behavior and accompanying evolutionary trends which have proven so useful in understanding the ungulates.

PRIMATE EVOLUTION

Fig. 28–1 Bush babies are among the most primitive of the primates.

The predominant evolutionary trends in primates, as in ungulates, are toward greater size, and toward increasing diurnality—more activity during the daylight hours. With the increasing importance of daylight in their lives comes a concomitant trend toward binocular color vision for better visual resolution and visual (gestural) communication. The most primitive primates are the small, nocturnal prosimians such as the tarsiers, lemurs, bush babies (Fig. 28–1), and tree shrews. Except for the tree shrew and the bush baby, which are solitary, most prosimians are monogamous pair-bonders and occupy small, exclusive territories, out of which they drive their offspring.

Out of this inauspicious beginning came the higher primates, whose social systems run the gamut from strict monogamy to unconstrained promiscuity. Except for the harem structure, this entire range can be observed in our nearest (and, we would like to think, most highly evolved) relations, the twelve species of anthropoid apes. Gibbons pair-bond on defended territories. Like tawny owls, they force their offspring to fend for themselves once they are mature. Gorillas, on the other hand, live in groups of many adults in which social stability is maintained by dominance. At the top of the hierarchy is usually the oldest "silverback" male, followed by ever-younger ones, females, and juveniles (Fig. 28–2). Orangutans, the "old men of the

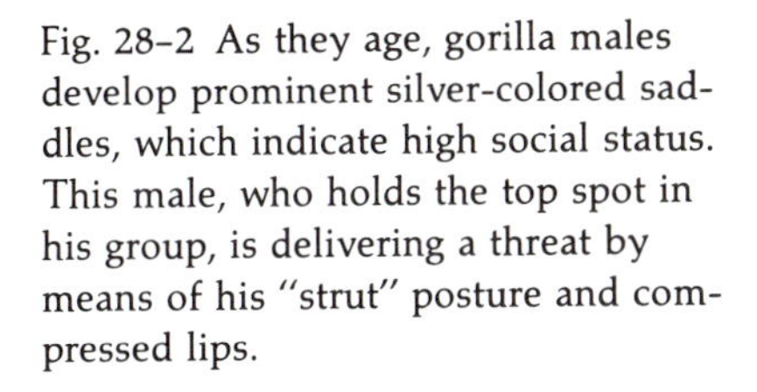

Fig. 28–2 As they age, gorilla males develop prominent silver-colored saddles, which indicate high social status. This male, who holds the top spot in his group, is delivering a threat by means of his "strut" posture and compressed lips.

forest," wander about in solitude on loose home ranges, stumbling upon each other from time to time. Native legend has it that the wise orangs could talk if they wanted to, but that they are too smart—someone would just put them to work if they did.

Chimpanzees have yet another system. Large groups occupy distinct home ranges, but the group is rarely together. Instead, the chimps form small subgroups which may gather, change, and disperse several times in the course of a day (Fig. 28-3). This casual, undisciplined system of loose associations is mirrored in their completely promiscuous sexual system. And yet underneath this egalitarian and libertarian facade is a strict hierarchy.

We cannot hope in one chapter to do justice to the vast body of data, at once massive and miserably incomplete, that we have accumulated on primates. The only lesson we learn from phylogenetic and ecological comparisons as they relate to social systems in primates is that essentially anything is possible and virtually nothing can be predicted with confidence. Our focus will be on social behavior, then, rather than on behavioral ecology, for out of this chaos common threads run through higher

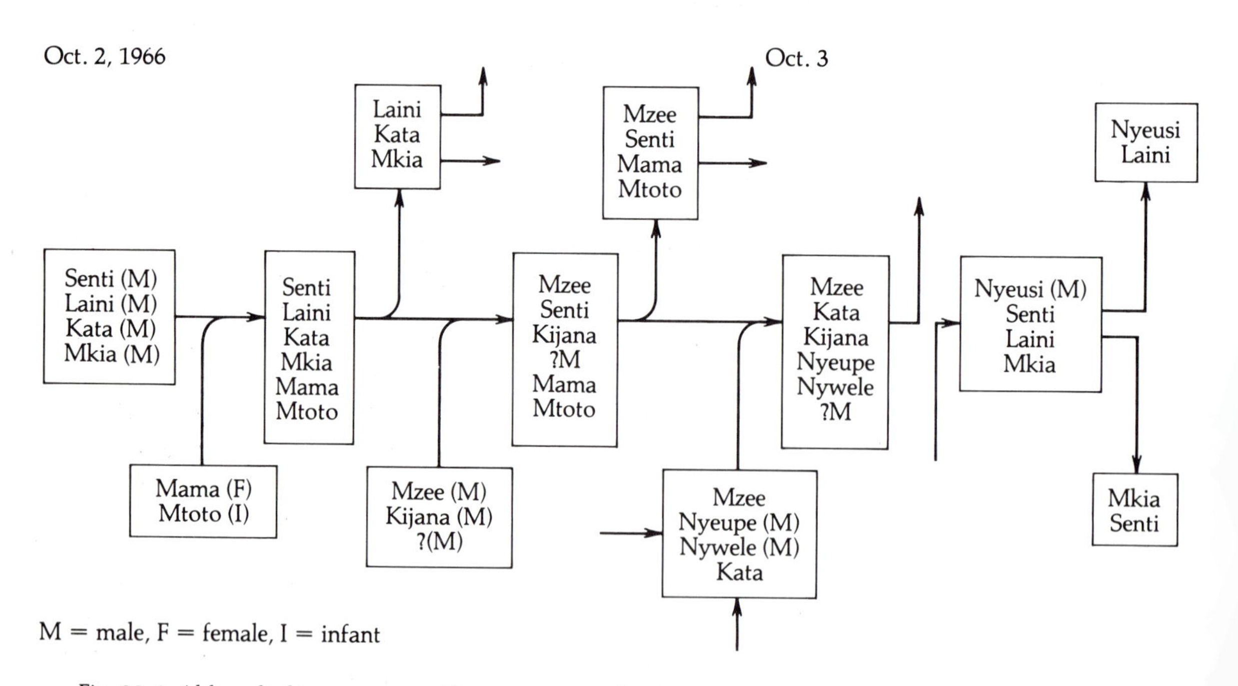

Fig. 28-3 Although chimpanzees reside as troops on discrete territories, their daily lives are spent in loose, ever-changing parties. Here are two days of social recombination in Budongo Forest.

primate sociobiology—threads which can distinguish our order from others.

PRIMATE HIERARCHIES

Most of these diagnostic similarities revolve around the various ways in which dominance may be established and maintained. The typical nonprimate systems, in which several males coexist, are all based on some sort of linear hierarchy. One familiar version is the territorial matrix in which males establish, defend, and are dominant on their own territories, which serve as places to mate and/or feed. Territories may be of differing qualities with the best ones subject to greater competition, and so a male hierarchy is inevitably established based on which has managed to acquire and maintain the most favorable spot. Excluded males often form bachelor herds with their own explicit dominance hierarchies. In another version, the "tending bond" typified by buffalo and mountain sheep, males engage in one-on-one ritualized tests of strength and endurance which establish their positions in the hierarchy. In any of these cases the females typically have no detectable hierarchy.

Primate hierarchies, in contrast, usually involve both sexes and are often based on an intricate system of cooperation. Some, like Japanese macaques and olive baboons, live in troops of forty to eighty animals which forage on the ground in the woodlands and grasslands. Troops are organized such that several dominant males are in the center with the females and juveniles, surrounded by other adult males. From behavioral observations it seems obvious that the function of this particular arrangement is defense. Rigidly enforced when the animals are out of the sheltering safety of their trees, the structure is relaxed in the forest. The troop is run by the few dominant males as a gang, which has an internal hierarchy of its own. The males, probably brothers, cooperate in gaining control of a troop and continue, Mafia-like, to maintain their ascendancy by defending each other. Although a particular adult male outside the ruling clique might well be able to defeat any of the dominant males singly, he stands little chance of challenging them successfully as a group. Hence this sort of gang rule is extraordinarily stable. Though old, lacking his canines, and blind in one eye, the alpha male of one troop of Japanese macaques remains

Fig. 28–4 Twenty-five-year-old "Arrowhead," though slightly built, missing his canines, and blind in one eye, is the alpha male of this troop of Japanese macaques. Despite his obvious weakness, he is kept in power by his allies, none of which has ever been observed to challenge him.

in power because of his small but fanatically devoted group of supporters, who perhaps prefer a known evil to the unknown (Fig. 28–4). In fact, the way to make it to the top in many primate societies is either to form your own gang of malcontents or to ingratiate yourself with the powers that be.

This latter strategy is illustrated by the hamadryas baboons. These savannah primates are organized into strict harem groups of perhaps three females per male. Depending on food availability, the harems forage independently or as parts of larger groups. But even in a group the harems do not mix (Fig. 28–5), and the pasha male not only defends his harem from other males, but punishes the slightest dawdling or mixing on the part of any of his harem (Fig. 28–6). Bachelor males assemble harems of their own either by kidnapping young females from established harems, a risky process unless the female is too young to reproduce, or by following a harem in a totally submissive way until the harem males begin to display some degree of toleration toward them. Later the subordinate male may be incorporated as a dependable ally of the ruling male and become eventual heir to the harem.

Females in primate societies also form dominance hierarchies. In multimale systems these hierarchies are based not on contests or rituals, but rather on male alliances: that is, not on what you know, but whom you know. A female who is tolerated

Fig. 28–5 Although many harems of hamadryas baboons will collect together in the open for the sake of defense, the harem units remain distinctly separate and tightly clustered.

Fig. 28–6 When a female hamadryas baboon strays too far from her harem she is soundly rebuked by the male, which bites her on the back of her neck. Her screams send females from other groups running to the protection of their males.

by the highest ranking male (a toleration indicated by his magnanimous indifference to her presence or his passive willingness to allow himself to be groomed by her) receives from these "favors" the conferred benefit of his status. When she is attacked or threatened, his reflected glory allows her to retreat into the dominant male's vicinity where her attacker dare not pursue her, and from this sanctuary to threaten the would-be attacker herself. Her antagonist cannot respond to such a "protected threat" for fear of reprisal from the male, should he chance to see the threat. The favor of a less dominant male confers a proportionate status on the female so blessed. Of course, when a female's dominant friend dies or falls in rank, her fall, in the grand tragic manner, is inevitable.

The intellectual capacity of the higher primates is the factor which enables every member of the troop to keep a running register of all current alliances as well as their practical consequences. In fact, social order in the simian world is even more wonderfully complex than this, with the female hierarchy in particular exhibiting many far-reaching consequences. Much of the dominance hierarchy of the next generation may be predicted from the present one: the sons and daughters of high-ranking females will usually "inherit" some of that status, while the offspring of low-ranking, friendless females are likely to wind up at the bottom of the social ladder (Fig. 28–7). At least part of the mechanism of this phenomenon is quite clear: mothers intervene in the inevitable squabbles which erupt between juveniles, and so the status of a parent often determines the outcome of peer conflict. This social asymmetry often persists into adulthood. Indeed, the relative adult status of a group of brothers and sisters is also strongly influenced by the way in which the mother chooses to intervene among her own young, with the intriguing result that older brothers dominate younger ones, while younger daughters tend to be "mothers' pets" and to dominate their older sisters.

The result of these bewildering relationships is a stability and social order almost medieval in character. "Place" is established by a genetically controlled cultural process, and at the surface we observe apparent friendship and social tranquility. In fact, close observation reveals that beneath this peaceful facade subordinate animals live in nervous terror of the wrath of their "betters," and look to them constantly for reassurance (Fig. 28–8). The peace and calm, the seemingly frictionless workings

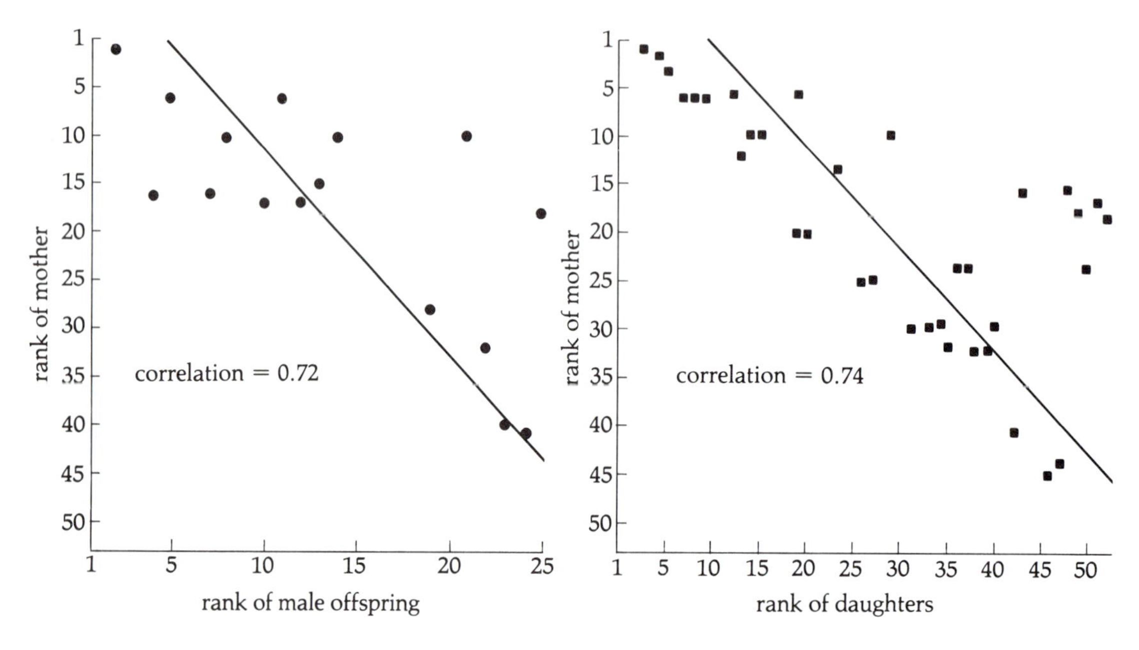

Fig. 28–7 Social status in many primates is strongly influenced by the rank of an individual's mother. The mother's rank is plotted here against those of her offspring. The regression line for both males and females shows a strong correlation with maternal rank. (1.00 would represent a perfect correlation, 0.00 no correlation, and −1.00 a negative correlation.)

Fig. 28–8 When they are upset, subordinate chimpanzees seek reassurance from one of the dominant clique. A. A female with a six-month-old infant is comforted by a touch from a male. B. An adolescent male reassures an adolescent female, again with a touch.

of the society, are actually the result of a continual sequence of subtle dominance interactions—passive displacements, averted glances, submissive postures, grooming ceremonies, and constant attention to the moods and movements of the leaders (Fig. 28–9).

Fig. 28–9 A group of male rhesus monkeys groom each other in order of increasing status. Grooming is a crucial element in dominance: high-ranking individuals constantly remind lower ranking ones of their superior status by presenting themselves for grooming. Subordinates, perhaps seeking to ingratiate themselves, solicit opportunities to groom their superiors.

One consequence of this slavish "enculturated" concern about the mood of the dominant individual, the enhanced stability of the gang-rule system, and the whole self-stabilizing network of social alliances, is that the behavior of the dominant individual can become quite idiosyncratic through the years without provoking a challenge to the hierarchy from below. Indeed, the other males in the ruling group must be continually repressing themselves in order to make their behavior conform because quite often, after the alpha male dies and the beta male succeeds him, the new alpha male begins to run the troop in a completely different manner, with, of course, the wholehearted support of the rest of the gang, be he saint or villain.

Another product of the close attention social primates must pay to even the most subtle indicators of mood and intention on the part of their leaders is reflected in the uncanny ability of chimpanzee males to coordinate their hunting efforts. One of the favorite targets of these, our nearest relatives, is juvenile baboons. A typical hunting sequence will begin as one chimp nonchalantly stops whatever he is doing and wanders, poker-faced, into a group of baboons. Silently, almost magically, other males descend from neighboring trees and begin in what seems to be a haphazard way to meander toward the same general spot. Suddenly they all close on a victim and in an instant the hunt is over. The chimps immediately divide up the spoil, and the enraged baboons soon become calm, apparently none the wiser in the long run. Infrequent as they are, such hunts illustrate how closely and continuously the chimps must be watching each other.

PRIMATE COMMUNICATION

We know that primates communicate by sound as well as by facial and bodily gestures, but understanding their vocal repertoire has proven difficult. Consider the advantages anthropologists make use of when they go about learning a new, unwritten language. First, they can get to know the culture and the way it works by living in it. Second, they may share a species-specific language structure (Chapters 30 and 31) with the speakers. Third, they understand instinctively many nonverbal gestures and expressions that convey much of the meaning of any human language. Finally, their brains automatically subdivide the wealth of acoustic stimuli into the same discrete categories as the senders of the messages do, as we will see in the next chapter. With primates, none of these advantages exists. Our evolution has best prepared us to comprehend ourselves, so that deciphering the communications of other species must be accomplished without the aid of nature. The relatively simple, straightforward, and obvious communication system of honey bees, for example, was observed by countless individuals throughout the course of history, and studied intensively by von Frisch for forty years, before even he could decipher its most basic message.

Imagine, then, the problems posed by primate communication. It used to be thought that the difficulty in finding the key to primate language lay in the stupidity of the apes, in their lack of any "real" or consistent vocal communication like ours. This view was fed by the failure of the heroic efforts of behaviorists to teach chimpanzees to speak English. We know now that the physical construction of the chimpanzee vocal tract prevents them from producing the necessary range of vowel sounds, while they may readily be taught to communicate in a simple language employing hand signs, plastic symbols, or illustrated computer buttons as words. In fact the most formidable barrier remaining in communication studies of the great apes may be that, unlike human infants who are innately driven to learn language, young monkeys simply have no interest in mastering this particular arbitrary form of communication.

Armed with a more thoughtfully calibrated respect for primate abilities, then, recent research has been able to turn up two preliminary clues. First, at least some of what primates are saying to each other is coded along auditory continua instead of being expressed as discrete sounds, rather like the difference between the way a bee signals 325 m at 18° right of the sun and the way we say the same thing. Bees can express any distance and direction, but the differences are graded, reflecting various locations on a continuum, whereas our words are discrete entities, not related acoustically to each other at all. A second observation is that at least some of the graded signals primates produce may actually be distinguished as if they *were* discrete, being recognized individually according to special, species-specific boundaries which we simply cannot perceive.

Japanese macaques, for instance, produce many classes of sounds, one of which is known to students as "coo" sounds. These all sound more or less the same to us, but careful sound-spectrograph analysis reveals the existence of seven subclasses. In one field study the behavior of the macaques was recorded along with their sounds, and later analyzed for correlations. Each of the seven aurally indistinguishable variations on the "coo" theme was correlated with a different behavioral situation. For example, certain coos awkwardly labeled "smooth-early" (for their peculiar acoustic characteristics) are produced by calm, isolated youngsters, while "smooth-late" coos are generated primarily by estrus females attempting to initiate courtship. We know now that the auditory processing of Japanese macaques is specialized for making these particular subtle distinctions. This was demonstrated by comparing how long it took the animals to begin distinguishing prerecorded coos on the basis of this acoustic characteristic in an experimental situation, as compared to two other species of macaque. On the average the two other species took five times as long. When asked to distinguish coos on the basis of the frequency at which the coo started, however, Japanese macaques were almost four times slower than the other species. Clearly, evolution has built even closely related species to attend to very different stimuli, and to respond to the same stimuli in very different ways. Although the evidence is not conclusive, there is reason to suspect that the Japanese macaques have innate "coo" filters, a set of acoustic feature detectors or sign stimuli which enable them to decode the signals of conspecifics automatically without having to worry about features that are irrelevant to their particular system.

PRIMATES AND HUMANS

Such experiments show that higher primates are equipped both with the mental wherewithal and the innate processing circuits necessary to deal with the world in terms of discrete signs. Both these lines of evidence argue against the old notion that our species is different from the animal world in kind rather than in degree—a line of argument which has been based on our previous inability to teach primates and our ignorance of their physiology and life histories. Long lists of "distinctions" used to be complacently drawn up featuring examples of "uniquely human" behaviors—tool making, cultural learning (Fig. 28-10), creative problem solving—which we now know to be going on daily in the wild, among even some of our "lower" animal

Fig. 28-10 Tool using and social learning are crucial elements in the chimpanzee social system. Here an adult has selected and pruned a stick which it then uses to "fish" for ants or termites. A juvenile male observes the process narrowly and shares in the spoils.

compatriots. The most precious distinction, however, has always been our elegant sort of species-specific language, with its use of arbitrary, culturally defined "words" to *name* things as opposed to the rudimentary expression of emotion of which we supposed other animals to be capable. Alas, even here we are not alone: at least one species of monkey, the vervet monkey, is now known to have apparently arbitrary, acoustically distinct warning calls for different kinds of predators, one each for leopards, eagles, and poisonous snakes, and probably a fourth for primates.

The final bulwark of those who want to believe *Homo sapiens* to be a special creation is our "self-awareness," our ability to have individual thoughts and feelings, personalities and intentions. This assumption, often espoused by people who have never observed higher animals carefully, puts us comfortably outside the evolutionary continuum. Ironically, much the same view once prevailed among Europeans who were confronted for the first time with the peoples of Africa, America, even Asia—and as far as we know, vice versa. As we have come to know the rest of the primate world, however, the view that we are different in kind has become more and more absurd. Higher primates clearly have personalities, intentions, thoughts, as well as "feelings." Anyone who has ever worked with them has come to feel this keenly, but evidence to convince those who can only learn about them through books has been lacking. What, after all, constitutes the "awareness of self" that philosophers and humanists consider so crucial to a "thinking" animal like ourselves? One criterion hitherto considered paramount has been that only we have mental "pictures" or concepts of ourselves in relation to the world around us, a sort of awareness which we have come to understand as individuality.

A preliminary laboratory demonstration of this "self-awareness" as it exists in our animal relations comes from the work of G. G. Gallup. He anesthetized chimpanzees and other higher primates familiar with mirrors and then marked the animals with spots on their heads. When allowed to look in a mirror, they each examined the spots curiously, and began immediately to rub at them in an effort to remove them—indicating that they knew the images in the mirror to be representations of themselves. Interestingly, gibbons and lower primates do not react in this way. Perhaps they simply cannot comprehend mirrors; alternatively, it is tempting to suppose that they may lack the ability to form the all-important "self-image" and all that this sort of awareness implies.

The realization that higher primates live in a world of thought and emotion must forever change our view of them, of ourselves, and of our evolution. A baboon mother who carries her dead infant for days, refusing to eat or to interact with her group, appears in a new light (Fig. 28–11), as does the discovery that hundreds of thousands of years ago our rough, brutal ancestors seem to have buried their dead with flowers. More than anything, an increasing familiarity with primates serves to remind us of our place on the evolutionary continuum, and reassures us that the thoughtfully anthropomorphic point of view which behaviorism traditionally deplores has its value in guiding our speculation. The hypotheses such thinking engenders need be consigned to the realms of fantasy only if we are unable to come up with convincing and critical ways of putting them to the test. Establishing the extent of our similarities to and dissimilarities from our nearest extant evolutionary ancestors provides a valuable opportunity for the exercise of our critical and imaginative faculties.

Fig. 28–11 When an infant baboon dies, the parent often carries the dead infant for days, occasionally attempting to wake it, and may refuse to eat or interact with the group. This oddly human reaction to death is most obvious in chimpanzees, and is in sharp contrast to the obliviousness or uncomprehending curiosity exhibited by most mammals.

SUMMARY

The primates, though our closest relatives, are difficult to study. As with ungulates, the basic trend in primate evolution has been toward increasing size, but in this case no corresponding gradient of social systems is evident. Several features of primate group structures are striking. Males often form gangs by which they dominate groups. Females establish a hierarchy as well, but it is based on male alliances rather than female fights. A degree of heritability of status enters into the calculations. Subtle dominance interactions are frequent: passive displacement, grooming, physical reassurance. This sophisticated a social system depends on small groups of intelligent individuals with good visual resolution. As we come to understand the social communication of higher primates more fully, the more similarities with our species become apparent and our relative positions on the evolutionary continuum begin to materialize.

STUDY QUESTIONS

1. Why don't Arrowhead's small gang of supporters oust him? What adaptive value might the tendency toward "respect for elders" in our order have?
2. In primates, as in most animal societies, females are generally treated by theorists as passive ciphers in the social systems. What active influences might they be generating which have escaped observation?
3. What might account for the existence of a hierarchy among chimpanzees which seems to have no influence on mating?

FURTHER READING

Eaton, G. G. "Social Order of Japanese Macaques." *Scientific American* 235, no. 4 (1976): 96–106.

Premack, A. J., and Premack, D. "Teaching Language to an Ape." *Scientific American* 227, no. 4 (1972): 92–99.

Sackett, G. P. "Monkeys Reared in Isolation." *Science* 154 (1966): 1468–73.

Teleki, G. "The Omnivorous Chimpanzee." *Scientific American* 228, no. 1 (1973): 32–42.

Washburn, S. L., and DeVore, I. "Social Life of Baboons." *Scientific American* 204, no. 6 (1961): 62–71.

Part VIII

Human Ethology

CHAPTER 29

Human Behavior

Modern ethology's greatest opportunity is also its most difficult to realize: we must use the ethological perspective to extend our knowledge of *Homo sapiens.* Our own athropocentric biases give us a curiously foreshortened perspective which makes it difficult not to consider the species a special creation, somehow exempt from the evolutionary continuum. We find it difficult to accept the idea that we too might be subject to the mindless releasers, drives, motor programs, and directed-learning routines that are the handmaidens of instinct, and which reveal the rest of the animal world to be the unwitting servants of their genes. We are often, as von Frisch taught us, blind to our own blindnesses. We cannot readily see what evolution did not design us to see, or perhaps to think or imagine beyond the built-in limits of the system. We assert that we are different because we alone of all nature's creatures know what we are doing and determine our own behavior. But the evidence that is accumulating suggests that our behavior too is to a surprising degree controlled by internal programming; and although we can perhaps overcome that programming to a greater extent than other animals, we are still more influenced by it than we have been taught to think.

INTROSPECTION

The basic difficulty we face in analyzing our own species is our admittedly natural tendency to trust introspection. After all, we have a tempting advantage in studying *Homo sapiens:* we can ask members of this species direct questions and they will give us answers. This straightforward process should, it would seem, short-circuit the need for tedious, well-controlled experimentation. And, of course, the simplest individual for an investigator to consult is himself. Alas, this methodological frontal attack on our species, the technique used by countless generations of philosophers, makes three unlikely assumptions: first, that we are honest with ourselves (much less with others); second, that what we are conscious of is all, or most, or at least the most important part, of what is going on in our CNS; and third, that the information being served up to our consciousness after several rounds of species-specific filtering and processing is an accurate representation of the objects and events around us.

In fact, there is no reason to suppose that we are aware of most of the "thinking" that is going on within us. We are blissfully unconscious for instance of the orders continually being issued to direct such vital activities as breathing, walking, digestion, or the pumping of blood. Our role as passive observers extends to surprisingly intellectual levels: we seem to be regularly handed "answers" without any meaningful access to the data under consideration. Patients who have lost all or part of their visual cortex are, of course, blind—completely unaware of anything in the affected part of the visual field. Yet if asked to guess what might be in front of them, they prove able to reach for objects with remarkable accuracy and to describe their shapes and orientations.

Similarly, a standard psychological test involves flashing one or more patterns on a screen for a fraction of a second and asking the subject what he has seen. Very short presentations are perceived unconsciously and can elicit any of a number of specific emotional and motor responses, though the subject will be consciously unaware of what was projected. Slightly longer presentations *are* consciously perceived, but the perception is indirect: patterns with missing elements—a flower missing one of its six petals—are seen as whole; a word with a missing or inappropriate letter is perceived as complete and correct; and so on. Indeed, patients with partial lesions, able to see and verbally report only, say, the right visual field, consciously perceive patterns as extending into the left field, whether or not the pattern being shown them exists in the left field. On a far smaller scale, our minds regularly fill in the visual gap—the blind spot—in each retina, a gap necessitated by evolution's curious "decision" to put the processing cells and the axons carrying their output *between* the lens and the retina. (The blind spot is the hole in the receptor array through which the axons pass on their way to the optic chiasma.) Our minds automatically extrapolate from contiguous patterns to fill in gaps in visual data, as part of that essential and unconscious processing which constructs our conscious experience.

All this is possible because the visual information received by the eyes has gone first to the optic tectum (called the LGN in mammals), the only internal projection of the visual world that fish, reptiles, and amphibians possess. It is here, completely below the level of conscious visual experience, that these relatively sophisticated discriminations are probably made. Who knows what other sorts of information processing may be accomplished outside of consciousness? Indeed, there is no reason to suppose that we are conscious of *any* of the routine sorts of sensory processing and decision-making that are programmed into lower animals, and probably into us.

In addition to this "predigestion" of sensory data is the powerful but unconscious programming of "mood" which goes on inside animals, humans included. We are all familiar with the workings of adrenalin, testosterone, and estrogen, but research has recently discovered that the brain employs a battery of additional chemicals to control attitude, attention, and temperament. Among the entries on this rapidly growing list of chemical mind-benders are LRF (lutenizing release factor, which apparently acts as an aphrodisiac), beta-endorphin (an internal pain killer), the enkephalins (involved in depression), factor S (a sleep controller), bombesin and somatostatin (which compete with each other to modulate the sympathetic nervous system), and so on. Each has sweeping effects but normally no failsafe provision for conscious control. What moderates the levels of these substances if not our programming, which cues itself off releasers as yet unknown, and what other bits of behavior-controlling chemical clockwork are yet to be discovered?

If we have as it seems little access to the real data of life and only partial control of mood, at least, we might argue, we have consciousness to monitor and analyze what is going on, to deliver informed and truthful answers to our inquiries, and to try to set things right when our programming begins leading us in directions we do not wish to go. But what is human consciousness, and how honest and reliable is it? Is consciousness an epiphenomenon, an accidental side effect of some increase in brain size? And if so, does it serve as a primarily passive observer of what is happening, able to comment but not to control, or has it seized some of the behavioral reins? If it is not an accident, might consciousness be an evolutionary strategy, selected for by nature and designed to deal with certain classes of problems which face our species for which more conventional programming fails?

As we saw in Chapter 18, all these questions arise in a greatly subdued form even in honey bees, and the answers are not easily to be found. The most important problem to appreciate, however, is that introspection simply cannot be trusted for the answers to the crucial question of the source, evolutionary purpose, and consequence of human behavior. For one thing, a surprising amount of what passes for conscious behavior in humans appears, upon investigation, to originate in the unconscious. One sobering example is what might be called "justification"—finding reason or purpose for our own behavior or for external events. Our species seems uniquely driven to make sense of—to justify—the senseless, the uncertain, or the unknown, while other species face such ambivalent circumstances with placid apathy or generalized fear. Consider something as obvious as the behavior of a small child when asked why he did something or how something works. The ready reply is most likely to be a stream of nonsense rather than the truthful "I don't know." We seem to come ready-equipped with the capacity to hide our true thought processes not only from others but from ourselves as well, and when logic fails we appear actually preprogrammed to substitute surmise and hypothesis for fact.

SPLIT-BRAIN RESEARCH

This behavior is compellingly illustrated by what are called "split-brain" patients—individuals who lack the cerebral com-

The human brain is a powerful, highly organized, well-programmed computer, designed from birth to deal with our species-specific neural processing needs. So specialized is this mental machine that damage to specific areas produces specific observable defects in neural processing. For example, a lesion (that is, any interruption in the flow of information, whether from an injury, a tumor, unavoidable surgery, or a blood clot) in the underside of the brain near the junction of the occipital and temporal lobes (see Fig. 29–1) results in the peculiar loss of our ability to assign names to faces. This deficit is both striking and highly specific—the victim cannot name even pictures of a spouse or a close friend. The names themselves are not lost since the patient can readily name familiar voices. The patient's ability to recognize faces remains unaffected, too: he can match profile shots with full-face photographs. Nor does the victim be-

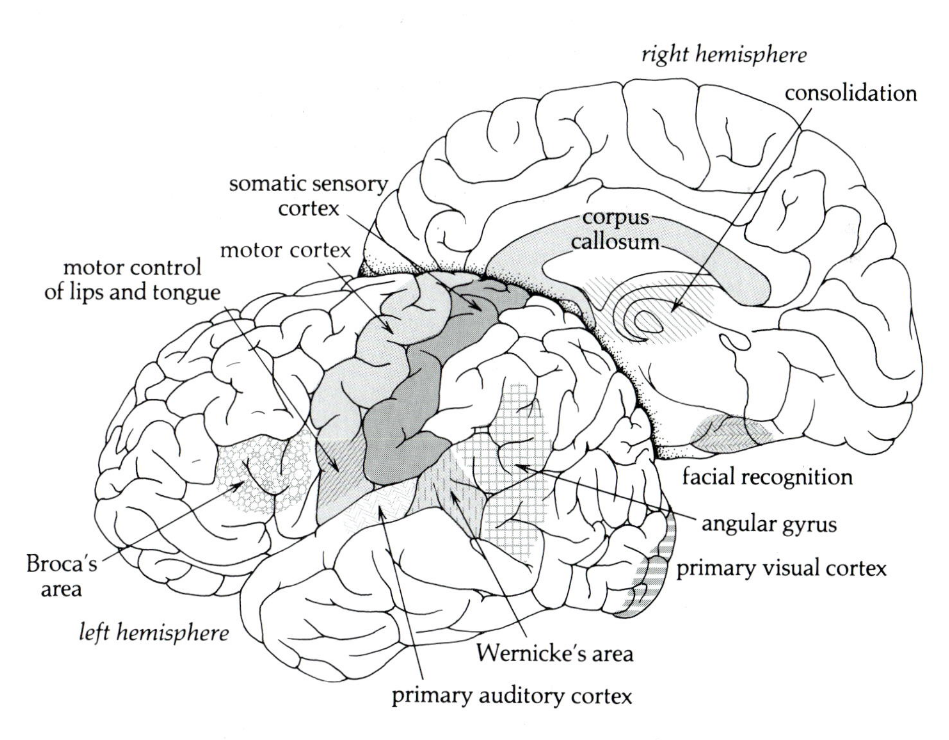

Fig. 29–1 The human brain is composed of a variety of highly specialized processing areas. Shown here are some of the neural centers involved in language and memory as described in the text. For purposes of illustration the corpus callosum, which connects the two hemispheres, has been cut in the manner of a split-brain operation, and the hemispheres separated to expose the inner surface of the right hemisphere. The cerebellum and brain stem have been removed.

come unable to name familiar objects or other animals. The small but critically important area of the human brain which has been affected has one task and one task only: linking faces and their names.

This pattern of preordained specificity is evident throughout the human brain. Lesions in the hippocampus and the adjacent part of the temporal lobe, for example, abolish memory consolidation. The patient can still retrieve previously stored information from long-term memory and can hold current thoughts in mind (in short-term memory) so long as he keeps running over them every minute or so, but following the lesion absolutely nothing new can be learned.

By far the most famous example of the brain's neural specialization concerns the two areas of the brain that process language: Broca's area and Wernicke's area (see Fig. 29–1). Both spoken and written language go to Wernicke's area to be decoded. A lesion here destroys our ability to comprehend either, while blocking the input from the visual cortex (which is processed in the angular gyrus en route) eliminates reading ability without disturbing speech recognition. (In fact, what the angular gyrus seems to do is to turn written language into spoken words which it then whispers to Wernicke's area.) Preliminary encoding of thoughts for linguistic expression also occurs in Wernicke's area, but these signals must be processed by Broca's area before they can be turned into grammatically correct language. A lesion in Broca's area leaves the patient knowing what he wants to say but unable to express it according to the linguistic conventions of proper tense, declension, number, gender, etc.; the use of other linguistic refinements such as pronouns, conjunctions, and prepositions is likewise severely affected. A lesion in Wernicke's area, on the other hand, leaves a patient talking perfectly grammatical nonsense.

Yet more striking, both these linguistic areas are found on only one side of the brain—the left in almost everyone. The other side can recognize a few simple nouns, but that's about it. Since both hemispheres share all the information they process with each other, the existence of a "mute" side comes to light only when one side is destroyed or when the connections between the two are cut (the so-called split brain procedure). Only with highly refined testing can the workings of the separate hemispheres of a normal person be dissected. Equipment and techniques have only recently been designed to detect the extra 0.2 sec required for information from the "wrong" hemisphere to be rerouted to the proper side.

After a lengthy series of comparisons, though, it has become clear that the left side is also specialized for analytical mathematics and contains the learned, hardwired programs for fine motor control. The right side, on the other hand, appears to focus on spatial and acoustic patterns. This leads to the odd result that although people with extensive *left* hemisphere damage cannot talk, they are able to sing, and despite their inability to understand the linguistic content of speech, they are able to recognize the patterns of inflection and shading which convey its emotional overtones. In contrast, a person with severe *right* hemisphere lesions may be perfectly able to decode the linguistic content of a statement without having a clue as to whether the remark was made in anger or in jest, with sorrow or delight.

Studies of split-brain patients tell us in addition something about the way memory is stored on the two sides and how associations are exchanged. If the pathological conditions which require such surgery allow only the posterior half of the "bridge" between the two hemispheres to be cut, the visual experiences of each of the two hemispheres are isolated, but no longer, as is often the case with split-brain subjects, does the conversational left hemisphere deny that anything has been seen when experimentors show the right side a word. Instead, the left side in these partial splits will claim to have glimpsed a picture, though only through a game of mental pantomime can it identify the word. When the right hemisphere of one such patient was shown the word "knight," for example, the left began a monologue about tents and banners, tournaments, men on horses, and finally, in this case, guessed the right word. The word had evoked a set of pictorial associations from the seemingly unconscious right hemisphere, associations which would normally accompany the visual representation of the word itself, and perhaps color its interpretation on the verbal left side.

This difference in cognitive method between the two hemispheres is a fascinating strategy. Our brains process the same input, spoken and written language in particular, in parallel by two entirely different programs to extract its "full" meaning. Could it be that this dual approach, this joint collaboration of two specialists on the same problem, is the trick which accounts for our species' intellectual preeminence? At the same time, could this two-headed strategy be the source of our species' ubiquitous anxiety and corrosive internal conflict? In any case, the human mind is gradually being revealed as a device whose overall operation is in no way left to chance, the logic of whose masterful design and programming stands as the single greatest challenge to us as students of behavior.

missure, the enormous nerve tract which connects the left and right halves of the brain. For years it was thought that the operation which "split" the brain had no effect other than the desirable one of reducing the frequency of epileptic seizures. Indeed, the patients noticed no difference in their own behavior after the surgery nor did their friends and family, their doctors, or most psychologists. Roger Sperry and his colleagues, however, have discovered that the patients end up, in effect, with two independent brains with no channel of communication between them. More important, only one of the hemispheres has sufficient proficiency in producing speech to communicate with the experimenters. This finding led to a remarkable series of discoveries about prewired lateralization and specialization of brain function: the two halves of the brain each handle entirely separate sorts of functions in our lives, functions which are still being discovered. Of special interest in this context is the intellectual behavior of a split-brain patient when forced into a mistake. Since the left hemisphere receives its input from the right hand and the right half of the visual field (Chapter 7) and vice versa, it is possible, for example, to show a picture to one hemisphere and to ask the hand of the other to select the corresponding object out of a box on the basis of feel alone (Fig. 29-2). Split-brain subjects will gamely search and select an object, but their choice is almost always wrong. When faced with this discrepancy between intent and action and asked why they have made the wrong choice, subjects respond with a wide range of rationalizations, from accusing the experimenter of having shown a picture of the wrong object to explaining how the incorrect object was, after all, similar to the picture in many respects. After a series of such obvious mistakes a patient will become agitated, but may be calmed by being reminded of his surgery and its effects, and of the very artificial testing conditions. Thus reassured, the patient continues the experiment but persistently produces the same fanciful explanations for mistakes as before without ever saying he doesn't know or referring to the neurosurgery.

In short, these patients resolve the intellectual dilemma through a quick, sincere, thoroughly unconscious fabrication. If different pictures are shown to a split-brain subject's two hemispheres simultaneously and the subject is asked to choose appropriate matches from a set of pictures, each hemisphere will direct its hand independently of the other to its own logical

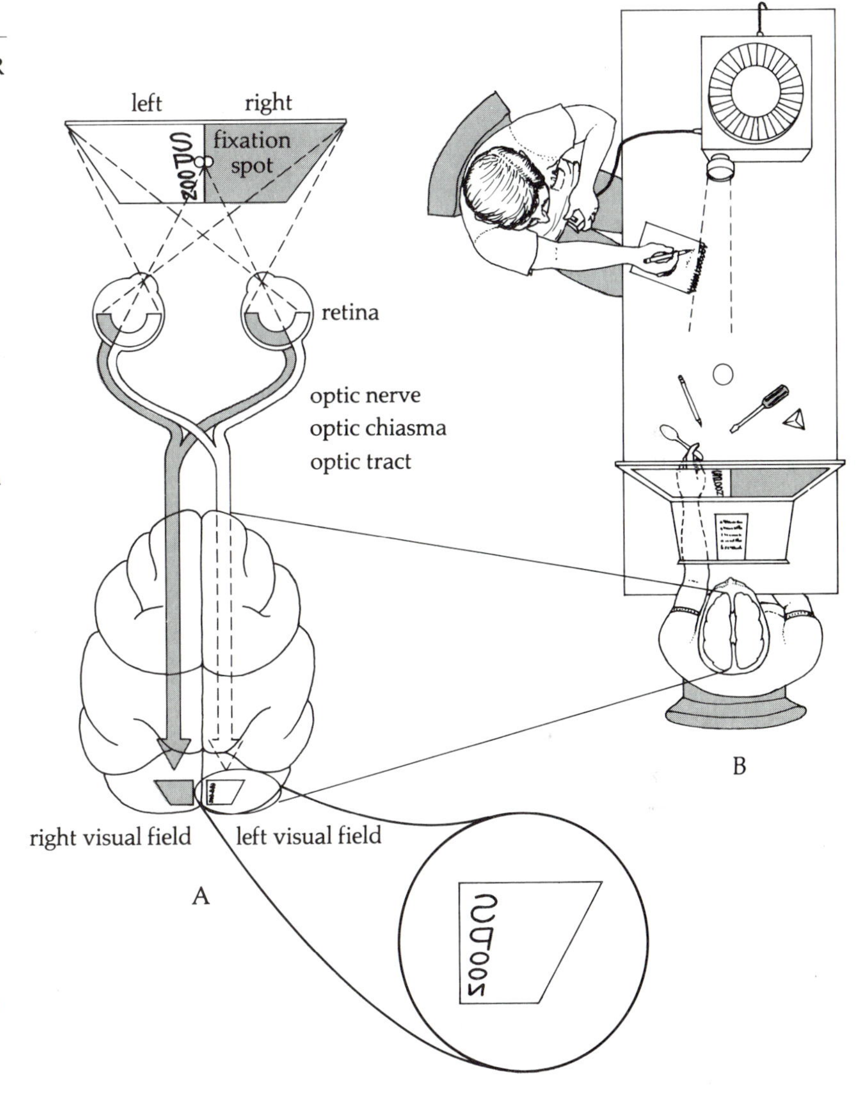

Fig. 29-2 Split-brain subjects lack any connection between the two hemispheres. Hence, when they fixate their eyes on a spot, each side perceives a visual field not seen by the other (A). In this example, the word "spoon" is flashed briefly into the left visual field and the image is projected exclusively to the right visual cortex. Although the subject's conversational left side will assure the experimenter that it saw nothing, the left hand (which is controlled by the right hemisphere) can, on command, pick out the spoon by feel (B). The right hand would fail this task.

choice—yet only the verbal side will be able to explain its actions. If the verbal side (usually the left) sees a bird's claw and the right a snowy landscape, the right hand will select a picture of a chicken and the left a snow shovel (Fig. 29-3). Asked why he chose as he did, the subject's conversational left hemisphere will unhesitatingly respond that the claw of course belongs to the chicken, and the shovel is for cleaning up after it. Confronted with a nonsensical situation, his brain strives valiantly to impose some order on it, to make comprehensible the inherently senseless.

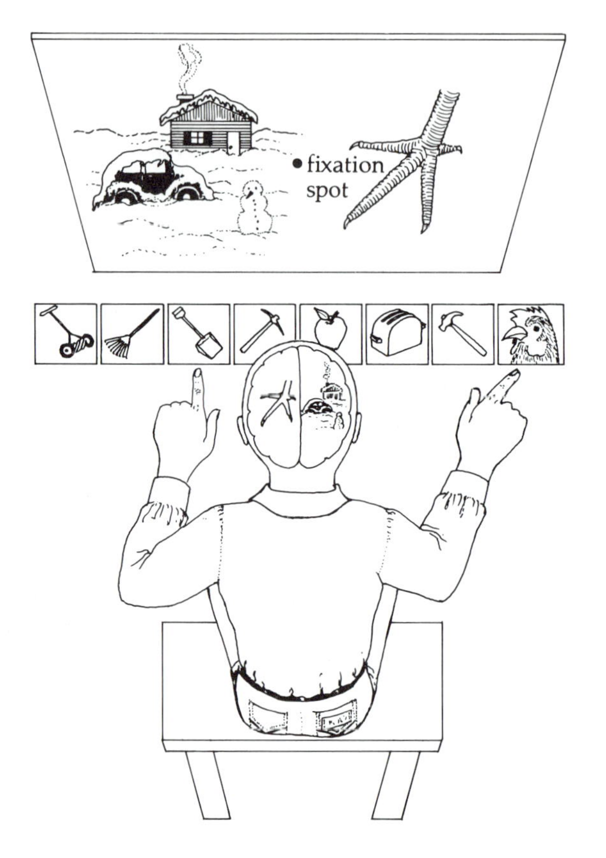

Fig. 29-3 When the two hemispheres view different pictures, the two hands select different best matches from an array of choices. Despite this obvious disagreement, the conversational left hemisphere has no difficulty in explaining the logic of both choices in regard to the one picture it saw.

Nor is this overwhelming urge to fabricate (and to do it automatically and well) restricted to split-brain patients and children. "Confabulation," as it is known formally, has been the subject of countless experiments on normal subjects. In a typical example, a table covered with perfectly identical pairs of pantyhose is presented to subjects who are then asked to choose a pair. After a subject has made her selection, she is asked to explain her choice. Such a subject has no difficulty in explaining her preference, usually as the softest, sheerest, most pleasingly colored, and so on. The answer "I don't know" is almost never recorded.

So quick and unconscious is this justification behavior—in the split-brain case the linguistic hemisphere's "coverup" for the unaccountable foibles of its mute companion, and thus its creation of order out of chaos—and so early is its appearance in children, that it is hard to believe it is not innate. The greatest moments in science, for example, have come when minds driven to solve a problem have taken a few confusing, even conflicting bits of data and have created from them a pattern—a theory—to explain the previously inexplicable. This ability and desire to imagine and create patterns is the basis of much of our aesthetics—design, art, and music—our science and superstition, such as early astronomy and astrology, Aristotelean and Newtonian physics, and of our amusements: countless parlor games, puzzles, and optical illusions.

Can such a species, so curiously gifted, so inherently sly, crafty, and devious that it will try to make itself and its world look good no matter what, be trusted to know—or even to inquire—the extent to which it is the master of its fate? Has our species alone come into the world unprogrammed and undirected to conquer the problem of survival, development, and reproduction solely by the force of brute unfettered intelligence? But from what, if not from deceitful introspection, can we hope to construct an answer to such a question?

HUMAN EVOLUTION

One potential source of knowledge might lie in the study of our own evolution. Although it seems obvious that we now live in a world very different from that in which and to which we evolved, the history of that evolution, could we decipher it, might hold important clues. There is precious little that either introspection or experimentation can really tell us about our niche or the selection pressures which got us into it and made us what we are, but the fossil record permits some plausible guessing. We do know that we evolved from arboreal omnivores with excellent binocular vision and well-coordinated hands. We used to be furry, and even now when we get anxious or angry the familiar phenomenon of "goosebumps" is the result of our bodies' attempt to raise fur, an imitation of the practical animal trick of increasing our apparent size when making a threat. When we lost our fur is not clear, but contemporary artists' representations to the contrary, it may have been shortly after our ancestors followed those first ungulates and carnivores into the challenging but virgin niches of the newly created savannahs. A naked primate with a profusion of sweat glands would be better able to regulate its body temperature while chasing down prey in the sweltering daytime heat of the savannahs.

Archeological evidence shows that seven million years ago we (*Australopithecus*, that is) were clubbing and eating lizards, snakes, tortoises, rodents, porcupines, and savannah baboons, and robbing other carnivores of their kills. We were still maintaining a peaceable steady diet of leaves, seeds, and berries, but the dawning opportunity for a carnivorous primate was being exploited too. Our present nutritional needs document this part of our evolution. Most of the amino acids, vitamins, and minerals we require represent metabolic pathways for synthesis or extraction discarded by selection. The ready availability in our diet of these substances rendered them unnecessary and inefficient biochemical baggage. This list of nutrients which our bodies have come to expect points unambiguously to an animal whose diet is well-supplied with meat, or with the complex combinations of grains and dairy products which vegetarians must consume to take its place.

By a million years ago we *(Homo erectus)* were the most dangerous animal on earth, using stone axes to kill animals which had no natural enemies, like elephants and giraffes, and other species of large mammals which we seem to have hunted to extinction. During this period our brain size was beginning to increase at an alarming rate (Fig. 29–4), indicating that being smart was of great selective advantage in this new life. We

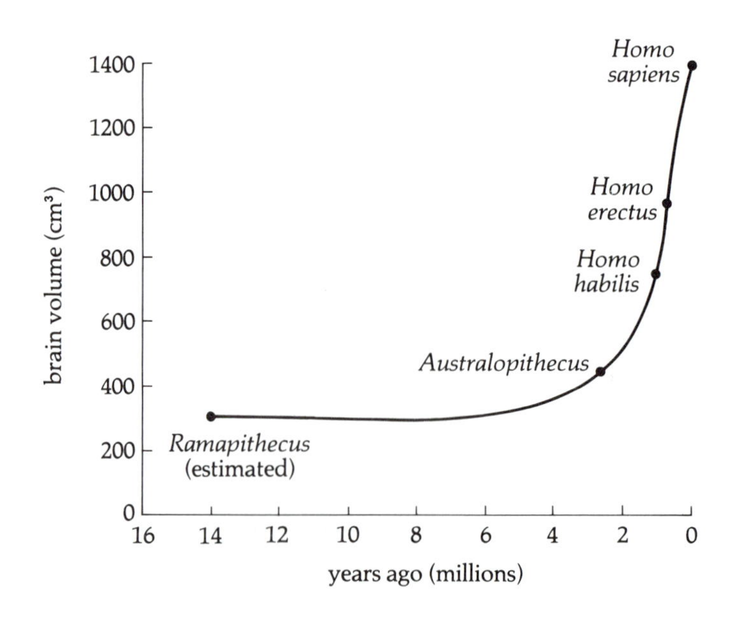

Fig. 29–4 The brain capacity of the evolutionary line which led to humans had been getting larger at a rapid rate—presumably the result of strong selection pressure—until it peaked about fifty thousand years ago and began a slow decline. [Redrawn with permission of Macmillan Publishing Co., Inc., from *The Ascent of Man: An Introduction to Human Evolution,* by David R. Pilbeam. Copyright © 1972 by David R. Pilbeam.]

began to paint pictures of ourselves and our prey on cave walls (Fig. 29–5), and *Homo sapiens* was born.

And then, ten to twenty thousand years ago, there appeared a more familiar, "civilized" version of *Homo sapiens.* It was the evolutionarily recent and sudden domestication, first of animals, and then of crops in the active planting and harvesting of grains, that gave rise to cities, division of labor, "wealth," and, of course, centralized authority, taxation, and large-scale war. And with this sudden, enforced shift to a very new environment with new, vastly larger groups of people with vastly smaller spaces between them, began for better or worse our rapid cultural evolution. With only 250 generations since the invention of large cities, our genes have had essentially no opportunity to

Fig. 29–5 Early man left a tangible record of the propensity and technology for hunting ungulates which made our species even then the most dangerous on earth.

adapt to whatever new selection pressures there must be. Our present environment is obviously artificial, some sort of compromise between what our genes expect and need and what the technology on which we so completely depend demands. Just how much can be learned from the current behavior of this species of clothed animal in its prefabricated cultural zoo is as yet uncertain.

But archeological evidence has its limits, particularly when we begin to ask the key questions of behavioral ecology. For example, our ancestors were clearly social, but what was their system? Were they monogamous pair-bonders or promiscuous? Were there one-male harems or multimale groups with dominance hierarchies? Was the limiting resource for our ancestors high-quality (animal) protein, or carbohydrates, or salt, or water? Or perhaps it was not food at all but mates, or territories for hunting or breeding. And whatever the crucial resource, how did the social structure serve to establish and limit access to it?

"PRIMITIVE" CULTURES

There exist today, however, rare and rapidly vanishing groups of primitive peoples untainted by the technology and civilization which are the necessary consequences of crop cultivation and herding, and which offer an unprecedented opportunity to step back through thousands of years into the world of our ancestors. Perhaps observations of these few existing tribes of hunter-gatherers and the analogies that we can draw from other animals may serve to guide our present speculation and future observation.

The first step must be to look for the ethological patterns of species-typical behavior throughout *Homo sapiens.* This quest is a difficult one. Our eyes, always alert for the qualities which set people apart, need rigorous schooling to perceive those characteristics which we share. When we look at the thousands of human cultures (or pseudospecies) with attentive eyes, however, there are a suspicious number of persistent species-specific characteristics—practices which, although sometimes submerged or metamorphosed almost beyond recognition, nevertheless persist throughout the world. E. O. Wilson lists seven: property rights, body adornment, incest taboos, sexual roles, rites of passage, intraspecific war, and that ultimate example of creating order out of chaos, belief in the supernatural. To what extent are those pervasive practices coincidental, and to what degree might they be part of our genetic inheritance from those uniquely intelligent but necessarily pragmatic creatures which began millions of years ago to carve out the niche we now enjoy?

Primitive peoples, then, share many of our species-specific behaviors, yet live in a niche which resembles much more closely that of our ancestors. Of these primitive tribes, the most interesting in many ways are the !Kung bushmen (the ! represents a "click" which characterizes their language) of the Kalahari, who until the last decade were still carving out a living on the cruel desert/savannah habitat of southern Africa. The very unattractiveness and apparent worthlessness of their habitat protected them until quite recently from the encroachments of other cultures. Their habitat, the most demanding on the forest/scrub/grassland/savannah scale, was first invaded by ungulates, followed closely by their predators: carnivores and our primate ancestors. The !Kung niche differs from those of other carnivores in a variety of ways, the two most obvious being their mixed diet of meat and vegetable products, and the use of technology.

How do niche and social behavior interact? !Kung adults form monogamous pair bonds and invest heavily in their offspring. But unlike the usual pattern among animals, the pairs live in bands of twenty to thirty. These groups occupy semiexclusive territories of about 500 km^2 centered on more-or-less reliable sources of water. Group interchange is frequent as individuals often visit other bands for days or even weeks. The bushmen hunt game and gather fruits and vegetables. This labor is divided along strictly sexual lines: women gather and men hunt in small groups (Fig. 29–6). The gathering is more efficient—women bring in about 60 percent of the protein and carbohydrates, yet work fewer hours than the men. The gatherers harvest more than a hundred species of plants, but their efforts are concentrated on the mongongo, a tasty fruit with an edible, nutritious nut. The game, of course, supplies essential amino acids, vitamins, and minerals.

Both food sources are seasonal, and it is the rhythm of the seasons which most directly controls the behavior of the !Kung. Unlike those of the ungulates, which were preadapted to the

A

B

Fig. 29-6 Labor among the !Kung bushmen is divided along strict sexual lines. The men hunt (A) while the women gather (B; mongongo nuts in this case).

arid savannah, the limiting resource for the bushmen is water. About 95 percent of the rain falls in the six months from late spring through early autumn. During this period the plants grow quickly, providing a plentiful supply of fruits and nuts for the bushmen and forage for the animals they hunt. Water is widespread. A group will settle in on a mongongo forest and eat their way out of it, clearing an ever-widening circle of food until the daily round trip for foraging may reach about ten miles. From these foraging bases the men hunt everything from birds to giraffe. As the dry season sets in, the !Kung groups begin to retreat to the more dependable water holes. As the winter drags on, food becomes less and less plentiful. The early spring brings hot, dry weather and the most severe test. Now groups may share a water hole if the year has been dry, and they must travel long distances for desirable foods. Requests for such sharing are made to the headman of the group owning the water, and permission is generally freely given. After this period of hardship, the rains come again.

An essential element in !Kung behavioral ecology is the variability in food supply. Through much of the year mongongo fruits/nuts are superabundant, and at its peak per capita meat consumption may reach an incredible 4½ lb a day. Even in the worst of the dry season, however, meat is still supplying about 20 percent of the diet, and vegetable food, although more distant and less desirable than in better seasons, is still adequate.

This kind of variability is predictable and serves to limit the carrying capacity. What is *not* predictable is exactly where and how much rain will fall, and therefore where the plants will be growing. In a drought year—and about two in every five are dry—some areas will get rain and others will not, and only some water holes will receive enough water to last through the dry season.

The low density of resources and the unpredictable variability in their distribution is the key to the !Kung social system. From a base camp, individuals must each day travel even farther for food, and a typical !Kung gatherer walks twelve hundred to two thousand miles a year. The larger the group size, the larger the area which must be harvested, and so the farther individuals must walk. The efficiency of gatherers, then, constitutes the downward pressure on the typical group size of twenty to thirty.

The upward pressure comes from the hunters. Although there are advantages to group hunting in terms of stalking, attacking, tracking, defending the kill, butchering, and transporting the meat back to camp, two hunters rather than the typical population of six to ten would be sufficient. But game, unlike fruits and nuts, is a chancy resource, and the more teams hunting, the likelier they are to make a kill. The probability that thirty people (of whom ten are active hunters) will go hungry is far less than the likelihood of starvation for six people relying on the luck of two hunters. One wildebeest provides 450 lb of meat, so the whole group can easily be fed from one such kill by one pair of hunters while the other eight hunt fruitlessly for a week or two. The !Kung social system, then, depends on an essential reciprocal altruism, both within and between groups. Within the group, animal kills are shared so that the successful feed the unsuccessful, on the implicit assumption that the next week the roles may be reversed. Between groups, that most important and patchy of all resources, water, is shared. For when the frequent droughts come, who is to say that a group lucky enough to have extra water on its land this year may not need to "borrow" it from another group the next?

The cement for this bond of reciprocal altruism is kinship. The bands are complex kin groups with the leadership role, such as it is in these unassertive people, determined largely by kinship. Visits to neighboring groups are always kin visiting kin. And requests to use part of another group's territory or its water are inevitably made through lines of kinship. The kinship ties among the !Kung are reinforced by strong cultural traditions. Hunting and gathering, both life-supporting social functions, are based on culturally transmitted technology. Gathering depends on the proper use of a carefully manicured digging stick and various well-designed containers—canteens made from ostrich eggs, intricately woven nets, pouches, and so on. And, of course, much training is required to recognize edible plants and to harvest and prepare food from them. Hunting is based on delicate snares, well-balanced spears, highly crafted bows with strings of sinew which require constant attention and maintenance, poisoned arrows of perennial grass stems tipped with carved bone shafts and heads (though the heads are now manufactured from scraps of wire fencing), knives (formerly of stone), and of course, fire. And then the use of all these implements and others (including oracle disks to divide the movement of prey), as well as the particular techniques of tracking to be employed, depend on cultural teaching.

Children, who must learn all these critical practices thoroughly, seem to be absorbing the essential knowledge even in their play. To all appearances no distinction is made between the sexes as children, but 86 percent of juvenile play groups are comprised of a single sex, and the differences in play preferences notorious in Western culture are present here. Anthropologists report that boys spend 45 percent of their time exploring "technical" equipment (i.e., using crude bows and arrows to "hunt" dung beetles and explore termite mounds), girls 4 percent; boys devote 17 percent of their time to rough-and-tumble play, girls 6 percent.

What ethological lessons can we draw from the !Kung? For one thing, their existence is far less tenuous than it seems: despite the harsh environment, !Kung only have to work about forty hours a week (housework included) to survive. Much has been made of the !Kung's selfless sharing and guileless gentleness, but both are, as we have seen, mythical: sharing is based on kinship and the expectation of repayment, and people who take but do not give are the objects of criticism and scorn. And as to their gentleness, they have about the same attempted homicide rate as urban America. Their disputes generally involve adultery or sharing, and are marked by long-lasting kinship feuds and grisly bouts of retribution. In fact, it is the social pressure of the dry winters when two or three groups must share a water hole which brings out the worst. The normal group of twenty to thirty requires no strong leader to keep the peace—social pressure among intimates is sufficient to modulate the behavior of others—but a winter gathering of 50–150 !Kung is another matter, and in the absence of a firm power structure a group of this size is inherently unstable. Other habitats which favor larger groups require—and display—more traditional organization. But with this increase in central authority and manpower come the elements which permit our species' characteristic intergroup conflict. It seems clear on close examination that the !Kung's highly touted lack of war may be the consequence of there being nothing much to fight over and too few adult males to mount a good conflict in any case.

The pattern emerges in primitive cultures again and again of group formation based on kinship, and group cohesion based on intergroup competition, discord, or even war (Fig. 29–7). In fact, Eibl-Eibesfeldt argues that our almost unique propensity for intraspecific conflict and war, a tendency which leads to typical male mortality rates of 25 percent in hotly contested habitats, is in fact *the* selection pressure which made our species so terribly and terrifyingly smart.

Unfortunately, it is virtually impossible to pinpoint innately

Fig. 29–7 The propensity for intraspecific war is not restricted to Western or "advanced" societies. Here two tribes in New Guinea perpetuate a bloody, well-organized war worthy of medieval Europe.

directed adult behavior from simple observation, even of primitive cultures. Behavior that results from programmed learning and is fueled by innate drives may be as dependent on external contingencies as behavior that finds its roots in plastic learning or enculturation. For however romantically primitive the !Kung may look to us, they have an extraordinary cultural technology orders of magnitude beyond that of any other species, and the interaction of this awesome cultural force with their programming is beyond our power to sort out as yet. Even when we see the repeated occurrence of similar cultural patterns among various primitive peoples in different habitats, we cannot with certainty distinguish the cases in which cultural training reinforces genetic predispositions and those in which it represses them. As the next chapter will argue, there is groundwork which must be done first. We must seek out wholly unenculturated humans and look at their behavior. Fortunately we have a ready source of suitable subjects for such an endeavor: our children.

SUMMARY

When we try to apply the ethological perspective to our species, to look for evidence of instinct and programming, we find most potential sources of evidence either incomplete or unreliable. Introspection is especially tricky since we seem to be driven to fabricate explanations for our own behavior. Evolutionary and archeological evidence indicates that our species moved into an open-savannah hunting niche enriched by vegetarian gathering, experienced an enormous increase in brain size, invented a primitive but powerful technology, and then quite recently domesticated plants and animals and began to live in cities. Anthropological studies of surviving hunter-gatherer groups reveal several relatively consistent species-specific characteristics: a familiar monogomous social system with a sexual division of labor, strong cultural traditions and technology, and a fair share of violence. As with other species, the details of the human social system seem fine-tuned to the ecology of the habitat, and kinship plays a prominent role. But the enormous cultural influences that shape human behavior underscore our need to study those truly naïve subjects: children.

STUDY QUESTIONS

1. If we were to treat *Homo sapiens* as just another species of mammal, a variety of curious patterns would demand explanation.

A. Weaning of infants, even after they have teeth and are perfectly capable of eating solid food, is usually difficult. During lactation, females are nearly infertile. Why should infants not want to be weaned, and why should parents want to wean them? (The term "want" is being used here in the evolutionary rather than the conscious sense.) Assume for the purposes of argument that solid food is more nutritious than milk, and that breast-feeding is no more or less convenient or physically demanding on parents or children than providing solid food. How could you test your answer? Are there other examples of parent-child conflict in the same or other species which you can cite and compare?

B. In most human societies grandparents take extraordinary interest in their children's children. Oddly enough, grandmothers are generally more enthusiastic than grandfathers. In the past this has been written off as "maternal instinct," although lately the notions that a sinister, male-imposed behavioral shaping is the cause. I doubt that this is the whole story, particularly since it fails to explain some of the other data. For example, grandparents are more altruistic to their daughters' offspring than to their sons'. Assume that all this behavior is genetically programmed. What could explain these asymmetrical interactions? How could you test your model? Does your model allow you to predict which married children—sons or daughters—are more likely to live with their spouses in their parents' home? It should. Can you cite similar examples in nature?

C. In virtually all human cultures females devote more time to childcare than do males, even after breast-feeding has ended. Many maintain that this is a cultural phenomenon perpetuated by the deliberate plotting of the male sex. There may be something to this explanation, but there also exist plausible evolutionary explanations. Why should one sex in a human pair invest more in a child than the other? Is there a way to test your model? Are there similar examples from nature?

D. Homosexuality is widespread among human males. For example, the Kinsey report of the 1940s found that more than 10 percent of American males had had significant homosexual relations within the preceding three years. Considering the social pressures and selection bias, the actual value could be somewhat higher or lower. Although psychologists and sociologists used to interpret homosexuality as a sort of neurosis brought about by early family or social experiences, current attitudes are less focused. Popular ethologists have interpreted homosexuality as a behavioral abnormality generated by the conflict between our present artificial, urban environment and our hunter-gatherer genes. Two recent observations weaken all these explanations: (1) homosexuality may have a genetic component and behave as a single-locus recessive with a probability of being expressed (i.e., "penetrance") of about 50 percent; and (2) according to some studies, homosexuality seems to occur at roughly the same frequency in primitive cultures.

Assume for the moment that there really is a homosexuality gene. Its frequency in the population must then be on the order of 30–35 percent and stable. Assume further that homozygous recessives leave no offspring. You might expect that such an unfit gene would rapidly disappear from the gene pool. This obviously is not happening. Indeed, there is every reason to suppose that the gene's frequency has been stable throughout the history of our species. Given these two assumptions, what could explain the stability of this hypothetical gene? (There are at least two explanations, but one will be sufficient.) How could you test your model? Are there examples of similarly stable, apparently unfit genes in nature?

2. It is frequently said that since we are hunters, our social system ought to be like those of carnivores rather than resembling those of the vegetarian ungulates and the other primates. How does this speculation hold up with hunter-gatherers?

3. Chimps and humans diverged five to fifteen million years ago and are classified in different genera. Yet the actual genetic differences are smaller than those normally occurring between two species in the same genus. Clearly the genes that gave rise to such blatant differences in behavior must have been very important. What do you think these genes might code for?

FURTHER READING

Binford, S. R., and Binford, L. R. "Stone Tools and Human Behavior." *Scientific American* 220, no. 4 (1969): 70–84.

Braidwood, R. J. "The Agricultural Revolution." *Scientific American* 203, no. 3 (1960): 130–48.

Gazzaniga, M. S. "The Split Brain in Man." *Scientific American* 218, no. 2 (1967): 24–29.

———. "One Brain—Two Minds?" *American Scientist* 60 (1972): 311–17.

Geschwind, N. "Specializations of the Human Brain." *Scientific American* 241, no. 3 (1979): 180–99.

Isaac, G. "Food-Sharing Behavior of Protohuman Hominids." *Scientific American* 238, no. 4 (1978): 90–108.

Kimura, D. "The Asymmetry of the Human Brain." *Scientific American* 228, no. 3 (1973): 70–80.

Sperry, R. W. "The Great Cerebral Commisure." *Scientific American* 210, no. 1 (1964): 42–52.

CHAPTER 30

Infants

Just as primitive hunter-gatherers, uncorrupted by our technological culture, provide us with a glimpse of our presumptive ancestors, so too do infants offer us a window into our natural unenculturated selves—a world of which each of us was once a part, but which some trick of our natures has put out of reach of our memories. Infants provide natural experiments, telling us, under proper test conditions, what they already "know": the inborn rules by which they structure their perceptual worlds, the innate behaviors which ensure their survival, and the prewired orchestration which guides their development. Our species' profound susceptibility to experience, though, whether through imprinting, programmed learning, or as we want to think, so-called plastic learning, taints truly inborn attributes at a very early age. The importance of the questions we are asking, however, makes it worth any amount of effort and ingenuity to find gentle, noncoercive ways to get reliable answers from these innocently recalcitrant experimental subjects.

NEWBORNS

What, if anything, can newborn humans do? From observation we know that they have a remarkable repertoire of behavior and innate knowledge about how to survive, some of which becomes useless with time and is therefore lost as they grow. For example, from birth babies are able to cry (Fig. 30–1). At first glance this behavior seems too obvious and simple to merit consideration, yet crying is a motor program requiring the coordinated movement of many muscles. It serves as a one-word language for survival, the infant's sole means of communicating to its parent that help is needed. The cry itself is a stimulus superbly designed to get a response, for it is inherently irritating to the human parent. Indeed, it must be attributed to the strength of our genes' drive to regenerate themselves that human parents are not more often tempted to deal with the noise in the most direct and forthright manner.

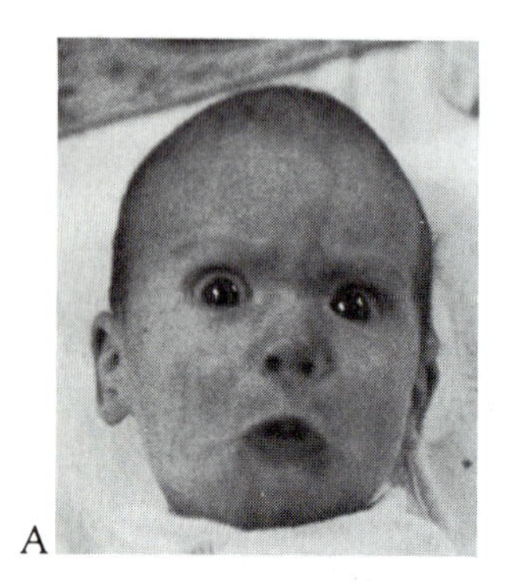

A

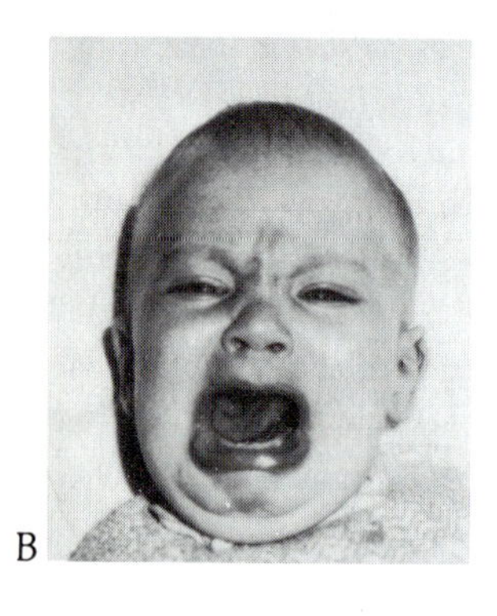

B

Fig. 30–1 One of the first well-coordinated and effective motor programs displayed by infants is crying, seen here in its ominous low-intensity form (A) as well as its disconcerting high-intensity version (B). As primal communication, a baby's crying is extraordinarily effective.

Despite the apparent weakness and helplessness of babies, it is startling to discover how strong they are. A touch on the palm or the soles of the feet releases a stereotyped grasping reflex more than sufficient to allow the newborn to support its own weight in a clinging position immediately after birth (Fig. 30–2). The most effective stimulus for this response seems to be fur, and it is difficult to avoid the interpretation that the behavior is a holdover from our very distant past. Like many other cases of sensory deprivation, however, this response in humans disappears with lack of use.

Newborns have more active behavior patterns at their disposal as well. Babies in their first week of life begin a stereotyped side-to-side searching movement when touched on the lips or the cheeks (Fig. 30–3). Only when the newborn discovers the nipple does this motor program end as the child opens its mouth, grasps the nipple with its lips and gums, and begins a powerful rhythmic sucking. After a few days this "rooting" behavior becomes more directed so that the infant searches only on the side touched. Whether this change involves learning or maturation of abilities already in the system is not known. During nursing babies gradually begin to seek strong, direct eye contact with whomever is feeding them, and it is difficult to resist the inference that programmed learning is taking place, though whether the student in this case is the baby or its adoring parents (or both) is unclear.

A newborn also displays several striking locomotor patterns. When held by the hands with its feet just touching the ground, it will perform a rhythmic walking motion. Similarly, when held so that hands and knees touch the ground it will begin a coordinated crawling movement with diagonally opposite limbs (right arm and left leg, for example) moving together in alternation with the other pair. Neither of these motor programs is of any use to an infant which is too weak to support its own weight in this posture, so they disappear until true attempts to crawl and walk begin, at which time these behaviors must be laboriously rediscovered.

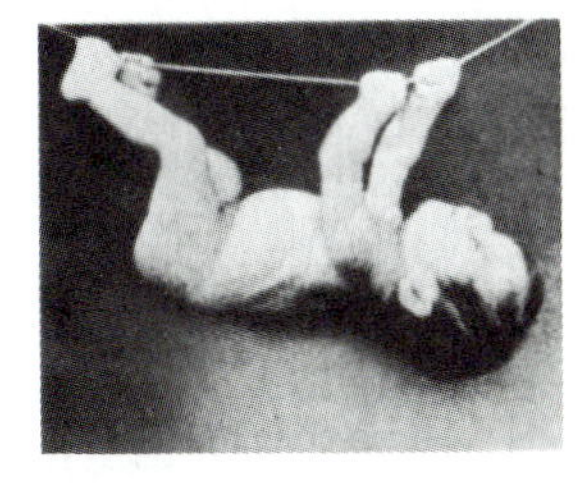

Fig. 30-2 Newborns are markedly strong, easily able to support their own weight by means of the powerful "grasping reflex." In our nearest primate relatives, this behavior permits a baby to cling to its mother.

Placing a newborn horizontally in the water and supporting its chin elicits a remarkable, coordinated swimming movement. This striking behavior is accompanied by dramatic reactions of the entire body to water known as the diving reflex. The apparent releaser is cold water on the face. Heart rate drops abruptly to lower blood pressure, peripheral vessels constrict to send all the blood to the brain, the automatic urge to breathe ceases, and brain metabolism switches into a poorly understood state which allows it to go without oxygen for several minutes—up to forty minutes in icy water. The result is an animal prepared for prolonged swimming or even temporary total immersion. These physiological reactions to cold water persist in an attenuated form into adulthood. Babies readily "learn" to swim at this age (Fig. 30-4), but in the absence of opportunity the motor program disappears and must be mastered again.

Less clearly functional, though perhaps revealing nonetheless, is the propensity of newborns still on the delivery table to move their limbs in synchrony with speech. Since they ignore

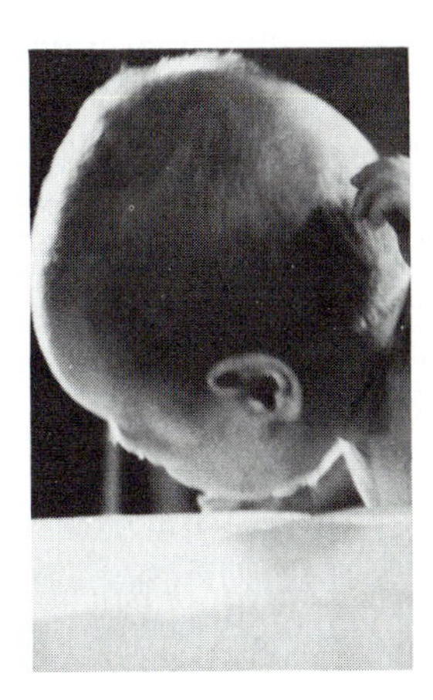
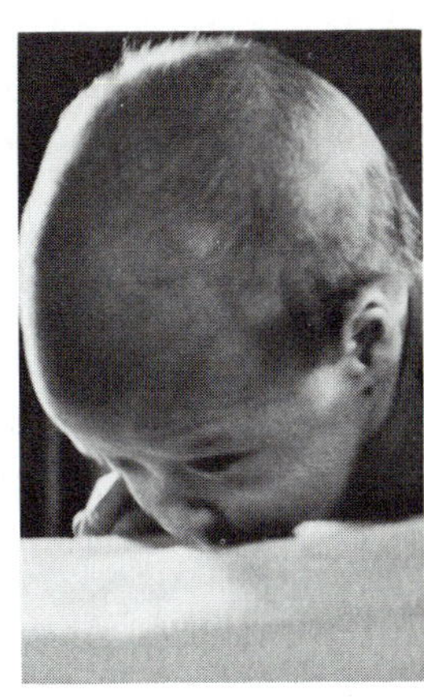
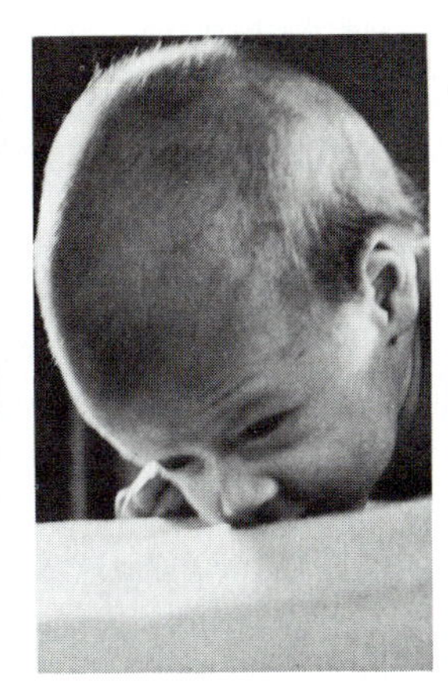
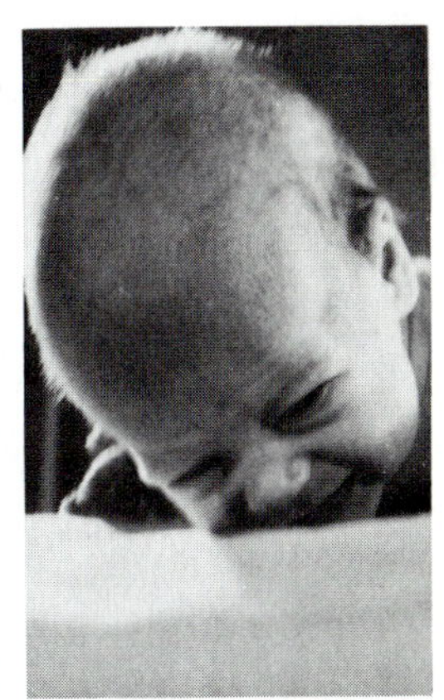

Fig. 30-3. When touched on the face or lips, newborns begin a rhythmic, coordinated search for a nipple, followed by stereotyped sucking behavior.

Fig. 30-4 Newborns take readily to the water, automatically altering their physiology and performing coordinated swimming movements. Experience at this age leads quickly to competent swimming, while deprivation means that children will have to go through a long and laborious learning process later if they are ever to swim.

music, a far more suitable source of acoustic rhythm, it seems likely that they recognize and attend to speech as something special from the beginning. In the dark the same newborn will turn its head in the direction of sounds, especially speech, as will blind children. This implies that at least some of the processing for sound localization is innate, as must be the appropriate motor responses. Perhaps it also hints that our senses of auditory and visual space might be prewired, calibrated at birth to work together to form a multimodal map of our sensory world.

Localization of odors, turning away abruptly from noxious smells, is also apparent in newborns only a few minutes old. Certainly there could be survival value in such an aversive reaction. As adults we localize olfactory stimuli by comparing "arrival times" at each of our two nostrils, discerning a minute interval threshold of 0.5 msec. Whether newborns use the same strategy with the same impressive resolution is not known.

THE FIRST MONTH

Since most of these behavioral tours de force have been observed right on the delivery table where the opportunity for prenatal learning and practice seems limited to say the least, there can be little doubt that the patterns are innate. By the time an infant is a week or two old, several significant additions have been made to its repertoire. Although critical experiments have not been performed, these new elements are probably derived from maturation rather than from learning. For example, two-week-old infants have been found to be able (that is, programmed) to recognize certain sorts of danger. A silhouette expanding symmetrically to suggest an approaching object releases a variety of protective behaviors. The baby's eyes widen as it pulls back its head and brings its hands and arms up to cover its face. The sequence and duration of the separate behavioral elements is always the same, implying that the whole behavior is innate and under the control of a single motor program. We may hope that two-week-old infants have had little opportunity to learn the necessity for defensive behavior. The processing behind this protective response must be relatively sophisticated, since an asymmetrically expanding silhouette which would be appropriate to an object that was not directly approaching produces no reaction.

Another apparent motor program, the reach-and-grasp response, is seen in infants when we present them with an object within their arms' reach. A child will extend its arm in the appropriate direction, but will fail to adjust the trajectory of the hand even if the object is noticeably moved. Grasping is triggered when the arm has been fully extended or when it touches an object. The time course and pressure of the grasp are independent of any feedback from the size of the object, or its texture, or even its existence. The sequence and duration of the reach and grasp is constant. Months later this motor program disappears, to be replaced by a feedback-sensitive grasp.

The clearly innate abilities of the newborn do not end with motor programs. From birth newborns show extraordinary interest in facelike arrangements of dots, and soon concentrate their attention on eyes or mouth to the exclusion of other equally obvious features. Their gaze fixates on the stimulus, and infants older than about four weeks often smile. This coordinated bit of facial muscle manipulation has a powerful effect on adults, and since blind children do it too, thereby ruling out any possibility of its being a conditioned response (Fig. 30–5), we may be fairly safe in guessing that smiling is an innate survival tactic employed by human genes to cement that powerfully irrational bond from parent to offspring that is so essential to the survival of such a helpless, noisy, dirty, and intrusive organism as a child.

Perhaps the most uncanny ability of newborns is the facility with which they mimic the facial expressions of adults. When we stop to think about it, it means that they must innately

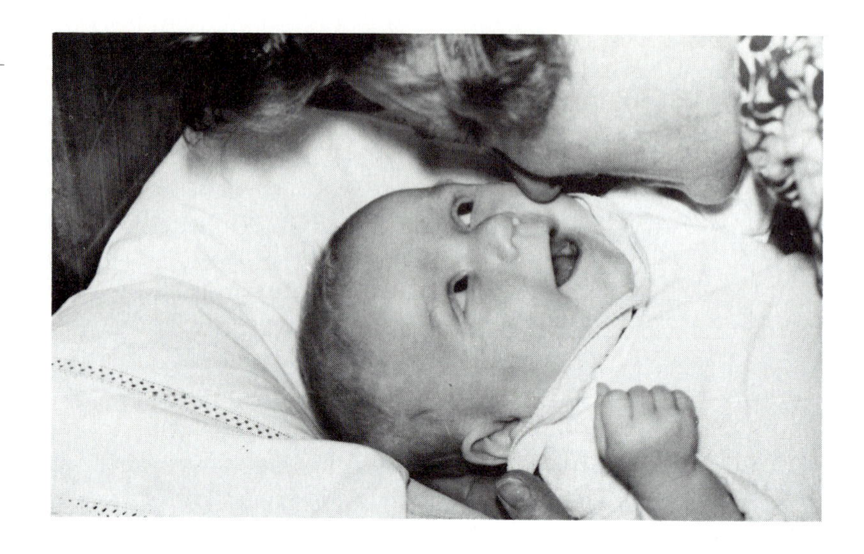

Fig. 30-5 Smiling appears in babies at about four weeks of age. This coordinated and stereotyped motor program is entirely innate. Here an eleven-week-old girl, blind from birth, has stopped the restless wandering of her sightless eyes to "fixate" on her mother's voice, and to smile. This behavior on the infant's part helps forge a strong and essential emotional attachment between parent and child.

recognize certain classes of facial expressions and then switch on the motor programs which control the dozens of facial muscles to produce the same expression in return. All this occurs without the help of so much as a mirror. There is every reason to believe that facial expressions work as sign stimuli, and indeed this may be the basis of the art of cartooning, in which only a specific fraction of the many lines, shapes, and shadings of a face are needed to communicate the desired meaning. Our easily recognized and stereotyped facial reactions to sweet, sour, and bitter are evident in newborns as well, as are the equally obvious facial expressions and coordinated movements we call smiling, laughing, crying, pouting, and frowning. These universal sign stimuli are present even in blind children, and are recognized by humans for what they are the world over.

Two- and three-week-old infants seem to have some innate preconceptions about inanimate objects as well. When a visual stimulus approaches its hand a newborn clearly expects to be able to touch it. When an object disappears behind an obstacle the same newborns, like honey bees tracking the sun, expect it to continue to exist and to continue to move at the same speed. They are visibly upset if it appears on the other side of the obstacle too soon, or if the obstacle is removed and the object is gone. Three-week-olds forget about objects which have been out of sight for more than 15 sec, while ten-week-olds remember longer.

THE SECOND MONTH

By the time the infant reaches four to six weeks of age, researchers can begin to use an extraordinarily powerful technique which capitalizes on the child's developing capacity to habituate. Infants are notorious for their ability to achieve rapid and frequent boredom. Their attention, as measured by continued eye contact or nipple sucking, is rapidly engaged by anything novel. As a result we can ask infants a series of same-or-different questions. By habituating (boring) an infant to one wavelength and then switching to another, this technique has revealed that newborns come programmed to classify the multitude of shades and hues that we can distinguish with our trichromatic vision into four general categories—red, yellow, green, and blue—with boundaries which correspond closely to those used by adults. To an infant all reds look alike, but the slightest step across the arbitrary perceptual boundary into "yellow" rivets its attention. A glance at the response characteristics of our three classes of cone cells indicates that this categorization is being accomplished by active processing (Fig. 30-6). We do not yet understand the advantage that this sort of selective categorization must give to species which use it, but perhaps it divides the objects in the world into categories that are—or were—useful, or facilitates learning by reducing the number of variables a young animal must deal with.

The same ingenious experimental technique has revealed that

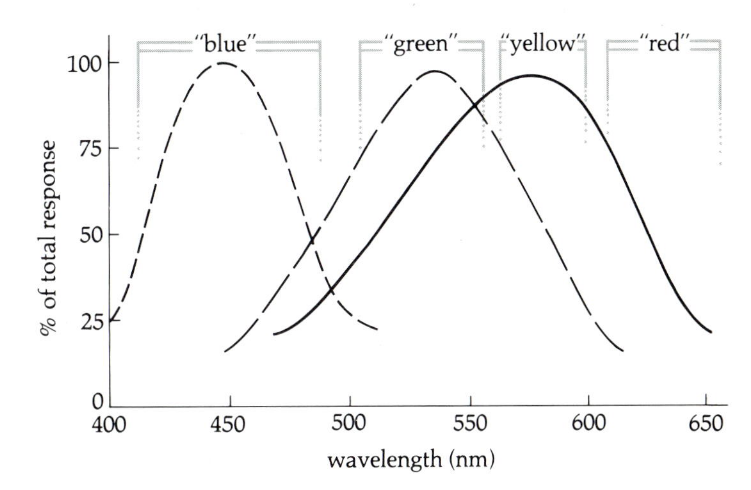

Fig. 30-6 Although humans have only three visual pigments, and those pigments respond best to blue, green, and yellow-green, infants categorize colors into four groups corresponding to red, yellow, green, and blue (as indicated by the bracketing over the pigment curves). Some sort of central processing must be involved to separate yellow and green, and to factor out red from yellow.

infants perceive geometric objects to be the same whether they have been inverted, rotated, or moved nearer or farther away. This innate ability to recognize an object regardless of its orientation would have been adaptive for infants in the hunter-gatherer society that ours sprang from, but must be consciously suppressed for many functions of life today, especially reading. Two-year-olds refuse to notice any difference between pictures that are right-side-up or upside-down. To a child just beginning to read, "b" and "d" and "6" and "9" are identical. He must work determinedly to overcome his mind's inborn tendency to gloss over differences of this sort if he is to read successfully.

By the time an infant is six weeks old a familiar childhood attitude toward problems begins to emerge. Infants seem innately motivated to solve problems. The curious thing is that, like so many examples of programmed learning, there seems to be no obvious reinforcement for the behavior other than the simple satisfaction of the pursuit itself. Indeed, success extinguishes all interest in the problem. Are we born with some inner "need" to solve puzzles, to satiate some intellectual appetite by the very act of solution?

LANGUAGE

More accessible to testing than this shadowy problem-solving "drive" is the ability of six-week-old infants to recognize the consonant sounds of human speech. Linguists have known for some time that the tens of thousands of words in each of the more than three thousand human languages are constructed of about forty phonemes, or speech sounds. Many more phonemes are physically possible, and even these forty are created out of combinations of only a few vocal gestures—specific movements of the lips, mouth, and tongue (Fig. 30–7). From analysis of sound spectrograph recordings of human speech and attempts to synthesize sound, it has become clear that the linguistic information in speech is carried by three "formants"—that is, by the frequency, amplitude, and temporal modulation of the lowest three frequencies emitted during speech. (The other formants are not without function: they convey much of the emotional tone so crucial to communication.) Vowel sounds are created by the size and therefore the resonance of the mouth and larynx (Fig. 30–8). The formants represent those reso-

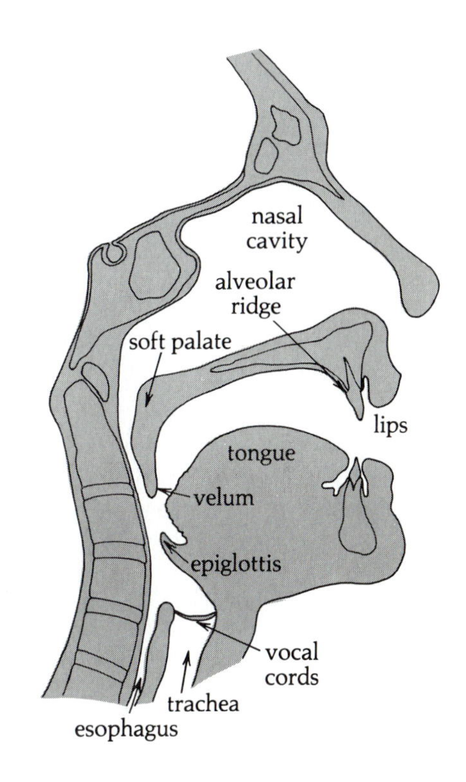

Fig. 30-7 The human vocal apparatus uses the passage of air over vocal cords to generate noise, and then employs cavities to emphasize resonances selectively: the throat, mouth, and nasal cavity. [From *The Origins and Development of the English Language*, 2d edition, by Thomas Pyles. Copyright © 1971 by Harcourt Brace Jovanovich, Inc. Reproduced by permission of the publisher.]

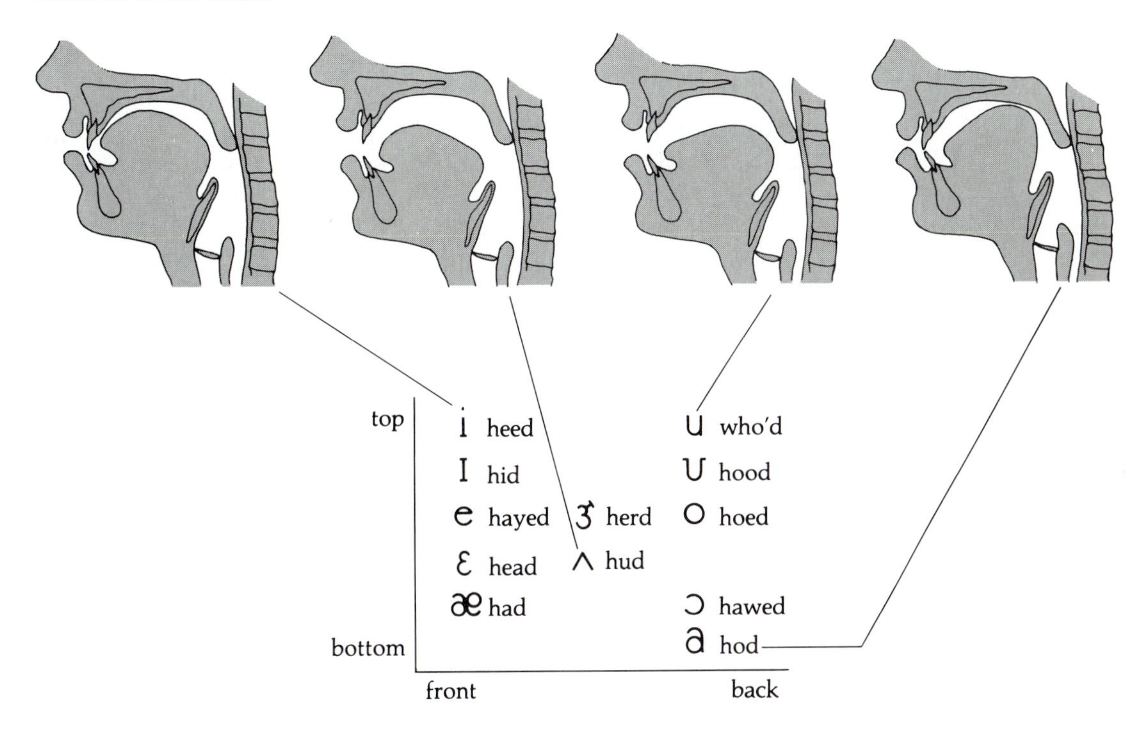

Fig. 30-8 The control of relative cavity size between mouth and throat (as well as access to the nasal cavity) is crucial to the formation of vowels. To a first approximation, the tongue plays the key role. The approximate location and position of the tongue at the point of airway constriction for the set of American English vowels is indicated in this matrix.

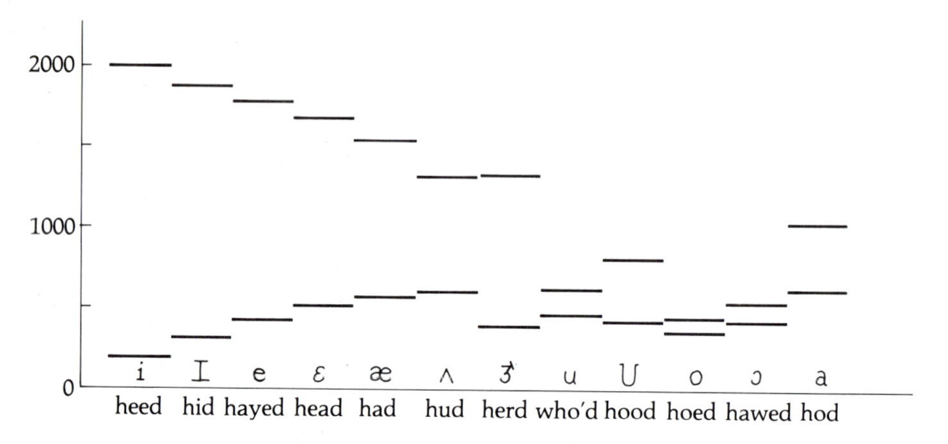

Fig. 30–9 The key acoustic feature in distinguishing most vowel sounds is the frequency ratio of the first two formants. Shown here are these formants for the dozen vowels of standard American English along with their linguistic representation according to the most common convention.

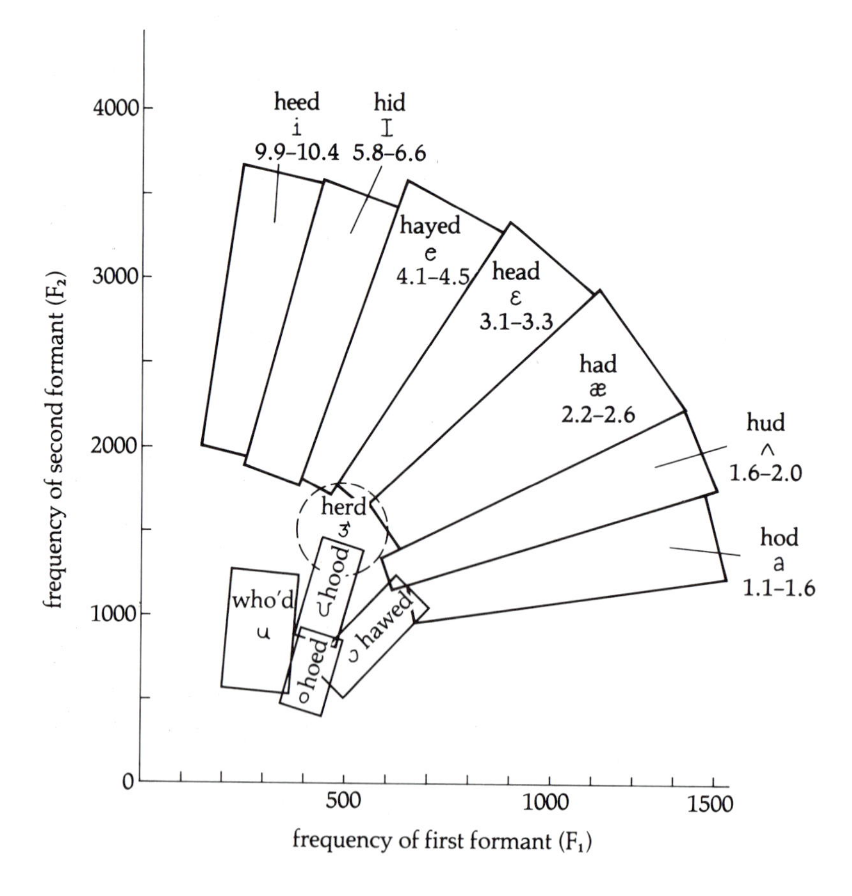

Fig. 30–10 The frequencies of the two lowest formants of the vowels of standard English. About 90 percent of vowels produced by native English-speakers falls within the boundaries shown.

nances, and we recognize vowels by the ratios of the frequencies of the first two formants (Fig. 30–9). The plasticity of this processing strategy allows us to recognize most vowels automatically, regardless of the pitch or absolute frequency of the speaker, by simply measuring the frequency ratios (Fig. 30–10). This ability is apparently innate.

LINGUISTICS

The human vocal apparatus (Fig. 30–7) employs a sloppy sound generator (the vocal cords) accompanied by three cavities (the adjustable pharynx and mouth, and the static nasal passage) and a trio of modulators (the lips, the tongue, and the velum, which opens and closes the nose cavity). Because the vocal cords are so poorly built and the cavities have so many possible resonances, speech is comprised of an almost endless jumble of frequencies. And yet we are able as humans to understand speech, even in a noisy environment. This feat is possible because there is a specialized sound decoder for language in our brains which is constructed along classical ethological lines to sort out sign stimuli and ignore everything else. Work with synthetic speech devised specifically to sort out the "lowest common denominators" in speech recognition makes it clear that the key diagnostic features of linguistic sounds are the relative frequencies, timing, and frequency modulations of the lowest three frequencies or "formants" of speech.

Relative frequency is crucial in the identification of those parts of speech that we perceive as vowels. Vowels are produced by vibrating the vocal cords and shaping the mouth cavity with the tongue and lips to create distinctive resonances. Figure 30–8 indicates the approximate position the tongue must adopt to generate the dozen vowels of standard English. The effect of the tongue's position on the first two formants of the various vowel sounds is shown in Figure 30–9 for a male voice. Figure 30–10 shows the range for formant frequency pairs for these same twelve vowels over many different voices: for almost any vowel, the *ratio* of its first two formant frequencies is roughly the same regardless of the pitch of the individual voice. As a result, our brains can identify each vowel by the sort of simple frequency comparison that is becoming increasingly familiar as our knowledge of biological information processing expands.

Although this processing strategy for vowels appears to be the same kind of innate acoustic analysis seen in other mammals, the precise definition of vowel boundaries is cultural. Not only do different languages place the divisions between vowel classes in different places (Spanish, for example, lumps the phonemes "i" and "I" together so that native Spanish speakers perceive "sit" and "seat" to be the same word), but different dialects of the same language may vary widely in this regard.

In contrast, we produce consonants by making mechanical "gestures" which change the pattern of airflow through our vocal apparatus, while at the same time we are controlling the time at which the vocal cords begin to vibrate. There are three general timing options open: the vocal cords may begin sounding *before* the mechanical consonant gesture which shapes the sound (the so-called prevoiced consonants), or *simultaneously* (voiced), or *afterward* (unvoiced). These correspond to negative, zero (or very short), and long *Voice Onset Times* (VOT). The consonant gesture itself can be performed with the lips, or with the tongue against the teeth, the alveolar ridge, the soft palate, or the throat.

The final variable which determines the consonant which we are producing is the *nature* of the gesture. It can be a complete blockage of air ("stop" consonants), a substantial but incomplete blockage ("fricative" consonants), a combination beginning as a stop but ending in a fricative ("affricates"), no blockage (semivowels), or the nose can be used as the sound path (nasal consonants).

The two dozen consonants of English occupy only some of the available compartments of this three-dimensional matrix. For example, among the stop consonants, the lips produce the voiced "ba" and unvoiced "pa," the tongue against the alveolar ridge generates the voiced "da" and unvoiced "ta," while when it is against the soft palate it creates the voiced "ga" and unvoiced "ka." The set of nasal consonants "ma," "na," and the familiar "ing" sound are all prevoiced and produced at the lips, alveolar ridge, and soft palate, respectively. These nine consonants are illustrated in Figure 30-11.

English has only three affricates, all produced on the alveolar ridge: the prevoiced "la," the voiced "j" of "jar," and the unvoiced "ch" of "char." By far the richest group in English are the fricatives. "Ha" is an unvoiced glottal fricative, the only glottal in standard English. Another pair is produced by the lower lip against the upper teeth: the voiced "va" and unvoiced "fa." The rest of the fricatives are generated on the alveolar ridge. One pair, the voiced "th" of "thy" and the unvoiced "th" of "thigh," are created with the tongue against the tips of the upper teeth. The voiced "za" and unvoiced "sa" are produced at the tooth/gum line. Our standard English prevoiced "ra" is created by constricting the flow of air with the tongue at the base of the alveolar ridge, while the voiced and unvoiced versions of the same gesture generate the "zh" of "vision" and "sh," respectively.

Each of these consonants has its own acoustic signature which we are wired to recognize. Infant studies show that we come pretuned to classify sounds according to whether VOT is earlier than about −35 msec (prevoiced), later than −35 msec (unvoiced), or in the middle (voiced; compare the rows in Fig. 30-11). Indeed, the auditory cortices of other mammals regularly make the same three distinctions. The place in the vocal tract that the consonant is produced, of course, determines the acoustic resonances at the beginning of the gesture, and

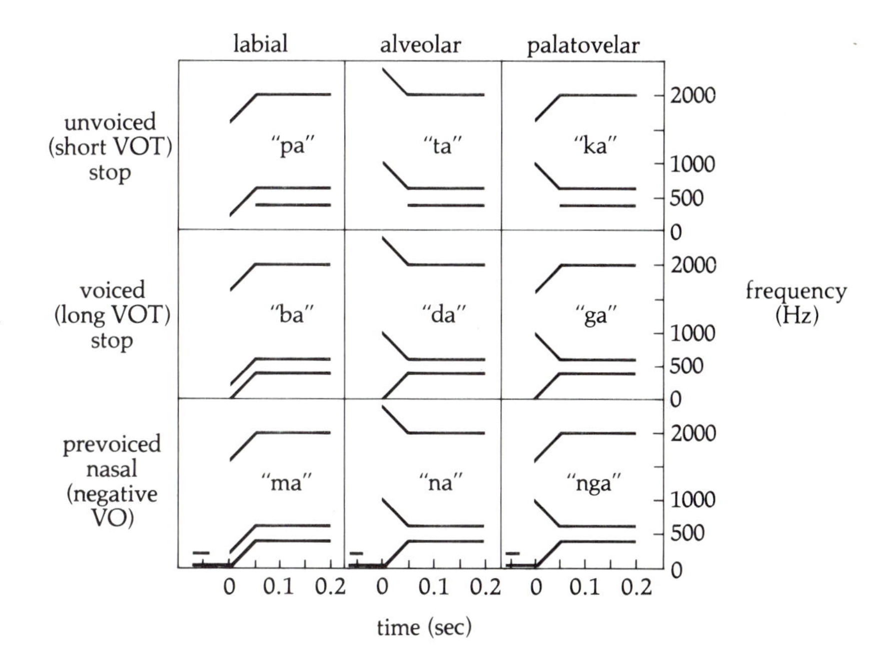

Fig. 30-11 Stop consonants can be formed not only with the lips (labial consonants), but with the tongue against either the alveolar ridge behind the upper teeth, or against the soft palate at the back of the mouth (palatovelar). Ignoring the nonnasal prevoiced consonants absent from English, and including the prevoiced nasal, a matrix of nine familiar consonants is created.

therefore the initial frequency ratios by which we differentiate consonants with the same voicing and gestural nature (compare the columns of Fig. 30-11). The more subtle distinctions between stops, nasals, fricatives, and affricates are all encoded in the initial frequency modulations (compare the first two rows of Fig. 30-11 with the third). FM detectors are well known in the mammalian auditory cortex, but appear to be species-specific in their tuning.

Hence, the whole wonderfully complex process by which the human brain decodes speech turns out to be a simple acoustic processing task whose feature detectors are so widespread in mammals that they probably predate the development of language. Indeed, the major linguistic limitation on our nearest relative the chimpanzee is probably that its vocal tract is not fitted out to produce the wide variety of resonance ratios necessary for generating vowels. Doubtless some quirk in the workings of Wernicke's and Broca's areas had already preadapted them for primitive language-like operations and these idiosyncratic peculiarities of the anthropoid brain and their subsequent evolution must have shaped the structure of our special system of communication. For all we know, the same areas lie ready but underutilized in the chimps which, even if they were somehow supplied surgically with our human vocal apparatus, would still lack the unique and powerful *drive to learn language* which is the one certain piece of programming which sets humans apart from the apes.

We produce consonants, on the other hand, by specific movements of tongue and lips, "shaping" sound by mechanical manipulation of the vowel resonances that accompany them. Consonants depend for their individual characteristics on the temporal relationships between the onsets and the frequency modulations of the formants. Consider the syllables "ba" and "pa," for example, both produced with the lips. Say them very slowly yourself. "Ba" begins with the consonant rush of air (the so-called plosive sound), and the voiced "ah" sound of the vowel occurs almost simultaneously. In "pa," on the other hand, not only is the plosive itself delayed a fraction of a second after the beginning of speech, but the "ah" follows the plosive somewhat later as well (Fig. 30–12). From tests with synthetic speech it is clear that as listeners we distinguish "ba" from "pa" on the basis of the differences in voice onset times (VOT). The same distinction applies to "da" (the "ah" comes quickly), and "ta" (the "ah" is delayed) and other pairs of "voiced" and "unvoiced" consonants. This can be seen in Figure 30–11, and the same principles underlie our distinction of other groups of consonant sounds.

Two curiosities emerge from the study of consonants. First, adult subjects, when presented with a series of different VOTs, do not classify them along a continuum but rather hear them as either one syllable or the other. There exists in our minds a distinct perceptual boundary where "b" suddenly becomes "p." Second, there is a curious correspondence of consonant sounds and their concomitant boundaries between languages. Peter Eimas sought to discover whether these boundaries might be innate or if they were the products of coincidence and learning. He used infants as young as six weeks of age, and took advantage of their maddening propensity for boredom that we mentioned earlier. Eimas could present a series of, say, "ba" sounds, and measure the babies' sucking rates on a microswitch-equipped pacifier. When the sounds began the babies would respond with vigorous sucking, but if the same sound continued the sucking waned. If Eimas then switched to synthetic "pa" sounds, the sucking rate would shoot up as the listlessness disappeared. The test, then, was to offer pairs of sounds differing 20 msec in VOT, to see whether the infants perceived them as different. The results demonstrate that all "ba's" sound the same, as do all "pa's," but when the pairs straddle the arbitrary adult barier, the infants recognize the sounds as different.

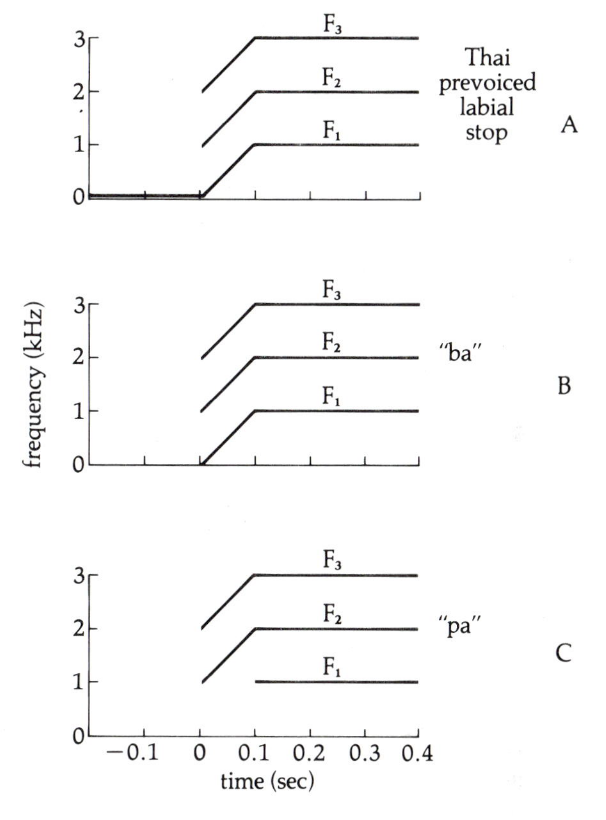

Fig. 30–12 Three classes of labial stop consonants exist. All stop the flow of air with the lips—the kind of mechanical vocal gesture which gives rise to consonants—but differ in the time at which the voicing of the vowel sound begins. In A, a consonant not found in English and beyond most of our imaginations, the "ah" sound precedes the "plosive" stop. In B, "ba," the stop and the vowel occur at about the same time, while for "pa" (C) the plosive stop is followed later by the vowel sound.

It required an elegant adumbration of this technique, however, to determine decisively that these perceptual boundaries are innate rather than something the precocious infant has already acquired from listening to its parents. Some languages use other VOTs to create consonants not found and never heard in English. Figure 30–12A for instance, shows a sound from Thai in which the voiced "ah" *precedes* the plosive. Adult English speakers do not use this consonant and cannot even distinguish it from sloppy versions of the nearest English ones. The infants of English-speaking parents nevertheless recognize the alien consonants as distinct, and set their boundaries almost precisely where they are found among native Thai speakers. A variety of these examples exist from several different languages, and the converse experiment—asking children from other language groups to distinguish English consonants which are absent from their parents' speech and perceptions—shows that we are all born to sort the cacophony of incoming auditory input automatically into discrete preexisting phoneme clusters on the basis of universally shared rules regarding VOT, frequency modulation, and so on. The ability to distinguish boundaries disappears when one of two neighboring categories of sounds is absent from the speakers' language—use it or lose it, it seems, like so many of our inborn abilities. It is therefore no wonder that small children can so readily learn different languages and their respective pronunciations, while English-speaking adults find it an uphill climb to hear and produce "un-English" consonant sounds.

Young children harbor a wealth of precious information about how we are wired, and new experimental procedures are allowing us brief glimpses of the riches to be found there. Using this new data on infant abilities and reinterpreting older observations, ethologists and cognitive psychologists are discovering clear evidence for genetic programming in many behaviors which the behaviorists who had studied them before had overlooked. Releasers, drives, motor programs, and directed learning all play major roles in our development of humans, just as they do in the lives of other animals. To ethologists infant behavior, like the behavior of other social animals, exhibits familiar patterns whose evolutionary significance is sometimes glaringly obvious, sometimes subtly hidden. The notion that we are born into the world with no internal preparation or guidance should now seem impossible, and the data that infant studies are yielding supports this contention. The youngest members of our species have a fascinating story to tell those who know what questions to ask and how to interpret the answers.

SUMMARY

The young of our species come equipped with innate processing rules and behaviors which structure their perceptions and aid in their survival. The catalogue discovered to date includes a variety of coordinated facial gestures—crying, smiling, flirting, and stereotyped reactions to tastes—and body movements including swimming, "walking," nursing, and clinging. Infants are able to recognize certain sorts of danger, localize sounds and odors, and sort the sensory continua of colors and speech sounds into arbitrary and appropriate categories. Doubtless the list is much longer. Our genes seem to have taken some care in arming and guiding their seemingly helpless porters in the business of survival. The critical question for ethologists is how much of this original programming persists into childhood, and what new sorts of internal direction might appear then.

STUDY QUESTIONS

1. Most crawling infants spend their time progressing from one object to another and putting everything in their paths into their mouths, be it a toy, a stone, or dirt on the floor. Despite the obvious inedibility of these objects, infants seem to take real pleasure in this behavior. Assuming they are not just doing it to torture their parents, what might be going on here?

2. One of the most obvious behaviors of babies is "flirting." The infant will look wide-eyed at a person, break into a shy smile, and hide its face against the person holding it. Later, however, flirting gives way to a strong fear-of-strangers phase and bouts of crying at the approach of new or rarely seen individuals. How would you interpret this sequence of behaviors against the background of human evolution?

FURTHER READING

Bower, T. G. R. "Visual World of Infants." *Scientific American* 215, no. 6 (1966): 80–92.

———. "The Object in the World of the Infant." *Scientific American* 225, no. 4 (1971): 30–38.

Fantz, R. L. "Origin of Form Perception." *Scientific American* 204 no. 5 (1961).

Gibson, E. G., and Walk, R. D. "The 'Visual Cliff'." *Scientific American* 202, no. 4 (1960): 64–71.

CHAPTER 31

Intellectual Development

Although our genes seem to have taken great care to prepare us for the first few months of life, to keep us alive and safe and to organize our perceptual world, is there any reason to suppose that they do anything more? Traditionally we have supposed that here, at the inception of that elusive quantity, intellect, genetic influences stop, or at least become so subjugated to the exigencies of natural and cultural environment that they are no longer forces to be dealt with. We allow that they might have some effect on certain "physical" attributes of our behavior, attributes such as temperament, but that their dominion over our behavior might exceed such clearly metabolic influences has always seemed unlikely. The difficulty in distinguishing between the products of learning and those of genetically legislated maturation in our species makes scientific exploration of this problem almost overwhelmingly complex. Can we ever hope to know if our genes' invisible hands are involved in predetermining or predisposing our continued development and—even if we should be able to find out—would knowledge of any genetic controls we might discover deep beneath our consciousness even be of use to a species so intent upon effecting cultural control and change?

In fact, human development does appear to proceed along

relatively predictable and adaptive lines with a remarkable will of its own—but genetic predisposition must not be equated with Calvinistic predestination. The more we learn about the various holds our biological inheritance has over us, the more tools we have for freeing us from the hold. If, for instance, we can pinpoint critical periods in the various aspects of our intellectual development, we will be free to counteract to some extent the crippling effects of deprivation, to provide the appropriate intellectual sustenance to young minds at the right time and at the rate they are best able to digest it. As a result, the ethological perspective may actually hold more promise for effecting cultural change and realizing our species' full potential than the prevailing viewpoint, which holds that our minds are blank slates, unshapen clay infinitely malleable by society and environment.

"INTELLIGENCE" TESTS

Where, then, shall we look for evidence of genetic orchestration of intellectual development? The most accessible and quantifiable data come from that statistical and political quagmire, IQ testing, which fuels the perennial debate over the relative contribution of genes (nature) and environment (nurture) to the development of human intelligence. What lessons can ethology draw from this most intriguing and potentially powerful part of the nature-nurture debate? In fact, the evidence brought forth in the IQ dispute for genetic effects on at least some aspects of human intelligence is clear. Figure 31–1 compares the test scores of identical twins, fraternal twins, and unrelated children over their first six years of life. The close similarity between identical twins and the complete lack of any between unrelated children is striking.

However, such "longitudinal" data (data derived from following the same individuals over time) are relatively scarce compared to the sorts of group statistics which fill the literature. The most abundant of these statistical sets are, of course, IQ scores. What does IQ measure? We do not know for sure, but whatever it measures, scores are relatively stable with age (Table 31–1) and do correlate well with future academic performance. Test administrators would like to believe that they are testing some sort of naïve and native, educational-indepen-

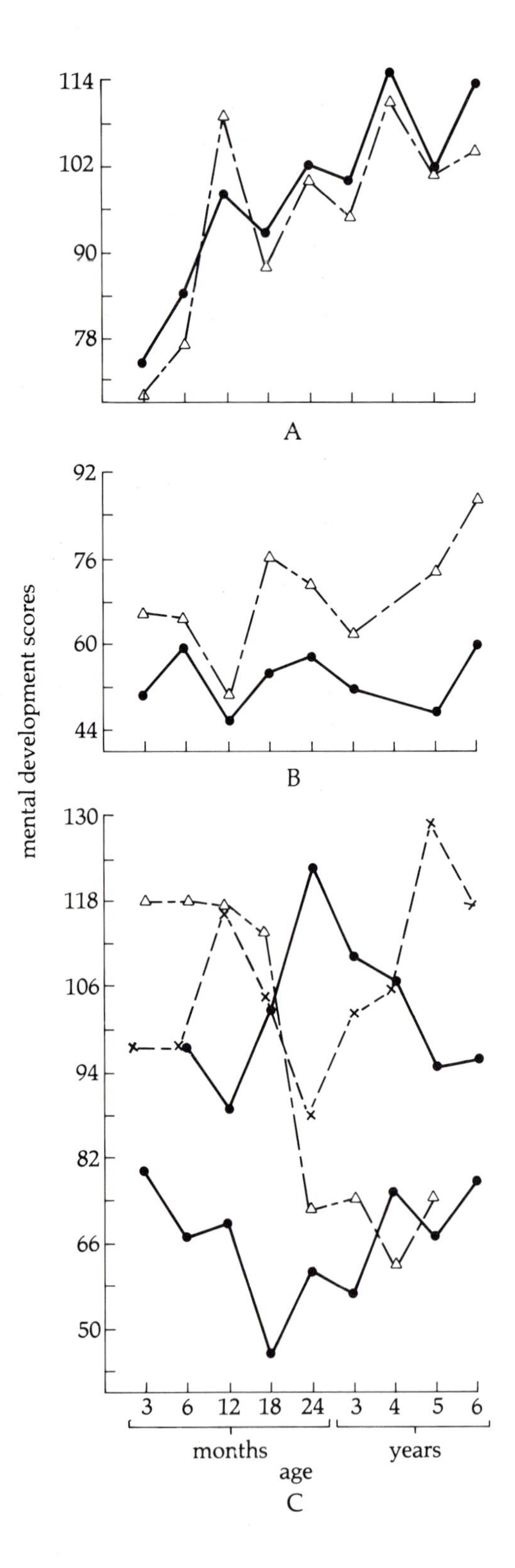

Fig. 31–1 Test scores of identical twins (A) track each other more closely than scores of fraternal twins (B), which in turn are more similar than the scores of unrelated children (C). These data suggest that the timing of the learning "spurts" so characteristic of children is under genetic control.

TABLE 31–1 *Correlation of IQ at Age 18 with IQ at Other Ages*[a]

Age of testing	*Correlation*
2	0.0
3	0.2
4½	0.5
6	0.65
9	0.75
12	0.80
15	0.85
50	0.85

Source: Christopher Jencks, *Inequality* (New York: Harper & Row, 1972).
[a] The data have been corrected for the variation in score normally seen between two administrations of the same test to the same individual (∼3 points).

dent, culture-free intelligence—a measure which would be of inestimable value to our ethological evaluation. We can be quite sure that this is *not* the case, though, since exposure to Western schooling dramatically raises the test scores of children around the world. As a result, we must take IQ for the moment simply as a measure of academic potential in ordinary American schools, a quantity which, given a homogeneous cultural background, probably would reflect individual mental differences of some sort. The data in Table 31–2 compare various degrees of genetic and environmental "relatedness," and when analyzed in any simple way, peg the genetic contribution to the observed range of scores at about 65 percent and indicate an environmental input of around 35 percent. This is not too surprising: the species-specific mental abilities of *Homo sapiens* are genetically controlled, but depend heavily (although less than has often been thought) on environmental input.

In itself, the genes-and-IQ debate is of little ethological interest, yet the data provide raw materials for ethological model making. An ethologist, of course, is interested in mechanisms, and so would want to know how and when environmental input is programmed to be selected and accepted. Looking back at Table 31–1, we can see why educators have traditionally assumed that most adult intellectual capacity is formed between

TABLE 31–2 *Correlation Between Test Scores (Corrected for Normal Variability) of People with Differing Genetic and Environmental Backgrounds*

	N	Subjects	Correlation
a.	4700	identical twins reared together	0.97
b.	3800	fraternal twins reared together	0.70
c.	122	identical twins reared apart	0.81
d.	2200	siblings reared together	0.59
e.	650	unrelated children reared together	0.36
f.	10,000	unrelated children reared apart	0.00
g.	10,000	parents and their natural children	0.55
h.	1200	parents and their adopted children	0.28

Source: Christopher Jencks, *Inequality* (New York: Harper & Row, 1972).

ages 3 and 12; and yet Table 31–3, which shows the net environmental effect on IQ of going to a "good" (upper 10 percent) vs. a "poor" (lowest 10 percent) school, suggests that even elementary schools have little effect on IQ per se. The data in Table 31–3 show, for example, that if the average entering IQ in "bad" junior highs was 90 and in "good" ones was 110, the average IQs at graduation might be 89 and 110, respectively. Hence, although students at "good" schools have higher IQs it is not because they have gone to "good" schools.

Where, then, does the environment's 35 percent input come from? Three hypotheses come to an ethological mind: (1) children might be programmed to accept environmental input only from parents; (2) environmental input to basic cognitive abilities

TABLE 31–3 *Effects on IQ of Going to a "Bad" (Lowest 10%) vs. a "Good" (Highest 10%) School*

Level of schooling	*Net change in IQ*
elementary	3 points
junior high	1 point
high school	1 point
college	0 points

Source: Christopher Jencks, *Inequality* (New York: Harper & Row, 1972).

might be mostly over before schooling begins at age 6; or (3) environmental input for IQ, as in birdsong learning, might come very early and lie dormant and untestable until the child is older. The second hypothesis has been the main focus of such ambitious social programs as "Head Start." Children who received this preschool training initially did much better on tests, but later the tangible advantages of the training disappeared from their test scores. In an essential control experiment researchers administered the tests *without* the program in case simple familiarity with the testing protocol itself might have been important. The control scores closely mimic those of the experimental group. Environmental "intervention" between ages 3 and 6 was, at least under these standard remedial programs, relatively ineffective.

No extensive tests have been done to distinguish between hypotheses 1 and 3, but since the few programs which focused their aim on the one- to three-year-old group *did* seem to have lasting effects, the latter seems the most likely explanation. Improvement under these programs is roughly "dosage dependent": the more the children were exposed to an enriched, non-home environment (or the more the parents were taught about how to enrich their infant's world), the better they did, even much later. If a birdsong model—which would predict that in order to be assimilated in any lasting way the information must be "fed" into the system at an early age, long before it can be detected by testing—is even partially relevant, a reanalysis of child adoption and separation data of the sort illustrated in Table 31-2 with close attention paid to *when* the source of environmental input was switched should provide powerful evidence. Only the children adopted before this hypothetical

---

IQ

When Benet developed the IQ test in France with the modest goal of identifying mentally retarded children, he could not have dreamed of its enormous use in the U.S. today and its central role in the emotional nature/nurture debate. How, from the correlations in Table 31-2, do we extract the conclusion that those intellectual abilities measured by IQ tests are mostly genetically determined, and what does the term "correlation" mean in the first place?

In statistics, correlation is a measure of the similarity between two test scores. For example, Table 31–1, which lists the correlation between IQ at age 18 and IQ at other ages, gives the correlation between ages 18 and 50 as 0.85. This means that, on the average, 85 percent of the deviation from the mean score of 100 measured at age 18 will be retained at age 50: people with scores of 140 would, on average, score 134 at age 50. A correlation of 0.0 would mean that the age 50 scores would bear no relation to the age 18 results, and would therefore average 100. Hence, the standard correlation of 0.80 between IQ and high school grades is very high.

When we look at the data in Table 31–2, we see that the IQs of unrelated children raised apart are, as we would expect, completely uncorrelated, but those of genetically identical twins reared together in the same environment are virtually the same. So far, so good. But now how do we separate the genetic from the environmental components of the similarity? To do this precisely, we would need to use the elegant model developed by Christopher Jencks (Fig. 1–3) which takes into account such subtleties as the tendency for parents to have similar IQs (the correlation is 0.57) and socioeconomic backgrounds (0.12 in the U.S.), and the possibility that high-IQ genes and intellectually stimulating environmental conditions might occur together more frequently than if they were randomly distributed, for adoptive parents and adopted children to come from restricted and very different samples of the population, and so on. As it turns out, however, even the naïve assumption that any observed correlation will represent the sum of the genetic and environmental correlations gives about the same result with far less effort, and that is the approach we will use.

Looking at the data in Table 31–2, then, we can immediately use that from identical twins raised apart (line c) as a potentially pure measure of the genetic component, 81 percent, while the correlation between genetically unrelated children reared together (line e) puts the environmental component at 36 percent. Unfortunately, the two numbers do not add up to 100 percent: the identical-twin correlation is too high for a variety of reasons (one unfortunate one being that, as researchers have recently discovered, some of it was fabricated by an overzealous psychologist). But fraternal twins (related genetically by one-half) reared together provide more reliable information. Since the correlation is the sum of the environmental component and half of the potential genetic input, and would be about 1.00 if the genes were the same, it follows that the missing half of the genetic component must be 30 percent and so the full genetic influence is around 60 percent. Whatever parameter we use to evaluate the data, every correlation tends toward a 55–80 percent genetic basis for IQ and a 20–45 percent environmental influence. The mean (excluding the identical twins) is 65 percent nature and 35 percent nurture.

Two caveats must accompany any analysis of IQ data. The first is that our conclusions about relative genetic and environmental influences are only average estimates reflecting some set of "ordinary" contingencies which will rarely describe any specific case. Moreover, it is more than likely that severe environmental deprivation can wreck a normal child's future IQ. The rare documented examples of "feral" children raised by animals and the all-too-frequent cases of infants closely confined by a deranged parent through their formative years are compelling instances of this tragic effect. Such children lose irretrievably the capacity for language and many other cognitive skills. To what extent the less obvious forms of mental deprivation may affect whatever "quotient" it is that IQ tests measure remains unknown, and cannot be deduced from these (or any other) figures.

The second caveat we must work under is that "heritability"—the precise term for the genetic component in such calculations—is a strange and distinctly nonbiological measure. Heritability specifies how much of the observed *variance* between individuals is the result of genetic differences. The remaining variance is assumed to be generated by environmental variation. We must take care with this measure, however, because the *proportion* of variance that results from genetic influence will vary with the degree of uniformity in the environment, while the proportion created by the environment will depend on the degree of naturally occurring genetic variation. This interaction can give rise to some remarkably counterintuitive conclusions. For example, the heritability of arms in humans is zero because all observed variation in this trait is environmental (i.e., wars, thalidomide, etc.). The consequence for test scores is that in relatively homogeneous social environments where little cultural variation exists between families, more of the observed difference between scores must be accounted for by genetic variance. Hence, heritability estimates for a culturally diverse country like the U.S. are generally lower than for, say, England.

In the end, the genes and IQ debate remains subject to myriad interpretations. For those whose political or emotional biases predispose them to the environmental viewpoint, we may quibble that IQ tests may not reflect "true" intellect, we may show that their applicability across cultural lines is dubious at best, and we may point out that the notion of heritability is often misleading. On the other side, genetic determinists can defend the data and point out that no matter what the measure, the genes always "win." Perhaps a more constructive reading of the data might reveal that the demonstrated high heritability of IQ reflects not so much the iron hand of the genes as our incompetence as teachers: our growing ethological understanding of what environmental cues are salient to the developing intellect, and when those cues can be interpreted, may serve to lower the "heritability" of IQ to levels the traditional critics of ethology will find comfortable.

critical period should reflect the environment of their adoptive parents, while the scores of those adopted any time after the age-3 cutoff should remain uninfluenced by the change. Ethological models argue, too, that to understand which of the many sorts of interactions the human child has with its world are of special significance to the developing mind, we should focus our attention on this apparent intellectual watershed.

PIAGET AND MONTESSORI

Indeed, the work of Jean Piaget suggests that the time before age 3 is a period of highly structured cognitive development. Piaget and his students have discovered in children the world over an ordered and reproducible course of intellectual events which seems to be self-generating and inflexible. Piaget discarded the behavioristic model in which learning was believed to *cause* intellectual development for one in which it is development that makes learning possible. This is a model familiar to ethologists: as we have seen, birds cannot "learn" their songs until their genes, through an inevitable process of maturation, make the learning possible.

The examples of our species' curious and stereotyped intellectual blossomings are almost endless. One well-known example is the steady progression in our grasp of the inexorable laws of the natural world. A child of two months grasps a rattle, but if his view of his hand is blocked he will forget what he cannot see and drop it in about 5 sec. Similarly, a child of this age will follow an object visually, but somehow lose sight of it when it is placed on or against a larger object. In fact the child's eyes continue to "follow" the now-stationary target as though it were still moving at the same speed and in the same direction as before.

When a child is four months old an object no longer "disappears" in front of a larger one, but does become invisible if it is even partially covered. Even if the child watches while the object is hidden, he will search for it not where it has been hidden but in whatever place he is used to seeing it. By six months babies can find partially hidden objects but not fully covered ones. By twelve months the child will search for an object where he saw it hidden, but until eighteen months he cannot master concealed movement of objects. The child can find a toy he sees being hidden in a box, but if the box with the object in it is moved—in full view of the child—the toy is lost.

The lesson here is that although deprivation at any time can stifle development, the steps which constitute the learning "program" cannot be rearranged or speeded up. Intensive studies have shown that we cannot teach a child that a wad of clay weighs the same whether rolled into a compact ball or stretched out into a long cylinder until the innate ability to grasp the concept of conservation of mass matures. On the same principle, no amount of coaching can convince even the brightest three-year-old that a small glassful of water does not become less when poured into a large glass.

Indeed, when any of these sophisticated stages in intellectual development appears, it is as though the abilities were drawn unsolicited from the very air children breathe. Maria Montessori's very successful school of education, in fact, maintains that mere *exposure* to well-designed learning tools during sensitive periods, rather than the usual laborious schooling, is the key to the development of cognitive abilities and skills.

LANGUAGE ACQUISITION

A striking demonstration of the strength of these concepts is seen in the effects of accidental deafening on the child's use of that all-important human tool, language. The analogy with birdsong in this instance is particularly intriguing: a chronically deaf infant will begin "babbling" at the usual age, but will never develop structured vocalizations. But after words have crystallized sufficiently in a normal child, deafening will have little effect, as though these elements in the individual's repertoire have become "hardwired" and independent of feedback. In fact, language acquisition shows every sign of being less an isolated intellectual feat and more like the innate development of species-typical behavior than we have heretofore wanted to believe.

There are two ways of thinking about language: either we are such a clever species that we invented it *ex nihilo* or like everything else in nature it owes its existence to genes and evolution. At first glance it seems that language must be cultural: we do learn it, and it comes in many different forms. But then we learn to walk and to crawl in different ways from sets of instructions that are obviously innate. Languages, it turns out, may not be all that different. Beginning with a scathing review of B. F. Skinner's explication of the behavioristic model for language learning *Verbal Behavior*, the eminent linguist Noam Chomsky has shown that standard contingency learning (even assuming there is some environmental reinforcement for the proper use of language in small children) simply cannot account for the syntactic structures recognized as grammatical and nongrammatical by a literate child. Chomsky has gone on to propose the existence of and gather evidence for a clear, basic, abstract, cross-cultural core to language which he calls its "deep structure." It is this deep structure, an innate ability to decode and encode basic linguistic thoughts (tucked away perhaps in Wernicke's area) which would account for the ability of children to recognize and generate syntax beyond their experience, and would explain the many grammatical similarities between unrelated languages.

The deep structure, the drive to learn language and its predictable time course, and now the demonstration that linguistic phoneme categories are innate, all combine to form what Chomsky calls an "innate language-acquisition device," and explains our species' facility at language learning. Indeed, the task of learning language as an adult is formidable, as most of us who have had to face a foreign-language requirement in high school or college can attest. And yet children, many of whom will turn out as adults to be painfully slow, manage this feat effortlessly. Our drive to communicate through some sort of language is so great that a group of chronically deaf children, with no model of language at all to draw from, may *invent* one among themselves with many of the syntactic features of spoken language.

Consider vocabulary size as a crude indicator of what is going on in language acquisition. From age 1 onward children begin to acquire a passive vocabulary—a dictionary of words that they recognize whether or not they actually use them in their own speech. This passive vocabulary increases from a rate of more than seven words a day at the beginning of the language-learning process to more than twenty a day by age eight (Table 31–4) until in a well-educated adult it reaches a total of at least forty thousand words. *The Shorter Oxford English Dictionary*, by comparison, has seventy-five thousand entries.

It is highly unlikely that this enormous array of words and meanings is acquired and originated without innate help, no matter what we would like to give ourselves credit for. Vocabularies are imbued with rules by which root words are used as nouns, verbs, adjectives, and even adverbs, and children seem programmed to look for these rules. There comes that day in the life of every parent when their two-year-old begins to make studied errors, correcting in mid-sentence "I ran . . ." to "I runned . . ." in blind obedience to the theoretical linguistic rules he is discovering to be true in other contexts. In this case linguistic learning takes dramatic and painful precedence over the child's previous longstanding and correct usage of even the most complex constructions.

TABLE 31–4 *Passive vs. Active Vocabulary Size at Various Ages*

Age	*Approximate passive vocabulary size*	*Active vocabulary*
1	~50	~5
6	~13,000	—
7	~22,000	—
8	~28,000	—
Well-educated adult	~40,000	~5000

Source: Peter Marler.

Language learning appears to be based on some innate structuring of the world and of language, as well as a "seek out the local conventions" program which makes irregular verbs and those crazy plurals like "geese," "larvae," and "data" so annoying. The path from speech to meaning begins with syllable sounds or "phonemes," which are organized through phonological syntax into words. Words are then organized according to "lexical syntax" into sentences which generate meanings in our minds. The basis of the system is our naming of objects, facilitated by a very strong tendency in our species to organize and classify. Psycholinguists organize our naming tendencies into three levels of organization or specificity—superordinate, basic, and subordinate—which correspond respectively to a general classification such as furniture, a basic one of chairs, tables, and lamps, and a very specific one such as rocking chair, kitchen tables, and desk lamps. The key seems to be the higher level categories, for once a generic (superordinate) concept is acquired, acquisition of particular names and their creative and meaningful generation in novel contexts is enhanced. Are these rules deduced or are they innate? Perhaps they are cleverly inferred from experience, but if so, it is intelligence which we share with the rest of the vertebrate hoi polloi. Even pigeons have been found to form the same sort of hierarchical concepts while merely observing slides. The ability, even readiness to generalize, to isolate and focus on diagnostic similarities in things of particular importance for an animal (food especially) seems to be a very old and widespread preadaptation for language which our species has put to good use.

Then, too, there seems every reason to suppose that words are stored in our brains in logical, organized arrays. Minor strokes or other sorts of modest brain damage can create strikingly specific aphasias—an inability to remember any more the names of plants, for example.

Thus the emerging picture of language acquisition, that greatest of human achievements, is being drawn along ethological lines. Our genes program us to classify the acoustic world into a few discrete consonant categories, and these categories doubtless provide the models against which we learn to produce the same sounds on our own. This innate tendency toward classification makes possible the manageable decoding of auditory input, as well as the ability to "name" things and to remember such arbitrary symbols as words. The brain comes already programmed to organize the world in a hierarchial way, a system crucial for linguistic thought and learning. We come programmed to seek out and learn the rules by which word forms are generated and combined, and if Chomsky is correct, the deep structure of grammar is prewired as well. The incredible mystery of how we as foolish, inattentive, undisciplined, and illogical children can manage to achieve language seems to have moved one step back and become instead a question of how evolution has spawned a set of genes which are such excellent teachers.

HUMAN ETHOLOGY

The patterns which have emerged from this ethological study of language and the other developmental phenomena in human children are not unique: they recur again and again throughout human development, revealing the hand of genetic programming at work even beyond infancy. Many ethologically inclined observers see in our behavioral ontogeny—in activities so pervasive as adolescent rebellion, courtship rituals, and parenting—similarities to the behavioral development of other species in the animal world. The myth of ourselves as completely separate creations, divorced from our biological inheritance, has created an egotistical blindness to analogies which might open the way to new and important discoveries about how we live and learn. The enormous opportunity before us is to ask the three ethological questions—what, how, and why—of *Homo sapiens,* on the assumption that we are a species whose genes are doing everything possible to help us survive and reproduce.

We cannot know where, during the course of evolution, our increasing mental capacities spawned the will that now battles with our genes for control of our behavior. What we can see, though, is that our moral and intellectual existence as well as our physical survival as a species depend both on our realization that our genes still have a powerful hand in our affairs, and on our continuing inquiry into what they are doing, how they are doing it, and why they have programmed or predisposed us to act in particular ways. We should be treating ourselves as one of many interesting species and looking hard for putative releasers, critical periods, drives, cases of imprinting, motor programming, and so on. Only in this way can we hope to make any effective progress toward understanding, not to mention changing, our behavior—be it moral, scholastic, or sociological. In this regard, observation and the psychological literature are our greatest resources.

Once we realize how much we have in common with our fellow species, the ethological models of animal behavior will be the best source of hypotheses to direct our analysis of human programming. Of particular interest to human ethologists should be the mechanisms and principles underlying information processing, critical periods, and programmed learning, for this new way of looking at the world may open some doors to the realization of human potential which have been closed to us before. It seems likely that current attitudes toward education—particularly the conviction that humans are infinitely plastic in all things at all times—is especially debilitating and open to ethological revision.

Somewhat more immediate perhaps is the terrible irony that our yet-to-be-understood instincts, painstakingly evolved for the propagation of our genes, together with our incredible, unadaptive technology, form an extraordinarily dangerous threat. If these fears prove valid, ethology's greatest challenge will lie in bringing its unique perspective into the traditional disciplines of human psychology, sociology, and anthropology with the object of discovering what those hundreds of thousands of selfish, pragmatic, utterly amoral genes in each of us are up to. We must learn more about the behavioral programs specified in ourselves in order to circumvent those that, in our present social environment, predispose us to inhumane actions, social chaos, or unmanageable wars. From the evidence of recorded history, it is more than probable that our posterity—our children, and our children's children—will require our educated intervention.

SUMMARY

When examined from a biological perspective the intellectual development of children appears to be more of a piece with animal development than with environmental conditioning, proceeding as it does in a relatively fixed order and schedule with a will of its own. Combined with genetic studies, the evidence from children supports the hypothesis that internal guidance plays a major role in the development of at least IQ, object concepts, and language acquisition, and points to likely roles for critical periods, drives, and sign stimuli. If cognitive development of humans is organized along the lines so evident in the rest of the animal kingdom, then ethology, with its mechanistic insights and models, may well be able to suggest useful remedies for current problems in the education, enculturation, and socialization of children.

STUDY QUESTIONS

1. Beginning with the "terrible twos," most children begin systematically to misbehave, testing their parents' patience and breaking previously established rules. How could this behavior be "fit" or adaptive, given the child's obvious inability to make it on its own should the parents abandon the increasingly frustrating enterprise of raising a child?
2. At or following puberty children normally rebel, rejecting their parents and their parents' values. This seems an enormous waste of

everyone's time since, in a few years, the children will most likely hold views very similar to the ones they rejected and be more or less reconciled to their parents. Psychologists refer to this as a "stage," but ethologists ought to be able to do more than assign a meaningless name to a widespread behavioral phenomenon. Develop a hypothetical model, suggest a test of it, and cite similar examples from nature.

3. There have been several cases of children growing up with wild animals. Although the best-known example is the wild boy of Avignon, the most informative case is that of the wolf children of Midnapore. For several years prior to 1920 natives had reported seeing a small, man-like animal in the jungle. A Christian missionary decided to investigate, and had a blind built where the sightings were most common. In the evening he saw three large wolves emerge from a hole under a termite mound, followed by two cubs and two human children running on all fours.

A few days later the missionary and his party returned and dug out the den. Two of the adult wolves fled, but one, a female, stayed to fight. The two children, both girls, eight and one and a half years old, were captured. They were both mute except for well-formed howls which they delivered punctually at 10 P.M., 1 A.M., and 3 A.M. The two girls huddled together in a corner and refused to interact with other children for several months. They ate from the ground, tore off any clothing, and were unable to stand. They flashed bared teeth at those who came near them, and would occasionally bite and scratch. The younger child learned a word for milk. The two girls played happily with each other, and sometimes would interact briefly with other crawling children.

After about eight months the two children began to accept a certain measure of protection from their adopted mother, but after eleven months the younger girl fell ill and died. The older turned her attention to other animals—kid goats, a hyena cub, etc. After fourteen months she began to interact with her adopted parents. After two years she could walk after a fashion, but never learned to run. When in a hurry, she would drop onto all fours. After three years she suddenly developed a modest vocabulary of forty-five poorly pronounced words. She fell ill and died in 1929.

No one knows how or when these children came to be living with wolves. Perhaps they were abandoned by their parents, were taken into the den as food, and found to taste bad. Some people speculate that they were abandoned because they were hopelessly retarded, but there is no reason to assume so. Analyze this story as an ethologist. Using what you know about human development, you should be able to make some good guesses about the age at which the girls came into the den. From that, you should be able to predict other stages of human development from this story, and suggest somewhat less dramatic ways of testing them. You should be able to predict what would have happened to the younger girl had she lived. How were the girls able to survive in the den at all? Were they "preadapted" in any way?

4. Whenever we attempt to test learning, we are usually left with some ambiguity over whether a failure is the result of an animal's not having learned or the consequence of its inability to recall the information quickly and reliably. Given the evidence from lesion studies (for example, the case of the patient with a relatively minor lesion in which the ability to recall plant names was entirely abolished) and the observation that the "learning" of the names of subordinate objects is greatly facilitated by the acquisition of a superordinate concept, formulate a hypothesis of human memory organization. How would you test it? How would you go about examining this hypothesis in animals?

FURTHER READING

Bower, T. G. R. "Repetitive Processes of Child Development." *Scientific American* 235, no. 5 (1976): 38–47.

Moskowitz, Breyne Arlene. "The Acquisition of Language." *Scientific American* 239, no. 5 (1978): 92–108.

Piaget, Jean. "How Children Form Mathematical Concepts." *Scientific American* 189, no. 5 (1953): 74–79.

APPENDIX A

[Lorenz has wondered why ethology did not begin in England just after Darwin's *Origin of the Species.* Actually, it did, only to die out. The author of the following article saw the implications of Darwin's theory on animal behavior, and began his own experiments. They were interrupted when the Church of England overturned the will which had given him the position of chief tutor to (then) young Bertrand Russell. Spalding died a few years later. Reprinted from *MacMillans Magazine* 27 (1873): 282–93.]

Instinct: With Original Observations on Young Animals BY DOUGLAS A. SPALDING

The exquisite skill and accurate knowledge observable in the lives of the lower animals, which men generally have regarded as instinctive—born with them—have ever been subjects of wonder. In the hands of the natural theologian, whose armoury has been steadily impoverished in proportion as mystery has given way before science, instinct is still a powerful weapon. When the divine expatiates on the innate wisdom and the marvelous untaught dexterity of beasts, birds, and insects, he is in little danger of being checked by the men of science. His learned enemies are dumb, when in triumph he asks the old question:

> "Who taught the nations of the field and wood
> To shun their poison and to choose their food?
> Prescient, the tides or tempests to withstand,
> Build on the wave, or arch beneath the sand?"

The very little that our psychologists have done for instinct may be told in a few words. The only theory of instinct, of the nature of an explanation, is that put forward by Mr. Herbert Spencer as part of his philosophy of evolution; but, as a theory, it is only beginning to be understood and appreciated among scientific men; while some eminent thinkers question the reality of the phenomena to be explained. Professor Bain, our other psychologist, and his able following of trained disciples, simply discredit the alleged facts of instinct. Unfortunately, however, instead of putting the matter to the test of observation and experiment, they have contented themselves with criticising the few accidental observations that have been recorded, and with arguing against the probability of instinctive knowledge. In defending the Berkeleian Theory of Vision, Professor Bain, in answer to the assertion that the young of the lower animals manifest an instinctive perception of distance by the eye, contends that "there does not exist a body of careful and adequate observations on the early movements of animals." Writing long ago on the same subject, Mr. Mill also, while admitting that "the facts relating to the young of the lower animals have been long felt to be a real stumbling block in the way of the theory," maintains that "our knowledge of the mental operations of animals is too imperfect to enable us to affirm positively that they have this instinct." Denying the facts, however, was not Mr. Mill's mode of saving the theory. He was rather of opinion that the "animals have to us an inexplicable facility both of finding and selecting the objects which their wants require." How very inexplicable, he conceives, their mental operations may possibly be, may be gathered from the fact of his suggesting an experiment to ascertain whether a blind duckling might not find the water as readily as one having sight. The position of psychologists of the too purely analytical school, however, is not that the facts of instinct are inexplicable; but that they are incredible. This view is set out most explicitly in the article on Instinct in "Chambers's Encyclopaedia." Thus: "It is likewise said that the chick recognizes grains of corn at first sight, and can so direct its movements as to pick them up at once; being thus able to know the meaning of what it sees, to measure the distance of objects instinctively, and to graduate its movements to that knowledge—all which is, in the present state of our acquaintance with the laws of mind, wholly incredible." And it is held, that all the supposed examples of instinct may be—for anything that has yet been observed to the contrary—nothing more than cases of rapid learning, imitation, or instruction.

Thus it would appear that with regard to instinct we have yet to ascertain the facts. With a view to this end, I have made many observations and experiments, mostly on chicken. The question of instinct, as opposed to acquisition, has been discussed chiefly in connection with the perceptions of distance and direction by the eye and the ear. Against the instinctive character of these perceptions it is argued, that as distance means movement, locomotion, the very essence of the idea is such as cannot be taken in by the eye or ear; that what the varying sensations and feelings of sight and hearing correspond to, must be got at by moving over the ground—by experience. On the other hand, it is alleged that, though as regards man the prolonged helplessness of infancy stands in the way of the observer, we have only to look at the young of the lower animals to see that as a matter of fact they do not require to go through the process of learning the meaning of their sensations in relation to external things; that chickens, for example, run about, pick up crumbs, and follow the call of their mother *immediately* on leaving the shell. For putting this matter to the test of experiment, chickens, therefore, are most suitable and convenient subjects. I have observed and experimented on more than fifty chickens, taking them from under the hen while yet in the eggs. But of these, not one on emerging from the shell was in a condition to manifest an acquaintance with the qualities of the outer world. On leaving the shell they are wet and helpless; they struggle with their legs, wings, and necks, but are unable to stand or hold up their heads. Soon, however, they may be distinctly seen and felt pressing against and endeavouring to keep in contact with any warm object. They advance very rapidly. I have seen them hold up their heads well, peck at objects, and attempt to dress their wings when only between four and five hours old. But there is no difficulty in conceiving that, with great spontaneity and a strong power of association, much might be learned in four or five hours. Professor Bain is of opinion, from observations of his own on a newly dropped lamb, that "a power that the creature did not at all possess naturally, got itself matured as an acquisition in a few hours." Accordingly, in the absence of precautions, the time that must elapse before chickens have acquired enough control over their muscles to enable them to give evidence as to their instinctive power of interpreting what they see and hear, would suffice to let in the contention that the eye and the ear may have had opportunities of being educated. To obviate this objection with respect to the eye, I had recourse to the following expedient. Taking eggs just when the little prisoners had begun to break their way out, I removed a piece of the shell, and before they had opened their eyes drew over their heads little hoods, which, being furnished with an elastic thread at the lower end, fitted close round their necks. The material of these hoods was in some cases such as to keep the wearers in total darkness; in other instances it was semi-transparent. Some of them were close at the upper end, others had a small aperture bound with an elastic thread, which held tight round the base of the bill. In this state of blindness—the blindness was very manifest—I allowed them to remain from one to three days. The conditions under which these little victims of human curiosity were first permitted to see the light were then carefully prepared. Frequently the interesting little subject was unhooded on the centre of a table covered with a large sheet of white paper, on which a few small insects, dead and alive, had been placed. From that instant every movement, with the date thereof, as shown by the watch, was put on record. Never in the columns of a Court Journal were the doings of the most royal personage noted with such faithful accuracy. This experiment was performed on twenty separate chickens at different times, with the following results. Almost invariably they seemed a little stunned by the light, remained motionless for several minutes, and continued for some time less active than before they were unhooded. Their behavior, however, was in every case conclusive against the theory that the perceptions of distance and direction by the eye are the result of experience, of associations formed in the history of each individual life. Often at the end of two minutes they followed with their eyes the movements of crawling insects, turning their heads with all the precision of an old fowl. In from two to fifteen minutes they pecked at some speck or insect, showing not merely an instinctive perception of distance, but an original ability to judge, to measure distance, with something like infallible accuracy. They did not attempt to seize things beyond their reach, as babies are said to grasp at the moon; and they may be said to have invariably hit the objects at which they struck—they never missed by more than a hair's breadth, and that too, when the specks at which they aimed were no bigger, and less visible, than the smallest dot to an *i*. To seize between the points of the mandibles at the very instant of striking seemed a more difficult operation. I have seen a chicken seize and swallow an insect at the first attempt; most frequently, however, they struck five or six times, lifting once or twice before they succeeded in swallowing their first food. The unacquired power of following by sight was very plainly exemplified in the case of a chicken that, after being unhooded, sat complaining and motionless for six minutes, when I placed my hand on it for a few seconds. On removing my hand the chicken immediately followed it by sight backward and forward and all round the table. To take, by way of example, the observations in a single case a little in detail:

A chicken that had been made the subject of experiments on hearing, was unhooded when nearly three days old. For six minutes it sat chirping and looking about it; at the end of that time it followed with its head and eyes the movements of a fly twelve inches distant; at ten minutes it made a peck at its own toes and the next instant it made a vigorous dart at the fly, which had come within reach of its neck, and seized and swallowed it at the first stroke; for seven minutes more it sat calling and looking about it, when a hive-bee coming sufficiently near was seized at a dart and thrown some distance, much disabled. For twenty minutes it sat on the spot where its eyes had been unveiled without attempting to walk a step. It was then placed on rough ground within sight and call of a hen with a brood of its own age. After standing chirping for about a minute, it started off towards the hen, displaying as keen a perception of the qualities of the outer world as it was ever likely to possess in after life. It never required to knock its head against a stone to discover that there was "no road that way." It leaped over the smaller obstacles that lay in its path and ran round the larger, reaching the mother in as nearly straight line as the nature of the ground would permit. This, let it be remembered, was the first time it had ever walked by sight.[1]

[1] Since writing this article, I see it stated in Mr. Darwin's new book, "The Expressions of the Emotions in Man and Animals," that "the wonderful power which a chicken possesses only a few hours after being hatched of picking up small particles of food, seems to be started into action through the sense of hearing; for, with chickens hatched by artificial heat, a good observer found that 'making a noise with a finger-nail against a board, in imitation of the hen-mother, first taught them to peck at their meat.' " My own observations give no countenance whatever to this view:—(1) I have frequently observed chickens finally hatched in a flannel nest over a jar of hot water and left undisturbed for a few hours, begin, immediately after the covering was removed, and while they still sat nestling together, to pick at each other's beaks and at specks of oatmeal when these were dropped on them, all noise being as far as possible avoided. (2) Each of the twenty chickens made subjects of the experiment described in the text, began to eat without any assistance from the sense of hearing; the greatest possible stillness being maintained and required during the experiment. (3) Chickens picked up food though rendered deaf while yet in the shell. One of these, deprived of both sight and hearing at its birth,

It would be out of place here to attempt to indicate the full psychological bearing of these facts. But this much may be affirmed, that they put out of court all those who are prepared only to argue against the instinctive perception by the eye of the primary qualities of the external world. When stripped of all superflous learning, the argument against this and every other alleged case of instinctive knowledge is simply that it is unscientific to assume an instinct when it is possible that the knowledge in question may have been *acquired* in the ordinary way. But the experiments that have been recounted are evidence that prior to experience chickens behave as if they already possessed an acquaintance with the established order of nature. A hungry chick that never tasted food is able, on seeing a fly or a spider for the first time, to bring into action muscles that were never so exercised before, and to perform a series of delicately adjusted movements that end in the capture of the insect. This I assert as the result of careful observation and experiment; and it cannot be answered but by observation and experiment at least as extensive. It is no doubt common for scientific men to discredit new facts, for no other reason than that they do not fit with theories that have been raised on too narrow foundations; but when they do this they are only geologists, or psychologists—they are not philosophers.

Before passing to the perceptions of the ear, it may be mentioned that, instead of hooding chickens, which had the advantage of enabling me to make many interesting observations on them when in a state of blindness, I occasionally put a few eggs, when just chipped, into a flannel bag made for the purpose. In this bag the hatching was completed artificially, and the chickens allowed to remain in the dark from one to three days. When placed in the light they deported themselves as regards sight in the manner already described. For the purpose of merely testing the perceptions of the eye or the ear this is by far the easier experiment. The hooding process requires considerable delicacy of manipulation, and the chickens are very liable to be injured.

With respect now to the space perceptions of the ear, which, in man at least, even Mr. Spencer regards as acquired by each individual. Chickens hatched and kept in the said bag for a day or two, when taken out and placed nine or ten feet from a box in which a hen with chicks were concealed, after standing for a minute or two, uniformly set off straight for the box in answer to the call of the hen, which they had never seen and never before heard. This they did, struggling through grass and over rough ground, when not yet able to stand steadily on their legs. Nine chickens were thus experimented upon, and each individual gave the same positive results, running to the box scores of times, and from every possible position. To vary the experiment I tried the effect of the mother's voice on hooded chickens. These, when left to themselves, seldom made a forward step, their movements were round and round, and backward; but when placed within five or six feet of the mother, they, in answer to her call, became much more lively, began to make little forward journeys, and soon followed her by sound alone, though, of course, blindly keeping their heads close to the ground and knocking against everything that lay in their path. Only three chickens were made subjects of this experiment. Another experiment consisted in rendering chickens deaf for a time by sealing their ears with several folds of gum paper before they had escaped from the shell. I tried at different times to stop the ears of a good many in this way, but a number of them got the papers off, others were found not quite deaf, and only three remained perfectly indifferent to the voice of the mother when separated from them by only an inch board. These had their ears opened when between two and three days old, and on being placed within call of the mother hidden in a box, they, after turning round a few times, ran straight to the spot whence came what must have been very nearly, if not actually, the first sound they had ever heard. It seems

---

was unhooded when three days old, and nine minutes after it vigorously pursued a large blue fly a distance of two feet, pecking at it several times: this bird proved perfectly deaf. Another with its ears similarly closed, was taken from the dark when a day and a half old, and when an experiment was being tried to ascertain whether it was perfectly deaf—which it turned out to be—it began to pick up and swallow small crumbs. What in this case really surprised me was that, the gum employed in closing its ears having also sealed up one of its eyes, it nevertheless picked up crumbs by sight of its one eye almost if not altogether as well as if it had had two.

scarcely necessary to make any comment on these facts. They are conclusive against the theory that, in the history of each life, sounds are at first but meaningless sensations; that the direction of the sounding object, together with all other facts concerning it, must be learned entirely from experience.

If now it be taken as established that in the perceptions of the eye and the ear, chickens at least manifest an instinctive knowledge of the relations and qualities of external things, the popular belief that the special knowledge, the peculiar art and skill, so marked in the various species of animals, come to them mostly without the labour of acquisition, is at once freed from all antecedent improbability. In the way of direct evidence, the little that I have been able to observe in this wide field goes to prove that the current notions are in accordance with fact. We have seen that chickens follow the call of their mother before they have had any opportunity of associating that sound with pleasurable feelings; and one or two observations, which must be taken for what they are worth, support the general opinion that they have an equally instinctive dread of their more deadly enemies. When twelve days old one of my little *proteges,* while running about beside me, gave the peculiar chirr whereby they announce the approach of danger. I looked up, and behold a sparrow-hawk was hovering at a great height over head. Having subsequently procured a young hawk, able to take only short flight, I made it fly over a hen with her first brood, then about a week old. In the twinkling of an eye most of the chickens were hid among grass and brushes. The hen pursued, and scarcely had the hawk touched the ground, about twelve yards from where she had been sitting, when she fell upon it with such fury that it was with difficulty that I was able to rescue it from immediate death. Equally striking was the effect of the hawk's voice when heard for the first time. A young turkey, which I had adopted when chirping within the uncracked shell, was on the morning of the tenth day of its life eating a comfortable breakfast from my hand, when the young hawk, in a cupboard just beside us, gave a shrill chip, chip, chip. Like an arrow the poor turkey shot to the other side of the room, and stood there motionless and dumb with fear, until the hawk gave a second cry, when it darted out at the open door right to the extreme end of the passage and there, silent and crouched in a corner, remained for ten minutes. Several times during the course of that day it again heard these alarming sounds, and in every instance with similar manifestations of fear. Unfortunately, my hawk coming to an untimely end, I was prevented from proceeding with observations of this class. But these few were so marked and unmistakeable in their character that I have thought them worth recording.

There are instincts, however, yet to be mentioned, concerning the reality of which I have thoroughly satisfied myself. The early attention that chickens give to their toilet is a very useful instinct, about which there can be no question. Scores of times I have seen them attempt to dress their wings when only a few hours old—indeed as soon as they could hold up their heads, and even when denied the use of their eyes. The art of scraping in search of food, which, if anything, might be acquired by imitation—for a hen with chickens spends the half of her time in scratching for them—is nevertheless another indisputable case of instinct. Without any opportunities of imitation, when kept quite isolated from their kind, chickens began to scrape when from two to six days old. Generally, the condition of the ground was suggestive; but I have several times seen the first attempt, which consists of a sort of nervous dance, made on a smooth table. As an example of unacquired dexterity, I may mention that on placing four ducklings a day old in the open air for the first time, one of them almost immediately snapped at and caught a fly on the wing. More interesting, however, is the deliberate art of catching flies practised by the turkey. When not a day and a half old I observed the young turkey already spoken of slowly pointing its beak at flies and other small insects without actually pecking at them. In doing this, its head could be seen to shake like a hand that is attempted to be held steady by a visible effort. This I observed and recorded when I did not understand its meaning. For it was not until after, that I found it to be the invariable habit of the turkey, when it sees a fly settled on any object, to steal on the unwary insect with slow and measured step until sufficiently near, when it advances its head very slowly and steadily till within an inch or so of its prey, which is then seized by a sudden dart. If all this can be proved to be instinct, few, I think, will care to maintain that *anything* that can be learned from experience *may* not also appear as an intuition. The evidence I have in this case, though not so abundant as could be wished, may yet, perhaps, be held sufficient. I have mentioned that this masterpiece of turkey cleverness when first observed, was in the incipient stage, and, like the nervous dance that precedes the actual scraping, ended in nothing. I noted it simply as an odd performance that I did not understand. The turkey, however, which was never out of my sight except when in its flannel bag, persisted in its whimsical pointing at flies, until before many days I was delighted to discover that there was more in it than my philosophy had dreamt of. I went at once to the flock of its own age. They were following a common hen, which had brought them out; and as there were no other turkeys about the place, they could not possibly learn by imitation. As the result, however, of their more abundant opportunities, I found them already in the full and perfect exercise of an art—a cunning and skilful adjusting of means to an end—bearing conspicuously the stamp of experience. But the circumstances under which those observations were made left me no room for the opinion that the experience, so visible in their admirable method of catching flies, was original, was the experience, the acquisition of those individual birds. To read what another has observed is not, however, so convincing as to see for oneself, and to establish a case so decisive more observation may reasonably be desired; at the same time, it can scarcely be attempted to set aside the evidence adduced, on the ground of improbability, for the *fact* of instinct: all that is involved in this more striking example, has, we venture to think, been sufficiently attested.

A few manifestations of instinct still remain to be briefly spoken of. Chickens as soon as they are able to walk will follow any moving object. And, when guided by sight alone, they seem to have no more disposition to follow a hen than to follow a duck, or a human being. Unreflecting on-lookers, when they saw chickens a day old running after me, and older ones following me miles and answering to my whistle, imagined that I must have some occult power over the creatures, whereas I simply allowed them to follow me from the first. There is the instinct to follow; and, as we have seen, their ear prior to experience attaches them to the right object. The advantage of this arrangement is obvious. But instincts are not conferred on any principle of supplying animals with arts very essential to them, and which they could not very well learn for themselves. If there is anything that experience would be sure to teach chickens, it would be to take care when they had got a piece of food not to let their fellows take it from them, and from the very first they may be seen to run off with a worm, pursued by all their companions. But this has been so stamped in their nature that, when they have never seen one of their kind, nor ever been disturbed in the enjoyment of a morsel, they nevertheless, when they get something larger than can be swallowed at once, turn round and run off with it.

Another suggestive class of phenomena that fell under my notice may be described as imperfect instincts. When a week old my turkey came on a bee right in its path—the first, I believe, it had ever seen. It gave the danger chirr, stood for a few seconds with outstretched neck and marked expression of fear, then turned off in another direction. On this hint I made a vast number of experiments with chickens and bees. In the great majority of instances the chickens gave evidence of instinctive fear of these sting-bearing insects; but the results were not uniform, and perhaps the most accurate general statement I can give is, that they were uncertain, shy, and suspicious. Of course to be stung once was enough to confirm their misgivings for ever. Pretty much in the same way did they avoid ants, especially when swarming in great numbers.

Probably enough has been said to leave no doubt in minds free from any bias on the subject, that in the more important concerns of their lives the animals are in great part guided by knowledge that they individually have not gathered from experience. But equally certain is it that they do learn a great deal, and exactly in the way that we are generally supposed to acquire all our knowledge. For example, every chicken, as far as my observations go, has to learn not to eat its own excrement. They made this mistake invariably; but they did not repeat it oftener than once or twice. Many times they arrested themselves when in the very act, and went off shaking their heads in disgust, though they had not actually touched the obnoxious matter. It also appeared that, though thirsty, they did not recognize water by sight, except perhaps in the form of dew-drops on the grass; and they had to some extent to learn to drink. Their first attempts were awkward; instead of dipping in their beaks, they pecked at the water, or rather at specks in the water, or at the edge of the water. All animals have a capacity to learn: each individual must learn the topography of its locality, and numerous other facts. Many dogs, horses, and elephants may be able to learn more than some men. But I have no doubt that observation will bear out the popular belief that what may be called the professional knowledge of the various species—those special manifestations of practical skill, dexterity, and cunning that mark them off from each other, no less clearly than do the physical differences whereon naturalists base their classifications—is instinctive, and not acquired. As we shall see, the creatures have not in a vast multitude of instances the opportunity to acquire these arts. And if they had the opportunity, they have not individually the capacity to do so, even by way of imitation. We have seen as a matter of fact that it is by instinct that the chicken, and, I may add, the turkey, scratch the surface of the earth in search of insects; also, that the turkey has a method of catching flies so remarkably clever that it cannot be witnessed without astonishment. Now chickens like flies no less than turkeys, and, though with less success, often try to catch them. But it is a significant fact that they do not copy the superior art. To give every opportunity of imitation, I placed a newly-hatched chicken with my turkey, when the latter was eleven days old. The two followed me about for several weeks, and when I deserted them they remained close companions throughout the summer, neither of them ever associating with the other poultry. But the chicken never caught the knowing trick of its companion—seemed, indeed, wholly blind to the useful art that was for months practised before its eyes.

Before passing to the theory of instinct, it may be worthy of remark that, unlooked for, I met with in the course of my experiments some very suggestive, but not yet sufficiently observed, phenomena; which, however, have led me to the opinion that not only do the animals learn, but they can also forget—and very soon—that which they never practised. Further, it would seem that any early interference with the established course of their lives may completely derange their mental constitution, and give rise to an order of manifestations, perhaps totally and unaccountably different from what would have appeared under normal conditions. Hence I am inclined to think that students of animal psychology should endeavour to observe the unfolding of the powers of their subjects in as nearly as possible the ordinary circumstances of their lives. And perhaps it may be because they have not all been sufficiently on their guard in this matter, that some experiments have seemed to tell against the reality of instinct. Without attempting to prove the above propositions, one or two facts may be mentioned. Untaught, the new-born babe can suck—a reflex action; and Mr. Herbert Spencer describes all instinct as "compound reflex action"; but it seems to be well known that if spoon-fed, and not put to the breast, it soon loses the power of drawing milk. Similarly, a chicken that has not heard the call of the mother until eight or ten days old then hears it as if it heard it not. I regret to find that on this point my notes are not so full as I could wish, or as they might have been. There is, however, an account of one chicken that could not be returned to the mother when ten days old. The hen followed it, and tried to entice it in every way; still it continually left her and ran to the house or to any person of whom it caught sight. This it persisted in doing, though beaten back with a small branch dozens of times, and indeed cruelly maltreated. It was also placed under the mother at night, but it again left her in the morning. Something more curious, and of a different kind, came to light in the case of three chickens that I kept hooded until nearly four days old—a longer time than any I have yet spoken of. Each of these on being unhooded evinced the greatest terror of me, dashing off in the opposite direction whenever I sought to approach it. The table on which they were unhooded stood before a window, and each in its turn beat against the glass like a wild bird. One of them darted behind some books, and squeezing itself into a corner, remained cowering for a length of time. We might guess at the meaning of this strange and exceptional wildness; but the odd fact is enough for my present purpose. Whatever might have been the meaning of this marked change in their mental constitution—had they been unhooded on the previous day they would have run to me instead of from me—it could not have been the effect of experience; it must have resulted wholly from changes in their own organization. . . .

So far good. But it will occur to every reader that the peculiar depths of animal psychology are not yet explored. Two classes of phenomena still lie in the dark. First, there are the many extraordinary and exceptional feats of dogs and other animals, which seem to be constantly falling under the observation of everybody except the few that are interested in these matters. Second, all the more wonderful instincts, especially those of insects, are such that it is hard, if at all possible, to conceive how they ever could have been derived from experience.

With regard to the first, it is not desirable to say much. Though volumes of marvellous stories have been written, I am not aware that any careful experiments have been tried, and, as the performances in question are of an exceptional character, it is perhaps but scientific caution not as yet to put too much stress on them. For my own part, though I have been very intimate with dogs, I have been singularly unfortunate in having never witnessed any of their more incomprehensible clairvoyant-like achievements. I have known them to do many surprising things, but I have always found that they had, or might have had, something to go upon—enough, coupled with quick intelligence, to account for their exploits. What may be said in this connection, if, indeed, it be prudent to say anything is that, while we certainly cannot have all the data of experience from without of all the vastly different living things which people the earth, the air, and the ocean—while we certainly can have no trace of many feelings that arise from changes in the organisms of the different creatures, and which, instinctively interpreted, start them on lines of action—a host of statements, generally accepted as fact, suggest the opinion that even such animals as dogs, are alive to, conscious, sensible of influences that scarcely affect us, or wholly escape our cognition. If this be so, they have a basis of experience from which to start in their calculations that we want, and, if so, well may their actions seem to us, as Mr. Mill said, hopelessly inexplicable. Take, not the most remarkable, but the best-authenticated example of this class—the frequently alleged fact of dogs and other animals returning in a straight line, or by the most direct routes, through districts they had never before traversed, to places from which they had been taken by devious tracks, and even shut up in close boxes. To most people this is a phenomenon sufficiently incomprehensible. They are certain they themselves could do nothing at all like it. But there is in some men what may be just a hint of this faculty. Most people that have lived only in cities are very soon lost in a strange and trackless district, and still sooner in a pathless wood; in the one case, after wandering this way and that for a few hours, in the other, after merely turning round a few times, they can tell nothing of the direction whence they came. But all men are not so easily lost; some, without consciously making notes, retain, after long wandering in such situations, a strong and often accurate impression, not of the ground they have gone over, but of the direction in which lies the place whence they started. Without attempting to throw any light on the mental chemistry of this perception, we would submit that in it may perhaps be found a clue to the mystery of those astonishing home-journeys of dogs, sheep, cats, pigeons, bees, &c., of which hundreds are on record.

It is, however, with the other dark enigma that we are more especially concerned. We do not think it necessary to examine the proof of the actuality of such marvellous instincts as those of bees and wasps. But for the too fond love of a theory we venture to think none would doubt the reality, or the instinctive character, of their "far-sighted," or, more correctly, blind provisions for the future. The problem before us is not whether for example, the male of the fish Arius does, and by instinct, hatch the eggs of the female in his mouth, but how such a singular mode of incubation ever has a beginning? Perhaps the most widely known instance of this class of instincts is the provision of the solitary wasp for the worm that will issue from her egg after her own death. She brings grubs—food that as a wasp she never tasted—and deposits them over the egg, ready for the larva she will never see. The life history of every insect exhibits instincts of this perplexing description. Witness the caterpillar, how at the proper time it selects a suitable situation and spins for itself a silken cocoon. It may be admitted at once that the creatures, *as we behold them,* never could have lived to acquire such instincts by any process of experience and inheritance of which we can conceive. Nor let it be supposed that it is only in the insect world, where all is so strange, that instincts are to be met with so essential to lives of the individuals or their progeny that without them the creatures in their present shape could never have existed. Of this kind are the first movements observable in the life of a bird, and which take place within the shell. I have often observed the self-delivery of the chicken. The prison wall is not burst in pieces by spontaneous, random struggles. By a regular series of strokes the shell is cut in two—chipped right round in a perfect circle, some distance from the great end. Moreover, the bird has a special instrument for this work, a hard, sharp horn on the top of the upper mandible, which being required for no other purpose disappears in a few days. Obviously each individual bird, no more acquires the art of breaking its way out than it furnishes itself with the little pick hammer used in the operation; and it is equally clear that a bird could have never escaped from the egg without this instinct. Again, how were eggs hatched before birds had acquired the instinct to sit upon them? Or who will throw light on the process of such an acquisition? Nor are the subsequent phenomena easier of explanation. A fowl that never before willingly shared a crumb with a companion, will now starve herself to feed her chickens, which she calls by a language she never before used—may have never even heard—but which they are born to understand. Once more, it is clearly because she cannot do otherwise that a she-rabbit, when with her first young, digs a hole in the earth away from her ordinary habitation, and there builds a nest of soft grass, lined with fur stripped from her own body. But how as to the origin of this habit?

We need not accumulate examples of seemingly unfathomable instincts. And it may be confessed at once, that in the present state of our knowledge it would be hopeless to attempt to guess at the kinds of experiences that may have originally, when the creatures wore different shapes and lived different lives, wrought changes in their nervous systems that, enduring and being modified through many changes of form, have given to the living races the physical organisations of which these wonderful instincts are the corresponding mental facts. Nor, perhaps, can it be confidently asserted that in experience and heredity we have all the terms of the problem. The little we can say is, that though in the dark we need not consider ourselves more in the dark as to the origin of those strange instincts than we are concerning the origin of those wonderful organs of astonishing and exquisite mechanism that, especially among the insects, are the instruments of those instincts. Nay, more, if the view we have put forward concerning the connection between mental manifestations and bodily organization be correct, the question of the origin of these mysterious instincts is not more difficult than, or different from, but is the same with, the problem of the origin of the physical structure of the creatures; for, however they may have come by their bodies, they cannot fail to have the minds that correspond thereto. When, as by a miracle, the lovely butterfly bursts from the chrysalis full-winged and perfect, and flutters off a thing of soft and gorgeous beauty, it but wakes to a higher life, to a new mode of existence, in which, strange though it may sound, it has, for the most part, nothing to learn, *because* its little life flows from its organisation like melody from a music box. But we need not enlarge on this a second time.

In seeking to understand the phenomena of instinct we of course get the full benefit of the law of Natural Selection, which, though it throws no light on the origin of anything, mental or physical—for, as Mr. Darwin says, it "has no relation whatever to the primary cause of any modification of structure"—nevertheless helps us to understand the existence of instincts far removed from the circumstances or conditions of life under which they could have been acquired. Suppose a Robinson Crusoe to take, soon after his landing, a couple of parrots, and to teach them to say in very good English, "How do you do, sir?"—that the young of these birds are also taught by Mr. Crusoe and their parents to say, "How do you do, sir?"—and that Mr. Crusoe, having little else to do, sets to work to prove the doctrine of Inherited Association by direct experiment. He continues his teaching, and every year breeds from the birds of the last and previous years that say "How do you do, sir?" most frequently and with the best accent. After a sufficient number of generations his young parrots, continually hearing their parents and a hundred other birds saying "How do you do, sir?" begin to repeat these words so soon that an experiment is needed to decide whether it is by instinct of imitation; and perhaps it is part of both. Eventually, however, the instinct is established. And though now Mr. Crusoe dies, and leaves no record of his work, the instinct will not die, not for a long time at least; and if the parrots themselves have acquired a taste for good English the best speakers will be sexually selected, and the instinct will certainly endure to astonish and perplex mankind, though in truth we may as well wonder at the crowing of the cock or the song of the skylark. Again, turkeys have an instinctive art of catching flies, which, it is manifest, the creatures in their present shape may have acquired by experience. But suppose the circumstances of their life to change; flies steadily become more abundant, and other kinds of food scarcer: the best fly-catchers are now the fittest to live, and each generation they are naturally selected. This process goes on, experience probably adding to the instinct in ways that we need not attempt to conceive, until a variety or species is produced that feeds on flies alone. To look at, this new bird will differ considerably from its turkey ancestors; for change in food and in habits of life will have affected its physical conformation, and every useful modification of structure will have been preserved by natural selection. My point however is, that thus, by no inconceivable steps, would be produced a race of birds depending for all their food on an instinctive art, which they, as then constituted, could never have acquired, because they never could have existed without it.

No doubt, to the many, who love more to gaze and marvel than to question and reflect, all this will seem miserably inadequate as a clue to one of the greatest mysteries of life. But enough, if I have indicated my view of how the most inexplicable of instincts may have had their origin; or rather, if I have shown how our utter inability to trace them back to their origin tells nothing against the probability that they all came into existence in accordance with those laws of acquisition and heredity that we now see operating before our eyes. We cannot tell how the pupa of the dragon-fly came by the instinct that prompts it to leave the water and hang itself up to dry. But we may be able to explain this quite as soon as to unveil the origin of the hooks by which it hangs itself up. And if ever human intelligence should so trace the evolution of living forms as to be able to say, "Thus was developed the bill-scale wherewith birds now break their way out of the shell," it will probably be able to add, "and these were the experiences to which we must trace the instinct that makes every little bird its own skillful accoucheur."

APPENDIX B

[Behaviorist theory suggests that almost any animal can be conditioned to perform any behavior of which it is physically capable. The Brelands' business enterprise provided a unique test of this theory. Reprinted with the permission of the authors and the American Psychological Association from *American Psychologist* 6 (1951): 202–4.]

A Field of Applied Animal Psychology

BY KELLER BRELAND AND MARIAN BRELAND

Recent developments in behavior theory have made possible a new field of applied psychology. This new field has yet to be finally christened. It might be called the field of applied animal psychology or the field of behavioral engineering. We consider it an excellent example of how the findings of "pure" research can be put to practical use.

The core of the field is the work of the neobehaviorists, which has so ordered the facts of behavior that many of their experimental data and those of earlier workers have become immediately applicable to the engineering of animal behavior. We have found most useful the systematic formulation presented by B. F. Skinner in *The Behavior of Organisms.* This body of theory has made it possible for us since the spring of 1947 to develop a flourishing and expanding business concerned with the mass production of conditioned operant behavior in animals.

Applied animal psychology brings together the two formerly unrelated fields of professional animal training and modern behavioral science. The field is new in that it represents, we believe, the first application of systematic behavior theory to the control of animal behavior. We are now in a position to outstrip oldtime professional animal trainers in speed and economy of training. In many instances we can use automatic training methods. We can apply to our training the data of comparative psychology, utilizing new tricks, new animals. We can turn out multiple units—200 "Clever Hanses" instead of one. Furthermore, the systematic nature of the theory puts us in a position to advance to new and more elaborate behavior patterns, to predict results and forestall difficulties.

So far, all our applications have been made for the purpose of advertising exhibits for General Mills, Inc. We developed first a series of trained chicken acts, which were used for county fair booth exhibits in the Midwest, for the purpose of advertising farm feeds. These acts were performed by a group of two-year-old hens which had been culled from a neighbor's flock and were destined for the stew pot. We used a hen-sized stage, some specially constructed props, and a solenoid-driven automatic feed hopper for dispensing reinforcements in the form of scratch grain.

One hen played a 5-note tune on a small piano, another performed a "tap dance" in costume and shoes, while a third "laid" wooden eggs from a nest box; the eggs rolled down a trough into a basket—the audience could call out any number of eggs desired, up to eight, and the hen would lay that number, nonstop.

The basic operation in all these acts was reinforcement at the proper moment in the behavior sequence, by presenting the chicken with a small amount of scratch grain from the solenoid-operated hopper. During the training period, successive approximations to the desired behavior, and component parts of the final pattern, were reinforced. During performances, longer ratios or more elaborate completed patterns were reinforced to keep the behavior at a high level of strength.

During the ensuing year, three sets of these acts were prepared and shipped all over the United States in the hands of men who had had only one or two days' training. The birds played thousands of performances without a single failure, except for an occasional sluggish performance due to ill health or overfeeding. The acts proved to be unprecedented crowd-stoppers at the fairs and feed-store "Open-house" events where they played, showing to as many as 5000 people in a day.

The success of these acts led to the development of a trained pig show, "Priscilla the Fastidious Pig," whose routine included turning on the radio, eating breakfast at a table, picking up the dirty clothes and putting them in a hamper, running the vacuum cleaner around, picking out her favorite feed from those of her competitors, and taking part in a quiz program, answering "Yes" or "No" to questions put by the audience, by lighting up the appropriate signs.

Priscilla was likewise shown at fairs and special feed-store events and conventions throughout the country. She also appeared on television. She was even more successful than the chicken acts at jamming fair booths and feed-stores with spectators. The pig act was in use almost steadily from the fall of 1948 to the summer of 1950. It was necessary to train a replacement about every 3 to 5 months, since the pigs rapidly became too large for easy shipping. After training, the pigs were turned over to their handler, usually a General Mills' feed salesman, who had had one or two days' instruction at our farm, or in the field under our supervision.

In addition to teaching handlers to manage the animals on the road, we have twice taught instructors to do the basic training of the animals and assist with the instruction of the handlers. Both experiences in training instructors were successful and demonstrated clearly that people with no special psychological background can learn the methods and theory behind our animal training procedures. One instructor was a woman college graduate who had taken her degree in statistics and sociology. The other was an average male high school graduate, whose only specialty had been radio repair work. Both acquired in a few weeks most of the techniques of training the existing acts, and enough of the theory and nature of the process to train new acts on their own.

Our next development was a baby chick act. Sixty to 100 chicks are trained for one show. Beginning at about one week of age, they are trained for about 10 days. The show is run with about 10 or 12 chicks on stage. Each runs up a ramp or inclined plane to a platform from which he can reach a feed hopper. He "roots" the top chick off, grabs a bit of feed, then in turn gets pushed off by the next in line. As he goes, he falls onto a tilting pan and is deposited onto the stage floor, accompanied by the sounding of a chime and flashing of a trade name sign. This sequence of behavior results in an endless chain of baby chicks running up the ramp and sliding off. When the group becomes sleepy, they are replaced by a fresh batch, and the show can thus go on indefinitely; it has actually been used about 12 hours in most cases.

This act is our first "packaged act." It is designed to run virtually automatically. It requires only the attendant to keep the feed hopper full and change the group of chicks on stage when they become sleepy. No special training is required for the attendant; mimeographed directions are shipped out with the chicks and provide the only necessary instruction. This act has been a perennial favorite and we have trained more than 2500 chicks to fill orders for this display.

We have developed two variations on the baby chick act. One uses a projector to present advertising copy, with an endless chain of baby chicks in motion around it. The other variation substitutes for the ramp a series of steps onto which the chicks must jump.

A calf was trained for the General Mills booth at the International Dairy Exposition at Indianapolis. "Larro Larry" took part in a quiz program by lighting up "Yes" and "No" signs, as did Priscilla the Pig, and played "Bull in the China Shop" by systematically upsetting an elaborate display of dishes, to the great alarm of the passing crowd.

A turkey act has been developed in which members of the audience play a game with the turkey. The bird is placed in a display case and has access through an opening to part of the miniature playing field. The turkey is trained to rake a steel ball off this field into his goal. The audience player is given a long pole with a magnet on one end and tries to guide the ball along the playing field into his goal before the turkey wins. Various barriers are placed along the playing field to make the game more difficult for both players.

Additional acts using grown chickens have been designed and used, two involving discrimination problems: the Card Sharp, who picks out a better poker hand than a member of the audience, and the Old Shell Game, in which the chicken picks out the shell with the bean under it; and two contests between two birds, a High Jump contest, and a Strength of Pull test. Another automatic act was created by training a hen, on a very high fixed ratio, to beat a toy drum for hours at a time. We also trained a hen in some bizarre contortions; the hen twisted her neck to one side and over her back so that she appeared to be looking frantically in all directions at once. This was billed as "The Civilian Aircraft Spotter" or "The Atom Bomb Neurosis." We have done a few experiments and some developmental work on rats, hamsters, guinea pigs, ducks, pigeons, rabbits, cats, dogs, and crows.

There are, obviously, innumerable other possibilities in the field of advertising exhibits. One is the perfection of the "packaged act," the fully automatic unit which can be shipped anywhere, set up in a store window or convention booth, and operated day in and day out with no more instructions than are necessary for the operation of any machine designed for such use. One adaptation of the automatic act is the animated display—show window advertising in which live animals take the place of puppets and robots.

However, probably the biggest applications exist in the entertainment world. Here we can take over the formal animal training involved in the standard animal act for stage, circus, and movies, and do it faster, cheaper, better, and in multiple units. It is possible to create new acts, whole new circuses, in fact, using unusual animals and unusual acts, and again do it cheaply, quickly, and in numbers limited only by time and production facilities. Television offers unusual opportunities. We can invade the field of night-club entertainment with novel small animals. We can sell or rent trained animal units to hospitals, doctors' offices, waiting rooms of various sorts, or even to private individuals, supplying instructions on care and maintenance.

Another important application of animal psychology is the training of farm animals. Farm dogs and horses could be rendered much more useful to the average farmer if they were given appropriate training. Farmers could themselves be instructed in training and handling their own animals.

The training of dogs for the blind could probably be done on a larger scale, more rapidly and efficiently. One of the big problems of the "Seeing Eye" institution was obtaining instructors. The difficulty was, apparently, that the first masters of the art did not have a sufficiently precise theoretical formulation in training the dogs and hence could not pass the information on to new instructors. They then encountered another problem in instructing the blind to handle the dogs and met numerous failures here in adapting client to dog. Many of these failures could now doubtless be avoided.

Dogs, of course, can be trained more readily with the new methods in all the traditional fields of canine service to mankind: hunting, guarding children and property, and detective work. Military use of dogs in such tasks as guard duty and carrying messages can also be made more effective.

This, then, seems to be the general outline of a promising new field which we have only begun to explore. It is so vast, we feel, that we cannot begin to develop one-tenth of the projects we have thought of. More psychologists, grounded in the theory, are needed to advance the technology and explore the undeveloped portions of this program.

Furthermore, once the technology gets under way and the business develops, there will be active need for academic psychologists to do the background research necessary for full development of the program. And, of course, as psychologists continue to do basic research using animals as subjects, one by-product will be new and better methods of applied animal psychology.

Two types of problems have cropped up repeatedly in our efforts. (1) Apparatus problems have consumed much more time than problems connected with the behavior. The apparatus must be suited to the physique of the particular animal, must be durable enough to stand up under cross-country shipment, and must be foolproof enough to be operated by relatively untrained personnel. (2) We need to know the answers to various "academic" problems, such as "What sort of fixed ratio will an animal sustain on a response made to a disappearing manipulandum, a key available, for example, only every three minutes?" "What would constitute an adequate reinforcement for a hamster, to sustain performances over several hours without satiation?" "What are the emotional characteristics of rabbits and guinea pigs—to what sort and magnitude of stimuli will they adapt, and what is the nature of the curve of recovery from such an adaptation?"

The study of these and related questions—in short, the reexamination of the whole field of comparative psychology in this new light—by psychologists who have available the facilities of an animal laboratory, would greatly speed up the development of the applied field.

In conclusion, we feel that here is a genuine field of applied psychology, old as "group living" and "parenthood" in its subject matter, but new in method and approach, which psychologists can enter with promise of financial reward and a sense of accomplishment and ultimate benefit to the science. For we all know that there is nothing as convincing to the layman of the worth of a discipline as achievement, and the present field offers the psychologist a fine opportunity to demonstrate control of his subject matter.

[Techniques of animal training have been known for centuries. The development of principles of operant conditioning, however, formalized and simplified these techniques. Keller and Marian Breland, of Animal Behavior Enterprises in Hot Springs, Arkansas, have demonstrated that the principles of operant conditioning can be used to train a variety of animals for commercial purposes. In the following reading, however, they report a peculiar "breakdown" of learned behavior, which they have repeatedly observed in their animals. *American Psychologist* 16 (1961): 681–84; reprinted by permission of the authors and the American Psychological Association.]

The Misbehavior of Organisms

BY KELLER BRELAND AND MARIAN BRELAND

There seems to be a continuing realization by psychologists that perhaps the white rat cannot reveal everything there is to know about behavior. Among the voices raised on this topic, Beach (1950) has emphasized the necessity of widening the range of species subjected to experimental techniques and conditions. However, psychologists as a whole do not seem to be heeding these admonitions, as Whalen (1961) has pointed out.

Perhaps this reluctance is due in part to some dark precognition of what they might find in such investigations, for the ethologists Lorenz (1950, p. 233) and Tinbergen (1951, p. 6) have warned that if psychologists are to understand and predict the behavior of organisms, it is essential that they become thoroughly familiar with the instinctive behavior patterns of each new species they essay to study. Of course, the Watsonian or neobehavioristically oriented experimenter is apt to consider "instinct" an ugly word. He tends to class it with Hebb's (1960) other "seditious notions" which were discarded in the behavioristic revolution, and he may have some premonition that he will encounter this bête noir in extending the range of species and situations studied.

We can assure him that his apprehensions are well grounded. In our attempt to extend a behavioristically oriented approach to the engineering control of animal behavior by operant conditioning techniques, we have fought a running battle with the seditious notion of instinct.* It might be of some interest to the psychologist to know how the battle is going and to learn something about the nature of the adversary he is likely to meet if and when he tackles new species in new learning situations.

Our first report (Breland & Breland, 1951) in the *American Psychologist*, concerning our experiences in controlling animal behavior, was wholly affirmative and optimistic, saying in essence that the principles derived from the laboratory could be applied

* In view of the fact that instinctive behaviors may be common to many zoological species, we consider *species specific* to be a sanitized misnomer, and prefer the possibly septic adjective *instinctive*.

to the extensive control of behavior under nonlaboratory conditions throughout a considerable segment of the phylogenetic scale.

When we began this work, it was our aim to see if the science would work beyond the laboratory, to determine if animal psychology could stand on its own feet as an engineering discipline. These aims have been realized. We have controlled a wide range of animal behavior and have made use of the great popular appeal of animals to make it an economically feasible project. Conditioned behavior has been exhibited at various municipal zoos and museums of natural history and has been used for department store displays, for fair and trade convention exhibits, for entertainment at tourist attractions, on television shows, and in the production of television commercials. Thirty-eight species, totaling over 6000 individual animals, have been conditioned, and we have dared to tackle such unlikely subjects as reindeer, cockatoos, raccoons, porpoises, and whales.

Emboldened by this consistent reinforcement, we have ventured further and further from the security of the Skinner box. However, in this cavalier extrapolation, we have run afoul of a persistent pattern of discomforting failures. These failures, although disconcertingly frequent and seemingly diverse, fall into a very interesting pattern. They all represent breakdowns of conditioned operant behavior. From a great number of such experiences, we have selected, more or less at random, the following examples.

The first instance of our discomfiture might be entitled, What Makes Sammy Dance? In the exhibit in which this occurred, the casual observer sees a grown bantam chicken emerge from a retaining compartment when the door automatically opens. The chicken walks over about 3 feet, pulls a rubber loop on a small box which starts a repeated auditory stimulus pattern (a four-note tune). The chicken then steps up onto an 18-inch, slightly raised disc, thereby closing a timer switch, and scratches vigorously, round and round, over the disc for 15 seconds, at the rate of about two scratches per second until the automatic feeder fires in the retaining compartment. The chicken goes into the compartment to eat, thereby automatically shutting the door. The popular interpretation of this behavior pattern is that the chicken has turned on the "juke box" and "dances."

The development of this behavioral exhibit was wholly unplanned. In the attempt to create quite another type of demonstration which required a chicken simply to stand on a platform for 12–15 seconds, we found that over 50% developed a very strong and pronounced scratch pattern, which tended to increase in persistence as the time interval was lengthened. (Another 25% or so developed other behaviors—pecking at spots, etc.) However, we were able to change our plans so as to make use of the scratch pattern, and the result was the "dancing chicken" exhibit described above.

In this exhibit the only real contingency for reinforcement is that the chicken must depress the platform for 15 seconds. In the course of a performing day (about 3 hours for each chicken) a chicken may turn out over 10,000 unnecessary, virtually identical responses. Operant behaviorists would probably have little hesitancy in labeling this an example of Skinnerian "superstition" (Skinner, 1948) or "mediating" behavior, and we list it first to whet their explanatory appetite.

However, a second instance involving a raccoon does not fit so neatly into this paradigm. The response concerned the manipulation of money by the raccoon (who has "hands" rather similar to those of the primates). The contingency for reinforcement was picking up the coins and depositing them in a 5-inch metal box.

Raccoons condition readily, have good appetites, and this one was quite tame and an eager subject. We anticipated no trouble. Conditioning him to pick up the first coin was simple. We started out by reinforcing him for picking up a single coin. Then the metal container was introduced, with the requirement that he drop the coin into the container. Here we ran into the first bit of difficulty: he seemed to have a great deal of trouble letting go of the coin. He would rub it up against the inside of the container, pull it back out, and clutch it firmly for several seconds. However, he would finally turn it loose and receive his food reinforcement. Then the final contingency: we put him on a ratio of 2, requiring that he pick up both coins and put them in the container.

Now the raccoon really had problems (and so did we). Not only could he not let go of the coins, but he spent seconds, even minutes, rubbing them together (in a most miserly fashion), and dipping them into the container. He carried on this behavior to such an extent that the practical application we had in mind—a display featuring a raccoon putting money in a piggy bank—simply was not feasible. The rubbing behavior became worse and worse as time went on, in spite of nonreinforcement.

For the third instance, we return to the gallinaceous birds. The observer sees a hopper full of oval plastic capsules which contain small toys, charms, and the like. When the S_D (a light) is presented to the chicken, she pulls a rubber loop which releases one of these capsules onto a slide, about 16 inches long, inclined at about 30 degrees. The capsule rolls down the slide and comes to rest near the end. Here one or two sharp, straight pecks by the chicken will knock it forward off the slide and out to the observer, and the chicken is then reinforced by an automatic feeder. This is all very well—most chickens are able to master these contingencies in short order. The loop pulling presents no problems; she then has only to peck the capsule off the slide to get her reinforcement.

However, a good 20% of all chickens tried on this set of contingencies fail to make the grade. After they have pecked a few capsules off the slide, they begin to grab at the capsules and drag them backward into the cage. Here they pound them up and down on the floor of the cage. Of course, this results in no reinforcement for the chicken, and yet some chickens will pull in over half of all the capsules presented to them.

Almost always this problem behavior does not appear until after the capsules begin to move down the slide. Conditioning is begun with stationary capsules placed by the experimenter. When the pecking behavior becomes strong enough, so that the chicken is knocking them off the slide and getting reinforced consistently, the loop pulling is conditioned to the light. The capsules then come rolling down the slide to the chicken. Here most chickens, who before did not have this tendency, will start grabbing and shaking.

The fourth incident also concerns a chicken. Here the observer sees a chicken in a cage about 4 feet long which is placed alongside a miniature baseball field. The reason for the cage is the interesting part. At one end of the cage is an automatic electric feed hopper. At the other is an opening through which the chicken can reach and pull a loop on a bat. If she pulls the loop hard enough the bat (solenoid operated) will swing, knocking a small baseball up the playing field. If it gets past the miniature toy players on the field and hits the back fence, the chicken is automatically reinforced with food at the other end of the cage. If it does not go far enough, or hits one of the players, she tries again. This results in behavior on an irregular ratio. When the feeder sounds, she then runs down the length of the cage and eats.

Our problems began when we tried to remove the cage for photography. Chickens that had been well conditioned in this behavior became wildly excited when the ball started to move. They would jump up on the playing field, chase the ball all over the field, even knock it off on the floor and chase it around, pecking it in every direction, although they had never had access to the ball before. This behavior was so persistent and so disruptive, in spite of the fact that it was never reinforced, that we had to reinstate the cage.

The last instance we shall relate in detail is one of the most annoying and baffling for a good behaviorist. Here a pig was conditioned to pick up large wooden coins and deposit them in a large "piggy bank." The coins were placed several feet from the bank and the pig required to carry them to the bank and deposit them, usually four or five coins for one reinforcement. (Of course, we started out with one coin, near the bank.)

Pigs condition very rapidly, they have no trouble taking ratios, they have ravenous appetites (naturally), and in many ways are among the most tractable animals we have worked with. However, this particular problem behavior developed in pig after pig, usually after a period of weeks or months, getting worse every day. At first the pig would eagerly pick up one dollar, carry it to the bank, run back, get another, carry it rapidly and neatly, and so on, until the ratio was complete. Thereafter, over a period of weeks the behavior would become slower and slower. He might run over eagerly for each dollar, but on the way back, instead of carrying the dollar and depositing it simply and cleanly, he would repeatedly drop it, root it, drop it again, root it along the way, pick it up, toss it up in the air, drop it, root it some more, and so on.

We thought this behavior might simply be the

dilly-dallying of an animal on a low drive. However, the behavior persisted and gained in strength in spite of a severely increased drive—he finally went through the ratios so slowly that he did not get enough to eat in the course of a day. Finally it would take the pig about 10 minutes to transport four coins a distance of about 6 feet. This problem behavior developed repeatedly in successive pigs.

There have also been other instances: hamsters that stopped working in a glass case after four or five reinforcements, porpoises and whales that swallow their manipulanda (balls and inner tubes), cats that will not leave the area of the feeder, rabbits that will not go to the feeder, the great difficulty in many species of conditioning vocalization with food reinforcement, problems in conditioning a kick in a cow, the failure to get appreciably increased effort out of the ungulates with increased drive, and so on. These we shall not dwell on in detail, nor shall we discuss how they might be overcome.

These egregious failures came as a rather considerable shock to us, for there was nothing in our background in behaviorism to prepare us for such gross inabilities to predict and control the behavior of animals with which we had been working for years.

The examples listed we feel represent a clear and utter failure of conditioning theory. They are far from what one would normally expect on the basis of the theory alone. Furthermore, they are definite, observable; the diagnosis of theory failure does not depend on subtle statistical interpretations or on semantic legerdemain—the animal simply does not do what he has been conditioned to do.

It seems perfectly clear that, with the possible exception of the dancing chicken, which could conceivably, as we have said, be explained in terms of Skinner's superstition paradigm, the other instances do not fit the behavioristic way of thinking. Here we have animals, after having been conditioned to a specific learned response, gradually drifting into behaviors that are entirely different from those which were conditioned. Moreover, it can easily be seen that these particular behaviors to which the animals drift are clear-cut examples of instinctive behaviors having to do with the natural food-getting behaviors of the particular species.

The dancing chicken is exhibiting the gallinaceous birds' scratch pattern that in nature often precedes ingestion. The chicken that hammers capsules is obviously exhibiting instinctive behavior having to do with breaking open of seed pods or the killing of insects, grubs, etc. The raccoon is demonstrating so-called washing behavior. The rubbing and washing response may result, for example, in the removal of the exoskeleton of a crayfish. The pig is rooting or shaking—behaviors which are strongly built into this species and are connected with the food-getting repertoire.

These patterns to which the animals drift require greater physical output and therefore are a violation of the so-called law of least effort. And most damaging of all, they stretch out the time required for reinforcement when nothing in the experimental setup requires them to do so. They have only to do the little tidbit of behavior to which they were conditioned—for example, pick up the coin and put it in the container—to get reinforced immediately. Instead, they drag the process out for a matter of minutes when there is nothing in the contingency which forces them to do this. Moreover, increasing the drive merely intensifies this effect.

It seems obvious that these animals are trapped by strong instinctive behaviors, and clearly we have here a demonstration of the prepotency of such behavior patterns over those which have been conditioned.

We have termed this phenomenon "instinctive drift." The general principle seems to be that wherever an animal has strong instinctive behaviors in the area of the conditioned response, after continued running the organism will drift toward the instinctive behavior to the detriment of the conditioned behavior and even to the delay or preclusion of the reinforcement. In a very boiled-down, simplified form, it might be stated as "learned behavior drifts toward instinctive behavior."

All this, of course, is not to disparage the use of conditioning techniques, but is intended as a demonstration that there are definite weaknesses in the philosophy underlying these techniques. The pointing out of such weaknesses should make possible a worthwhile revision in behavior theory.

The notion of instinct has now become one of our basic concepts in an effort to make sense of the welter of observations which confront us. When behaviorism tossed out instinct, it is our feeling that some of its power of prediction and control were lost with it. From the foregoing examples, it appears that although it was easy to banish the Instinctivists from the science during the Behavioristic Revolution, it was not possible to banish instinct so easily.

And if, as Hebb suggests, it is advisable to reconsider those things that behaviorism explicitly threw out, perhaps it might likewise be advisable to examine what they tacitly brought it—the hidden assumptions which led most disastrously to these breakdowns in the theory.

Three of the most important of these tacit assumptions seem to us to be: that the animal comes to the laboratory as a virtual *tabula rasa,* that species differences are insignificant, and that all responses are about equally conditionable to all stimuli.

It is obvious, we feel, from the foregoing account, that these assumptions are no longer tenable. After 14 years of continuous conditioning and observation of thousands of animals, it is our reluctant conclusion that the behavior of any species cannot be adequately understood, predicted, or controlled without knowledge of its instinctive patterns, evolutionary history, and ecological niche.

In spite of our early successes with the application of behavioristically oriented conditioning theory, we readily admit now that ethological facts and attitudes in recent years have done more to advance our practical control of animal behavior than recent reports from American "learning labs."

Moreover, as we have recently discovered, if one begins with evolution and instinct as the basic format for the science, a very illuminating viewpoint can be developed which leads naturally to a drastically revised and simplified conceptual framework of startling explanatory power (to be reported elsewhere).

It is hoped that this playback on the theory will be behavioral technology's partial repayment to the academic science whose impeccable empiricism we have used so extensively.

APPENDIX C

Highly Recommended Films

Signals for Survival. 55 min, color, sound. Available from McGraw-Hill Films (600 Grand Ave., Ridgefield, NJ 07654).

Tinbergen's classic studies of gull communication and behavior are reviewed. An excellent example of pure observational ethology.

The Mussel Specialist. 30 min, color, sound. Available from McGraw Hill Films (address above).

Through an interplay of releasers and selective learning, young oystercatchers master the difficult art of opening shellfish.

Big Horn. 15 min., color, sound, no narration. Available from McGraw-Hill Films (address above).

The life cycle of bighorn sheep, complete with imprinting, selective play, dominance fighting, and mating, is depicted in this overwhelmingly beautiful film.

The Hunters. 75 min, color, sound, #50320. Available from Penn State Audio-Visual Services (Penn State University, University Park, PA 16802).

The lifestyle of !Kung bushmen is graphically presented, mainly as a chronicle of a week-long hunt.

Japanese Macaques—Food preparation. 4 min, color, silent, #E1466. Available from Penn State Audio-Visual Services (address above).

Illustrates potato and grain washing.

Competition Elimination by a Newly Hatched Cuckoo. 5 min, black & white, silent, #E721. Available from Wisconsin Bureau of Audio-Visual Instruction (University of Wisconsin, 1327 University Ave., Madison, WI 53706).

Shows the blind baby cuckoo using its stereotyped motor program to eject its host's eggs and chicks.

Castles of Clay. 55 min, color, sound. Available from Benchmark Films (145 Scarborough Rd., Briarcliff Manor, NY 10501), and Penn State Audio-Visual Services (address above).

Impressive study of the life cycle and ecology of African termites.

Year of the Wildebeest. 55 min, color, sound. Available from Benchmark Films (address above).

Follows a herd of wildebeest through one year of their lives, looking at the behavior, ecology, and predators.

Life on a Silken Thread. 55 min, color, sound. Available from Time-Life Films (100 Eisenhower Dr., Paramus, NJ 07652).

Records the wide variety of lifestyles exhibited by spiders, including time-lapse photography of orb weaving.

Bibliography

1. THE BIOLOGY OF BEHAVIOR

Dawkins, Richard. *The Selfish Gene.* New York: Oxford University Press, 1976.

Klinghammer, Eric, and Hess, Eckhard H. "Parental Feeding in Ring Doves." In *Control and Development of Behavior: An Historical Sample from the Pens of Ethologists,* edited by Peter H. Klopfer and Jack P. Hailman. Reading, Mass.: Addison-Wesley, 1972.

Morgan, Clifford T., and King, Richard A. *Introduction to Psychology.* New York: McGraw-Hill, 1951.

Nottebohm, Fernando, and Nottebohm, Marta E. "Vocalization and Breeding Behavior of Surgically Deafened Ring Doves." *Animal Behavior* 19 (1971): 313–27.

Pfungst, Oskar. *Clever Hans: The Horse of Mr. von Osten.* Edited by Robert Rosenthal. New York: Holt, Rinehart, and Winston, 1965.

Silver, Rae. "The Parental Behavior of Ring Doves." *American Scientist* 66 (1978): 209–15.

Watson, John B. *Behaviorism.* New York: W. W. Norton, 1930.

2. EARLY ETHOLOGY

Evans, Howard E. *The Comparative Ethology and Evolution of the Sand Wasps.* Cambridge, Mass.: Harvard University Press, 1966.

Frisch, Karl von. *The Dance Language and Orientation of Bees.* Translated by Leigh E. Chadwick. Cambridge, Mass.: Harvard University Press, The Belknap Press, 1967.

———. *Bees: Their Vision, Chemical Senses, and Language.* Ithaca, N.Y.: Cornell University Press, 1971.

Hess, Eckhard H. "Space Perception in the Chick." *Scientific American* 195 (1956): 71–80.

———. *Imprinting: Early Experience and the Developmental Psychobiology of Attachment.* New York: Van Nostrand Reinhold, Behavioral Science Series, 1973.

Hoogland, R.; Morris, Desmond; and Tinbergen, Niko. "The Spines of Sticklebacks *(Gasterosteus* and *Pygosteus)* as a Means of Defense Against Predators *(Perca* and *Esox).*" *Behavior* 10 (1957): 207–36.

Lorenz, Konrad. "Der Kumpan in der Umwelt des

Vogels." *Journal für Ornithologie* 80 (1935). Translated and reprinted in Lorenz, Konrad, *Studies in Animal and Human Behavior.*

———. *King Solomon's Ring: New Light on Animal Ways.* New York: Thomas Y. Crowell, 1952.

———. *Studies in Animal and Human Behavior.* 2 vols. Translated by Robert Martin. Cambridge, Mass.: Harvard University Press, 1970, 1971.

Malthus, Thomas. *An Essay on the Principle of Population.* 6th ed. 2 vols. London, 1826.

Pavlov, Ivan P. *Conditioned Reflexes.* Translated by G. V. Anrep. London: Oxford University Press, 1927.

Tinbergen, Niko. "Über die Orientierung des Bienenwolfes *(Philanthus triangulum* Fabr.*)*." *Zeitschrift für Vergleichende Physiologie* 16 (1932): 305–34.

———. "Über die Orientierung des Bienenwolfes *(Philanthus triangulum* Fabr.). II. Die Bienenjagd." *Zeitschrift für Vergleichende Physiologie* 21 (1935): 699–716.

———. *The Study of Instinct.* New York: Oxford University Press, 1951.

———. *Curious Naturalists.* New York: Doubleday, Anchor Books, The Natural History Library, 1958.

———. *The Animal in Its World: Explorations of an Ethologist, 1932–1972.* 2 vols. Cambridge, Mass.: Harvard University Press, 1972.

———, and Kruyt, W. "Über die Orientierung des Bienenwolfes (*Philanthus triangulum* Fabr.). III. Die Bevorzugung bestimmter Wegmarken." *Zeitschrift für Vergleichende Physiologie* 25 (1938): 292–334.

Uexküll, Jakob von. "A Stroll Through the World of Animals and Men: A Picture Book of Invisible Worlds." In idem, *Instinctive Behavior: The Development of a Modern Concept,* translated and edited by Claire H. Schiller. New York: International Universities Press, 1957.

3. PRINCIPLES OF EARLY ETHOLOGY

Beer, Colin. "Incubation and Nest-building by the Black-headed Gull." Ph.D. dissertation, Oxford University, 1960.

Brower, Lincoln P.; Brown, J. V.; and Cranston, F. P. "Courtship Behavior of the Queen Butterfly, *Danaus gilippus berenice* (Cramer)." *Zoologica* 50 (1965): 1–39.

Crane, Jocelyn. "Combat, Display and Ritualization in Fiddler Crabs (Ocypodidae, genus *Uca*)." *Proceedings of the Royal Society of London, Series B: Biological Sciences* 1966: 459–472.

Darwin, Charles. *The Descent of Man, and Selection in Relation to Sex.* 2 vols. New York: Appleton, 1871.

Fabricius, Eric. "On the Ethology of Young Waterfowl." In *Control and Development of Behavior: An Historical Sample from the Pens of Ethologists,* edited by Peter H. Klopfer and Jack P. Hailman. Reading, Mass.: Addison-Wesley, 1972.

Green, M.; Green, R.; and Carr, W. F. "The Hawk-Goose Phenomenon: A Replication and an Extension." *Psychonometric Science* 4 (1966): 185–86.

Green, R.; Carr, W. J.; and Green, M. "The Hawk-goose Phenomenon: Further Confirmation and a Search for the Releaser." *Journal of Psychology* 69 (1968): 271–76.

Greenspan, Beverly N. "Male Size and Reproductive Success in the Fiddler Crab." *Animal Behavior* 28 (1980): 387–92.

Lack, David. "The Display of the Blackcock." *British Birds* 32 (1939): 290–303.

Lorenz, Konrad J. "Der Kumpan in der Umwelt des Vogels." *Journal für Ornithologie* 80 (1935).

Melzak, R.; Penick, E.; and Beckett, A. "The Problem of 'Innate Fear' of the Hawk Shape: An Experimental Study with Mallard Ducks." *Journal of Comparative and Physiological Psychology* 52 (1959): 694–98.

Mueller, Helmut C., and Parker, Patricia G. "Naïve Ducklings Show Different Cardiac Response to Hawk than to Goose Models." *Behavior* 74 (1980): 101–13.

Smith, S. M. "Innate Recognition of Coral Snake Pattern by a Possible Avian Predator." *Science* 187 (1975): 759–60.

———. "Coral Snake Pattern Recognition and Stimulus Generalization by Naive Great Kiskadees (Aves: Tyrannidae)." *Nature* 265 (1977): 535–36.

Solomon, Michael. "Coastal Distribution, Display and Sound Production by Florida Fiddler Crabs (genus *Uca*)." *Animal Behaviour* 15 (1967): 449–59.

Tinbergen, Niko. "Social Releasers and the Experimental Method Required for Their Study." *Wilson Bulletin* 60 (1948): 6–52.

———. *The Study of Instinct.* New York: Oxford University Press, 1951 (reprinted 1969).

———. "The Shell Menace." *Natural History* 72 (1963): 28–35.

———; Broekhuysen, G. J.; Feekes, F.; Houghton, J. C. W.; Kruuk, H.; and Szulc, E. "Egg Shell Removal by the Black-headed Gull, *Larus ridibundus* L.: A Behavior Component of Camouflage." *Behaviour* 19 (1963): 74–117.

Tinbergen, Niko; Kruuk, H.; and Paillette, M. "Egg Shell Removal by the Black-headed Gull (*Larus r. ridibundus* L.) II. The Effects on the Response to Colour." *Bird Study* 9 (1962): 123–31.

———; and Stamm, R. "How do Black-headed Gulls Distinguish Between Eggs and Egg-shells?" *British Birds* 55 (1962): 120–29.

Tinbergen, Niko; Meeuse, B. J. D.; Boerema, L. K.; and Varossieau, W. "Die Balz des Samtfalters, *Eumenis semele* (L.)" *Zeitschrift für Tierpsychologie* 5 (1942): 182–226.

4. THE EMERGENCE OF MODERN ETHOLOGY

Hailman, Jack P. "The Ontogeny of an Instinct." *Behaviour Supplements* 15 (1967): 1–159.

Klinghammer, Erich, and Hess, Eckhard H. "Parental Feeding in Ring Doves (*Streptopelia roseogrisea*): Innate or Learned?" In *Control and Development of Behavior: An Historical Sample From the Pens of Ethologists,* edited by Peter H. Klopfer and Jack P. Hailman. Reading, Mass.: Addison-Wesley, 1972.

Lehrman, Daniel S. "The Physiological Basis of Parental Feeding in the Ring Dove (*Streptopelia risoria*)." In *Control and Development of Behavior: An Historical Sample from the Pens of Ethologists,* edited by Peter H. Klopfer and Jack P. Hailman. Reading, Mass.: Addison-Wesley, 1972.

Miller, D. B., and Gottlieb, Gilbert. "Acoustic Features of Wood Duck (*Aix sponsa*) Maternal Calls." *Behaviour* 57 (1976): 260–80.

———. "Maternal Vocalizations of Mallard Ducks (*Anas platyrhynchos*)." *Animal Behaviour* 26 (1978): 1178–94.

Tinbergen, Niko, and Kuenen, D. J. "Über die ausloesenden und die richtung gebenden Reizsituationen der Sperrbewengung von jungen Drosseln (*Turdus m. merula* L. und *T. e. ericetorum* Turton)." *Zeitschrift für Tierpsychologie* 3 (1939): 37–60.

5. NERVES AND CIRCUITS

Albertson, D. B., and Thompson, J. N. "The Pharynx of *C. elegans.*" *Philosophical Translations of the Royal Society of London* 275 (1976): 299–325.

Bullock, Theodore H. *Introduction to Nervous Systems.* San Francisco: W. H. Freeman, 1977.

Delcomyn, Fred. "Neural Basis of Rhythmic Behavior in Animals." *Science* 210 (1980): 492–98.

Dethier, Vincent G. *The Hungry Fly.* Cambridge, Mass.: Harvard University Press, 1976.

Getting, P. A. "*Tritonia* Swimming: Triggering of a Fixed-Action Pattern." *Brain Research* 96 (1975): 128–33.

Goldschmidt, R. "Das Nervensystem von *Ascaris.*" *Zeitschrift für Wissenschaftliche Zoologie, Abteilung A* 92 (1909): 306–57.

Katz, Bernard. *Nerve, Muscle, and Synapse.* New York: McGraw-Hill, 1966.

Kennedy, Donald. "Nerve Cells and Behavior." *American Scientist* 59 (1971): 36–42.

———; Evoy, W. H.; and Hanawalt, J. T. "Release of Coordinated Behavior in Crayfish by Single Central Neurons." *Science* 154 (1966): 917–19.

Ward, Samuel; Thompson, N.; White, S. G.; and Brenner, Sidney. "Electron Microscopal Reconstruction of the Anterior Sensory Anatomy of *C. elegans.*" *Journal of Comparative Neurology* 160 (1975): 313–38.

White, J. G.; Southgate, E.; Thompson, J. N.; and Brenner, Sidney. "Structure of the Ventral Nerve Cord of *C. elegans.*" *Philosophical Transactions of the Royal Society of London* 275 (1976): 327–48.

Willows, A. O. Dennis, and Hoyle, Graham. "Neuronal Network Triggering a Fixed-Action Pattern." *Science* 166 (1969): 1549–51.

Wilson, Donald M. "Central Nervous Control of Flight in a Locust." *Journal of Experimental Biology* 38 (1961): 401–90.

6. VISUAL DESIGN

Fox, Robert; Lehmkuhle, Stephen W.; and Westendorf, David H. "Falcon Visual Acuity." *Science* 192 (1976): 263–65.

Fraenkel, G. S., and Gunn, D. L. *The Orientation of Animals.* New York: Dover, 1961.

Frisch, Karl von. *Bees: Their Vision, Chemical Senses, and Language.* Ithaca, N.Y.: Cornell University Press, 1971.

Gibson, K. S., and Tyndall, E. P. T. "Visibility of Radiant Energy." *Scientific Papers of the Bureau of Standards* 19 (1923): 1–47.

Hailman, Jack. *Optical Signals.* Bloomington, Ill.: Indiana University Press, 1977.

Hewitt, Paul G. *Conceptual Physics.* Boston: Little, Brown, & Co., 1974.

Land, M. F. "Compound Eyes: Old and New Mechanisms." *Nature* 287 (1980): 681–86.

Perez-Miravete, A. *Behavior of Micro-Organisms.* New York: Plenum, 1973.

7. VISUAL PROCESSING

Allman, John M., and Kaas, J. H. "Representation of the Visual Field on the Medial Wall of Occipital-Parietal Cortex in the Owl Monkey." *Science* 191 (1976): 572–75.

Banks, M. S.; Aslin, R. N.; and Letson, R. D. "Sensitive Period for the Development of Human Binocular Vision." *Science* 190 (1975): 675–77.

Browne, L. Barton, ed. *Experimental Analysis of Insect Behavior.* New York: Springer-Verlag, 1974.

Cynader, M., and Chernenko, G. "Abolition of Direction Selectivity in the Visual Cortex of the Cat." *Science* 193 (1976): 504–5.

Daw, N. W.; Berman, N. E. J.; and Ariel, M. "Interaction of Critical Periods in the Visual Cortex of Kittens." *Science* 199 (1978): 565–66.

Dews, P., and Weisel, Törsten N. "Consequences of Monocular Deprivation on Visual Behavior in Kittens." *Journal of Physiology* 206 (1970): 437–55.

Ewert, Jörg-Peter. *Neuro-Ethology.* New York: Springer-Verlag, 1980.

Hirsch, H. V. B., and Spinelli, D. N. "Visual Experience Modifies Distribution of Horizontally and Vertically Oriented Receptive Fields in Cats." *Science* 168 (1970): 869–71.

Hubel, David H., and Wiesel, Törsten N. "Shape and Arrangement of Columns in Cat Striate Cortex." *Journal of Physiology* 165 (1963): 559–68.

———. "Receptive Fields of Cells in Striate Cortex of Very Young, Visually Inexperienced Kittens." *Journal of Neurophysiology* 26 (1963): 994–1002.

———. "Receptive Fields and Functional Architecture in Two Non-striate Visual Areas (18 and 19) of the Cat." *Journal of Neurophysiology* 28 (1968): 229–89.

Leehey, Susan Cohen; Moskowitz-Cook, Anne; Brill, Sarah; and Held, Richard. "Orientational Anisotropy in Infant Vision." *Science* 190 (1975): 900–2.

Leventhal, Audie Gene, and Hirsch, Helmut V. B. "Cortical Effect of Early Selective Exposure to Diagonal Lines." *Science* 190 (1975): 902–4.

Mansfield, R. J. W. "Neural Basis of Orientation Perception in Primate Vision." *Science* 186 (1974): 1133–35.

Matsumoto, S. G., and Murphey, R. K. "Sensory Deprivation During Development Decreases the Responsiveness of Cricket Giant Interneurons." *Journal of Physiology* 268 (1977): 533–48.

Murphey, R. K., and Matsumoto, S. G. "Experience Modifies the Plastic Properties of Identified Neurons." *Science* 191 (1976): 564–66.

Pettigrew, John D., and Konishi, Masakazu. "Neurons Selective for Orientation and Binocular Disparity in the Visual Wulst of the Barn Owl." *Science* 193 (1976): 675–77.

Ratliff, Floyd. *Mach Bands: Quantitative Studies on Neural Networks in the Retina.* New York: Holden-Day, 1965.

Shinkman, P. G., and Bruce, C. J. "Binocular Differences in Cortical Receptive Fields of Kittens after Rotationally Disparate Binocular Experience." *Science* 197 (1977): 285–87.

Stryker, M. P., and Sherk, H. "Modification of Cortical Orientation Selectivity in the Cat by Restricted Visual Experiences: A Reexamination. *Science* 190 (1975): 904–6.

Wehner, Rudiger, ed. *Information Processing in the Visual System of Arthropods.* New York: Springer-Verlag, 1972.

8. AUDITORY DESIGN

Autrum, H., and Schneider, W. "Vergleichende Untersuchungen über den Erschütterungssin der Insekten." *Zeitschrift für Vergleichende Phy-*

siologie 31 (1948): 77–88.

Bennet-Clark, H. C. "The Mechanism and Efficiency of Sound Production in Mole Crickets." *Journal of Experimental Biology* 52 (1970): 619–52.

Capranica, Robert R. *The Evoked Vocal Response of the Bullfrog.* Cambridge, Mass: MIT Press, 1965.

Gerhardt, H. Carl. "Significance of Two Frequency Bands in Long-Distance Vocal Communication in the Green Tree Frog." *Nature* 261 (1976): 692–94.

Greenewalt, Crawford H. "How Birds Sing." *Scientific American* 221, no. 5 (1969): 126–39.

Griffin, Donald R. *Listening in the Dark: The Acoustic Orientation of Bats and Men.* New Haven, Conn: Yale University Press, 1958.

Haskell, P. T. "Sound Production." In *The Physiology of Insects,* vol. 2, edited by M. Rockstein. New York: Academic Press, 1974.

Heran, H. "Wahrnehmung und Regelung der Flugeigengeschwindigkeit bei *Apis mellifica.*" *Zeitschrift für Vergleichende Physiologie* 42 (1959): 102–63.

Manley, G. A. "Some Aspects of the Evolution of Hearing in Vertebrates." *Nature* 230 (1971): 506–9.

Prozesky-Schulze, L.; Prozesky, O. P. M.; Anderson, F.; and van der Merwe, G. J. J. "Use of a Self-made Sound Baffle by a Tree Cricket." *Nature* 255 (1975): 142–43.

Tautz, Jürgen, and Markl, Hubert. "Caterpillars Detect Flying Wasps by Hairs Sensitive to Airborne Vibration." *Behavioral Ecology and Sociobiology* 4 (1978): 101–10.

von Békésy, G. *Experiments in Hearing.* New York: McGraw-Hill, 1960.

———. "Mechanisms of Hearing." *Symposia of the Society for Experimental Biology* 16 (1962): 267–88.

———, and Rosenblith, W. A. "The Mechanical Properties of the Ear." In *Handbook of Experimental Psychology,* edited by S. S. Stevens. New York: John Wiley and Sons, 1951.

Wiley, R. Haven, and Richards, Douglas G. "Physical Constraints on Acoustic Communication in the Atmosphere: Implications for the Evolution of Animal Vocalizations." *Behavioral Ecology and Sociobiology* 3 (1978): 69–94.

9. AUDITORY PROCESSING

Batteau, D. Wayne. "The Role of the Pinnae in Human Localization." *Proceedings of the Royal Society of London, Series B: Biological Sciences* 168 (1967): 158–80.

Brownell, P. H. "Compressional and Surface Waves in Sand: Used by Desert Scorpions to Locate Prey." *Science* 197 (1977): 479–82.

———, and Farley, Roger D. "Prey-Localizing Behaviour of the Nocturnal Desert Scorpion, *Paruroctonus mesaensis:* Orientation to Substrate Vibrations." *Animal Behavior* 27 (1979): 185–93.

Goldman, L. S., and Henson, O. W., Jr. "Prey Recognition and Selection by the Constant Frequency Bat *Pteronotus p. parnelli.*" *Behavioral Ecology and Sociobiology* 2 (1977): 411–19.

Griffin, Donald R. *Listening in the Dark.* New Haven, Conn.: Yale University Press, 1958.

———; Webster, Frank A.; and Michael, C. R. "The Echolocation of Flying Insects by Bats." *Animal Behaviour* 8 (1960): 141–54.

Knudsen, Eric I.; Konishi, Masakazu; and Pettigrew, John D. "Receptive fields of Auditory Neurons in the Owl." *Science* 198 (1977): 1278–80.

Kuwada, Shigeyuki; Yin, Tom C. T.; and Wickesberg, Robert E. "Response of Cat Inferior Colliculus Neurons to Binaural Beat Stimuli: Possible Mechanisms for Sound Localization." *Science* 206 (1979): 586–88.

Lang, Horst H. "Surface Wave Discrimination Between Prey and Nonprey by the Back Swimmer *Notonecta glauca* L. (Hemiptern, Heteroptera)." *Behavioral Ecology and Sociobiology* 6 (1980): 233–46.

Mills, A. W. "Auditory Localization." In *Foundations of Modern Auditory Theory,* vol. 2, edited by J. V. Tobias. New York: Academic Press, 1972.

Rheinlaender, Jurgen, and Morchen, Alfred. "Time-Intensity Trading in Locust Auditory Interneurons." *Nature* 281 (1979): 672–74.

Roeder, Kenneth D. "Auditory System of Noctuid Moths." *Science* 154 (1966): 1515–21.

———, and Treat, A. E. "The Detection and Evasion of Bats by Moths." *American Scientist* 49 (1961): 135–48.

Simmons, James A. "Echolocation in Bats: Signal Processing of Echoes for Target Range." *Science* 171 (1971): 925–28.

——— et al. "Target Structure and Echo Spectral Discrimination by Echolocating Bats." *Science*

186 (1974): 1130–32.

Suga, Nobuo, and Jen. P. H-S. "Disproportionate Tonotopic Representation for Processing CF-FM Sonar Signals in the Mustache Bat Auditory Cortex." *Science* 194 (1976): 542–44.

Suga, Nobuo; O'Neill, William E.; and Manabe, Toshiki. "Cortical Neurons Sensitive to Combinations of Information-Bearing Elements of Biosonar Signals in the Mustache Bat." *Science* 200 (1978): 778–81.

———. "Harmonic-Sensitive Neurons in the Auditory Cortex of the Mustache Bat." *Science* 203 (1979): 270–74.

10. OTHER SENSES

Bastian, Joseph, and Heiligenberg, Walter. "Phase-Sensitive Midbrain Neurons in *Eigenmannia:* Neural Correlates of the Jamming-Avoidance Response." *Science* 209 (1980): 828–31.

Burghardt, Gordon M. "Chemical-Cue Preferences of Inexperienced Snakes." *Science* 157 (1967): 718–21.

———. "Chemical Prey Preference Polymorphism in Newborn Garter Snakes." *Behaviour* 52 (1975): 202–25.

Dethier, Vincent. *The Hungry Fly.* Cambridge, Mass.: Harvard University Press, 1976.

Ewert, Jörg-Peter. *Neuroethology.* New York: Springer-Verlag, 1980.

Hopkins, C. D. "Electric Communication in Fish." *American Scientist* 62 (1974): 426–37.

Kaib, Manfred. "Die Fleisch- und Blumenduftrezeptoren auf der Antonne der Schmeissfliege." *Journal of Comparative Physiology* 95 (1974): 105–21.

Schneider, D. "Insect Olfaction: Deciphering System for Chemical Messages." *Science* 163 (1969): 1031–36.

11. MOTOR PROGRAMS

Bentley, David R., and Hoy, Ronald R. "Post-Embryonic Development of Adult Motor Programs in Crickets." *Science* 170 (1970): 1409–11.

Delcomyn, Fred. "Neural Basis of Rhythmic Behavior in Animals." *Science* 210 (1980): 492–98.

Doty, R. W. "Neural Organization in Deglutition." In *Handbook of Physiology: Alimentary Canal,* edited by W. Heidel. Washington, D.C.: American Physiological Association, 1968.

Elsner, Norbert. "Neural Economy: Bifunctional Muscles and Common Central Pattern Elements in Leg and Wing Stridulation of the Grasshopper." *Journal of Comparative Physiology* 89 (1974): 227–36.

Evarts, Edward V. "Motor Cortex Reflexes Associated with Learned Movement." *Science* 179 (1973): 501–3.

Ewer, R. F. "Food Burying in the African Ground Squirrel, *Xerus erythropus.*" *Zeitschrift für Tierpsychologie* 22 (1965): 321–27.

Ewert, Jörg-Peter. *Neuroethology.* New York: Springer-Verlag, 1980.

Fentress, John C. "Development and Patterning of Movement Sequences in Inbred Mice." In *The Biology of Behavior,* edited by J. A. Kiger. Corvallis, Ore.: Oregon State University Press, 1972.

———. "Development of Grooming in Mice with Amputated Forelimbs." *Science* 179 (1973): 704–5.

Ferron, J. "Solitary Play of the Red Squirrel." *Canadian Journal of Zoology* 53 (1975): 1495–99.

Frisch, Karl von. *Animal Architecture.* New York: Harcourt Brace Jovanovich, 1974.

Gazzaniga, M. S., and Le Doux, Joseph E. *The Split Brain and the Integrated Mind.* New York: Plenum, 1976.

McFarland, David, and Sibly, R. " 'Unitary Drives' Revisited." *Animal Behaviour* 20 (1972): 548–63.

Nottebohm, Fernando, and Nottebohm, Marta E. "Vocalizations and Breeding Behavior of Surgically Deafened Ring Doves." *Animal Behaviour* 19 (1971): 313–27.

Pearson, K. G. "Central Programming and Reflex Control of Walking in the Cockroach." *Journal of Experimental Biology* 56 (1972): 173–93.

Smith, Andrew W. "Investigation of the Mechanisms Underlying Nest Construction in the Mud Wasp." *Animal Behaviour* 26 (1978): 232–40.

Waldron, I. "Mechanisms for the Production of the Motor Output Pattern in Flying Locusts." *Journal of Experimental Biology* 47 (1967): 201–12.

Willows, A. O. D.; Dorsett, D. A.; and Hoyle, G. "The Neuronal Basis of Behavior in *Tritonia.* I. Functional Organization of the Central Nervous System." *Journal of Neurobiology* 4 (1973): 207–37.

Willows, A. O. D., and Hoyle, G. "Neuronal Network Triggering a Fixed-Action Pattern." *Science* 166 (1969): 1549–51.

Wilson, Donald M. "Central Nervous Control of Flight in a Locust." *Journal of Experimental Biology* 38 (1961): 401–90.

———. "Genetic and Sensory Mechanisms for Locomotion and Orientation in Animals." *American Scientist* 60 (1972): 358–65.

12. MOTIVATION AND DRIVE

Aschoff, J. "Circadian Rhythms in Man." *Science* 148 (1965): 1427–32.

———, ed. *Circadian Clocks.* Amsterdam: North Holland Publishing Co., 1965.

Bloom, F.; Segal, D.; Ling, N.; and Guillemin, R. "Endorphins: Profound Behavioral Effects in Rats Suggest New Etiological Factors in Mental Illness." *Science* 194 (1976): 630–32.

Dethier, Vincent G. "Insects and the Concept of Motivation." *Nebraska Symposium on Motivation 1966* (1966): 105–36.

———. *The Hungry Fly.* Cambridge, Mass.: Harvard University Press, 1976.

Emlen, Steven. "Bird Migration: Influence of Physiological State upon Celestial Orientation." *Science* 165 (1969): 716–18.

Enright, J. T. "Influences of Seasonal Factors on the Activity Onset of the House Finch." *Ecology* 47 (1966): 662–66.

Farner, D. S. "The Photoperiodic Control of Reproductive Cycles in Birds." *American Scientist* 52 (1964): 137–56.

———, and Lewis, R. A. "Photoperiodism and Reproductive Cycles in Birds." *Photophysiology* 6 (1971): 325–70.

Harding, Cheryl F., and Follett, Brian, K. "Hormone Changes Triggered by Aggression in a Natural Population of Blackbirds." *Science* 203 (1979): 918–20.

Lombardi, J. R., and Vandenbergh, J. G. "Pheromonally Induced Sexual Maturation in Females: Regulation by the Social Environment of the Male." *Science* 196 (1977): 545–46.

Lorenz, Konrad Z. "The Comparative Method in Studying Innate Behavior Patterns." *Symposia of the Society for Experimental Biology* 4 (1950): 221–68.

Meier, A. H. "Daily Hormone Rhythms in the White-Throated Sparrow." *American Scientist* 61 (1973): 184–87.

Myers, R. D., and McCaleb, M. L. "Feeding: Satiety Signal from Intestine Triggers Brain's Noradrenergic Mechanism." *Science* 209 (1980): 1035–37.

Pengelley, E. T., ed. *Circannual Clocks: Annual Biological Rhythms.* New York: Academic Press, 1974.

Toates, F. M., and Archer, J. "Comparative Review of Motivational Systems Using Classical Control Theory." *Animal Behaviour 26* (1978): 368–80.

Truman, James W. "Hormonal Release of Differentiated Behavior Patterns." In *Simpler Networks and Behavior,* edited by J. C. Fentress. Sunderland, Mass.: Sinauer Associates, 1976.

———; Fallon, A. M.; and Wyatt, G. R. "Hormonal Release of Programmed Behavior in Silk Moths." *Science* 194 (1976): 1432–33.

Truman, James W., and Sokolove, P. G. "Silk Moth Eclosion: Hormonal Triggering of a Centrally Programmed Pattern of Behavior. *Science* 175 (1972): 1491–93.

Wise, R. A. "Hypothalamic Motivational Systems: Fixed or Plastic Neural Circuits?" *Science* 162 (1968): 377–79.

13. INVERTEBRATE ORIENTATION AND NAVIGATION

Blakemore, Richard P. "Magnetotactic Bacteria." *Science* 190 (1975): 377–79.

———; Frankel, Richard B.; and Kalmijn, Adrianus J. "South-seeking Magnetotactic Bacteria in the Southern Hemisphere." *Nature* 286 (1980): 384–85.

Brower, Lincoln P. "Monarch Migration." *Natural History* 86 (1977): 40–53.

Farkas, S. R., and Shorey, H. H. "Chemical Trail-Following by Flying Insects: A Mechanism for Orientation to a Distant Odor Source." *Science* 178 (1972): 67–68.

Fraenkel, F. S., and Gunn, D. L. *The Orientation of Animals.* New York: Dover, 1961.

Frankel, Richard B.; Blakemore, Richard P.; and

Wolfe, R. S. "Magnetite in Freshwater Magnetotactic Bacteria." *Science* 203 (1979): 1355–57.

Frisch, Karl von. *The Dance Language and Orientation of Bees.* Translated by Leigh E. Chadwick. Cambridge, Mass.: Harvard University Press, The Belknap Press, 1967.

Gould, J. L. "Sun Compensation by Bees." *Science* 207 (1980): 545–47.

———; Kirschvink, Joseph L.; and Deffeyes, Kenneth S. "Bees Have Magnetic Remanence." *Science* 201 (1978): 1026–28.

Kalmijn, Adrianus, and Blakemore, Richard P. "Magnetic Behavior of Mud Bacteria." In *Animal Migration, Navigation and Homing,* edited by Klaus Schmidt-Koenig and William T. Keeton. Berlin: Springer-Verlag, 1978.

Kennedy, J. S.; Ludlow, A. R.; and Sanders, C. J. "Guidance System Used in Moth Sex Attraction." *Nature* 288 (1980): 475–77.

Kennedy, J. S., and Marsh, D. "Pheromone-regulated Anemotaxis in Flying Moths." *Science* 184 (1974): 999–1001.

Larkin, Ronald P. "Transoceanic Bird Migration: Evidence for Detection of Wind Direction." *Behavioral Ecology and Sociobiology* 6 (1980): 229–32.

Lindauer, Martin. "Recent Advances in the Orientation and Learning of Honeybees." In *Proceedings of the Fifteenth International Congress on Entomology.* College Park, Maryland: The Entomological Society of America, 1977.

———, and Martin, Hermann. "Magnetic Effects on Dancing Bees." In *Animal Orientation and Navigation,* edited by S. R. Galler et al. Washington, D.C.: U.S. Government Printing Office, 1972.

Martin, Hermann, and Lindauer, Martin. "Der Einfluss der Erdmagnetfelds und die Schwereorientierung der Honigbiene." *Journal of Comparative Physiology* 122 (1977): 145–87.

Papi, Floriano. "Orientation by Night: The Moon." *Cold Spring Harbor Symposium in Quantitative Biology* 25 (1960): 475–80.

Pardi, L. "Innate Components in the Solar Orientation of Littoral Amphipods." *Cold Spring Harbor Symposium of Quantitative Biology* 25 (1960): 395–401.

———, and Papi, Floriano. "Die Sonne als Kompass bei *Talitrus saltator* (Montagu) (Amphipoda, Talitridae)." *Naturwissenschaften* 39 (1952): 262–63.

Ruttner, H. "Untersuchungen über die Flugaktivität und das Paarungverhalten der Dronen. VI. Flug auf und über Höhenrücken." *Apidologie* 7 (1976): 331–41.

Schmidt-Koenig, Klaus. "Directions of Migrating Monarch Butterflies in Some Parts of the Eastern United States." *Behavioral Processes* 4 (1979): 73–78.

Zmarlicki, C., and Morse, Robert A. "The Mating of Aged Virgin Queen Honeybees." *Journal of Apicultural Research* 2 (1963): 62–63.

14. VERTEBRATE NAVIGATION

Carr, Archibald. "Adaptive Aspects of the Scheduled Travel of *Cholonia.*" In *Animal Orientation and Navigation,* edited by R. M. Storm. Corvallis, Ore.: Oregon State University Press, 1967.

DeRosa, Christopher T., and Taylor, Douglas H. "Homeward Orientation Mechanisms in Three Species of Turtles *(Trionyx spinifer, Chrysemys picta,* and *Terrapene carolina).*" *Behavioral Ecology and Sociobiology* 7 (1980): 15–23.

Emlen, Steven T. "Bird Migration: Influence of Physiological State upon Celestial Orientation." *Science* 165 (1969): 716–18.

———; Demong, R.; Wiltschko, W.; Wiltschko, R.; and Bergman, S. "Magnetic Direction Finding: Evidence for Its Use in Migratory Indigo Buntings." *Science* 193 (1976): 505–8.

Gwinner, Eberhard, and Wiltschko, Wolfgang. "Circannual Changes in Migration Orientation of the Garden Warbler, *Sylvia borin.*" *Behavioral Ecology and Sociobiology* 7 (1980): 73–78.

Kalmijn, Adrianus. "Experimental Evidence of Geomagnetic Orientation in Elasmobranch Fishes." In *Animal Migration, Navigation, and Homing,* edited by Klaus Schmidt-Koenig and William T. Keeton. Berlin: Springer-Verlag, 1978.

Keeton, William T. "Orientation by Pigeons: Is the Sun Necessary?" *Science* 165 (1969): 922–28.

———. "Do Pigeons Determine Latitudinal Displacement from the Sun's Altitude?" *Nature* 227 (1970): 626–27.

———. "Magnets Interfere with Pigeon Homing." *Proceedings of the National Academy of Sciences of the U.S.A.* 68 (1971): 102–6.

———. "Release Site Bias as a Possible Guide to the "Map" Component in Pigeon Homing." *Journal of Comparative Physiology* 86 (1973): 1–16.

———. "The Orientational and Navigational Basis of Homing in Birds." *Advances in the Study of Behavior* 5 (1974): 47–132.

———. "Avian Orientation and Navigation: A Brief Overview." *British Birds* 72 (1979): 451–70.

———; Larkin, Timothy S.; and Windsor, Donald M. "Normal Fluctuations in the Earth's Magnetic Field Influence Pigeon Orientation." *Journal of Comparative Physiology* 95 (1974): 95–103.

Kiepenhener, Jakob. "Pigeon Homing: Deprivation of Olfactory Information Does Not Affect the Deflector Effect." *Behavioral Ecology and Sociobiology* 6 (1979): 11–22.

Kirschvink, Joseph L., and Gould, James L. "Biogenic Magnetite as the Basis of Magnetic Field Sensitivity in Animals." *BioSystems* 13 (1981): 181–201.

Kramer, Gustav. "Die Sonnenorientierung der Vögel." *Verhandlungen der Deutschen Zoologischen Gesellschaft, Freiburg* (1952): 72–84.

———. "Recent Experiments on Bird Orientation." *Ibis* 101 (1959): 399–416.

Kreithen, Melvin L., and Keeton, William T. "Detection of Polarized Light by the Homing Pigeon, Columba Livia." *Journal of Comparative Physiology* 89 (1974): 83–92.

Kreithen, Melvin L., and Quine, Douglas B. "Infrasound Detection by the Homing Pigeon: A Behavioral Audiogram." *Journal of Comparative Physiology* 129 (1979): 1–4.

Matthews, G. V. T. *Bird Navigation.* London: Cambridge University Press, 1955.

Michener, Martin, and Walcott, Charles. "Homing of Single Pigeons—an Analysis of Tracks." *Journal of Experimental Biology* 47 (1967): 99–131.

Papi, Floriano; Keeton, William T.; Brown, A. Irene; and Benvenuti, Silvano. "Do American and Italian Pigeons Rely on Different Homing Mechanisms?" *Journal of Comparative Physiology* 128 (1978): 303–17.

Schmidt-Koenig, Klaus, and Keeton, William T., eds. *Animal Migration, Navigation, and Homing.* Berlin: Springer-Verlag, 1978.

Schmidt-Koenig, Klaus, and Walcott, Charles. "Tracks of Pigeons Homing with Frosted Lenses." *Animal Behaviour* 26 (1978): 480–86.

Scholz, A. T.; Horrall, R. M.; Cooper, J. C.; and Hasler, A. D. "Imprinting to Chemical Cues: The Basis for Home Stream Selection in Salmon." *Science* 192 (1976): 1247–49.

Walcott, Charles. "The Homing of Pigeons." *American Scientist* 62 (1974): 542–52.

———; Gould, James L.; and Kirschvink, Joseph L. "Pigeons Have Magnets." *Science* 205 (1979): 1027–29.

Walcott, Charles, and Green, Robert P. "Orientation of Homing Pigeons Altered by a Change in the Direction of an Applied Magnetic Field." *Science* 184 (1974): 180–82.

Wallraff, H. G. "Über die Heimfinde vermogen von Brieftauben nach Haltung in verschiedenartig abgeschirmten Volieren." *Zeitschrift für Vergleichende Physiologie* 52 (1966): 215–59.

———. "Weitere Volienversuche mit Brieftanben: Wahrschienlicher Einfluss dynamischer Faktosen der Atmosphare auf die Orientierung." *Zeitschrift für Vergleichende Physiologie* 68 (1970): 182–201.

Wiltschko, Wolfgang, and Wiltschko, Roswitha. "Magnetic Compass of European Robins." *Science* 176 (1972): 62–64.

Yodlowski, Marilyn L.; Kreithen, Melvin L.; and Keeton, William T. "Detection of Atmospheric Infrasound by Homing Pigeons." *Nature* 265 (1977): 725–26.

15. ANIMAL COMMUNICATION

Bennet-Clark, H. C. "The Courtship Song of *Drosophila.*" *Behaviour* 31 (1968): 288–301.

———, and Ewing, A. W. "Pulse Interval as a Critical Parameter in the Courtship Song of *Drosophila.*" *Animal Behaviour* 17 (1969): 755–59.

Bentley, David R., and Hoy, Ronald R. "Genetic Control of Cricket Song Patterns." *Animal Behaviour* 20 (1972): 478–92.

Birch, M. C. *Pheromones.* Amsterdam: North-Holland Publishing Co., 1974.

Cade, W. "Acoustically Orienting Parasitoids: Fly

Phonotaxis to Cricket Song." *Science* 190 (1975): 1312–13.

Capranica, Robert K. *The Evoked Vocal Response of the Bullfrog: A Study of Communication by Sound.* Cambridge, Mass.: MIT Press, 1965.

———; Frishkopf, L. S.; and Nevo, E. "Encoding of Geographic Dialects in the Auditory System of the Cricket Frog." *Science* 182 (1973): 1272–75.

Ewing, Arthur W. "Complex Courtship Songs in the *Drosophila funebris* Species Group: Escape from an Evolutionary Bottleneck." *Animal Behaviour* 27 (1979): 343–49.

Hopkins, Carl D. "Sex Difference in Electric Signalling in an Electric Fish." *Science* 176 (1972): 1035–37.

Hoy, Ronald R.; Hahn, J.; and Paul, R. C. "Hybrid Cricket Auditory Behavior: Evidence for Genetic Coupling in Animal Communication." *Science* 195 (1977): 82–83.

Hoy, Ronald R., and Paul, R. C. "Genetic Control of Song Specificity in Crickets." *Science* 180 (1973): 82–83.

Jensson, T. A. "Evolution of Anoline Lizard Display Behavior." *American Zoologist* 17 (1977): 302–15.

Lloyd, James E. "Studies on the Flash Communication System in *Photinus* Fireflies." *Miscellaneous Publications of the Museum of Zoology, University of Michigan* 130 (1966): 1–95.

———. "Aggressive Mimicry in *Photuris* Fireflies: Signal Repertoires by Femmes Fatales." *Science* 187 (1975): 452–53.

———. "Male *Photuris* Fireflies Mimic Sexual Signals of Their Females' Prey." *Science* 210 (1980): 669–71.

Lorenz, Konrad Z. "Die Entwicklung der Vergleichenden Verhaltensforschung in den letzten 12 Jahren." *Zoologischer Anzeiger Supplement* 16 (1953): 36–58.

Meltzoff, A. R., and Moore, M. K. "Imitation of Facial and Manual Gestures by Human Neonates." *Science* 198 (1977): 75–78.

Rudinsky, J. A. "Masking of the Aggregation Pheromone in *Dendroctonus pseudotsugae.*" *Science* 166 (1969): 884–85.

———, and Ryker, L. C. "Sound Production in Scolytidas: Rivalry and Premating Stridulation of Male Douglas Fir Beetles." *Journal of Insect Physiology* 22 (1976): 997–1003.

Schoonhoven, L. M. "Chemosensory Bases of Host Plant Selection." *Annual Review of Entomology* 13 (1968): 115–36.

Sebeok, Thomas, ed. *How Animals Communicate.* Bloomington, Ind.: Indiana University Press, 1977.

Tinbergen, Niko. "Comparative Studies of the Behavior of Gulls." *Behaviour* 15 (1959): 1–70.

Walker, Thomas J. "Specificity in the Response of Female Tree Crickets to Calling Songs of the Males." *Annals of the Entomological Society of America* 50 (1957): 626–36.

Wilcox, R. S. "Communication by Surface Waves: Mating Behavior of a Water Strider *(Gerridae).*" *Journal of Comparative Physiology* 80 (1972): 255–66.

———. "Sex Discrimination in Water Striders: Role of a Surface Wave Signal." *Science* 206 (1979): 1325–27.

16. LEARNING AND INSTINCT

Alloway, Thomas M. "Learning and Memory in Insects." *Annual Review of Entomology* 17 (1972): 43–56.

Aristotle. *History of Animals.* Book IX, Chap. 40.

Bogdany, F. J. "Linkage of the Learning Signals in Honey Bee Orientation." *Behavioral Ecology and Sociobiology* 3 (1978): 323–36.

Erber, Joachim. "The Dynamics of Learning in the Honey Bee *(Apis mellifera carnica).* I. The Time Dependence of the Choice Reaction." *Journal of Comparative Physiology* 99 (1975): 231–42.

———. "The Dynamics of Learning in the Honey Bee *(Apis mellifera carnica).* II. Principles of Information Processing." *Journal of Comparative Physiology* 99 (1975): 243–55.

———; Masuhr, Thomas; and Menzel, Randolf. "Localization of Short-term Memory in the Brain of the Bee, *Apis mellifera.*" *Physiological Entomology* 5 (1980): 343–58.

Erber, J., and Schildberger, K. "Conditioning of an Antennal Reflex to Visual Stimuli in Bees." *Journal of Comparative Physiology* 135 (1980): 217–25.

Fabre, J. H. *The Hunting Wasps.* New York: Dodd, Mead & Co., 1915.

———. *More Hunting Wasps.* New York: Dodd, Mead & Co., 1921.

Forel, Auguste. *Das Sinnesleben der Insekten.* Munich, 1910.

Horridge, G. A. "Learning of Leg Position by Headless Insects." *Nature* 193 (1962): 697–98.

Kerr, W. E.; Duarte, F. A. M.; and Oliveria, R. S. "Genetic Component in Learning Ability in Bees." *Behavioral Genetics* 5 (1975): 331–37.

Menzel, Randolf. "Behavioral Access to Short-term Memory in Bees." *Nature* 281 (1979): 368–69.

———; Erber, Joachim; and Masuhr, Thomas. "Learning and Memory in the Honey Bee." In *Experimental Analysis of Insect Behaviour,* edited by L. Barton Browne. New York: Springer-Verlag, 1974.

Opfinger, Elisabeth. "Über die Orientierung der Biene an der Futterquelle." *Zeitschrift für Vergleichende Physiologie* 15 (1931): 431–87.

17. PROGRAMMED LEARNING

Baerends, G. P. "Fortpflanzungsverhalten und Orientierung der Grabwespe." *Tijdschrift voor Entomologie* 84 (1941): 68–275.

Bateson, P. P. G. "The Imprinting of Birds." In *Ethology and Development,* edited by S. A. Barnett. London: Heinemann Medical Books Ltd., 1974.

Bernstein, Ilene, and Sigmundi, Ronald A. "Tumor Anorexia: A Learned Food Aversion?" *Science* 209 (1980): 416–18.

Brower, J. V. "Experimental Studies of Mimicry in Some North American Butterflies. I. The Monarch, *Danaus plexippus,* and Viceroy, *Limenitis archippus archippus.*" *Evolution* 12 (1958): 32–47.

Brower, L. P., and Brower, J. V. "Birds, Butterflies, and Plant Poisons: A Study in Ecological Chemistry." *Zoologica* 49 (1964): 137–59.

Burghardt, Gordon M., and Hess, Eckhard H. "Food Imprinting in the Snapping Turtle *Chelydra serpentina.*" *Science* 151 (1966): 108–9.

Collier, G.; Hirsch, E.; and Hamlin, P. H. "The Ecological Determinants of Reinforcement in the Rat." *Physiology and Behavior* 9 (1972): 705–16.

Cullen, E. "Adaptations in the Kittiwake to Cliff Nesting." *Ibis* (1957): 275–302.

Emlen, Steven J. "The Development of Migratory Orientation in Young Indigo Buntings." *The Living Bird* 8 (1969): 113–26.

Gallagher, J. E. "Sexual Imprinting: A Sensitive Period in Japanese Quail." *Journal of Comparative Physiology* 91 (1977): 72–78.

Garcia, J.; McGowan, B. K.; and Green, K. F. "Biological Constraints on Conditioning." In *Classical Conditioning II,* edited by A. H. Black and W. F. Prokasy. New York: Appleton-Century-Crofts, 1972.

Grier, J. B.; Counter, S. A.; and Shearer, W. M. "Prenatal Auditory Imprinting in Chickens." *Science* 155 (1967): 1692–93.

Hess, Eckhard H. "The Relationship Between Imprinting and Motivation." *Nebraska Symposium on Motivation* (1959): 44–77.

———. *Imprinting.* New York: Van Nostrand Reinhold, 1973.

Immelmann, Klaus. "Ecological Significance of Imprinting in Early Learning." *Annual Review of Ecology and Systematics* 6 (1975): 15–37.

Janzen, D. H. "The Deflowering of Central America." *Natural History* 83 (1974): 48–53.

Jenkins, P. F. "Cultural Transmission of Song Patterns and Dialect Development in a Free-Living Bird Population." *Animal Behaviour* 26 (1977): 50–78.

Konishi, Masakazu. "The Role of Auditory Feedback in the Control of Vocalization in the White-Crowned Sparrow." *Zeitschrift für Tierpsychologie* 22 (1965): 770–83.

———, and Nottebohm, Fernando. "Experimental Studies on the Ontogeny of Avian Vocalization." In *Bird Vocalization,* edited by R. A. Hinde. New York: Cambridge University Press, 1969.

Lorenz, Konrad. "Imprinting." *Auk* 54 (1937): 245–73.

Marler, Peter, and Tamura, M. "Culturally Transmitted Patterns of Vocal Behavior in Sparrows." *Science* 146 (1964): 1483–86.

Miller, D. B., and Gottlieb, Gilbert. "Acoustic Features of Wood Duck Maternal Calls." *Behaviour* 57 (1976): 260–80.

———. "Maternal Vocalizations of Mallard Ducks." *Animal Behaviour* 26 (1978): 1178–94.

Peters, Susan S.; Searcy, William A.; and Marler, Peter. "Species Song Discrimination in Choice Experiments with Territorial Male Swamp and Song Sparrows." *Animal Behaviour* 28 (1980): 393–404.

Pusey, Anne E. "Inbreeding Avoidance in Chimpanzees." *Animal Behaviour* 28 (1980): 543–52.

Quinn, William B., and Dudai, Yadin. "Memory Phases in *Drosophila*." *Nature* 262 (1976): 576–77.

Rothstein, Stephen I. "Mechanisms of Avian Egg Recognition: Additional Evidence for Learned Components." *Animal Behaviour* 26 (1978): 671–77.

Rozin, P., and Kalat, J. W. "Specific Hungers and Poison Avoidance as Adaptive Specializations of Learning." *Psychological Review* 78 (1971): 459–86.

Seawright, J. E.; Kaiser, P. E.; Dame, D. A.; and Lofgren, C. S. "Learned Taste Aversions in Children Receiving Chemotherapy." *Science* 200 (1978): 1302–4.

Slotnick, B. M., and Katz, H. M. "Olfactory Learning-Set Formation in Rats." *Science* 185 (1974): 796–98.

Tinbergen, Niko. *The Curious Naturalists.* New York: Doubleday, 1958.

Vidal, Jean-Marie. "Relations Between Filial and Sexual Imprinting in Domestic Fowl." *Animal Behaviour* 28 (1980): 880–91.

Wilcoxon, H. C.; Dragoin, W. B.; and Kral, P. A. "Illness-Induced Aversions in Rat and Quail: Relative Salience of Visual and Gustatory Cues." *Science* 171 (1971): 826–28.

Yasukawa, K.; Blank, J. L.; and Patterson, C. B. "Song Repertoires and Sexual Selection in the Red-winged Blackbird." *Behavioral Ecology and Sociobiology* 7 (1980): 233–38.

18. CULTURAL AND PLASTIC LEARNING

Bonner, John Tyler. *The Evolution of Culture in Animals.* Princeton, N.J.: Princeton University Press, 1980.

Curio, Eberhard; Ernst, Ulrich; and Vieth, Willy. "Cultural Transmission of Enemy Recognition in Blackbirds: Effectiveness and Constraints." *Zeitschrift für Tierpsychologie* 48 (1978): 184–202.

———. "Cultural Transmission of Enemy Recognition: One Function of Mobbing." *Science* 202 (1978): 899–901.

Douglas-Hamilton, Ian. *Among the Elephants.* New York: Viking Press, 1975.

Frisch, Karl von. *The Dance Language and Orientation of Bees.* Cambridge, Mass.: Harvard University Press, 1967.

Gould, James L., and Gould, Carol G. "The Insect Mind: Physics or Metaphysics?" In *Animal Mind—Human Mind,* edited by Donald R. Griffin. Berlin: Springer-Verlag, 1981.
1981.

Hall, K. R. L. "Observational Learning in Monkeys and Apes." *British Journal of Psychology* 54 (1963): 201–26.

———, and Goswell, M. J. "Aspects of Social Learning in Captive Patas Monkeys." *Primates* 5 (1964): 59–70.

Hinde, Robert A., and Fisher, J. "Further Observations on the Opening of Milk Bottles by Birds." *British Birds* 44 (1951): 393–96.

Kawai, M. "Newly Acquired Pre-cultural Behavior of a Natural Troop of Japanese Monkeys." *Primates* 6 (1965): 1–30.

Klopfer, Peter H. "Observational Learning in Birds." *Behaviour* 17 (1961): 71–80.

Kovach, J. K. "Interaction of Innate and Acquired Color Preferences and Early Exposure Learning in Chicks." *Journal of Comparative and Physiological Psychology* 75 (1971) 386–98.

Leger, Daniel W.; Owings, Donald H.; and Gelfand, Deborah L. "Single-Rate Vocalizations of California Ground Squirrels: Graded Signals and Situation-Specificity of Predator and Socially Evoked Calls." *Zeitschrift für Tierpsychologie* 52 (1980): 227–46.

Lindauer, Martin. "Schwarmbienen auf Wohnungssuche." *Zeitschrift für Vergleichende Physiologie* 37 (1955): 263–324.

Lovell, H. B. "Sources of Nectar and Pollen." In *Hive and Honey Bee,* edited by R. A. Grout. Hamilton, Ill.: Dadant and Sons, 1963.

Marler, Peter. "Developments in the Study of Animal Communication." In *Darwin's Biological Work,* edited by R. P. Bell. Cambridge: Cambridge University Press, 1959.

———. "The Drive to Survive." In *Marvels of Animal Behavior.* Washington, D.C.: National Geographic Society, 1972.

———, and Tamura, M. "Culturally Transmitted Patterns of Vocal Behavior in Sparrows." *Science* 146 (1964): 1483–86.

Moore, Bruce R. "Role of Directed Pavlovian Reactions in Simple Instrumental Learning in the

Pigeon." In *Constraints on Learning*, edited by R. A. Hinde and J. Stevenson-Hinde. New York: Academic Press, 1973.

Norton-Griffiths, M. N. "Organization, Control and Development of Parental Feeding in the Oystercatcher." *Behaviour* 34 (1969): 55–114.

Pankiw, P. "Studies of Honey Bees on Alfalfa Flowers." *Journal of Apicultural Research* 6 (1967): 105–12.

Reinhardt, J. F. "Responses of Honey Bees to Alfalfa Flowers." *American Naturalist* 86 (1952): 257–75.

Seeley, Thomas D. "Life History Strategy of the Honey Bee." *Oecologia* 32 (1978): 109–18.

———, and Morse, Roger A. "Nest-Site Selection by the Honey Bee." *Insectes Sociaux* 25 (1978): 323–37.

Trager, James G. *The Enriched, Fortified, Concentrated, Country-Fresh, Lip-Smacking, Finger-Licking, International, Unexpurgated FOODBOOK*. New York: Grossman, 1970.

Vieth, Willy; Curio, Eberhard; and Ernst, Ulrich. "Cultural Transmission of Enemy Recognition in Blackbirds: Cross-Species Tutoring and Properties of Learning." *Animal Behaviour* 28 (1980): 1217–29.

19. CLASSICAL BEHAVIORAL GENETICS

Bentley, David R. "Genetic Control of an Insect Neuronal Network." *Science* 174 (1971): 1139–41.

———, and Hoy, Ronald. "Genetic Control of the Neuronal Network Generating Cricket (Teleogryllus gryllus) Song Patterns." *Animal Behaviour* 20 (1972): 478–92.

Bovet, D.; Bovet-Nitti, F.; and Oliverio, A. "Genetic Aspects of Learning and Memory in Mice." *Science* 163 (1969): 139–49.

Capranica, Robert R.; Frishkopf, L. S.; and Nevo, E. "Encoding of Geographic Dialects in the Auditory System of the Cricket Frog." *Science* 182 (1973): 1272–75.

Dilger, William C. "The Interaction Between Genetic and Experimental Influences in the Development of Species-Typical Behavior." *American Zoologist* 4 (1964): 155–60.

Entrikin, R. K., and Erway, L. C. "Genetic Investigation of Roller and Tumbler Pigeons." *Journal of Heredity* 63 (1972): 351–54.

Fuller, J. L., and Thompson, R. W. *Behavior Genetics*. New York: Wiley, 1960.

Gerhardt, H. Carl. "Vocalizations of Some Hybrid Treefrogs: Acoustic and Behavioral Analyses." *Behaviour* 49 (1979): 130–51.

Gonçalves, Lionel S., and Stort, Antonio C. "Honey Bee Improvement Through Behavioral Genetics." *Annual Review of Entomology* 31 (1978): 197–213.

Gould, James L. "Genetics and Molecular Ethology." *Zeitschrift fur Tierpsychologie* 36 (1974): 267–92.

Hinde, Robert A. "The Behaviour of Certain Cardueline F_1 Inter-Species Hybrids." *Behaviour* 9 (1956): 202–13.

Hoy, Ronald R., and Paul, R. C. "Genetic Control of Song Specificity in Crickets." *Science* 180 (1973): 82–83.

Kerr, W. E.; Duarte, F. A. M.; and Oliveria, R. S. "Genetic Component of Learning Ability in Bees." *Behavioral Genetics* 5 (1975): 331–37.

Lade, B. I., and Thorpe, William H. "Dove Songs as Innately Coded Patterns of Specific Behaviour." *Nature* 202 (1964): 366–68.

Manning, Aubrey. "The Place of Genetics in the Study of Behavior." In *Growing Points in Ethology*, edited by P. P. G. Bateson and R. A. Hinde. Cambridge: Cambridge University Press, 1976.

McGrath, T. A.; Shalter, Michael D.; Schleidt, Wolfgang M.; and Sarvella, Patricia. "Analysis of Distress Calls of Chicken × Pheasant Hybrids." *Nature* 237 (1972): 47–48.

Nye, W. P., and Mackensen, D. "Selective Breeding of Honey Bees for Alfalfa Pollination." *Journal of Apicultural Research* 7 (1968): 21–27.

Rothenbuhler, Walter C. "Behavior Genetics of Nest Cleaning in Honey Bees." *American Zoologist* 4 (1964): 111–23.

Sharpe, R. S., and Johnsgard, P. A. "Inheritance of Behavioral Characters in F_2 Mallard × Pintail Hybrids." *Behaviour* 27 (1966): 259–72.

Tryon, R. C. "Genetic Differences in Maze-Learning Ability in Rats." *Yearbook of the National Society for the Study of Education* 39 (1940): 111–19.

20. MOLECULAR ETHOLOGY

Bastock, M. "A Gene Mutation Which Changes a Behavior Pattern." *Evolution* 10 (1956): 421–39.

Bentley, David. "Single-Gene Cricket Mutations: Effects on Behavior, Sensilla, Sensory Neurons, and Identified Interneurons." *Science* 187 (1975): 760–64.

Bruce, Victor G. "Clock Mutants in *Chlamydomonas*." *Genetics* 70 (1972): 537–48.

Byers, Duncan; Davis, R. L.; and Kiger, J. A., Jr. "A Defect in Phosphodiesterase Due to the *dunce* Mutation of Learning in *Drosophila*." *Nature* 289 (1981): 79–81.

Caviness, V. S. "Altered Visual Connectivity in Reeler Mice." In *Approaches to the Cell Biology of Neurons,* edited by M. J. Cowan and J. A. Ferrendelli. Baltimore, Md.: Society for Neuroscience, 1977.

Delong, G. R., and Sidman, R. L. "Alignment Defect in Reeler Mouse Brain Cells." *Developmental Biology* 22 (1970): 584–600.

Gould, James L. "Genetics and Molecular Ethology." *Zeitschrift für Tierpsychologie* 36 (1974): 267–92.

Guillery, R. W., and Casagrande, V. A. "Studies of the Modifiability of the Visual Pathways in Midwestern Siamese Cats." *Journal of Comparative Neurology* 174 (1977): 15–46.

Hazelbauer, G. L., and Adler, Julius. "Role of Galactose Binding Protein in Chemotaxis of *E. coli* Toward Galactose." *Nature New Biology* 230 (1971): 101–4.

Hotta, Y., and Benzer, Seymour. "Mapping of Behaviour in *Drosophila* Mosaics." *Nature* 240 (1972): 527–35.

———. "Courtship in *Drosophila* Mosaics: Sex-Specific Foci for Sequential Action Patterns." *Proceedings of the National Academy of Sciences of the U.S.A.* 73 (1976): 4154–58.

Konopka, Ronald J., and Benzer, Seymour. "Clock Mutants of *Drosophila melanogaster*." *Proceedings of the National Acaaemy of Sciences of the U.S.A.* 68 (1971): 2112–16.

Kung, C.; Chang, S-Y.; Satow, Y.; van Houten, J.; and Hansma, H. "Genetic Dissection of Behavior in *Paramecium*." *Science* 188 (1975): 898–904.

Lopresti, Victor; Macagno, Eduardo R.; and Levinthal, Cyrus. "Structure and Development of Neuronal Connections in *Daphnia* Optic Lamina." *Proceedings of the National Academy of Sciences of the U.S.A.* 70 (1973): 433–37.

Macagno, Eduardo R. "Mechanism for the Formation of Synaptic Projections in the Arthropod Visual System." *Nature* 275 (1978): 318–20.

———; Lopresti, Victor; and Levinthal, Cyrus. "Variation of the Neuronal Connections in the Optic System of *Daphnia*." *Proceedings of the National Academy of Sciences of the U.S.A.* 70 (1973): 57–61.

Macnab, R. M., and Koshland, Daniel E. "The Gradient-Sensing Mechanism in Bacterial Chemotaxis." *Proceedings of the National Academy of Sciences of the U.S.A.* 69 (1972): 2509–12.

Purves, Dale. "Neuronal Competition." *Nature* 287 (1980): 585–86.

Quinn, William G., and Dudai, Y. "Memory Phases in *Drosophila*." *Nature* 262 (1976): 576–77.

Quinn, William G., and Gould, James L. "Nerves and Genes." *Nature* 278 (1979): 19–23.

Sidman, R. L.; Green, M. C.; and Appel, S. H. *Neurological Mutants of the Mouse.* Cambridge, Mass.: Harvard University Press, 1968.

Springer, Martin S.; Goy, Michael F.; and Adler, Julius. "Protein Methylation in Behavioral Control Mechanisms and in Signal Transduction." *Nature* 280 (1979): 279–84.

Sulston, J. E. "Post-Embryonic Development in the Ventral Cord of *C. elegans*." *Philosophical Transactions of the Royal Society of London* 275 (1976): 287–97.

———, and Horvitz, H. R. "Post-Embryonic Cell Lineages of *C. elegans*." *Developmental Biology* 56 (1977): 110–56.

White, J. G.; Albertson, D. G.; and Anness, M. A. R. "Connectivity Changes in a Class of Motoneuron During the Development of a Nematode." *Nature* 271 (1978): 764–66.

21. BEHAVIORAL ECOLOGY

Aneshansley, D.; Eisner, Thomas; Widom, J. M.; and Widom, B. "Biochemistry at 100°C: The Explosive Discharge of Bombardier Beetles."

Science 165 (1969): 61–63.
Blest, A. D. "Relations Between Moths and Predators." *Nature* 197 (1963): 1046–47.
Janzen, D. H. "Coevolution of Mutualism Between Ants and Acacias in Central America." *Evolution* 20 (1966): 249–75.
Lack, David. *Darwin's Finches.* New York: Cambridge University Press, 1947.
MacArthur, Robert H. *Geographical Ecology: Patterns in the Distribution of Species.* New York: Harper and Row, 1972.
———, and Wilson, Edward O. *The Theory of Island Biogeography.* Princeton, N.J.: Princeton University Press, 1967.
Mech, L. David. *The Wolf: Ecology and Behavior of an Endangered Species.* Garden City, N.Y.: Doubleday, 1970.
Moore, Peter D. "How Plants Exploit Animals." *Nature* 283 (1980): 428–30.
Slobodchikoff, C. R. "Experimental Studies of Tenebrionid Beetle Predation by Skunks." *Behaviour* 66 (1978): 312–22.
Thomas, Lewis. "The Mimosa Girdler." *Science 80,* no. 1 (1980): 83–84.
Tinbergen, Niko. *Curious Naturalists.* New York: Doubleday, 1958.

22. MATING AND TERRITORIALITY

Crook, J. H. "Social Organization and Environment: Aspects of Contemporary Social Ethology." *Animal Behaviour* 18 (1970): 197–209.
Davies, N. B. "Territorial Defence in the Speckled Wood Butterfly: The Resident Always Wins." *Animal Behaviour* 26 (1978): 138–47.
Emlen, S. T., and Oring, L. W. "Ecology, Sexual Selection, and the Evolution of Mating Systems." *Science* 198 (1977): 215–23.
Geist, Valerius. *Mountain Sheep.* Chicago: University of Chicago Press, 1971.
Kessel, E. L. "Mating Activities of Balloon Flies." *Systematic Zoology* 4 (1955): 97–104.
Krebs, John R. "Sexual Selection and the Handicap Principle." *Nature* 261 (1976): 192.
Lack, David. *Population Studies of Birds.* Oxford: Clarendon Press, 1966.
———. *Ecological Adaptation for Breeding in Birds.* London: Methuen, 1968.
Lorenz, Konrad. *On Aggression.* New York: Bantam, 1967.
Massey, Adrianne, and Vandenburgh, John G. "Puberty Delay by a Urinary Cue from Female House Mice in Feral Populations." *Science* 209 (1980): 821–22.
Orians, Gordon H. "Ecology of Blackbird Social Systems." *Ecological Monograph* 31 (1961): 285–312.
———. "Evolution of Mating Systems in Birds and Mammals." *American Naturalist* 103 (1969): 589–603.
Southern, H. N. "The Natural Control of a Population of Tawny Owls *(Strix aluco).*" *Journal of Zoology (London)* 162 (1970): 197–285.
Thornhill, Randy. "Sexual Selection and Paternal Investment in Insects." *American Naturalist* 110 (1976): 153–63.
Tinbergen, Niko. "On War and Peace in Animals and Man." *Science* 160 (1968): 1411–18.

23. INDIVIDUAL SELECTION

Bradbury, Jack W. "Lek Mating Behavior in the Hammer-headed Bat." *Zeitschrift für Tierpsychologie* 45 (1977): 225–55.
Brockmann, H. Jane; Grafen, Alan; and Dawkins, Richard. "Evolutionarily Stable Nesting Strategy in a Digger Wasp." *Journal of Theoretical Biology* 77 (1979): 473–96.
Brown, Jerram L. "Alternate Routes to Sociality in Jays." *American Zoologist* 14 (1974): 63–80.
Charnov, Eric L., and Krebs, John R. "The Evolution of Alarm Calls: Altruism or Manipulation?" *American Naturalist* 109 (1975): 107–12.
Dunford, C. "Kin Selection for Ground Squirrel Alarm Calls." *American Naturalist* 111 (1977): 782–85.
Greenberg, Les. "Genetic Component of Bee Odor in Kin Recognition." *Science* 206 (1979): 1095–97.
Hamilton, William B. "Genetical Evolution of Social Behavior." *Journal of Theoretical Biology* 7 (1964): 1–16.
Koenig, Walter D., and Pitelka, Frank A. "Relatedness and Inbreeding Avoidance: Counterploys in the Communally Nesting Acorn Woodpecker." *Science* 206 (1979): 1103–5.
Orians, Gordon H. "On the Evolution of Mating Systems in Birds and Mammals." *American Naturalist* 103 (1969): 589–603.
Sherman, Paul W. "Nepotism and the Evolution of

Alarm Calls." *Science* 197 (1977): 1246–53.
Smith, John Maynard. "The Evolution of Alarm Calls." *American Naturalist* 94 (1965): 59–63.
———. "The Theory of Games and the Evolution of Animal Conflict." *Journal of Theoretical Biology* 47 (1974): 209–21.
———. "Parental Investment: A Prospective Analysis." *Animal Behaviour* 25 (1977): 1–9.
Trivers, Robert L. "The Evolution of Reciprocal Altruism." *Quarterly Review of Biology* 46 (1971): 35–57.
———. "Parent-Offspring Conflict." *American Zoologist* 14 (1974): 249–64.
———, and Hare, H. "Haplodiploidy and the Evolution of Social Insects." *Science* 191 (1976): 249–63.
Wade, M. J., and Breden, F. "The Evolution of Cheating and Selfish Behavior." *Behavioral Ecology and Sociobiology* 7 (1980): 167–72.
Watts, C. R., and Stokes, A. W. "The Social Order of Turkeys." *Scientific American* 224, no. 6 (1971): 112–18.
West Eberhard, M. J. "The Evolution of Social Behavior by Kin Selection." *Quarterly Review of Biology* 50 (1975): 1–33.
Wickler, Wolfgang. "Kin Selection and Effectiveness in Social Insect Workers and Other Helpers." *Zeitschrift für Tierpsychologie* 48 (1978): 100–3.
Wiley, R. Haven. "Territoriality and Non-Random Mating in Sage Grouse, *Centrocercus urophasianus*." *Animal Behaviour Monographs* 6 (1973): 87–169.
Woolfenden, G. E. "Florida Scrub Jay Helpers at the Nest." *Auk* 92 (1975): 1–15.
———, and Fitzpatrick, J. W. "The Inheritance of Territory in Group Breeding Birds." *BioScience* 28 (1978): 104–8.

24. HONEY BEES

Anderson, R. H. "The Laying Worker in the Cape Honey Bee." *Journal of Apicultural Research* 2 (1963): 85–92.
Brines, Michael L., and Gould, James L. "Bees Have Rules." *Science* 206 (1979): 571–73.
Butler, Charles. *The Feminine Monarchie.* Oxford: Joseph Barnes, 1609 (reprinted New York: Da Capo Press, 1969).
Free, John B. *The Social Organization of Honey Bees.* London: Edward Arnold, 1977.
Frisch, Karl von. *The Dance Language and Orientation of Bees.* Cambridge, Mass.: Harvard University Press, 1967.
Gould, James L. "Honey Bee Recruitment," *Science* 189 (1975): 685–93.
———; Henerey, Michael; and MacLeod, Michael C. "Communication of Direction by the Honey Bee." *Science* 169 (1970): 544–54.
Grout, Roy A. *The Hive and the Honey Bee.* Hamilton, Ill.: Dadant and Son, 1963.
Lindauer, Martin. *Communication Among Social Bees.* Cambridge, Mass.: Harvard University Press, 1978.
Maeterlinck, Maurice. *The Life of the Bee.* New York: Dodd, Mead & Co., 1901.
Michener, Charles D. *The Social Behaviour of Bees.* Cambridge, Mass.: Harvard University Press, 1974.
Ribbands, C. Ronald. *Social Life of Honey Bees.* London: Bee Research Association, 1953.
Seeley, Thomas D., and Morse, Roger A. "Nest Site Selection by the Honey Bee." *Insectes Sociaux* 25 (1978): 323–37.
Wenner, Adrian M. *The Bee Language Controversy.* Boulder, Colo.: Educational Programs Improvement Corporation, 1971.
Wilson, Edward O. *The Insect Societies.* Cambridge, Mass.: Harvard University Press, 1971.

25. ANTS

Chauvin, Rémy. *The World of Ants.* New York: Hill and Wang, 1970.
Hölldobler, Bert. "Recruitment Behavior, Home Range Orientation, and Territoriality in Harvester Ants." *Behavioral Ecology and Sociobiology* 1 (1976): 3–44.
Regnier, F. E., and Wilson, Edward O. "Chemical Communication and 'Propaganda' in Slavemaker Ants." *Science* 172 (1971): 267–69.
Schneirla, T. C. *Army Ants.* San Francisco: W. H. Freeman, 1971.
Sudd, John H. *An Introduction to the Study of Ants.* London: Edward Arnold, 1967.
Weber, N. A. "The Attines: The Fungus-Culturing Ants." *American Scientist* 60 (1972): 448–56.
Wheeler, William Morton. *Ants.* New York: Columbia University Press, 1910.

Wilson, Edward O. *The Insect Societies.* Cambridge, Mass.: Harvard University Press, 1971.

26. AFRICAN UNGULATES

Bigalke, R. C. "The Springbok." *Natural History* 75 (1966): 20–25.
Buechner, H. K. "Territorial Behavior in Uganda Kob." *Science* 133 (1961): 698–99.
———, and Roth, H. D. "The Lek System in Uganda Kob Antelope." *American Zoologist* 14 (1974): 145–62.
Floody, Owen R., and Arnold, Arthur P. "Uganda Kob: Territoriality and the Spatial Distributions of Sexual and Agonistic Behaviors at a Territorial Ground." *Zeitschrift für Tierpsychologie* 37 (1975): 192–212.
Geist, Valerius. *Mountain Sheep.* Chicago: University of Chicago Press, 1971.
———, and Walther, F. R., eds. *The Behavior of Ungulates.* Morges, Canada: IUCN Publications New Series No. 24, 1974.
Gwynne, M. O., and Bell, R. H. U. "Selection of Vegetation Components by Grazing Ungulates." *Nature* 220 (1968): 390–93.
Jarman, P. J. "The Social Organization of Antelope in Relation to Their Ecology." *Behaviour* 48 (1974): 215–67.
———, and Jarman, M. V. "Social Behaviour, Population Structure and Reproductive Potential in Impala." *East African Wildlife Journal* 11 (1973): 329–38.
Leuthold, Walter. *African Ungulates.* New York: Springer-Verlag, 1977.
Owen-Smith, R. N. "Territoriality in the White Rhinoceros." *Nature* 231 (1971): 294–96.
Ralls, Katherine. "Agonistic Behavior in Maxwell's Duikers." *Science* 171 (1971): 443–49.
———. "Mammalian Scent Marking." *Mammalia* 39 (1975): 241–249.
Schaller, George B. *The Serengeti Lion.* Chicago: University of Chicago Press, 1972.

27. AFRICAN CARNIVORES

Bertram, Brian C. R. "Kin Selection in Lions and Evolution." In *Growing Points in Ethology,* edited by P. P. G. Bateson and R. A. Hinde. New York: Cambridge University Press, 1976.
Bygott, J. David; Bertram, Brian C. R.; and Hanby, Jeannette P. "Male Lions in Large Coalitions Gain Reproductive Advantages." *Nature* 282 (1979): 839–41.
Eaton, Randall L. *The Cheetah.* New York: Van Nostrand Reinhold, 1974.
Ewer, R. F. *The Carnivores.* Ithaca, N.Y.: Cornell University Press, 1973.
Klinghammer, Erich. *The Behavior and Ecology of Wolves.* New York: Garland STPM, 1979.
Kruuk, Hans. *The Spotted Hyena.* Chicago: University of Chicago Press, 1972.
Leyhausen, Paul. *Cat Behavior.* New York: Garland STPM, 1979.
Schaller, George B. *The Serengeti Lion.* Chicago: University of Chicago Press, 1972.
Van Läwick, Hugo, and van Läwick-Goodall, Jane. *Innocent Killers.* Boston: Houghton-Mifflin, 1971.
Walther, F. "Flight Behaviour and Avoidance of Predators in Thomson's Gazelle (*Gazella thomsoni* Günther 1884)." *Behaviour* 34 (1969): 184–221.

28. PRIMATES

Altmann, Stuart A., ed. *Social Communication Among Primates.* Chicago: University of Chicago Press, 1967.
Clutton-Brock, T. H., and Harvey, P. H. "Primate Ecology and Social Organization." *Journal of Zoology* 183 (1977): 1–40.
Dunbar, R. I. M. "Determinants and Evolutionary Consequences of Dominance Among Female Gelada Baboons." *Behavioral Ecology and Sociobiology* 7 (1980): 253–65.
Fouts, Roger S. "Acquisition and Testing of Gestural Signs in Four Young Chimpanzees." *Science* 180 (1973): 978–80.
Gallup, G. G. "Chimpanzees: Self-Recognition." *Science* 167 (1969): 86–87.
Gardiner, R. A., and Gardiner, B. T. "Teaching Sign Language to a Chimpanzee." *Science* 165 (1969): 664–72.
———. "Early Signs of Language in Child and Chimpanzee." *Science* 187 (1975): 752–53.
Goodall, Jane. "The Behaviour of Free-Living Chimpanzees in the Gombe Stream Reserve." *Animal Behaviour Monographs* 1 (1968): 165–301.
Krebs, John R. "Primate Social Structure and Ecol-

ogy." *Nature* 270 (1977): 99–100.

Kummer, H. *Social Organization of Hamadryas Baboons: A Field Study.* Chicago: University of Chicago Press, 1968.

Lee, P. C., and Oliver, J. I. "Competition, Dominance and the Acquisition of Rank in Juvenile Yellow Baboons *(Papio cynocephalus).*" *Animal Behaviour* 27 (1979): 576–85.

Michael, R. P.; Keverne, E. B.; and Bonsall, R. W. "Pheromones: Isolation of Male Sex Attractants from a Female Primate." *Science* 172 (1971): 964–66.

Premack, David. "Language in the Chimpanzee?" *Science* 172 (1971): 808–22.

Rumbaugh, Duane M., ed. *Language Learning by a Chimpanzee.* New York: Academic Press, 1977.

Rumbaugh, Duane M.; Gill, T. V.; and von Glasersfeld, E. C. "Reading and Sentence Completion by a Chimpanzee." *Science* 182 (1973): 731–33.

Sackett, G. P. "Monkeys Reared in Isolation with Pictures as Visual Input: Evidence for an Innate Releasing Mechanism." *Science* 154 (1966): 1468–73.

Schaller, George B. *The Mountain Gorilla: Ecology and Behavior.* Chicago: University of Chicago Press, 1963.

Seyfarth, Robert M., and Cheney, Dorothy L. "The Ontogeny of Vervet Monkey Alarm Calling Behavior: A Preliminary Report." *Zeitschrift für Tierpsychologie* 54 (1980): 37–56.

———; and Marler, Peter. "Vervet Monkey Alarm Calls: Semantic Communication in a Free-Ranging Primate." *Animal Behaviour* 28 (1980): 1070–94.

Snowdon, Charles T., and Pola, Yvonne V. "Interspecific and Intraspecific Responses to Synthesized Pygmy Marmoset Vocalizations." *Animal Behaviour* 26 (1978): 192–206.

Tutin, Caroline E. G. "Mating Patterns and Reproductive Strategies in a Community of Wild Chimpanzees *(Pan troglodytes schweinfurthii).*" *Behavioral Ecology and Sociobiology* 6 (1979): 29–38.

Wilson, Edward O. *Sociobiology.* Cambridge, Mass.: Harvard University Press, 1975.

29. HUMAN BEHAVIOR

Eibl-Eibesfeldt, Irenäeus. *The Biology of Peace and War.* New York: Viking, 1980.

Ekman, Paul; Sorenson, E. Richard; and Friesen, Wallace U. "Pan-Cultural Element in Facial Displays of Emotion." *Science* 164 (1969): 86–88.

Galaburda, A. M.; LeMay, M.; Kemper, T. L.; and Geschwind, Norman. "Right-Left Asymmetries in the Brain." *Science* 199 (1978): 852–56.

Gazzaniga, Michael S., and LeDoux, Joseph E. *The Integrated Mind.* New York: Plenum, 1977.

Geschwind, Norman. "The Organization of Language and the Brain." *Science* 170 (1970): 940–44.

Gurin, Joel. "Chemical Feelings." *Science 80,* no. 1 (1980): 28–35.

Hinde, Robert A. *Biological Bases of Human Social Behavior.* New York: McGraw-Hill, 1974.

Klein, R., and Armitage, R. "Rhythms in Human Performance: 1½-Hour Oscillations in Cognitive Style." *Science* 204 (1979): 1326–28.

Lee, Richard B. *The !Kung San.* Cambridge: Cambridge University Press, 1979.

Wieskrantz, L.; Warrington, Elizabeth K.; Sanders, M. D.; and Marshall, J. "Visual Capacity in the Hemianopic Field Following a Restricted Occipital Ablation." *Brain* 97 (1974): 709–28.

Wilson, Edward O. *Sociobiology.* Cambridge, Mass.: Harvard University Press, 1975.

———. *On Human Nature.* Cambridge, Mass.: Harvard University Press, 1978.

30. INFANTS

Aronson, E., and Rosenbloom, S. "Space Perception in Early Infancy: Perception Within a Common Auditory-Visual Space." *Science* 172 (1971): 1161–63.

Bornstein, Mark H. "Perceptual Development: Stability and Change in Feature Perception." In *Psychological Development from Infancy,* edited by M. H. Bornstein and W. Kessen. Hillside, N.J.: Erlbaum, 1979.

———; Kessen, W. K.; and Weiskopf, S. "The Categories of Hue in Infancy." *Science* 191 (1976): 201–2.

Bower, T. G. R. *Development in Infancy.* San Francisco: W. H. Freeman, 1974.

Condon, W. S., and Sander, L. W. "Neonate Movement Is Synchronized with Adult

Speech: Interactional Participation and Language Acquistion." *Science* 183 (1974): 99–101.
Cutting, J. E., and Rosner, B. "Categories and Boundaries in Speech and Music." *Perception and Psychophysics* 16 (1974): 564–70.
DeCasper, Anthony J., and Fifer, William P. "Of Human Bonding: Newborns Prefer Their Mothers' Voices." *Science* 208 (1980): 1174–76.
Eilers, R. E., and Minifie, F. D. "Fricative Discrimination in Early Infancy." *Journal of Speech and Hearing Research* 18 (1975): 158–69.
Eilers, R. E.; Wilson, W. R.; and Moore, J. M. "Development Changes in Speech Discrimination in Infants." *Journal of Speech and Hearing Research* 20 (1977): 766–80.
Eimas, Peter D. "Auditory and Linguistic Processing of the Cues for Place of Articulation by Infants." *Perception and Psychophysics* 16 (1974): 513–21.
———. "Speech Perception in Early Infancy." In *Infant Perceptions from Sensation to Cognition,* edited by L. B. Cohen and P. Salapatek. New York: Academic Press, 1975.
———, and Corbit, J. D. "Selective Adaptation of Linguistic Feature Detectors." *Cognitive Psychology* 4 (1973): 99–109.
Eimas, Peter D., and Miller, Joanne L. "Contextual Effects in Infant Speech Perception." *Science* 209 (1980): 1140–41.
Eimas, Peter D.; Siqueland, E. R.; Jusczyk, P.; and Vigorito, J. "Speech Perception in Infants." *Science* 171 (1971): 303–6.
Ekman, Paul, and Friesen, W. C. *Unmasking the Face.* Englewood Cliffs, N.J.: Prentice-Hall, 1975.
Fagan, J. F. "Infant's Recognition of Invariant Features of Faces." *Child Development* 47 (1976): 627–38.
Freedman, D. G. "Smiling in Blind Infants and the Issue of Innate vs. Acquired." *Journal of Child Psychology and Psychiatry and Allied Disciplines* 5 (1964): 171–84.
Goren, C. C.; Sarty, M.; and Wu, P. "Visual Following and Pattern Discrimination of Face-like Stimuli by Newborn Infants." *Pediatrics* 56 (1975): 544–49.
Haith, M. M.: Bergman, T.; and Mare, M. J. "Eye Contact and Face Scanning in Early Infancy." *Science* 198 (1977): 853–55.
Klaus, M. H., et al. "Maternal Attachment: The Importance of the First Postpartum Days." *New England Journal of Medicine* 28 (1972): 460–63.
Kuhl, P. K. "Speech Perception in Early Infancy." In *Hearing and Davis,* edited by S. K. Hirsh, D. H. Eldredge, I. J. Hirsh, and S. R. Silverman. St. Louis: Washington University Press, 1976.
Lasky, R.; Syrdal-Lasky, A.; and Klein, R. "VOT Discrimination by Four- to Six-month Old Infants from Spanish Environments." *Journal of Experimental Child Psychology* 20 (1975): 215–25.
Mattingly, I. G. "Speech Cues and Sign Stimuli." *American Scientist* 60 (1972): 327–37.
Meltzoff, Andrew N., and Barton, Richard W. "Intermodal Matching by Human Neonates." *Nature* 282 (1979): 403–4.
Meltzoff, A. R.; and Moore, M. K. "Imitation of Facial and Manual Gestures by Human Neonates." *Science* 198 (1977): 75–78.
Morse, P. "The Discrimination of Speech and Nonspeech Stimuli in Early Infancy." *Journal of Experimental Child Psychology* 14 (1972): 477–92.
———, and Snowdon, C. T. "Categorical Speech Discimination by Rhesus Monkeys." *Perception and Psychophysics* 17 (1975): 9–16.
Pyles, Thomas. *The Origins and Development of the English Language.* New York: Harcourt Brace Jovanovich, 1971.
Ojemann, George, and Mateer, Catherine. "Human Language Cortex: Localization of Memory, Syntax, and Sequential Motor-Phoneme Identification Systems." *Science* 205 (1979): 1401–3.
Schwartz, Joyce, and Tallal, Paula. "Rate of Acoustic Change May Underlie Hemispheric Specialization for Speech Perception." *Science* 207 (1980): 1380–81.
Strange, W., and Jenkins, J. J. "Role of Linguistic Experience in the Perception of Speech." In *Perception and Experience,* edited by R. D. Walk and H. L. Pick, Jr. New York: Plenum, 1978.
Streeter, L. A. "Language Perception of Two-month Old Infants Shows Effects of Both Innate Mechanisms and Experience." *Nature* 259 (1976): 39–41.
Studdert-Kennedy, M. "Universals in Phonetic Structure and Their Role in Linguistic Communication." In *Recognition of Complex Acoustic Signals,* edited by Theodore H. Bullock. Berlin: Dahlem Konferenzen, 1977.

Waters, R. S., and Wilson, W. A. "Speech Perception by Rhesus Monkeys: The Voicing Distinction in Synthesized Labial and Velar Stop Consonants." *Perception and Psychophysics* 19 (1976): 285–89.

Zelazo, P. R.; Zelazo, N. A.; and Kolb, S. "Walking in the Newborn." *Science* 176 (1972): 314–15.

31. CHILD DEVELOPMENT

Belmont, Lillian; Stein, Zena; and Zybert, Patricia. "Child Spacing and Birth Order: Effect on Intellectual Ability in Two-Child Families." *Science* 202 (1978): 995–96.

Chomsky, Noam. "On the Biological Basis of Language Capacities." In *The Neuropsychology of Language,* edited by R. W. Rieber. New York: Plenum, 1976.

Cioffi, Joseph, and Kandel, Gillray L. "Laterality of Stereognostic Accuracy of Children for Words, Shapes, and Bigrams: A Sex Difference for Bigrams." *Science* 204 (1979): 1432–34.

Elkind, D. *Children and Adolescents: Interpretive Essays on Jean Piaget.* Oxford: Oxford University Press, 1974.

Gesell, A. *Wolf Child and Human Child.* New York: Harper and Bros., 1940.

Goldin-Meadow, Susan, and Feldman, Heidi. "Development of Language-like Communication Without a Language Model." *Science* 197 (1977): 401–3.

Herrnstein, R. J.; Loveland, D. H.; and Cable, C. "Natural Concepts in Pigeons." *Journal of Experimental Psychology: Animal Behavioral Processes* 2 (1976): 285–302.

Holden, Constance. "Identical Twins Reared Apart." *Science* 207 (1980): 1323–28.

Jencks, Christopher. *Inequality.* New York: Harper and Row, 1972.

Kagan, Jerome, and Klein, R. E. "Cross-Cultural Perspectives on Early Development." *American Psychologist* 28 (1973): 947–61.

Lenneberg, E. H., and Lenneberg, E., eds. *Foundation of Language Development.* New York: Academic Press, 1975.

Marler, Peter. "Birdsong and Speech Development: Could There Be Parallels?" *American Scientist* 58 (1970): 669–83.

Mendlewicz, J., and Ranier, J. D. "Adoption Study Supporting Genetic Transmission of Manic-Depressive Illness." *Nature* 268 (1977): 327–29.

Montessori, Maria. *The Absorbent Mind.* New York: Dell, 1967.

Rose, Richard J.; Harris, E. L.; Christian, J. C.; and Nance, W. E. "Genetic Variance in Nonverbal Intelligence: Data from the Kinships of Identical Twins." *Science* 205 (1979): 1153–55.

Wilson, R. S. "Twins: Early Mental Development." *Science* 176 (1972): 914–17.

———. "Synchronies in Mental Development: An Epigenetic Perspective." *Science* 202 (1978): 939–47.

Wittig, M. A., and Peterson, A. C., eds. *Sex-Related Differences in Cognitive Functioning.* New York: Academic Press, 1979.

Index

Credits

Cover Wildlife Photographers/Bruce Coleman, Inc.
Part I David Overcash/Bruce Coleman, Inc.

1-1 From R. Silver, *American Scientist* 66 (1978): 209-15. Reprinted by permission of *American Scientist,* journal of Sigma Xi, The Scientific Research Society.
1-2 From N. Tinbergen and H. Falkus, *Signals for Survival,* published by Oxford University Press. © Oxford University Press 1970.
1-3 From C. Jencks, *Inequality: A Reassessment of the Effect of Family and Schooling in America.* © 1972 by Basic Books, Inc. Reprinted by permission.
1-4 From W. R. Miles, *Journal of Comparative Psychology* 10 (1930): 237-61. © 1930 The Williams and Wilkins Co., Baltimore, Md.
1-5 From O. Pfungst, *Clever Hans: The Horse of Mr. von Osten,* edited by R. Rosenthal (New York: Holt, Rinehart and Winston, 1965).
1-6 Courtesy Pfizer Inc.
1-7 From R. Sevenster, *Behaviour* Supplement 9 (1961): 1-170.
2-1A Wallace Kirkland, *Life* magazine. © 1954 Time, Inc.
2-1B E. H. Hess, *Scientific American* 195, no. 1 (1956): 71-80.
2-2 Sovfoto.
2-3 K. von Frisch, *Tanzsprache und Orientierung der Bienen.* Berlin: Springer-Verlag, 1965.
2-4 From R. Menzel and J. Erber, *Scientific American* 239, no. 1 (1978): 102-10. © 1978 Scientific American, Inc. All rights reserved.
2-5 Courtesy Thomas Eisner, Cornell University.
2-6 Nina Leen, *Life* magazine. © 1964 Time, Inc.
2-8 Courtesy N. Tinbergen and Addison-Wesley Publishing Co.
2-9, 2-10 Based on N. Tinbergen, *Zeitschrift für vergleichende Physiologie* 21 (1935): 699-716.
2-11 Based on N. Tinbergen, *Zeitschrift für vergleichende Physiologie* 16 (1932): 305-34.
2-12 From N. Tinbergen and W. Kruyt, *Zeitschrift für vergleichende Physiologie* 25 (1938): 292-334.
2-13 Based on R. Hoogland, D. Morris, and N. Tinbergen, *Behaviour* 10 (1957): 207-36.
3-1 From K. Lorenz and N. Tinbergen, *Zeitschrift für Tierpsychologie* 2 (1938): 1-29.
3-2 From C. G. Beer, unpublished Ph. D. dissertation, Oxford University, 1960.
3-3 From N. Tinbergen, *De Levende Natuur* 51 (1948): 65-69.
3-4 From C. Darwin, *The Descent of Man* (New York: Collier & Son, 1871).

3-5 From J. Crane, *Zoologica* 42 (1957): 69–82. By permission of the New York Zoological Society.

3-6 From N. Tinbergen, *Inleiding tot de Diersociologie* (Gorinchem: J. Noorduijn en Zoon N. V., 1946).

3-7, 3-8 From N. Tinbergen, *Wilson Bulletin* 60 (1948): 6–52.

3-9 From L. P. Brower, J. V. Z. Brower, and F. P. Cranston, *Zoologica* 50 (1965): 1–39. By permission of the New York Zoological Society.

4-1 Based on N. Tinbergen and D. J. Kuenen, *Zeitschrift für Tierpsychologie* 3 (1939): 37–60.

4-2, 4-3, 4-4, 4-5 From N. Tinbergen and A. C. Perdeck, *Behaviour* 3 (1950): 1–39.

4-6, 4-7 Based on J. P. Hailman, *Behaviour* Supplement 15 (1967): 1–159.

Part II S. C. Bisserot/Bruce Coleman, Inc.

5-5 From A. O. D. Willows, *Scientific American* 224, no. 2 (1971): 68–78. © 1971 Scientific American, Inc. All rights reserved.

5-6 Kim Taylor/Bruce Coleman, Inc.

5-7 Based on A. Gelperin, *Annual Review of Entomology* 16 (1971): 365–78. With permission; © Annual Reviews, Inc.

5-8 From R. Goldschmidt, *Zeitschrift für Wissenschaftliche Zoologie* 92 (1909): 306–57.

6-1 Based on *Iscotables* (7th ed.; Lincoln, Neb.: Instrumentation Specialties Co., 1974); G. Wald, *Science* 145 (1964): 1007–17, © 1964 American Association for the Advancement of Science (hereafter A.A.A.S.); and O. von Helversen, *Journal of Comparative Physiology* 80 (1972): 439–72.

6-3 From B. Diehn, *Science* 181 (1973): 1009–15. © 1973 A.A.A.S.

7-1 Based on H. Autrum and V. von Zwehl, *Zeitschrift für vergleichende Physiologie* 48 (1964): 357–84; and P. K. Brown and G. Wald, *Science* 144 (1964): 45–52; © 1964 A.A.A.S.

7-2 From S. Skoglund, *Acta Physiologica Scandinavica* 36, Supplement 124 (1956): 1–101.

7-7 Based on data from John Allman, California Institute of Technology.

7-9 From S. LeVay, D. H. Hubel, and T. N. Wiesel, *Journal of Comparative Neurology* 159 (1975): 559–76.

8-2 Courtesy Philip S. Callahan.

8-3 From L. Prozesky-Schulze, O. P. M. Prozesky, F. Anderson, and G. J. J. van der Merwe, *Nature* 255 (1975): 142–43.

8-4 From F. Nottebohm and M. E. Nottebohm, *Journal of Comparative Physiology* 108 (1976): 171–92.

8-5 Based on H. Autrum and W. Schneider, *Zeitschrift für vergleichende Physiologie* 31 (1948): 77–88; and H. Heran, *Zeitschrift für vergleichende Physiologie* 42 (1959): 103–63.

9-1 Based on G. von Békésy, *Symposia of the Society for Experimental Biology* 16 (1962): 267–88; G. von Békésy, *Experiments in Hearing*, © 1960 McGraw-Hill, used with permission of McGraw-Hill Book Co.; M. S. Gordon, G. Bartholomew, A. D. Grinnell, C. B. Jørgensen, and F. N. White, *Animal Function: Principles and Adaptations* (New York: Macmillan, 1968), © 1968 M. S. Gordon.

9-4 Based on E. J. Knudsen and M. Konishi, *Science* 200 (1978): 795–97. © 1978 A.A.A.S.

9-5 Based on J. A. Simmons, D. J. Howell, and N. Suga, *American Scientist* 63 (1975): 204–15. Reprinted by permission of *American Scientist*, the journal of Sigma Xi, The Scientific Research Society.

9-6 Based on T. Manabe, N. Suga, and J. Ostwald, *Science* 200 (1978): 339–42, © 1978 A.A.A.S.; N. Suga, W. E. O'Neill, and T. Manabe, *Science* 200 (1978): 778–81, © 1978 A.A.A.S.; and N. Suga and W. E. O'Neill, *Science* 206 (1979): 351–53, © 1979 A.A.A.S.

10-1 From V. G. Dethier, *American Scientist* 59 (1971): 706–715. Reprinted by permission of *American Scientist*, journal of Sigma Xi, The Scientific Research Society.

10-2 From P. H. Hartline, L. Kass, and M. S. Loop, *Science* 199 (1978): 1225–29. © 1978 A.A.A.S.

10-3 From P. S. Enger and T. Szabo, *Journal of Neurophysiology* 28 (1965): 800–18; and H. Scheich and T. H. Bullock, in *Handbook of Sensory Physiology III/3*, edited by A. Fessard (Berlin: Springer-Verlag, 1974).

Part III Jen and Des Bartlett/Bruce Coleman, Inc.

11-1 From A. O. D. Willows, *Scientific American* 224, no. 2 (1971): 68–78. © 1971 by Scientific American, Inc. All rights reserved.

11-2 From H. A. Dade, *Anatomy and Dissection of the*

Honey Bee (London: International Bee Research Association, 1978).

11-3 Reprinted from D. M. Wilson, *Neural Theory and Modeling: Proceedings of the 1962 OJAI Symposium,* edited by Richard F. Reiss, with the permission of the publishers, Stanford University Press. © 1964 by the Board of Trustees of the Leland Stanford Jr. University.

11-4A From A. O. D. Willows, *Neurobiology of Invertebrates,* edited by J. Salanki (New York: Plenum Publishing Corporation, 1973).

11-4B From G. S. Stent, W. B. Kristan, Jr., W. O. Friesen, C. A. Ort, M. Poon, and R. L. Calabrese, *Science* 200 (1978): 1348–57. © 1978 A.A.A.S.

11-5 From R. W. Doty and J. F. Bosma, *Journal of Neurophysiology* 19 (1956): 44–60. Courtesy of Charles C Thomas, Publisher, Springfield, Ill.

11-6A From N. E. Collias and E. C. Collias, *Proceedings of the XIIIth International Ornithological Congress* 1 (1963): 518–30.

11-6B,C Courtesy of Turid Hölldobler.

11-6D, E From G. Linhardt; courtesy of Naturhistorisches Museum, Braunschweig.

11-7 Courtesy of Anthony Janetos, University of Utah.

11-8 From H. W. Levi, *American Scientist* 66 (1978): 734–42. Reprinted by permission of *American Scientist,* journal of Sigma Xi, The Scientific Research Society.

11-9 Courtesy of Grant Frazier Gould.

12-1 From K. Lorenz, *Symposia of the Society for Experimental Biology* 4 (1950): 221–68.

12-2 From P. DeCoursey, *Zeitschrift für vergleichende Physiologie* 44 (1961): 331–54.

13-3 Redrawn from K. von Frisch, *Tanzsprache und Orientierung der Bienen* (Berlin: Springer-Verlag, 1965).

13-8 From R. Wehner, *Scientific American* 235, no. 1 (1976): 106–14. © 1976 by Scientific American, Inc. All rights reserved.

13-9 From E. M. N. Parker, *American Scientist* 59 (1971): 578–85. Reprinted by permission of *American Scientist,* journal of Sigma Xi, The Scientific Research Society.

13-10 From M. Lindauer and H. Martin, in *Animal Orientation and Navigation,* edited by S. R. Galler et al. (Washington, D.C.: U.S. Government Printing Office, 1972).

13-11 From M. Lindauer, *Verhandlungen der Deutschen Zoologischen Gesellschaft* 69 (1976): 156–83.

13-12 From R. B. Frankel, R. P. Blakemore, and R. S. Wolfe, *Science* 203 (1979): 1355–57. © 1979 A.A.A.S.

14-1 From E. Gwinner and W. Wiltschko, *Journal of Comparative Physiology* 125 (1978): 267–73.

14-2 From W. T. Keeton, *Science* 165 (1969): 922–28. © 1969 A.A.A.S.

14-3A From B. Elsner, in *Animal Migration, Navigation, and Homing,* edited by K. Schmidt-Koenig and W. T. Keeton (New York: Springer-Verlag, 1978).

14-3B From M. Michener and C. Walcott, *Journal of Experimental Biology* 47 (1967): 99–131.

14-3C From K. Schmidt-Koenig and C. Walcott, *Animal Behaviour* 26 (1978): 480–86.

14-4 From W. Wiltschko, *Zeitschrift für Tierpsychologie* 25 (1968): 537–58.

14-5 From W. T. Keeton, *Proceedings of the National Academy of Sciences* 68 (1971): 102–6.

14-6 Courtesy of Charles Walcott.

14-7 From C. Walcott and R. P. Green, *Science* 184 (1974): 180–82. © 1974 A.A.A.S.

14-10A From W. T. Keeton, T. S. Larkin, and D. M. Windsor, *Journal of Comparative Physiology* 95 (1974): 95–103.

14-10B, 14-11A From C. Walcott, in *Animal Migration, Navigation, and Homing,* edited by K. Schmidt-Koenig and W. T. Keeton (New York: Springer-Verlag, 1978, pp. 194–98).

14-11B Based on data from USGS aeromagnetic surveys and from Charles Walcott.

15-1 Courtesy of J. T. Bonner, Princeton University.

15-2 Based on J. E. Lloyd, *Miscellaneous Publications of the Museum of Zoology, University of Michigan* 130 (1966): 1–95; and A. D. Carlson and J. Copeland, *American Scientist* 66 (1978): 340–46. Reprinted by permission of *American Scientist,* journal of Sigma Xi, The Scientific Research Society.

15-3 From F. Schaller and H. Schwalb, *Zoologischer Anzeiger* 24 (1961): 154–66.

15-4 Courtesy of James E. Lloyd, University of Florida.

15-5 From M. Salmon and S. P. Atsaides, *American Zoologist* 8 (1968): 623–39.

15-6 From D. Hunsaker II, *Evolution* 16 (1962): 62–74.

15-7 From D. R. Bentley, *Science* 174 (1971): 1139–41. © 1971 A.A.A.S.

15-8 From A. W. Ewing and H. C. Bennett-Clark, *Behaviour* 31 (1968): 288–301.

15-9 From K. Lorenz, *Zoologischer Anzeiger* 17 (1953): 36–58.

15-10 Based on data from Carl Hopkins, University of Minnesota.

Part IV Eric Hosking/Bruce Coleman, Inc.

16-1, 16-2 From R. Menzel, J. Erber, and T. Masuhr, in *Experimental Analysis of Insect Behaviour,* edited by L. Barton Browne (New York: Springer-Verlag, 1974).

16-3 From R. Menzel and J. Erber, *Scientific American* 239, no. 1 (1978): 102–10. © 1978 Scientific American, Inc. All rights reserved.

17-1A J. Erber, *Journal of Comparative Physiology* 99 (1975): 231–42.

17-1B,C R. Menzel, J. Erber, and T. Masuhr, in *perimental Analysis of Insect Behaviour,* edited by L. Barton Browne (New York: Springer-Verlag, 1974).

17-2 Based on D. H. Janzen, *Natural History* 83, no. 4 (1974): 48–53.

17-3 Courtesy of Lincoln P. Brower, University of Florida.

17-4 From A. O. Ramsay and E. H. Hess, *Wilson Bulletin* 66 (1954): 196–206.

17-5 Based on P. Marler and S. Peters, *Science* 198 (1977): 519–21. © 1977 A.A.A.S.

17-6 Based on data from Peter Marler, Rockefeller University.

17-7 Based on P. Marler and M. Tamura, *Science* 146 (1964): 1483–86. © 1964 A.A.A.S.

18-1 From M. Lindauer, *Zeitschrift für vergleichende Physiologie* 37 (1955): 263–324.

18-2 From R. A. Grout, ed., *The Hive and the Honey Bee* (Hamilton, Ill.: Dadant & Sons, 1975).

18-3 From P. Marler, in *Darwin's Biological Work: Some Aspects Reconsidered,* edited by P. R. Bell (Cambridge: Cambridge University Press, 1959).

18-4 J. Markham/Bruce Coleman, Inc.

18-5 C. R. Carpenter Papers, Penn State Room, courtesy Pennsylvania State University Library.

Part V Courtesy Ginny Fonte, University of Colorado.

19-2 From T. A. McGrath, M. D. Shalter, W. M. Schleidt, and P. Sarvella, *Nature* 237 (1972): 47–48.

19-3 From B. I. Lade and W. H. Thorpe, *Nature* 202 (1964): 366–68.

19-4 Courtesy of William C. Dilger, Cornell University.

19-5 From D. R. Bentley, *Science* 174 (1971): 1139–41. © 1971 A.A.A.S.

19-6 From D. Bovet, F. Bovet-Nitti, and A. Oliverio, *Science* 163 (1969): 139–49. © 1969 A.A.A.S.

19-7 Based on R. C. Tryon, *Yearbook of the National Society for the Study of Education* 39 (1940): 111–19.

19-8 From W. C. Rothenbuhler, *American Zoologist* 4 (1964): 111–23.

20-2 From H. S. Jennings, *Behavior of the Lower Organisms.* © 1962, 1976 by Indiana University Press. All rights reserved.

20-3 From J. E. Sulston and H. R. Horvitz, *Developmental Biology* 56 (1977): 110–56. © 1977 by Academic Press, New York.

20-5 From R. J. Konopka and S. Benzer, *Proceedings of the National Academy of Sciences* 68 (1971): 2112–16.

20-6 Based on Y. Hotta and S. Benzer, *Nature* 240 (1972): 527–35.

20-7 Based on W. G. Quinn and Y. Dudai, *Nature* 262 (1976): 576–77.

Part VI Courtesy of David Bygott and Jeanette Hanby.

21-1 Courtesy of Irenäus Eibl-Eibesfeldt, Max Planck Institute for Behavioral Physiology.

21-2 Based on M. D. F. Udvardy, *The Audubon Society Field Guide to North American Birds: Western Region* (New York: Knopf, 1977); and J. Bull and J. Farrand, *The Audubon Society Field Guide to North American Birds: Eastern Region* (New York: Knopf, 1977).

21-3 From C. Darwin, *Journal of Researches* (New York: Collier, 1902).

21-5 From B. Cobb, *Field Guide to the Ferns.* © 1956, 1963 by Boughton Cobb. Reprinted by permission of the publisher, Houghton Mifflin Co.

21-7 Courtesy of Thomas Eisner, Cornell University.

21-8 Courtesy of Lincoln Brower, University of Florida.

21-9, 21-10 Courtesy of Carl Rettenmeyer, University of Connecticut.

21-11 M. P. L. Fogden/Bruce Coleman, Inc.

21-12 Courtesy of Thomas Eisner and David Aneshansley, Cornell University.

21-13 Courtesy of L. D. Mech.

21-15 David Overcash/Bruce Coleman, Inc.

21-16A Courtesy of Robert Silberglied, Harvard University.

21-16B Courtesy of Daniel H. Janzen, University of Pennsylvania.

22-1, 22-2, 22-3 Based on H. N. Southern, *Journal of Zoology* 162 (1970): 197–285.

22-4A Courtesy of Doris M. Gould.

22-4B Courtesy of Walter Leuthold.

22-5 Courtesy of Randy Thornhill, University of New Mexico.

22-6 Courtesy of Robert Silberglied, Harvard University.

22-7 From V. Geist, *Mountain Sheep* (Chicago: University of Chicago Press, 1971). © 1971 University of Chicago Press. All rights reserved.

23-1 From C. H. Holm, *Ecology* 54 (1973): 356–65. © 1973 Ecological Society of America.

23-2 Courtesy of Haven Wiley, University of North Carolina.

23-4 Courtesy of George D. Lepp.

Part VII Courtesy of Richard D. Estes.

24-1 Courtesy of Kenneth Lorenzen, University of California at Davis.

24-2 Courtesy of Bernd Heinrich, University of Vermont.

24-3 From S. F. Sakagami, M. J. Montenegro, and W. E. Kerr, *Journal of the Faculty of Science, Hokkaido University,* Series VI (Zoology) 15 (1965): 578–607.

24-4 Courtesy of Thomas Seeley, Yale University.

24-5, 24-6 Courtesy of Kenneth Lorenzen.

24-7 From R. A. Grout, ed., *The Hive and the Honey Bee* (Hamilton, Ill.: Dadant & Sons, 1963).

24-8 Based on A. M. Wenner, *Science* 138 (1962): 446–48. © 1962 A.A.A.S.

24-9 Courtesy of Norman E. Gary, University of California at Davis.

24-10 Courtesy of Kenneth Lorenzen.

24-11 Based on E. P. Jeffree, *Journal of Economic Entomology* 48 (1955): 723–26.

24-12 Based on M. Lindauer, *Zeitschrift für vergleichende Physiologie* 34 (1952): 299–345.

24-14A,B From K. von Frisch, *Tanzsprache und Orientierung der Bienen* (Berlin: Springer-Verlag, 1965).

24-14C Courtesy of Kenneth Lorenzen.

24-15 Based on K. von Frisch, *Osterreichische Zoologische Zeitschrift* 1 (1946): 1–48.

24-16, 24-17, 24-19, 24-20, 24-21 Based on K. von Frisch, *Tanzsprache und Orientierung der Bienen* (Berlin: Springer-Verlag, 1965).

24-22 From M. Lindauer, *Communication Among Social Bees* (Cambridge, Mass.: Harvard University Press, 1961). Reprinted by permission.

24-23 Courtesy of Kenneth Lorenzen.

24-24 From M. Lindauer, *Zeitschrift für vergleichende Physiologie* 37 (1955): 263–324.

25-1 Based on C. R. Ribbands, *The Behaviour and Social Life of Honey Bees* (London: Bee Research Association, 1953); and M. Pavan and G. Ronchetti, *Atti della Societa Italiana di Scienze Naturali, Milano* 94 (1955): 379–477.

25-2 From W. Hangartner, *Zeitschrift für vergleichende Physiologie* 62 (1969): 111–20.

25-3 Courtesy of Carl Rettenmeyer, University of Connecticut.

25-4 Courtesy of Bert Hölldobler, Harvard University.

25-5 Courtesy of R. W. Taylor.

25-6 Courtesy of Carl Rettenmeyer.

25-7 From S-P. Jozsef, *Termeszettudomanyi Kozlony* 1928 (1928): 215–219.

25-8 From A. Reyne, *Zoologische Mededelingen* 32 (1954): 233–57.

25-9A Courtesy of Howard Wildman, Princeton University.

25-9B Courtesy of Carl Rettenmeyer.

25-10A *(top)* From J. H. Sudd, *Discovery* 24 (1963): 15–19; *(bottom)* based on photograph by Turid Hölldobler, courtesy of Bert Hölldobler.

25-10B Courtesy of Bert Hölldobler.

25-10C Based on photographs by Turid Hölldobler, courtesy of Bert Hölldobler.

25-11A,B, 25-12, 25-13, 25-14 Courtesy of Carl Rettenmeyer.

25-11B, 25-12, 25-13, 25-14 Courtesy of Carl Rettenmeyer.

25-15 Based on T. C. Schneirla, *Army Ants: A Study*

in Social Organization (San Francisco: W. H. Freeman, 1971). © 1971 W. H. Freeman.

25–16 Courtesy of Carl Rettenmeyer.

25–17, 25–18 From H. Kutter, *Neujahrsblatt herausgegeben von der Naturforschenden Gesellschaft in Zurich* 171 (1969): 1–62.

26–2 Courtesy of David Bygott and Jennette Hanby.

26–3, 26–4 Courtesy of Richard D. Estes.

26–5, 26–6 Courtesy of David Bygott and Jennette Hanby.

26–7 Ted Grant, National Film Board of Canada.

26–8 From V. Geist, *Mountain Sheep* (Chicago: University of Chicago Press, 1971), © 1971 University of Chicago Press. All rights reserved.

26–9, 26–10 From W. Leuthold, *African Ungulates* (Berlin: Springer-Verlag, 1977).

27–1 Courtesy of David Bygott and Jennette Hanby.

27–2, 27–3 Courtesy of George Schaller, New York Zoological Society.

27–4 George Schaller/Bruce Coleman, Inc.

27–5 Based on G. Schaller, *The Serengeti Lion* (Chicago: University of Chicago Press, 1972), © 1972 University of Chicago Press. All rights reserved.

27–6 Courtesy of George Schaller.

27–7 Norman Myers/Bruce Coleman, Inc.

27–8 John Dominis, *Life* magazine. © 1967 Time, Inc.

27–9 Courtesy of George Schaller.

27–10 From R. C. Eaton, *The Cheetah: The Biology, Ecology, and Behavior of an Endangered Species* (New York: Van Nostrand Reinhold, 1974). © 1974 by Litton Educational Publishing Co. Reprinted by permission of Van Nostrand Reinhold Co.

27–11 Courtesy of David Bygott and Jennette Hanby.

27–12 Courtesy of George Schaller.

27–13 Norman Myers/Bruce Coleman, Inc.

27–14, 27–15 From H. Kruuk, *The Spotted Hyena* (Chicago: University of Chicago Press, 1972), © 1972 University of Chicago Press. All rights reserved.

27–16 George Schaller/Bruce Coleman, Inc.

27–17 Courtesy of George Schaller.

27–18 Courtesy of David Bygott and Jennette Hanby.

27–19 From H. Kruuk, *The Spotted Hyena* (Chicago: University of Chicago Press, 1972), © 1972 University of Chicago Press. All rights reserved.

27–20 Based on G. Schaller, *The Serengeti Lion* (Chicago: University of Chicago Press, 1972), © 1972 University of Chicago Press, all rights reserved; and F. Walther, *Behaviour* 34 (1969): 184–221.

28–1 Irven DeVore, Anthro-Photo.

28–2 A. H. Harcourt, Anthro-Photo.

28–3 With permission from Y. Sugiyama in R. P. Michael and J. H. Crook, eds., *Comparative Ecology and Behaviour of Primates* (London: Academic Press, 1973). © Academic Press, Inc. (London) Ltd.

28–4 Courtesy of Kurt B. Modahl, Oregon Regional Primate Research Center.

28–5 Joseph Popp, Anthro-Photo.

28–6 From H. Kummer, *Social Organization of Hamadryas Baboons* (Chicago: University of Chicago Press, 1968). Reprinted with permission of S. Karger AG, Basel.

28–7 Based on data from G. Gray Eaton, Oregon Regional Primate Research Center.

28–8A Nancy Nicolson, Anthro-Photo.

28–8B Richard Wrangham, Anthro-Photo.

28–9 From J. H. Kaufmann, in S. A. Altman, ed., *Social Communication Among Primates* (Chicago: University of Chicago Press, 1964). © 1964 University of Chicago Press. All rights reserved.

28–10 James Moore, Anthro-Photo.

28–11 Joseph Popp, Anthro-Photo.

Part VIII From R. Lee, Anthro-Photo.

29–2, 29–3 From M. S. Gazzaniga and J. E. LeDoux, *The Integrated Mind* (New York: Plenum, 1977).

29–4 Reprinted with permission of Macmillan Publishing Co., Inc., from D. R. Pilbeam, *The Ascent of Man: An Introduction to Human Evolution.* Copyright © 1972 by David R. Pilbeam.

29–5 Courtesy of the American Museum of Natural History.

29–6A From J. Tanaka, Anthro-Photo.

29–6B From M. Shostak, Anthro-Photo.

29–7 Courtesy of the Film Study Center, Harvard University.

30–1 Courtesy of Clare Holly Gould.

30-2 From A. Peiper, *Zeitschrift für Tierpsychologie* 8 (1951): 449–56.

30-3 From a film by H. F. R. Prechtl, Institut für den wissenschaftlichen Film, Göttingen, F.R.G., 1953.

30-4 From B. Prudden, *Your Baby Can Swim* (Stockbridge, Mass.: The Institute for Physical Fitness, 1974).

30-5 Reprinted with permission of the author and the publishers from D. G. Freedman, *Human Infancy: An Evolutionary Perspective* (Hillsdale, N.J.: Lawrence Erlbaum Associates, 1975).

30-6: From G. Wald, *Science* 145 (1964): 1007–17, © 1964 A.A.A.S.; and M. H. Bornstein, W. Kessen, and S. Weiskopf, *Science* 191 (1976): 201–2, © 1976 A.A.A.S.

30-7 From T. Pyles, *The Origins and Development of the English Language*, 2d ed. © 1971 Harcourt Brace Jovanovich, Inc. Reproduced by permission of the publisher.

30-8 From P. Lieberman, E. S. Crelin, and D. H. Klatt. Reproduced with permission of the American Anthropological Association from the *American Anthropologist* 74 (1972): 287–307.

30-9 From P. Ladefoged, *Elements of Acoustic Phonetics* (Chicago: University of Chicago Press, 1962). © 1962 University of Chicago Press. All rights reserved.

30-10 From G. E. Peterson and H. L. Barney, *Journal of the Acoustical Society of America* 24 (1952): 175–84.

30-11 From W. Strange and J. J. Jenkins, in *Perception and Experience*, edited by R. D. Walk and H. L. Pick, Jr. (New York: Plenum, 1978, pp. 125–29).

30-12 From L. Lisker and A. S. Abramson, *WORD* 20 (1964): 384–422.

31-1 From R. S. Wilson, *Science* 202 (1978): 939–47. © 1978 A.A.A.S.